# Automation, Production Systems, and Computer-Aided Manufacturing

# Automation, Production Systems, and Computer-Aided Manufacturing

**MIKELL P. GROOVER**

*Department of Industrial Engineering*
*Lehigh University*

PRENTICE-HALL, INC. Englewood Cliffs, New Jersey 07632

201/592-2000

*Library of Congress Cataloging in Publication Data*

Groover, Mikell P.
Automation, production systems, and computer-aided manufacturing.

Includes bibliographies and index.
1. Automation. 2. Manufacturing processes—Data processing. 3. Group technology. I. Title
T59.5.G76 629.8 79-23492
ISBN 0-13-054668-2

Editorial/production supervision and interior design by Barbara A. Cassel
Cover design by Mario Piazza
Manufacturing buyer: Gordon Osbourne



Printed in the United States of America

10 9 8 7 6 5 4 3 2

PRENTICE-HALL INTERNATIONAL, INC., *London*
PRENTICE-HALL OF AUSTRALIA PTY. LIMITED, *Sydney*
PRENTICE-HALL OF CANADA, LTD., *Toronto*
PRENTICE-HALL OF INDIA PRIVATE LIMITED, *New Delhi*
PRENTICE-HALL OF JAPAN, INC., *Tokyo*
PRENTICE-HALL OF SOUTHEAST ASIA PTE. LTD., *Singapore*
WHITEHALL BOOKS LIMITED, *Wellington, New Zealand*

*To*

*W. R. Groover*

*G. E. Kane*

# Contents

## 2 Production Operations and Automation Strategies 12



## 3 Economic Analysis in Production 40



# part II HIGH VOLUME DISCRETE PARTS PRODUCTION SYSTEMS 63

## 4 Detroit Type Automation 65



## 5 Analysis of Automated Flow Lines 93

# Preface

This book is intended to provide a survey of the various topics in production automation and related systems. These topics include flow-line production (Detroit automation), numerical control, industrial robots, computer-aided manufacturing, process monitoring and control, group technology, flexible manufacturing systems, and material requirements planning. It is virtually an encyclopedia of modern manufacturing systems: Everything you always wanted to know about production automation but could not find it in one book.

The book has been written primarily for engineers and engineering students who wish to learn about automation and computer-assisted manufacturing methods in modern production plants. Managers and students of graduate business schools will also find the book useful for its emphasis on economic as well as technical issues related to automation.

As a text, the book should be suitable for advanced undergraduates and graduate students in industrial, mechanical, and manufacturing engineering curricula. More and more courses in manufacturing systems and computer-aided manufacturing are being introduced in engineering schools throughout the country, and this book should be appropriate for these courses; it may also be appropriate for technical production courses in graduate schools of business administration. There are probably more topics in the book than most instructors would want to pack into a single-semester course. The text is designed to fit various course requirements, and the chapters are largely self-contained. Accordingly, different instructors can concentrate on those portions of the book which they want to emphasize in their particular courses.

*Automation, Production Systems, and Computer-Aided Manufacturing* is also intended for the practicing engineer and manager of a production function who must keep abreast of the ever-changing state of technology in manufacturing. Examples of actual systems are described to illustrate applications in industrial practice. Application principles and relative advantages and disadvantages of the various automation techniques are presented to help the practicing engineer and manager make rational decisions about possible installations in their companies.

## ACKNOWLEDGEMENTS

I want to thank those persons who helped me in various ways during the preparation of this book. For their helpful reviews of selected portions of the text, I would like to thank Dr. Emory Zimmers, one of my colleagues at Lehigh and Director of our Computer-Aided Manufacturing Lab; Dr. John Buzacott of the University of Toronto; Mr. Cliff Marshall of Ingersoll Rand Company; Dr. James Ries of E. I. duPont; Mr. Oliver Wight; and Dr. Kenneth Tzeng.

I am also indebted to a significant number of industrial concerns and individuals for providing me with subject material, case studies, and other forms of technical assistance. Among this group, I am especially grateful to Mr. Ron LoVetri, currently a graduate student at Lehigh but destined for better things; Mr. Thomas Nolan of the Bethlehem Steel Corp.; Mr. Alex Houtzeel of TNO; Mr. George Burke, Jr., of Burron Medical Products; and Mr. James O'Brien of the MTM Association.

Finally, I appreciate the care and diligence with which Mrs. Nan Fahy typed the manuscript for the book.

MIKELL P. GROOVER

# List of Symbols

| | |
|---|---|
| $a$ | Parameter used in the solution of second order differential equation. $a = K_d/2M$. |
| $B$ | Parameter used in the computation of $h(b)$ for a two stage flow line. |
| $B_{1,2}$ | Status of binary digits in a binary register (0 or 1). |
| $b$ | Buffer capacity of a storage buffer in an automated flow line with internal parts storage. |
| $C$ | Constant (e.g., $C$ in the Taylor tool life equation). |
| $C(s)$ | Controller transfer function. |
| $C_{as}$ | Cost per time for an automatic workstation in a flow line. |
| $C_{at}$ | Cost per time for the automatic transfer mechanism used on a flow line. |
| $C_L$ | Cost per time to operate the flow line. This includes labor, overhead, and allocation of capital cost over the expected service life. |
| $C_m$ | Cost of raw materials per product produced on a flow line. Also, cost per unit of the components of an assembly produced on an automated assembly machine. |
| $C_o$ | Operator cost per time for manual workstation or stand-alone machine tool. |
| $C_{pc}$ | Cost per workpiece on average. |
| $C_t$ | Cost of disposable tooling per workpiece on a flow line. |

$C_t$ Tool cost per cutting edge in a machining operation.
$D$ Proportion downtime of an automated flow line.
$D_k$ Proportion downtime of stage $k$ in a multistage flow line.
$\mathbf{D}_p$ Direction of a gradient—a vector quantity.
$D_w$ Weekly demand rate.
$d$ Depth of cut in a machining operation.
$d$ Balance delay on a flow line, usually used for assembly lines.
$E$ Line efficiency or proportion uptime of an automated flow line.
$E_k$ Line efficiency of stage $k$ in an automated flow line with internal parts storage.
$E_o$ Line efficiency of an automated line with no internal parts storage.
$E_\infty$ Overall line efficiency of an automated flow line with internal parts storage where the stages are separated by storage buffers of infinite capacity.
$e$ Electrical voltage.
$e$ Error signal. $e=e(s)$.
$e_i$ Error used in least squares computations where $i$ stands for the number of data sets.
$e_n$ Discrete value of the error signal in a feedback control system.
$F$ Frequency of line stops per cycle.
$F$ Force.
$F_c$ Cutting force in a machining operation.
$f$ Feed rate of a machine tool.
$f_b$ Feed rate of a moving belt flow line.
$f_p$ Pulse rate (frequency of pulse train).
$f_r$ Feed of a drill bit per revolution.
$f_t$ Feed of a milling cutter per tooth.
$G$ Forward transfer function. $G=G(s)$.
$\mathbf{G}_p$ Gradient—a vector quantity used in some search procedures.
$H$ Feedback transfer function. $H=H(s)$.
$H_s$ Hours of operation per shift.
$h$ Head of fluid in a tank.
$h(b)$ Ideal proportion of the downtime of stage 1 in a two-stage line when stage 2 could be operating within the limits imposed by capacity $b$.
$i$ Subscript used to identify work station.
$j$ Subscript used to denote the reason for a line breakdown.
$j$ Subscript used to identify work element in line balancing.
$K$ Parameter defined in Eq. (5.46).
$K$ Controller gain.
$K_d$ Damping coefficient of a dashpot.
$K_s$ Spring constant.
$k$ Subscript used to identify the stages in an automated flow line with internal parts storage.

$L$ Number of units by which $b$ exceeds $BT_d/T_c$.

$L_s$ Length of each workstation on a manual flow line.

$M$ Mass.

$M_p$ Magnitude of the gradient—a scalar quantity.

$m$ Probability that a defective assembly component will cause stoppage of a workstation. $m_i$ denotes the probability at station $i$.

$N$ Number of bits (binary digits).

$n$ Number of periods (usually years) in engineering economy calculations.

$n$ Number of workstations on a flow line.

$n$ Discrete variable used in root locus definitions.

$n$ Exponent in the Taylor tool life equation.

$n_a$ Number of automatic workstations in a partially automated flow line.

$n_e$ Number of minimum rational work elements in a job in line balancing.

$n_m$ Number of machines through which a workpart must be routed.

$n_o$ Number of manually operated workstations on a partially automated flow line.

$n_p$ Number of poles in a transfer function.

$n_s$ Number of separate step angles in one complete revolution of a stepping motor.

$n_t$ Number of teeth in a milling cutter.

$n_z$ Number of zeroes in the transfer function.

$P$ Number of pulses in a pulse train.

$P$ Numerical value of a pole in root locus.

$P_{ap}$ Proportion or yield of assemblies with no defective components. Proportion of acceptable product.

$P_{qp}$ Proportion of assemblies produced which contain at least one defect.

$P(s)$ Process transfer function.

$PC$ Production capacity in pieces per week.

$p$ A point on a response surface defined by its coordinates $x_1$, $x_2$.

$p$ Probability that a workpart will jam at a workstation in an automated flow line. $p_i$ is the probability at station $i$.

$Q$ Quantity of workparts produced on a flow line.

$Q$ Batch size or number of parts produced over a certain time period.

$q$ Fraction defective rate.

$R$ Resistance to flow.

$R_{ap}$ Rate of production of acceptable parts.

$R_c$ Theoretical production rate of a flow line.

$R_p$ Actual average production rate of the flow line.

$r$ Ratio of breakdown rates of stage 2 to stage 1 in a two-stage automated flow line.

$r$ Sample correlation coefficient in regression analysis.

$S$ Spindle speed of a machine tool.

$S$ Motor speed of a stepping motor.

$S_w$ Number of shifts of plant operation per week.
$s_e$ Standard error of estimate used in regression analysis.
$s_p$ Spacing between parts on a moving belt line.
$s$ Differential operator and variable of the Laplace transform.
$T$ Tool life of a cutting tool.
$T_c$ Ideal or theoretical cycle time of a flow line.
$T_d$ Average downtime when a line breakdown occurs.
$T_{ej}$ Time required to perform minimum rational work element $j$.
$T_h$ Workpiece handling time.
$T_m$ Machining time per workpiece.
$T_{no}$ Average nonoperation time per part associated with each operation.
$T_o$ Average operation time per part.
$T_p$ Average production time per workpiece.
$T_{si}$ Workstation process time at station $i$.
$T_{su}$ Setup time for a production machine.
$T_t$ Tolerance time on a moving belt line.
$T_{tc}$ Tool change time per cutting edge.
$\mathrm{T}_{th}$ Tool handling time per workpiece at each machine.
$\mathrm{T}_{wc}$ Time of total work content on the line.
$t$ Time variable. Also used as independent variable time in differential equations.
$V$ Surface speed in a machining operation.
$V(t)$ Voltage as a function of time $t$.
$V_b$ Belt speed of a moving belt line.
$V_{\min}$ Cutting speed for minimum cost.
$V_o$ Output voltage of the DAC conversion process.
$V_{\text{ref}}$ Reference voltage.
$W$ Number of work centers in the plant.
$x$ Input variable or independent variable.
$x$ Displacement variable.
$x_p$ Input to process and output of controller unit.
$x_{pn}$ Discrete value of $x_p$.
$y$ Dependent variable or output variable of the system.
$Z$ Numerical value of a zero in root locus.
$z$ Objective function or index of performance.
$\alpha_s$ Step angle in a stepping motor.
$\beta$ $\mathrm{Cos}^{-1}\zeta$ in root locus.
$\zeta$ Damping ratio.
$\omega$ Imaginary part of a complex number of $\sigma + j\omega$.
$\omega_d$ Damped natural frequency.

$\omega_n$ Natural frequency.

$\sigma$ Real part of a complex number $\sigma + j\omega$.

$\sigma_b$ Breakaway point or break-in point in root locus.

$\sigma_c$ Center of asymptotes in a root locus plot.

$\tau$ Time constant.

$\tau$ Time interval between sampling instants.

# Automation, Production Systems, and Computer-Aided Manufacturing

part I

# Fundamental Concepts in Manufacturing and Automation

## 1.1 AUTOMATION DEFINED

This is a book about production automation. New terms have evolved over the years to describe this technology. These terms, such as numerical control, manufacturing systems, and computer-aided manufacturing, were unknown 30 years ago. Old words such as mechanization have virtually disappeared from the technical vocabulary. The development of new automation technology represents a continuous evolutionary process which has existed for many decades. Some might even argue that the process began with the industrial revolution (circa 1770) when machines began to take over the work previously performed by manual labor. Automation is a process of technological development that will proceed into the foreseeable future.

Automation is the technology concerned with the application of complex mechanical, electronic, and computer-based systems in the operation and control of production. This technology includes:

Automatic machine tools for processing workparts.

Automatic materials handling systems.

Automatic assembly machines for assembling components into products.

Continuous-flow processes.

Feedback control systems.

Computer process control systems.

Computerized systems for data collection, planning, and decision making to support manufacturing activities.

The scope of this text will be primarily limited to automation of discrete-parts manufacturing of the type found in the metalworking industries.

Automated production systems can be classified into two basic types:

1. Fixed automation.
2. Programmable automation.

*Fixed automation* occurs when the sequence of processing operations in the system is fixed by the equipment configuration. The typical features of fixed automation systems are:

High initial investment cost.

High production rates, suitable for high-volume demand.

Basic operations in the sequence are usually simple. It is the integration and coordination of many such operations into one piece of equipment that makes the system complex.

Inflexible—changes in the process to accommodate product changeovers are difficult and costly.

Examples of fixed automation are found in transfer lines, automatic assembly lines, oil refineries, and certain chemical processes.

In *programmable automation,* the production equipment is designed to be flexible. The distinguishing features of a programmable automation system are:

High investment cost (but not as high as for a fixed automation system).

Capability to change the sequence of operations to adapt to different product configurations. The sequence of operations is controlled by a *program* (of instructions). The system is reprogrammable to change the operation sequence.

Flexibility makes the system suitable for low-quantity production runs of different products.

Low production rates relative to fixed automation systems.

A good example of programmable automation is a numerically controlled machine tool. A "part program," coded on punched tape, controls the sequence of operations to machine the workpart. New workpart designs are accommodated by means of new part programs.

In recent years there have been several trends in automation, not only in the automated machines but also in the thinking of manufacturing people who are responsible for the planning, design, and operation of these machines. One of the trends is to envision the separate machine tools, part transfer devices, and human operators as components of integrated production systems. This viewpoint provides a conceptual framework for determining the interrelationships and parameter values of the components which achieve optimum performance of the overall system. Another trend, based on the production systems concept noted above, is to attempt to achieve the best features of fixed automation and flexible automation in one highly automated system. The name given to this type of automation is "flexible manufacturing system," and we will be discussing this type of system in the final chapter of this book. A third trend in manufacturing automation is the tremendous increase in the use of digital computers to control production equipment. Related to this, computerized manufacturing data bases and information systems are being developed for overall planning, scheduling, and coordination of plant operations. These trends, taken together, lead us to the ultimate conclusion that we will see, probably by the year 2000, the computer-integrated automatic factory. On the basis of these trends, all of which are discussed herein, we have titled the book *Automation, Production Systems, and Computer-Aided Manufacturing.*

## 1.2 REASONS FOR AUTOMATING

A number of economic and social factors provide the motivation for automation in manufacturing. These reasons for automating include the following:

1. *Increased productivity.* Automation of manufacturing operations holds the promise of increasing the productivity of labor. This means greater output per hour of labor input. Higher production rates (output per hour) are achieved with automation than with the corresponding manual operations.

2. *High cost of labor.* The trend in the industrialized societies of the world has been toward ever-increasing labor costs. As a result, higher investment in automated equipment has become economically justifiable to replace manual operations. The high cost of labor is forcing business leaders to substitute machines for human labor. Because machines can produce at higher rates of output, the use of automation results in a lower cost per unit of product.

3. *Labor shortages.* In many advanced nations there has been a general shortage of labor. West Germany, for example, has been forced to import labor to augment its own labor supply. Labor shortages also stimulate the development of automation as a substitute for labor.

4. *Trend of labor toward the service sector.* This trend has been especially prevalent in the United States. At this writing (1979), the proportion of the work force employed in manufacturing stands at about 23%. In 1947, this percentage was 30%. By the year 2000, some estimates put the figure as low as 2% [6].[1] Certainly, automation of production jobs has caused some of this shift. However, there are also social and institutional forces that are responsible for the trend. The growth of government employment at the federal, state, and local levels has consumed a certain share of the labor market which might otherwise have gone into manufacturing. Also, there has been a tendency for people to view factory work as tedious, demeaning, and dirty. This view has caused them to seek employment in the service sector of the economy (government, insurance, personal services, legal, sales, etc.).

5. *Safety.* By automating the operation and transferring the operator from an active participation to a supervisory role, work is made safer. The safety and physical well-being of the worker has become a national objective with the enactment of the Occupational Safety and Health Act of 1970 (OSHA). It has also provided an impetus for automation.

6. *High cost of raw materials.* The high cost of raw materials in manufacturing results in the need for greater efficiency in using these materials. The reduction of scrap is one of the benefits of automation.

7. *Improved product quality.* Automated operations not only produce parts at faster rates than do their manual counterparts, but they produce parts with greater consistency and conformity to quality specifications.

8. *Reduced manufacturing lead time.* For reasons that we shall examine in subsequent chapters, automation allows the manufacturer to reduce the time between customer order and product delivery. This gives the manufacturer a competitive advantage in promoting good customer service.

9. *Reduction of in-process inventory.* Holding large inventories of work-in-process represents a significant cost to the manufacturer because it ties up capital. In-process inventory is of no value. It serves none of the purposes of raw materials stock or finished product inventory. Accordingly, it is to the manufacturer's advantage to reduce work-in-progress to a minimum. Automation tends to accomplish this goal by reducing the time a workpart spends in the factory.

All of these factors act together to make production automation a viable and attractive alternative to manual methods of manufacture.

[1]Numbers in brackets refer to the References at the end of the chapter.

## 1.3 ARGUMENTS FOR AND AGAINST AUTOMATION

Since the time when production automation became a national issue in the late 1950s and early 1960s, labor leaders and government officials have debated the pros and cons of automation technology. Even business leaders, who generally see themselves as advocates of technological progress, have on occasion questioned whether automation was really worth its high investment cost. There have been arguments to limit the rate at which new production technology should be introduced into industry. By contrast, there have been proposals that government (federal and state) should not only encourage the introduction of new automation, but should actually finance a portion of its cost. (The Japanese government does it.) In this section, we discuss some of these arguments for and against automation.

### Arguments against automation

First, the arguments against automation include the following:

1. Automation will result in the subjugation of the human being by the machine. This is really an argument over whether workers' jobs will be downgraded or upgraded by automation. On the one hand, automation tends to transfer the skill required to perform work from human operators to machines. In so doing, it reduces the need for skilled labor. The manual work left by automation requires lower skill levels and tends to involve rather menial tasks (e.g., loading and unloading workparts, changing tools, removing chips, etc.). In this sense, automation tends to downgrade factory work. On the other hand, the routine monotonous tasks are the easiest to automate, and are therefore the first jobs to be automated. Fewer workers are thus needed in these jobs. Tasks requiring judgment and skill are more difficult to automate. The net result is that the overall level of manufacturing labor will be upgraded, not downgraded.

2. There will be a reduction in the labor force, with resulting unemployment. It is logical to argue that the immediate effect of automation will be to reduce the need for human labor, thus displacing workers. Because automation will increase productivity by a substantial margin, the creation of new jobs will not occur fast enough to take up the slack of displaced workers. As a consequence, unemployment rates will accelerate.

3. Automation will reduce purchasing power. This follows from argument 2. As machines replace workers and these workers join the unemployment ranks, they will not receive the wages necessary to buy the products brought by automation. Markets will become saturated with products that people cannot afford to purchase. Inventories will grow. Production will stop. Unemployment will reach epidemic proportions. And the result will be a massive economic depression.

## Arguments in favor of automation

Some of the arguments against automation are perhaps overstated. The same can be said of some of the declarations that advocate the new manufacturing technologies. The following is a sampling of the arguments for automation:

1. Automation is the key to the shorter workweek. There has been and is a trend toward fewer working hours and more leisure time. (College engineering professors seem excluded from this trend.) Around the turn of the century, the average workweek was about 70 hours per week. The standard is currently 40 hours (although many in the labor force work overtime). The argument holds that automation will allow the average number of working hours per week to continue to decline, thereby allowing greater leisure hours and a higher quality of life.
2. Automation brings safer working conditions for the worker. Since there is less direct physical participation by the worker in the production process, there is less chance of personal injury to the worker.
3. Automated production results in lower prices and better products. It has been estimated that the cost to machine one unit of product by conventional general-purpose machine tools requiring human operators may be 100 times the cost of manufacturing the same unit using automated mass-production techniques [3]. Examples abound. The machining of a V-8 automobile engine block by transfer line techniques (to be discussed in Chapters 4 and 5) may cost $25 to $35. If conventional techniques were used on reduced quantities (and the quantities would indeed be much lower if conventional methods were used), the cost would increase to around $3000. The electronics industry offers many examples of improvements in manufacturing technology that have significantly reduced costs while increasing product value (e.g., color TV sets, stereo equipment, hand-held calculators, and computers).
4. The growth of the automation industry will itself provide employment opportunities. This has been especially true in the computer industry. As the companies in this industry have grown (IBM, Burroughs, Digital Equipment Corp., Honeywell, etc.), new jobs have been created. These new jobs include not only workers directly employed by these companies, but also computer programmers, systems engineers, and others needed to use and operate the computers.
5. Automation is the only means of increasing our standard of living. Only through productivity increases brought about by new automated methods of production will we be able to advance our standard of living. Granting wage increases without a commensurate increase in productivity will result in inflation. In effect, this will reduce our standard of living. To afford a better society, we must increase productivity faster than we increase wages and salaries. Therefore, as this argument proposes, automation is a requirement to achieve the desired increase in productivity.

No comment is offered on the relative merits of these arguments for and against automation. This book will be concerned principally with the technical and engineering aspects of automated production systems. Included within the engineering analysis is, of course, consideration of the economic factors that determine the feasibility of an automation project.

## 1.4 ORGANIZATION OF THIS BOOK

This book is organized into six parts. Part I is introductory and covers some fundamental concepts in manufacturing and automation. This chapter has attempted to define the term "automation" and to explore various reasons why automation projects should be considered in manufacturing. Chapter 2 is a description of the way production is carried out in industry. The functional organization that typically exists in a manufacturing firm, as well as the different types of production plants, are considered. Types of production operations and general mathematical models of these operations are presented in Chapter 2. Finally, Chapter 2 develops a list of nine "strategies for automation," based primarily on an analysis of the mathematical models. Chapter 3 deals with the fundamentals of economic analysis required in manufacturing. In this chapter we consider the different types of costs in manufacturing (fixed costs, variable costs, overhead, etc.), and how to analyze them for potential automation projects.

Part II contains three chapters and is concerned with high-volume discrete-parts production systems. This type of automation is often called "Detroit automation" because of its beginnings in the automobile industry. Chapter 4 describes the various types of equipment that are used, such as transfer lines and automatic assembly machines. Chapter 5 is concerned with the mathematical modeling of these automated flow line systems. Chapter 6 presents several methods of line balancing. Line balancing is important not only in automated lines, but also in manual assembly line production systems.

Part III includes three chapters on numerical control (NC). Detroit automation is for high-volume production, while numerical control represents automation for low-quantity production. Chapter 7 presents the fundamentals of NC. In Chapter 8, we consider how to program for NC. This is called "part programming," and we consider the two types of NC part programming: manual and computer-assisted. Numerical control was first demonstrated in 1952. Since that time there has been substantial progress in the development of new NC systems. In Chapter 9, we consider some of these new advances in NC technology. They include the use of the computer in NC machines, adaptive control, and industrial robots.

Part IV, entitled "Computer-Aided Manufacturing," contains six chapters. Chapter 10 is introductory. It defines computer-aided manufacturing (CAM) and presents a framework for dealing with topics in CAM.

Chapter 11 covers the fundamentals of computers and how to connect the computer to the manufacturing process. We divide the applications of the computer into two categories:

1. Direct monitoring and control.
2. Indirect applications in which the computer is used to support plant operations.

Chapters 12 through 15 are concerned with direct monitoring and control. Chapter 12 covers computer process monitoring. In Chapter 13, we present the fundamentals of process control—linear control theory. This is a basic concept in automation. Although Chapter 13 is not directly concerned with computer process control, it is prerequisite material for a comprehensive understanding of computer control. Chapter 14 is on direct digital control (DDC), which is based on the principles of linear control theory (Chapter 13). Chapter 15 deals with supervisory computer control, which is distinguished from DDC by virtue of its objective to optimize the performance of the process. Topics include steady-state optimization, adaptive control, and on-line search strategies.

Part V, "Production Systems for Manufacturing Support," consists of two chapters and considers many of the indirect applications of the computer in manufacturing. Chapter 16 covers systems for manufacturing support at the process level. This includes computer generation of work standards and computerized machinability data systems. Chapter 17 is concerned with production systems at the plant level. Material requirements planning is the principal topic in this chapter.

Part VI has two chapters on group technology and flexible manufacturing systems. Chapter 18 discusses group technology, parts classification and coding, and computer-automated process planning. Chapter 19 presents a description of the flexible manufacturing system, a concept that combines automated flow lines with NC machines under computer control. The final section of the chapter discusses the computer-automated factory of the future.

## REFERENCES

[1] Bowen, H.R., and Mangum, G.L. (Editors), *Automation and Economic Progress,* Prentice-Hall, Inc., Englewood Cliffs, N.J., 1966.

[2] Buckingham, W., *Automation,* Harper & Row, Publishers, Inc., New York, 1961.

[3] Cook, N.H., "Computer-Managed Parts Manufacture," *Scientific American,* February, 1975, pp. 22–29.

[4] Harrington, J., *Computer Integrated Manufacturing,* Industrial Press, Inc., New York, 1973.

[5] LUKE, H.D., *Automation for Productivity,* John Wiley & Sons, Inc., New York, 1972.

[6] MERCHANT, M.E., "The Inexorable Push for Automated Production," *Production Engineering,* January, 1977, pp. 44–49.

[7] SILBERMAN, C.E., and the Editors of *Fortune, The Myths of Automation,* Harper & Row, Publishers, New York, 1966.

[8] TERBORGH, G., *The Automation Hysteria,* W.W. Norton & Co., Inc., New York, 1966.

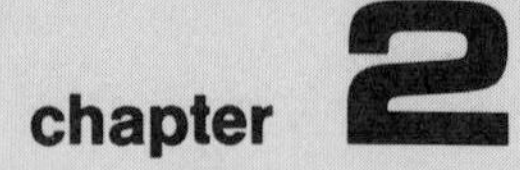

# Production Operations and Automation Strategies

## 2.1 WHAT IS PRODUCTION?

Production is a transformation process in which raw materials are converted into goods that have value in the marketplace. The goods are produced by a combination of manual labor, machinery, special tools, and energy. The transformation process usually involves a sequence of steps, each step bringing the materials closer to the desired final state. The individual steps are referred to as *production operations*.

Production operations occur in so many different forms that it is difficult to generalize the activity into types. In this chapter, we will explore the various basic industries, types of production plants, and production operations that exist in conventional manufacturing. This will include the derivation of a general mathematical model to represent production operations. From this foundation, a list of nine "automation strategies" will be developed. These are approaches for automating not only the individual operations but the entire manufacturing plant. The strategies form the basis for the specific topics in automation, production systems, and computer-aided manufacturing to be presented throughout the book.

## 2.2 BASIC INDUSTRIES

There is a wide variety of basic industries, including not only manufacturing but all others as well. By examining the publicly held corporations whose shares are traded on the major stock exchanges, it is possible to compile a list of industry types. Such a list is presented in Table 2.1. This list includes all types of industrial corporations, banks, utilities, and so on. Our interest in this book is on industrial firms that are engaged in production. Table 2.2 is a list of basic industries that produce goods, together with examples of companies that are members of these industries.

The companies represented by Table 2.2 can be divided into two types, depending on the nature of their production operations. The two types are the manufacturing industries and the process industries. *Manufacturing companies* are typically identified with discrete-item production: cars, computers, machine tools, and the components that go into these products. The *process industries* are represented by chemicals and plastics, petroleum products, food

**TABLE 2.1 Basic Industries: General**

| |
|---|
| Advertising |
| Aerospace |
| Automotive (cars, trucks, buses) |
| Beverages |
| Building materials |
| Cement |
| Chemicals |
| Clothing (garments, shoes) |
| Construction |
| Drugs, soaps, cosmetics |
| Equipment and machinery |
| Financial (banks, investment companies, loans) |
| Foods (canned, dairy, meats, etc.) |
| Hospital supplies |
| Hotel/motel |
| Insurance |
| Metals (steel, aluminum, etc.) |
| Natural resources (oil, coal, forest, etc.) |
| Paper |
| Publishing |
| Radio, TV, motion pictures |
| Restaurant |
| Retail (food, department store, etc.) |
| Shipbuilding |
| Textiles |
| Tire and rubber |
| Tobacco |
| Transportation (railroad, airlines, trucking, etc.) |
| Utilities (electric power, natural gas, telephone) |

TABLE 2.2 Basic Industries: Manufacturing and Process Industries

| *Basic industry* | *Representative company* |
|---|---|
| Aerospace | Boeing Co. |
| Automotive | General Motors |
| Beverages | Coca-Cola |
| Building materials | U.S. Gypsum |
| Cement | Lone Star Industries |
| Chemicals | E.I. du Pont |
| Clothing | Hanes Corp. |
| Drugs, soaps, cosmetics | Procter & Gamble |
| Equipment and machinery | |
| Agricultural | Deere |
| Construction | Caterpillar Tractor |
| Electrical | General Electric |
| Electronics | Hewlett-Packard |
| Household appliances | Maytag |
| Industrial | Ingersoll-Rand |
| Machine tools | Cincinnati Milacron |
| Office equipment, computers | IBM |
| Railroad equipment | Pullman |
| Steam generating | Combustion Engineering |
| Foods | |
| Canned foods | Green Giant |
| Dairy products | Borden |
| Meats | Oscar Mayer |
| Packaged foods | General Mills |
| Hospital supplies | American Hospital Supply |
| Metals | |
| Aluminum | Alcoa |
| Copper | Kennecott |
| Steel | U.S. Steel |
| Natural resources | |
| Coal | Pittston |
| Forest | Georgia-Pacific |
| Oil | Exxon |
| Paper | Kimberly Clark |
| Textiles | Burlington Industries |
| Tire and rubber | Goodyear |

processing, soaps, steel, and cement. Our focus in this book will be on manufacturing.

There are other ways to classify companies. One alternative would be to place a company into one of three categories:

1. Basic producer.
2. Converter.
3. Fabricator.

The three types form a connecting chain in the transformation of natural resources and basic raw materials into goods for the consuming public. The *basic producers* take the natural resources and transform these into the raw materials used by other industrial manufacturing firms. For example, steel producers transform iron ore into steel ingots. The *converter* represents the intermediate link in the chain. The converter takes the output of the basic producer and transforms these raw materials into various industrial products and some consumer items. For example, the steel ingot is converted into bar stock or sheet metal. Chemical firms transform petroleum products into plastics for molding. Paper mills convert wood pulp into paper. A distinguishing charactertistic of the converter is that its products are uncomplicated in physical form. The products are not assembled items. The production processes used to make the products may be complex but the products themselves are not.

The third category of manufacturing firms is the *fabricator*. These firms fabricate and assemble final products. The bar stock and sheet metal are transformed into machined engine components and automobile body panels. The plastics are molded into various shapes. Then these parts are assembled into final products, such as trucks, automobiles, appliances, garments, and machine tools. Fabricators include both the firms that produce the components and those which assemble the components into consumer goods.

There are several complicating factors in this classification. Some firms possess a high degree of *vertical integration*, which means that their operations include all three categories. The major oil firms are examples of vertical integration. They convert natural resources into finished petroleum products and then market these products directly to the consumer. Another complicating factor is that some companies—the conglomerates—are in so many different types of business that it is difficult to classify them. Some of their operations are in the basic producer category; others are converters; and still other lines of business fall into the fabricator category.

Let us next consider how these various industrial firms are organized to carry out their production functions.

## 2.3 ORGANIZATION IN MANUFACTURING FIRMS

In this section, we shall consider the organizational functions within a manufacturing firm classified as a fabricator. That is, it manufactures discrete components and assembles these components into final products for sale to its customers. Many medium-size and large corporations fall into this category. For some firms, hundreds of different products are manufactured. Each product is made up of individual components, sometimes numbering in the thousands. The task of organizing and coordinating the activities of the company to perform its production function is complex.

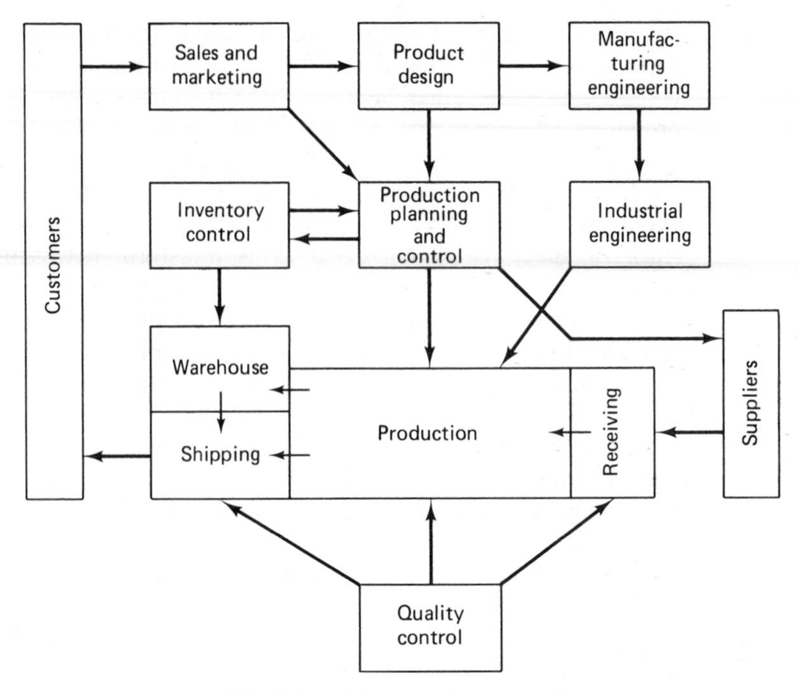


FIGURE 2.1 The manufacturing cycle.

Rather than resort to the traditional organization charts which show the typical line and staff structure, we shall consider the creation of the product from its beginning until the time the product is shipped. The sequence of events is sometimes called the *manufacturing cycle* and is pictured in Figure 2.1.

The steps in the manufacturing cycle will be different depending on the type of industry, product, size of the company, and management style. In general, though, the following functions will be performed within the firm in order to manufacture a product:

1. Sales and marketing.
2. Product design and engineering.
3. Manufacturing engineering.
4. Industrial engineering.
5. Production planning and control.
6. Manufacturing.
7. Quality control.
8. Shipping and inventory control.

The traditional responsibilities included in these functions will be discussed in the following subsections:

### Sales and marketing

The order to produce the product will typically originate from the sales and marketing department of the firm. It will be in one of three forms:

1. A customer order to manufacture an item to the customer's specifications.
2. A customer order to buy one (or more) of the manufacturer's proprietary products.
3. An order based on a forecast of future demand for a proprietary product. The forecast will be made by the marketing staff in coordination with the production control and inventory control departments.

### Product design and engineering

If the item is to be manufactured to specifications, the design will have been prepared by the customer. In this case, the manufacturer's design work will be nil. On the other hand, if the product is proprietary, the manufacturer is responsible for its development and design.

The product design is documented by means of component drawings, specifications, and a bill of materials that defines how many of each component go into the product. A prototype is often built for testing and demonstration purposes. The manufacturing engineering department is sometimes consulted to lend advice on matters of produceability. What changes in design could be made to reduce production costs without sacrificing function? Cost estimates are prepared to establish an anticipated price for the product.

Upon completion of the design and fabrication of the prototype, the top company management is invited in for a "show-and-tell" presentation. The design engineer in charge gives a presentation and demonstration of the product so that management can decide whether to manufacture the item. This decision is often a two-step procedure. The first is a decision by engineering management that the design is approved. Many companies call this an "engineering release." The second step is a decision by corporate management as to the general suitability of the product. This second decision represents an authorization to produce the item.

### Manufacturing engineering

The manufacturing engineering function consists typically of four responsibilities. First, as mentioned above, manufacturing engineers provide advice to the product design department on produceability. Ideally, the manufacturing engineer should be introduced to the product design near its inception so that recommendations can be offered which reflect the interests of the production departments.

Second, the principal purpose of manufacturing engineering is process planning. *Process planning* consists of determining the sequence of individual manufacturing operations needed to produce the part. The document used to specify this process sequence is called a *route sheet*. The route sheet is a listing of production operations and associated machine tools for each workpart.

The third responsibility of manufacturing engineering is the specification and design of tools, jigs, and fixtures used to produce the product. Decisions on new machine tools and other equipment are made by the manufacturing engineering department in many companies.

Fourth, after the product is in production, problems invariably arise. Out-of-tolerance materials are received from vendors, fixtures work improperly, the production shops make mistakes, components cannot be assembled, production workers on incentive sacrifice quality for the sake of their piece rate, and a host of other problems are encountered. It is the responsibility of manufacturing engineering to solve these problems or to seek out the people who can solve them. This part of the manufacturing engineer's job is sometimes called "troubleshooting."

### Industrial engineering

The purpose of industrial engineering in the manufacturing cycle is to determine the work methods and time standards for the individual production operations. The purpose of determining work methods is to find the best way to perform the task and to standardize and document that method. The purpose of setting time standards is to determine how much time the task should take using the standard method. Time standards may be used by the company for wage incentive and other purposes.

Additional functions undertaken by the industrial engineering department include cost reduction and productivity improvement, plant layout studies, equipment justification, and operations research projects.

### Production planning and control

The authorization to manufacture the product must be translated into a *master schedule* which specifies how many units of each product are to be delivered and when. In turn, this master schedule must be converted into purchase orders for raw materials, orders for components from outside vendors, and production schedules for parts made in the shop. These events must be timed and coordinated to allow delivery of the final product according to the master schedule. All of these activities are the responsibility of the production planning and control department. Production planning and control consists of several functions, as described in the paragraphs below.

MASTER SCHEDULE. The master schedule is a listing of the products to be produced, when they are to be delivered, and in what quantities. The usual scheduling periods in the master schedule are months. The master schedule must be consistent with the plant's production capacity. It should not list more quantities of products than the plant is capable of producing with its given number of machines and manpower.

REQUIREMENTS PLANNING. Based on the master schedule, the individual components and subassemblies that make up each product must be planned. Raw materials must be ordered to make the various components. Purchased parts must be ordered. And all of these items must be planned so that the components and assemblies are available when needed. This whole task is called *requirements planning* or *material requirements planning*.

SCHEDULING. Based on the outcome of the requirements planning activity, the next task is production scheduling. This involves the assignment of start dates and due dates for the various components to be processed through the factory. Several factors make the scheduling job complex. First, the number of individual parts and orders to be scheduled may run into the thousands. Second, each part has its own individual process routing to be followed. Some parts may have to routed through dozens of separate machines. Third, the number of machines in the shop is limited, and the machines are different. They perform different operations and they have different features and capacities. In effect, the orders compete with one another for time on the machines.

DISPATCHING. Based on the production schedule, the dispatching function is concerned with issuing the individual orders to the machine operators. This involves giving out order tickets, route sheets, part drawings, and job instructions. The dispatching function in some shops is performed by the shop foremen, in other shops by a person called a dispatcher.

EXPEDITING. Even with the best plans and schedules, things go wrong. It is the expediter's job to compare the actual progress of the order against the production schedule. For orders that fall behind schedule, the expediter takes corrective action. This may involve rearranging the sequence in which orders are to be done on a certain machine, coaxing the foreman to tear down one setup so that another order can be run, or hand-carrying parts from one department to the next just to keep production going. There are many reasons why things go wrong in production: parts-in-process have not yet arrived from the previous department, machine breakdowns, proper tooling is not available, quality problems, and so on.

### Manufacturing

The various production departments in the factory perform the operations that transform raw materials into finished goods. We shall discuss the various types of production shops in Section 2.4 and the manufacturing operations that are performed by them in Section 2.5. Included within the activities of the factory are such functions as materials handling (moving raw materials out of the warehouse and transporting work-in-process between the production departments) and machine maintenance (repairing the equipment when breakdowns occur, preventive maintenance, etc.).

### Quality control

The quality control department is responsible for assuring that the quality of the product and its components meets the standards specified by the designer. This function must be accomplished at various points throughout the manufacturing cycle. Materials and parts purchased from outside suppliers must be inspected when they are received. Parts fabricated inside the company must be inspected, usually several times during processing. Final inspection of the finished product is performed to test its overall functional and appearance quality.

### Shipping and inventory control

The final step in the manufacturing cycle involves shipping the product directly to the customer or stocking the item in inventory. The purpose of inventory control is to ensure that enough products of each type are available to satisfy customer demand. Competing with this objective is the desire that the company's financial investment in inventory be kept at a minimum. Inventory control interfaces with marketing and production control since there must be coordination between the various products' sales, production, and inventory level. None of these three functions can operate effectively without knowledge of what the others are doing. Inventory control is often included within the production control department.

The inventory control function applies not only to the company's final products. It also applies to raw materials, purchased components, and work-in-process within the factory. In each case, planning and control are required to achieve a balance between the danger of too little inventory (with possible stock-outs of raw materials) and the expense of too much inventory.

## 2.4 PRODUCTION PLANTS

There are several ways of classifying production shops. We will examine two classification schemes: according to the volume and rate of production, and according to type of plant layout.

### Plants for different production quantities

In terms of production volume, manufacturing plants can be classified into three types:

1. Job shop production.
2. Batch production.
3. Mass production.

This classification is normally associated with discrete-product manufacture, but it can also serve for plants used in the process industries. For example, some chemicals are produced in batches (batch production) whereas others are produced by continuous-flow processes (mass production). The three types of production shops are related to production volume as shown in Figure 2.2.

JOB SHOPS. The distinguishing feature of job shop production is low volume. The manufacturing lot sizes are small, often one of a kind. Job shop production is commonly used to meet specific customer orders, and there is a great variety in the kind of work the plant must do. Therefore, the production equipment must be flexible and general-purpose to allow for this variety of work. Also, the skill level of job shop workers must be relatively high so that they can perform a range of different work assignments. Examples of products manufactured in a job shop would be space vehicles, aircraft, machine tools, special tools and equipment, and prototypes of future products.

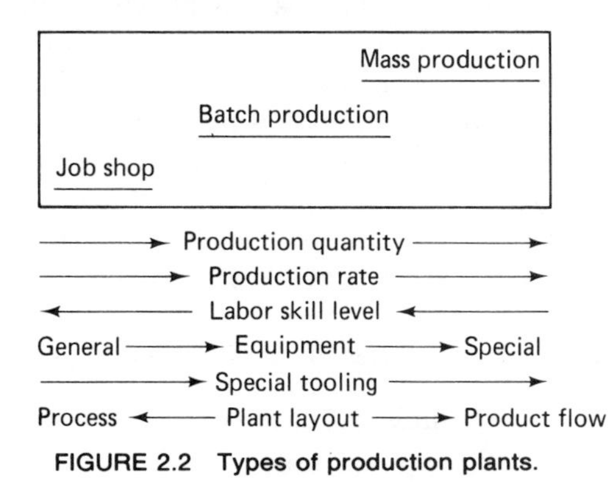


**FIGURE 2.2 Types of production plants.**

Construction work and shipbuilding are not normally identified with the job shop category, even though the quantities are in the appropriate range. Although these two activities involve the transformation of raw materials into finished products, the work is not performed in a factory.

BATCH PRODUCTION. This category involves the manufacture of medium-size lots of the same item or product. The lots may be produced only once, or they may be produced at regular intervals. The purpose of batch production is often to satisfy continuous customer demand for an item. However, the plant is capable of a production rate that exceeds the demand rate. Therefore, the shop produces to build up an inventory of the item. Then it changes over to other orders. When the stock of the first item becomes depleted, production is repeated to build up the inventory again.

The manufacturing equipment used in batch production is general-purpose but designed for higher rates of production. For example, turret lathes capable of holding several cutting tools are used rather than engine lathes. The machine tools used in batch manufacture are often combined with specially designed jigs and fixtures which increase the output rate. Examples of items made in batch-type shops include many types of industrial equipment, furniture, textbooks, and component parts for many assembled consumer products (household appliances, lawn mowers, etc.). Batch production plants include machine shops, casting foundries, plastic molding factories, and pressworking shops. Some types of chemical plants would also be included in this general category.

It has been estimated that perhaps as much as 75% of all parts manufacturing is in lot sizes of 50 pieces or less. Hence, batch production and job shop production constitute an important portion of total manufacturing activity.

MASS PRODUCTION. This is the continuous specialized manufacture of identical products. Mass production is characterized by very high production rates, equipment that is completely dedicated to the manufacture of a particular product, and very high demand rates for the product. Not only is the equipment dedicated to one product, but the entire plant is often designed for the exclusive purpose of producing the particular product. The equipment is special-purpose rather than general-purpose. The investment in machines and specialized tooling is high. In a sense, the production skill has been transferred from the operator to the machine. Consequently, the skill level of labor in a mass production plant tends to be lower than in a batch plant or job shop.

Two categories of mass production can be distinguished:

1. Quantity production.
2. Flow production.

Quantity production involves the mass production of single parts on fairly standard machine tools such as punch presses, injection molding machines, and automatic screw machines. These standard machines have been adapted to the production of the particular part by means of special tools—die sets, molds, and form cutting tools, respectively—designed for the part in question. The production equipment is devoted full time to satisfy a very large demand rate for the item. In mass production, the demand rate and the production rate are approximately equal. Examples of items in quantity production include components for assembled products that have high demand rates (automobiles, some household appliances, light bulbs, etc.), hardware items (such as screws, nuts, and nails), and many plastic molded products.

Flow production is the other category of mass production. The term suggests the physical flow of the product in oil refineries, continuous chemical process plants, and food processing. While these are examples of flow production, the term also applies to the manufacture of either complex single parts (such as automotive engine blocks) or assembled products. In these cases, the items are made to "flow" through a sequence of operations by material handling devices (conveyors, moving belts, transfer devices, etc.). Examples of flow production would include automated transfer machines for the production of complex discrete parts, and manual assembly lines for the assembly of complex products.

Figure 2.2 summarizes some of the important characteristics of these different types of production plants. It will be noted that the production ranges of the three major categories overlap to some degree. The reason is simply that it is difficult to draw a clear dividing line between the different types of shop.

## Types of plant layout

The term *plant layout* refers to the physical arrangement of the facilities within the production plant. A plant layout that is most suited to flow-type mass production would be quite impractical for job shop production, and vice versa. There are three types of plant layout that are associated with traditional production shops:

1. Fixed-position layout.
2. Process layout.
3. Product-flow layout.

We will see that there is a considerable correlation between the type of plant layout and the type of production shops as previously classified according to quantity.

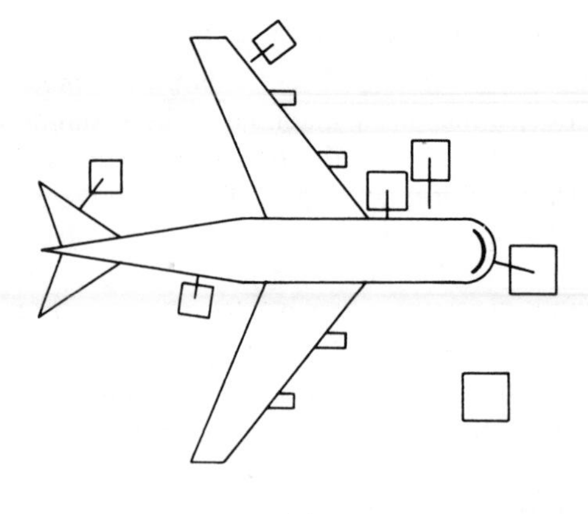

(a)

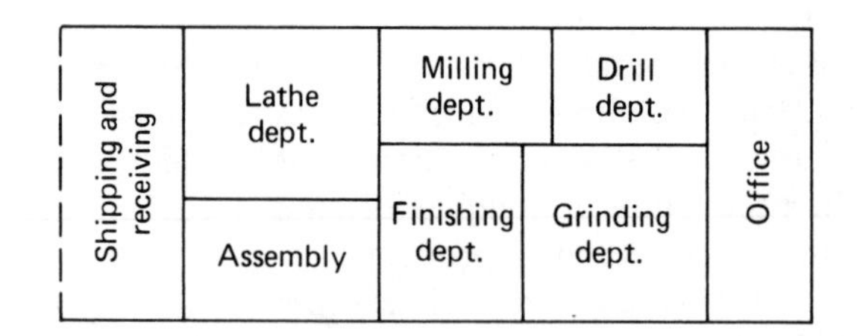


(b)

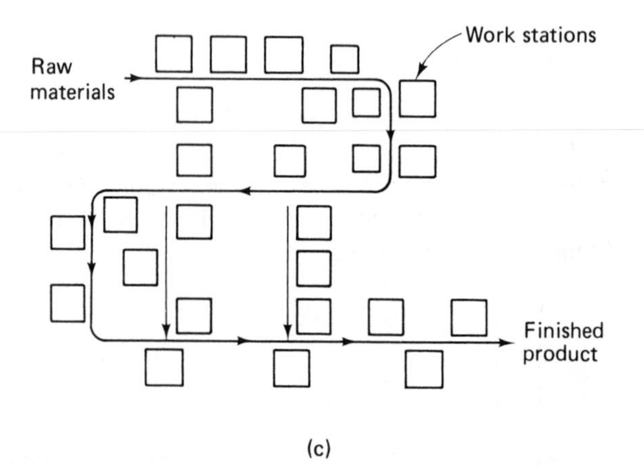


(c)

FIGURE 2.3 Types of plant layout: (a) fixed-position layout, (b) process layout, (c) product flow layout.

Fixed-Position Layout. In this layout, the term "fixed-position" refers to the product. Because of its size and weight, the product remains in one location and the equipment used in its fabrication is brought to it. Large aircraft assembly and shipbuilding would be examples of where the fixed-position layout would be utilized. The fixed-position layout is illustrated in Figure 2.3(a).

Another arrangement of facilities, similar to the fixed-position type, is the *project layout*. This is used for construction jobs such as buildings, bridges, and dams. As with the fixed-position layout, the product is large and the construction equipment and manpower must be moved to the product. Unlike the fixed-position arrangement, when the job is completed, the equipment is removed from the construction site. In a fixed-position layout, the product is eventually moved out of the plant and the plant remains for the next job. This type of arrangement is often associated with job shops in which complex products are fabricated in very low quantities.

Process Layout. In the process layout, the production machines are arranged into groups according to general type of manufacturing process. The lathes are in one department, drill presses are in another, plastic molding in still another department, and so forth. The advantage of this layout is its flexibility. Different parts, each requiring its own unique sequence of operations, can be routed through the respective departments in the proper order. Forklift trucks and hand carts are used to move materials from one work center to the next.

The process layout is typical in job shops and batch production. It is also used in quantity-type mass production. The process layout is illustrated in Figure 2.3(b).

Product-Flow Layout. If a plant specialized in the production of one product or one class of product in large volumes, the plant facilities should be arranged to produce the product as efficiently as possible. For complex assembled products, or items requiring a long sequence of operations, this efficiency is usually best achieved with the product-flow layout. With this type of layout, the processing and assembly facilities are placed along the line of flow of the product. The work-in-progress is moved by conveyor or similar means from one workstation to the next. The product is progressively fabricated as it flows through the sequence of workstations. As the name implies, this type of layout is appropriate for flow-type mass production. The arrangement of facilities within the plant is relatively inflexible and is warranted only when the production quantities are large enough to justify the investment. Figure 2.3(c) illustrates the product-flow layout.

These three layouts (fixed-position, process, and product-flow) are the conventional types found in manufacturing plants today. As we shall see in

Chapter 18, a fourth type, called the group technology layout, represents an attempt to combine the efficiency of the flow layout with the flexibility of the process layout.

## 2.5 PRODUCTION OPERATIONS

The number of different production processes represented by the manufacturing industries of Table 2.2 is immense. It is not the purpose of this section to survey all these processes. We leave the subject of manufacturing process technology to other texts, a few of which are listed in the references ([1], [5], [7]). The objective of this section is to examine some of the general characteristics of industrial processes.

The picture of a manufacturing shop developed in previous sections of this chapter is one in which a sequence of production operations is required to transform raw materials into finished product. We shall divide these operations into two types: processing operations and assembly operations. This classification is most applicable in discrete-product manufacturing.

### Processing operations

Processing operations include those manufacturing activities which transform a workpart from one state of completion into a more advanced state of completion. No materials or components are added to accomplish the transformation. Instead, energy is added (e.g., mechanical energy, heat, electrochemical energy) to change the shape of the workpart, remove material from it, change its physical properties, or other function. Processing operations can usually be classified into one of the following four categories:

1. Basic processes.
2. Secondary processes.
3. Operations to enhance physical properties.
4. Finishing operations.

Basic processes are those which give the work material its initial form. Metal casting and plastic molding would be examples. In both cases, the raw materials are converted into the basic geometry of the desired product. It is common for additional processing to be required to achieve the final shape and size of the workpart.

Secondary processes follow the basic process and are performed to give the workpart its final desired geometry. Examples in this category include machining (turning, drilling, milling, etc.) and pressworking operations (blanking, forming, drawing, etc.).

Operations to enhance physical properties do not perceptibly change the physical geometry of the workpart. Instead, the physical properties of the material are improved in some way. Heat-treating operations to strengthen metal parts and preshrinking used in the garment industry would be examples in this category.

Finishing operations are the final processes to be performed on the workpart. Their purpose is to improve the appearance, to provide a protective coating on the part, or some similar purpose. Examples in this fourth category include polishing, painting, and chrome plating.

### Assembly operations

Assembly and joining processes constitute the second major type of manufacturing operation. In assembly, the distinguishing feature is that two or more separate components are joined together. Included in this category are mechanical fastening operations, which make use of screws, nuts, rivets, adhesives, and so on, and joining processes, such as welding, brazing, and soldering. In the fabrication of a product, the assembly operations follow the processing operations.

As with any classification scheme, this classification of manufacturing operations leads to some ambiguities. There are processes that do not fit neatly into one single category. For example, the grinding operation might be considered as a finishing process since it is commonly used to provide a good finish and it does not perceptibly alter the final geometry of the workpiece. However, the grinding process does remove material from the workpiece. It is therefore classified as a machining process which is a secondary processing operation. In spite of such possible ambiguities, this classification scheme provides a perspective as to the types of manufacturing processes that take place in a production plant.

## 2.6 GENERAL MODELS OF PRODUCTION OPERATIONS

In this section, we examine several general production models, including both descriptive models and elementary mathematical ones.

### Operation sequence model

Before examining an individual production operation, let us review what happens to a typical workpart during processing. As mentioned previously, a series of operations are performed on the raw materials to produce the desired final form. In the routing of the workpart through the shop, it must be transported between successive operations. Delays often occur while the workpart waits in a queue for its turn to be processed on each machine. The

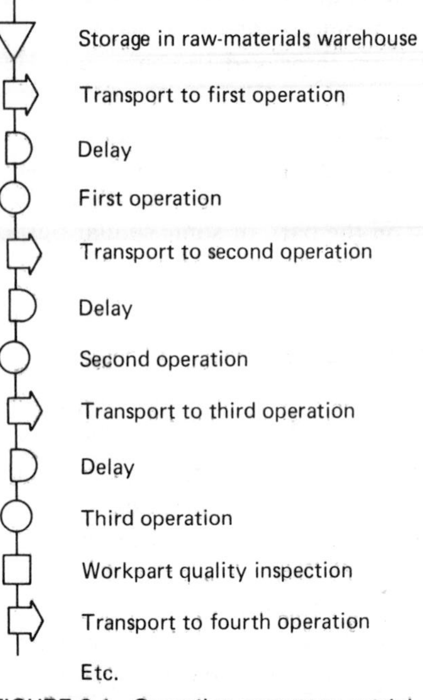


**FIGURE 2.4 Operation sequence model.**

series of events can be pictured as shown in Figure 2.4. This diagram represents a gross model of the sequence of production operations on a workpart. It can be seen that there is considerable nonproductive time during the sequence. Transports, unavoidable delays, and occasional inspections all represent events that take time but do not add value to the workpart.

It has been estimated[1] that of the total time an average part spends in a batch-type machine shop, only about 5% is spent on machine tools. In Figure 2.4, this 5% is symbolized by the circles (production operations are circles). Hence, 95% of the time spent in the shop is nonproductive. The obvious implication is that productivity could be improved if this waste could be reduced.

Of the 5% time spent on the machines, only 30% ($1\frac{1}{2}$% of the total) is occupied in actual metal cutting. The remaining 70% ($3\frac{1}{2}$% of the total) is nonproductive time on the machine. Again, this represents an opportunity for productivity gains.

It should be noted that these estimated percentages apply to batch-production machine shops. The percentages would be different for other types of production. Presumably, the corresponding nonproductive percentages would be significantly lower for flow-type mass production. Nevertheless, Figure

[1] M.E. Merchant, "The Inexorable Push for Automated Production," *Production Engineering*, January, 1977, pp. 45–46.

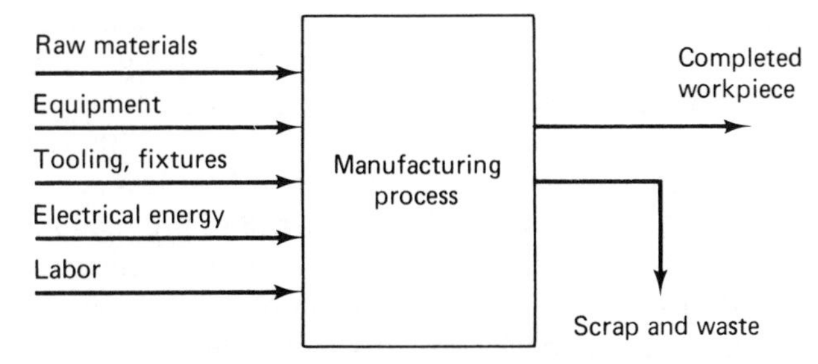


FIGURE 2.5 Manufacturing process model.

2.4 is a valid general model of production in which a sequence of operations are performed on the workpart.

## Manufacturing process model

Now let us examine a model for an individual production operation in the sequence. Figure 2.5 illustrates the input/output framework of a typical manufacturing process. Most production operations require five inputs:

1. Raw materials.
2. Equipment (machine tool).
3. Tooling and fixtures.
4. Energy (electrical energy).
5. Labor.

The manufacturing process adds value to the raw materials (or work-in-progress) by transforming them into a more desirable state. The process is usually carried out on production equipment that reflects a capital investment by the firm. The equipment is typically general-purpose and must be adapted to the particular workpart by the use of tools, fixtures, molds, die sets, and so on. This tooling must often be designed specifically for the given workpart. Electrical energy is required to operate the production equipment. Finally, labor is required to operate the equipment, load the raw workpart, unload the piece when the process is completed, check for malfunctions of the machine, and so forth.

The manufacturing process produces two outputs:

1. The completed workpiece.
2. Scrap and waste.

The term *completed workpiece* refers to the desired output of the particular manufacturing process. Since the workpart must be routed through several operations, only the last operation yields the finished part. The other operations produce work-in-progress for succeeding processes. As a by-product of all manufacturing processes, some scrap material and waste results. The scrap is in the form of metal chips (machining operations), skeleton (sheet metal press working operations), sprue and runner (plastic molding), and so on. The waste is represented by tools consumed in the operation, the mechanical inefficiency of the machine tool, and heat losses.

Having considered the sequence of production operations as well as the individual manufacturing processes in the sequence, we are now in a position to develop a mathematical model of production activity. The model will be simple, it will be based on production activity times, and it will serve to demonstrate the various automation strategies to be explored in Section 2.7.

### Mathematical models of production activity

The process sequence model illustrated in Figure 2.4 consists of many elements for a typical workpart. We shall reduce the complexity of the situation by dividing the elements into two main types: the operation elements and the nonoperation elements. An *operation element* is involved whenever the workpart is on the machine. The *nonoperation elements* include transportations, delays, and inspections. We shall not count the time the raw workpart spends in storage before its turn in the production schedule begins. Let us use $T_o$ to denote the operation time per machine or per operation, and $T_{no}$ to represent the nonoperation time associated with each process. If the number of machines through which the part must be routed is $n_m$, the total time to process the workpart through the plant (called the *manufacturing lead time*) is given by

$$\text{manufacturing lead time} = n_m(T_o + T_{no}) \tag{2.1}$$

This covers the case where there is only one workpart to be routed through the shop. If there are $Q$ workparts per batch, the throughput time is increased to reflect the operation times for all $Q$ workparts:

$$\text{manufacturing lead time} = n_m(QT_o + T_{no}) \tag{2.2}$$

This assumes that the nonoperation times per process occur simultaneously for all workparts in the batch. That is, all $Q$ parts are moved together, they wait in the queue next to the machine together, and so on. Equations (2.1) and (2.2) ignore the possibility of machine setup time. Nearly all production machines must be set up to process a particular workpiece. As indicated in Figure 2.5, special tools (e.g., fixtures, molds, die sets, cutting tools, etc.) must be used to adapt the machine to the given job. Fetching these tools, attaching them to the machine, and checking out the setup—all this takes time, which we shall call $T_{su}$. Adding this term to Eq. (2.2) we get

$$\text{manufacturing lead time} = n_m(T_{su} + QT_o + T_{no}) \tag{2.3}$$

All of the terms in Eqs. (2.1) through (2.3) are considered constants. In an actual job shop or batch plant, to which these equations are supposed to apply, the terms in the equations would vary. The batch size $Q$ would vary for different orders, the setup time $T_{su}$ and operation time $T_o$ would be different for different machines, the nonoperation time $T_{no}$ would tend to vary, and the process routings would not be the same for all parts; hence, $n_m$ would vary from part to part. In spite of this variability, Eq. (2.3)[2] can still be used if properly weighted average values of the terms $T_{su}$, $T_{no}$, and $T_o$ are utilized in the equation. Straight arithmetic averages are used for $Q$ and $n_m$. Specifically, the terms in Eq. (2.3) are computed as follows from plant data.

First, let $n_Q$ equal the number of batches of various parts to be considered. This number might represent the batches processed through the shop during a certain time period (e.g., month or year), or it might be a sample of batches to be used in the analysis. Given $n_Q$, the average batch quantity $Q$ is calculated by

$$Q = \frac{\sum_{i=1}^{n_Q} Q_i}{n_Q} \tag{2.4}$$

where $Q_i$ represents the batch quantity of batch $i$ among $n_Q$ batches. The value of $n_m$ for use in Eq. (2.3) is also an arithmetic average:

$$n_m = \frac{\sum_{i=1}^{n_Q} n_{mi}}{n_Q} \tag{2.5}$$

where $n_{mi}$ represents the number of operations (or machines) in the process routing for batch $i$. The setup time is given by

$$T_{su} = \frac{\sum_{i=1}^{n_Q} n_{mi} \overline{T}_{sui}}{\sum_{i=1}^{n_Q} n_{mi}} \tag{2.6}$$

where $\overline{T}_{sui}$ represents the average setup time for batch $i$. The nonoperation time is calculated similarly:

$$T_{no} = \frac{\sum_{i=1}^{n_Q} n_{mi} \overline{T}_{noi}}{\sum_{i=1}^{n_Q} n_{mi}} \tag{2.7}$$

[2] Equations (2.1) and (2.2) are special cases of Eq. (2.3).

where $\overline{T}_{noi}$ represents the average nonoperation time for batch $i$. Finally, the average operation time for use in Eq. (2.3) must be computed as the following weighted average:

$$T_o = \frac{\sum_{i=1}^{n_Q} n_{mi} Q_i \overline{T}_{oi}}{n_m Q n_Q} \tag{2.8}$$

where $\overline{T}_{oi}$ represents the average operation time per operation for batch $i$, and $n_m$ and $Q$ are the average values calculated by Eqs. (2.4) and (2.5).

Let us consider next those terms in Eq. (2.3) which represent the setup and process time at each machine. These two terms give the total batch time per machine:

$$\text{total batch time/machine} = T_{su} + QT_o \tag{2.9}$$

Dividing the total batch time per machine by the batch size gives the average production time per part for the machine. We shall symbolize this average production time $T_p$:

$$T_p = \frac{T_{su} + QT_o}{Q} \tag{2.10}$$

The average production rate $R_p$ at each machine is given by the reciprocal of $T_p$:

$$R_p = \frac{1}{T_p} \tag{2.11}$$

Equation (2.10) shows that the average production time is reduced as the batch size $Q$ is increased. When $T_p$ is reduced, the average production rate $R_p$ is increased.

We will next examine the components of the operation time $T_o$. The operation time is the time an individual workpart spends on a machine, but not all of this time is productive. Let us try to relate the operation time to a specific process. To illustrate, we will use a machining operation, as machining is common in discrete-parts manufacturing. Operation time for a machining operation is composed of three elements: the actual machining time $T_m$, the workpiece handling time $T_h$, and any tool handling time per workpiece $T_{th}$. Hence,

$$T_o = T_m + T_h + T_{th} \tag{2.12}$$

The tool handling time represents all the time spent in changing tools when they wear out, changing from one tool to the next for successive operations performed on a turret lathe, changing between the drill bit and tap in a drill and tap sequence performed at one drill press, and so on. $T_{th}$ is the average time per workpiece for any and all of these tool handling activities.

Each of the terms $T_m$, $T_h$, and $T_{th}$ has its counterpart in many other types of discrete-item production operations. There is a portion of the operation cycle, when the material is actually being worked ($T_m$), and there is a portion of the cycle when either the workpart is being handled ($T_h$) or the tooling is being adjusted or changed ($T_{th}$). We can therefore generalize upon Eq. (2.12) to cover many other manufacturing processes in addition to machining.

Let us consider two examples to illustrate the concepts embodied in these equations.

## EXAMPLE 2.1

Part number 090839 requires eight operations during its processing through the machine shop. The part is produced in batch sizes of 50. Average setup time per operation is 3.0 h. Average operation time per machine is 6 min. Average nonoperation time due to batch handling, delays and temporary storage, inspections, and so on, is 7.0 h. Compute the total time required to get the batch through the shop, assuming one 8-h shift per day.

The terms in Eq. (2.3) are as follows:

$$n_m = 8 \text{ machines} \qquad T_{su} = 3.0 \text{ h}$$

$$Q = 50 \text{ parts} \qquad T_o = 6/60 = 0.1 \text{ h}$$

$$T_{no} = 7.0 \text{ h}$$

Therefore, the manufacturing lead time is

$$8[3.0 + 50(0.1) + 7.0] = 120 \text{ h}$$

At 40 h/week, this amounts to 3 weeks from beginning to end of processing.

## EXAMPLE 2.2

The setup time for one particular operation during the process routing for part number 090839 of Example 2.1 is 3.0 h. The actual machining time is 2.5 min, workpart handling is 3.0 min, and tool handling time is 30 s. How much will the production rate be increased by increasing the batch size from 50 to 100 parts?

The terms in Eq. (2.12) are

$$T_m = 2.5 \text{ min}$$

$$T_h = 3.0 \text{ min}$$

$$T_{th} = 0.5 \text{ min}$$

Therefore,

$$T_o = 2.5 + 3.0 + 0.5 = 6.0 \text{ min/part}$$

Using Eq. (2.10), the average production time for a batch size of 50 parts is

$$T_p = \frac{3.0 + 50(6)/60}{50} = 0.16 \text{ h}$$

The corresponding production rate is 6.25 parts/h by Eq. (2.11).

By comparison, the average production time per part for a batch size of 100 is

$$T_p = \frac{3.0 + 100(6)/60}{100} = 0.13 \text{ h}$$

The corresponding production rate is 7.69 parts/h. This is a 23% increase.

## 2.7 AUTOMATION STRATEGIES

In this section, we summarize our previous discussion of production, and propose nine fundamental strategies for automation in manufacturing.

### Production summarized

Production has been characterized as the transformation of raw materials into finished goods by a sequence of processing and assembly operations. To carry out the transformation, many supporting functions must be performed in the plant. These include product design, manufacturing engineering, industrial engineering, quality control, production planning, scheduling, and inventory control. Manufacturing can be summarized as consisting of four essential functions:

1. Materials processing and assembly.
2. Materials handling and storage.
3. Control—both at the operations level and the plant level.
4. Development of a manufacturing data base and information system to support the foregoing three functions.

The subject of materials processing and assembly was discussed in Section 2.5. The materials handling and storage function was illustrated in the operation sequence model of Section 2.6. A significant portion of the total time a workpart spends in the factory is consumed in moving the parts between operations and temporary storage of the parts.

The control function in manufacturing includes both the regulation of the individual processes as well as supervision over the aggregate operations at the plant level. Control at the process level involves the achievement of certain performance objectives by properly manipulating the process inputs. It also includes quality control of the process output. Control at the plant level includes effective use of labor, proper utilization of machines, shipping products of good quality on schedule, and keeping plant operating costs at a minimum possible level.

A manufacturing data base is required to effectively organize the production functions of a modern factory. Two centuries ago, before the industrial revolution, all the information required to run the primitive shops of those days was contained in the mind of the person who owned the business. The operations of today's manufacturing plant are much too complex for one individual to cope with. Accordingly, a manufacturing data base is required to manage the plant. This conventional manufacturing data base includes:

Part drawings, material specifications, and bills of material (product design).

Route sheets which specify the process plans for the parts, tool lists for the route sheets, tool inventory records (manufacturing engineering).

Methods descriptions, time standards, equipment justification documents (industrial engineering).

Master schedules, production schedules, exception reports (production planning and control).

Inventory records (inventory control).

All of these elements of the manufacturing data base must be generated and maintained by the various support departments to run a modern production plant.

## Fundamental strategies in production automation

There are certain fundamental strategies that can be used to improve productivity in manufacturing operations. Since these strategies are commonly implemented by means of automation technology, we refer to them as *automation strategies*.

Table 2.3 presents a list of the nine strategies, together with an identification of the manufacturing functions discussed above which are affected by each strategy. The table also shows how the various terms of the production models—Eqs. (2.1) through (2.12)—are influenced. We discuss each of the nine schemes in the paragraphs on the following pages.

TABLE 2.3 Nine Automation Strategies

| *Strategy* | *Manufacturing function** | *Objective†* |
|---|---|---|
| 1. Specialization of operations | 1 | Reduce $T_m, T_o$ |
| 2. Combined operations | 1,2 | Reduce $T_h, T_{th}, n_m$ |
| 3. Simultaneous operations | 1,2 | Reduce $T_h, T_m, T_{th}, n_m$ |
| 4. Integration of operations | 1,2 | Reduce $T_h, n_m$ |
| 5. Reduce setup time | 1,3 | Reduce $T_{su}$ |
| 6. Improved materials handling | 2 | Reduce $T_{no}$ |
| 7. Process control and optimization | 1,3 | Reduce $T_m$ |
| 8. Computerized manufacturing data base | 4 | Reduce $n_m, T_{no}$ |
| 9. Computerized manufacturing control | 3,4 | Reduce $T_{no}$ |

*The four manufacturing functions to which the strategies might be applied:

1. Materials processing and assembly.
2. Materials handling.
3. Control-process level and plant level.
4. Manufacturing data base development.

†Production models to which objectives apply:

$$\text{Manufacturing lead time} = n_m(T_{su} + QT_o + T_{no}) \tag{2.3}$$

$$T_o = T_m + T_h + T_{th} \tag{2.12}$$

1. *Specialization of operations.* One way to increase productivity is to use a special-purpose machine designed to perform one operation with the greatest possible efficiency. This, of course, is analogous to the concept of labor specialization. This strategy is employed at each workstation of automatic transfer machines, to be discussed in Chapters 4 and 5.

2. *Combined operations.* Production occurs as a sequence of operations. Some complicated workparts may require 50 distinct processing steps. The objective of combining operations is to reduce the number of distinct production machines through which the part must be routed. This is accomplished by performing more than one operation at a given workstation, thereby reducing the total number of workstations needed. Since each operation typically requires a setup, the total number of setups are reduced by combining operations. The combined operations strategy means that the processes are performed in sequence at the same machine. This concept is utilized in numerically controlled (NC) machines, discussed in Chapters 7 and 8.

3. *Simultaneous operations.* A logical extension of the combined operations strategy is not only to perform multiple operations at the same workstation, but to perform these operations at the same time. Multiple spindle drill presses utilize this concept for drilling more than one hole simultaneously.

4. *Integration of operations.* This is a strategy in which a series of workstations are linked together as a single integrated mechanism by automatic work handling devices. With several stations, more than one part can be processed by the integrated system at one time, thereby increasing the overall production output. This concept reduces work-in-process and total throughput time. The system described by this strategy is a transfer line.

5. *Reduce setup time.* Whereas the strategy of combined operations reduces the number of setups, the fifth strategy is concerned with reducing the time of each setup. This can be accomplished by scheduling of similar workparts through the production machine, use of common fixtures for different but similar parts, and designing greater flexibility into the manufacturing system. Such goals can be achieved by means of group technology, to be discussed in Chapter 18.

6. *Improved materials handling.* A great opportunity for reducing nonproductive time exists through the use of mechanized and automated materials handling methods. The typical benefits of improved handling are reduced work-in-process and shorter throughput times. These benefits are generally achieved by minimizing the distances over which the workparts must be moved, providing for smooth flow of work through adjacent stations, handling the parts in larger unit loads, and other common sense principles.

7. *Process control and optimization.* This includes a whole range of control strategies intended to operate the process and equipment more efficiently. It includes conventional analog feedback control or computer control, which allows more flexibility in the type of control exercised over the process. Control strategies that can be implemented by computer include feedback control, optimal control, sequencing control, and adaptive control. We shall present the traditional process control principles in Chapter 13, and computer control in Chapters 14 and 15.

8. *Computerized manufacturing data base.* The computer is becoming an essential tool in manufacturing, just as it became an essential tool in business data processing 20 years ago. In addition to its use for process control, the computer is used to automate the steps in generating the manufacturing data base. The computer is used in product design, process planning, work measurement, and production planning and control. These areas will be covered in various portions of the book.

9. *Computerized manufacturing control.* Strategy 7 was concerned with control at the process level. Computerized manufacturing control is concerned with control at the plant level. This involves collecting data from throughout the factory and using those data to better manage the factory. The name sometimes used to describe such a system is "shop floor control," to be discussed in Chapter 17.

Let us use a numerical example to illustrate very simply how one of these automation strategies can improve productivity.

### EXAMPLE 2.3

Suppose in Example 2.1 that it were possible to apply a new automated machine based on numerical control. By using this machine, it is possible to combine four operations at one workstation. This reduces to five the number of machines ($n_m$) through which the part must be routed. The total operation time on the NC machine is 15 min. The setup time for the NC operation is still 3.0 h. Compute the total throughput time.

The terms in Eq. (2.3) from Example 2.1 are $Q=50$ parts, $T_{su}=3.0$ h, and $T_{no}=7.0$ h. The new value of $n_m=5$. We must compute a new value for $T_o$. Assume that the four operations which were not combined still have an average operation time of 6 min. Together with the new operation time of 15 min, the weighted average is

$$T_o=\frac{4(6)+1(15)}{5}=7.8 \text{ min}=0.13 \text{ h}$$

The new total throughput time is

$$5[3.0+50(0.13)+7.0]=82.5 \text{ h}$$

This is a 31% reduction from the previous 120 h.

## REFERENCES

[1] DOYLE, L. E., et al., *Manufacturing Processes and Materials for Engineers*, Prentice-Hall, Inc., Englewood Cliffs, N.J., 1969.

[2] EARY, D. F., and JOHNSON, G. E., *Process Engineering for Manufacturing*, Prentice-Hall, Inc., Englewood Cliffs, N.J., 1962.

[3] HARRINGTON, J., *Computer Integrated Manufacturing*, Industrial Press, Inc., New York, 1973.

[4] International Business Machines, *The Production Information and Control System*, Publication GE20-0280-2.

[5] LINDBERG, R. A., *Processes and Materials of Manufacture*, 2nd ed., Allyn and Bacon, Inc., Boston, 1977.

[6] LUKE, H. D., *Automation for Productivity*, John Wiley & Sons, Inc., New York, 1972.

[7] NIEBEL, B. W., and DRAPER, A. B., *Product Design and Process Engineering*, McGraw-Hill Book Company, New York, 1974.

[8] WILD, R., *Mass-Production Management*, John Wiley & Sons Ltd., London, 1972.

## PROBLEMS

**2.1.** A part is routed through six operations in a batch production shop (process-type plant layout). The setup times and operation times are given in the table. If the batch size is 100 parts and the average nonoperation time associated with each operation is 12 h, compute the total throughput time for the part.

| Operation | Setup time (h) | Operation time (min) |
|---|---|---|
| 1 | 4.0 | 5.0 |
| 2 | 2.0 | 3.5 |
| 3 | 8.0 | 10.0 |
| 4 | 3.0 | 1.9 |
| 5 | 3.0 | 4.1 |
| 6 | 4.0 | 2.5 |

**2.2.** For operation 3 in Problem 2.1, compute the average production rate by Eqs. (2.10) and (2.11) using batch sizes of 1, 10, 100, and 1000.

**2.3.** The workpart of Problem 2.1 can be machined on a new computer-controlled flexible manufacturing line recently installed in the plant. The changeover time (setup) to adapt the line to this part is 1.0 h. Total operation time is 16.5 min. If the entire 100 parts in the batch can be processed through the line one right after the other (with no interference from other workparts which are processed on the line), and the nonoperation time for the batch is 5 h, compute the manufacturing lead time.

**2.4.** In an operation on an engine lathe, the operator performs a series of cuts as given in the table. The delay time is related to handling, setting, and positioning the tools for each of the cuts. The setup time for this operation is negligible. The workpart handling time is 3.0 min/piece. Find the operation time and production rate for this part.

| Number | Cut | Machining time (min) | Delay time (min) |
|---|---|---|---|
| 1 | Face end | 0.5 | 2.0 |
| 2 | Turn | 2.5 | 1.5 |
| 3 | Drill | 1.5 | 2.0 |
| 4 | Ream | 1.0 | 1.5 |
| 5 | Tap | 0.5 | 1.5 |
| 6 | Cut off | 0.5 | 2.0 |

**2.5.** The part in Problem 2.4 is produced in batches of 1000. An automatic turret lathe is being considered to perform the cutting operations. The machining times will remain the same as in Problem 2.4, but the total delay time will be reduced from 10.5 min to 1.5 min. If the setup time for the automatic lathe is estimated at 5 h for this job, calculate the following:

(a) Total time to process the batch for the engine lathe operation of Problem 2.4.

(b) Total time to process the batch for the automatic turret lathe.

(c) Average production rate for the new method.

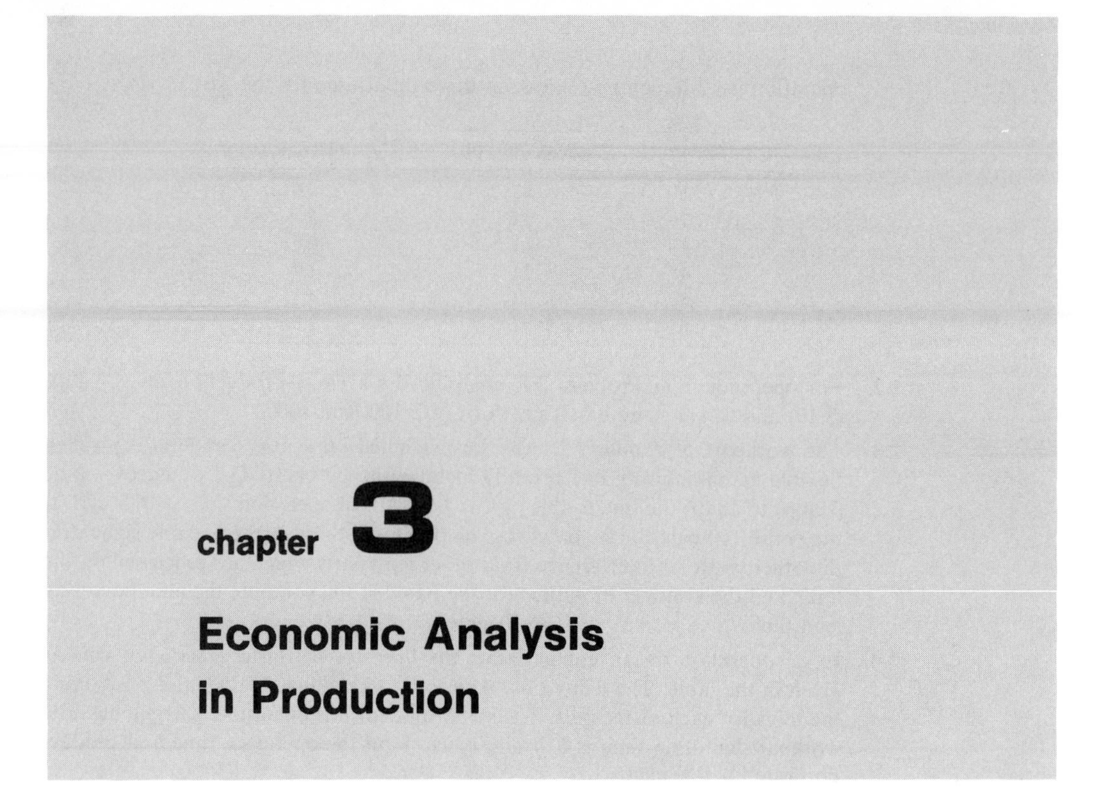

# chapter 3 Economic Analysis in Production

## 3.1 INTRODUCTION

Chapter 1 presented a number of reasons for considering automation. In the final analysis, the decision regarding a proposed automation project should be evaluated against the same economic criteria used to assess any other investment opportunity: Will the investment pay for itself and contribute to the profits of the firm? (In our discussion, we exclude consideration of investments that the firm is obligated to make to comply with government regulations. Examples would include capital expenditures to satisfy requirements of OSHA, antipollution, etc.)

In this chapter we consider the subject of engineering economy, with particular emphasis on economic analysis in manufacturing. The reader familiar with engineering economy will find the coverage in this chapter to be superficial. We omit such topics as depreciation, income taxes, accounting, risk analysis, and much more. Our objectives in the presentation are the following:

1. To present the fundamentals of investment analysis.

2. To identify the types of costs that are encountered in manufacturing.
3. To explain break-even analysis.

This will provide a sufficient base to deal with the problems in manufacturing and automation economics which will be encountered in subsequent chapters. The reader who would like to pursue the subject of engineering economy beyond what we cover in this chapter may wish to use one of the references listed ([1], [2], or [3]).

## 3.2 FUNDAMENTALS OF INVESTMENT ANALYSIS

### Basic concepts

Money is considered to possess a *time value* because when money is borrowed for a period of time, it is expected that the amount paid back will be greater than that which was borrowed. The difference is referred to as *interest*. The amount of interest is determined by three factors: the length of time the money was borrowed, the interest rate, and whether simple interest or compound interest was used to compute the amount. To explain these factors, let us concern ourselves with the difference between simple and compound interest.

SIMPLE INTEREST. The interest rate is generally expressed in terms of an annual rate. To compute the interest for a certain amount of money borrowed for exactly 1 year, the amount is multiplied by the interest rate. We can reduce this to the following formula:

$$I = Pi$$

where $i$ = annual interest rate
$P$ = principal (the starting amount)
$I$ = interest

With simple interest, when an amount of money is borrowed for a period greater than 1 year, the interest charge is determined by multiplying the yearly interest charge by the number of years:

$$I = Pni$$

where $n$ represents the number of years. The amount of money paid back at the end of $n$ years can be determined from the formula

$$F = P + I = P(1 + ni) \tag{3.1}$$

where $F$ represents the future amount to be paid, in this case under simple interest.

Compound Interest. With simple interest, the amount $I$ is directly proportional to the length of time $n$. If interest is compounded during the length of time, the final amount $F$ (principal plus interest) grows at a faster rate. To illustrate compound interest, consider the manner in which a savings account might grow if the interest were compounded annually. Using an initial deposit of $1000 and an interest rate of 5%, the savings account at the end of the first year would be worth

$$F_1 = \$1000(1+0.05) = \$1050$$

This amount is used to compute the interest for the second year. Thus, the savings would have grown by the end of the second year to

$$F_2 = \$1050(1+0.05) = \$1102.50$$

The general equation for calculating the future equivalent of some present value $P$ can be determined by the equation

$$F = P(1+i)^n \tag{3.2}$$

In engineering economy calculations, compound interest is almost always used, because it more accurately reflects the time value of money.

## Interest factors

Equation (3.2) represents one of six common interest-rate problems: the problem of computing the future worth of some present value, given the rate of return $i$ and the number of years $n$. There are a total of six of these problems. They occur frequently enough that an interest factor has been defined to cover each problem. In the paragraphs below the six interest factors are described.

1. *Single-payment compound amount factor* (*SPCAF*). This is the case we have previously considered: finding the future value $F$ of a present sum $P$. As the reader can deduce from Eq. (3.2), the formula used to calculate the single-payment compound amount factor is

$$\text{SPCAF} = (1+i)^n \tag{3.3}$$

We shall adopt the following notation for the SPCAF, which will be easier to remember and use than the name of the factor:

$$\text{SPCAF} = (F/P, i\%, n) \tag{3.4}$$

The terms in parentheses can be read: find $F$, given $P$, $i$, and $n$. This general form will be used for the interest factors that follow.

2. *Single-payment present worth factor (SPPWF).* This is the inverse of the previous interest problem. The SPPWF is used to compute the present worth of some future value.

$$\text{SPPWF} = (P/F, i\%, n) = \frac{1}{(1+i)^n} \tag{3.5}$$

3. *Capital recovery factor (CRF).* Instead of paying off a borrowed sum with a single future payment, another common method is to make uniform annual payments at the end of each of $n$ years. The amount of each payment is figured to yield the required interest rate $i$. The capital recovery factor is designed specifically for this case.

$$\text{CRF} = (A/P, i\%, n) = \frac{i(1+i)^n}{(1+i)^n - 1} \tag{3.6}$$

where $A$ represents the amount of the annual payment.

4. *Uniform series present worth factor (USPWF).* This solves the preceding problem in reverse: finding the present value of a series of $n$ future end-of-year equal payments.

$$\text{USPWF} = (P/A, i\%, n) = \frac{(1+i)^n - 1}{i(1+i)^n} \tag{3.7}$$

5. *Sinking fund factor (SFF).* "Sinking fund" refers to the situation in which we want to put aside a certain sum of money at the end of each year so that after $n$ years, the accumulated fund, with interest compounded, will be worth $F$. The sinking fund factor allows us to determine the amount $A$ to be put aside each year.

$$\text{SFF} = (A/F, i\%, n) = \frac{i}{(1+i)^n - 1} \tag{3.8}$$

6. *Uniform series compound amount factor (USCAF).* The reverse of the preceding problem arises when it is desired to know how much money has accumulated after $n$ years of uniform annual payments at interest rate $i$.

$$\text{USCAF} = (F/A, i\%, n) = \frac{(1+i)^n - 1}{i} \tag{3.9}$$

TABLES OF INTEREST FACTORS. Instead of calculating the value of the interest factor needed in a given problem, values are tabulated for a wide variety of interest rates and years. In the Appendix to this chapter, the interest factors are given for interest rates equal to 10%, 12%, 15%, and 20%. Although this list is not nearly complete, these values cover the range of annual rates of return that seem to prevail during the period in which this book is being written.

COMMENTS ON THE USE OF THE INTEREST FACTORS. In using the interest factors, we must be clear as to when the various cash flow transactions represented by $P$, $F$, and $A$ occur during the year. The present worth transaction, $P$, occurs at the beginning of the year. $F$ and $A$ transactions are assumed to be end-of-year cash flows.

Our definitions of the six interest factors were based on 1-year intervals or periods. Actually, interest can be compounded more frequently than annually. Savings accounts are often compounded quarterly. The foregoing interest factors can be adapted to periods other than annual periods. However, for our purposes it will be sufficient and convenient to maintain the annual compounding convention.

## Methods of evaluating investment opportunities

There are three basic methods of evaluating investment opportunities:

1. Present worth method (PW method).
2. Uniform annual cost method (UAC method).
3. Rate-of-return method (RR method).

All of these methods begin with the assumption that a minimum attractive rate of return has been defined as the criterion for deciding whether or not the investment opportunity has merit. If the investment proposal meets the criterion, the investment should be made. If not, the investment should not be made. In discussing the three methods, we ignore tax considerations.

PRESENT WORTH METHOD. The PW method uses the equilavent present value of all current and future cash flows to evaluate the investment opportunity. An example will illustrate use of this method.

**EXAMPLE 3.1**

A new production machine will cost \$83,000 and is expected to generate revenues of \$50,000 per year for 7 years. It will cost \$30,000 per year to operate the machine. At the end of the 7 years, the machine will be scrapped at zero salvage value. Use a minimum attractive rate of return of 10% to evaluate the investment.

To analyze this problem we will use the convention that costs or expenditures are positive cash flows and revenues are negative cash flows. We want to determine the equivalent present worth of all cash flows in this problem.

$$\begin{aligned} PW &= 83{,}000 - 50{,}000(P/A, 10\%, 7) + 30{,}000(P/A, 10\%, 7) \\ &= 83{,}000 - (20{,}000)(4.8684) \\ &= -14{,}368 \end{aligned}$$

Since the equivalent PW is negative, it means that a net profit is made on the investment. We can interpret this to mean that the actual rate of return is greater than 10%.

UNIFORM ANNUAL COST METHOD. The UAC method converts all current and future cash flows into their equivalent uniform annual cost over the period of the investment. This will be illustrated by using the data of Example 3.1.

### EXAMPLE 3.2

The same production machine as in Example 3.1 can be evaluated using the UAC method.

$$\begin{aligned} \text{UAC} &= 83{,}000(A/P, 10\%, 7) - 50{,}000 + 30{,}000 \\ &= 83{,}000(0.20541) - 20{,}000 \\ &= -2951 \end{aligned}$$

Again, since the net UAC is negative, the investment must be considered attractive. This should be no surprise in view of the results of Example 3.1. The PW of −$14,368 is equivalent to the UAC of −$2951 at 10% rate of return for 7 years. Let us check this.

$$\begin{aligned} \text{UAC} &= 14{,}368(A/P, 10\%, 7) \\ &= -14{,}368(0.20541) \\ &= -2951 \end{aligned}$$

RATE OF RETURN METHOD. The RR method goes slightly beyond either the PW or the UAC methods by actually calculating the rate of return. If the calculated rate of return exceeds the criterion, the investment is acceptable.

### EXAMPLE 3.3

The data from Examples 3.1 and 3.2 will be used. To find the actual rate of return for the production machine, an equation must be set up with the rate of return $i$ as the unknown. Either the equivalent PW or the equivalent UAC can be used to establish the equation. Then the value of $i$ is determined which drives the PW or UAC to zero. A PW equation will be used in our solution.

$$\text{PW} = 83{,}000 - 50{,}000(P/A, i\%, 7) + 30{,}000(P/A, i\%, 7)$$

$$= 83{,}000 - 20{,}000(P/A, i\%, 7) = 0$$

In this problem, the value of $(P/A, i\%, 7)$ can be solved.

$$(P/A, i\%, 7) = \frac{83{,}000}{20{,}000} = 4.150$$

Scanning the $P/A$ values in the interest tables at $n=7$ years for different values if $i$, we find that

$$(P/A, 15\%, 7) = 4.1604$$

which is very close to our computed value of 4.150. This means that the actual rate of return is very close to 15%. By linear interpolation, the calculated value is $i = 15.1\%$.

## Comparison of investment alternatives

Any of the three methods can be used to compare investment alternatives. Each method has its relative advantages. The present worth method has the advantage of being relatively easy to understand. It is easy to compare one alternative to another in terms of present dollar value. The UAC method is convenient to use when the service lives of the alternatives are different. Awkward adjustments must be made to the PW method when comparing alternatives with different service lives. The advantage of the rate of return method is that it provides a reading on what the anticipated rate of return will be. With the PW and UAC methods, we simply compare the investment alternatives to the criterion (10% in the examples). The disadvantage of the rate of return method is that a trial-and-error approach is usually required for most problems (not in our example 3.3, though).

In our work we will use the uniform annual cost method to compare alternatives. This will also be helpful for future chapters, when we must know cost rate per time to apply to a particular automated manufacturing system.

**EXAMPLE 3.4**

Two investment proposals for alternative production machines are to be compared. One is a manually operated machine. The other is automatic. Data are given in the table.

| | *Manual* | *Automatic* |
|---|---|---|
| First cost | \$50,000 | \$100,000 |
| Annual operating cost | \$35,000 | \$10,000 |
| Salvage value | \$5,000 | 0 |
| Service life (years) | 10 | 7 |

Use a rate of return of 12% to determine which alternative should be selected if it is assumed that both are equivalent in terms of revenues generated.

For the manual machine,

$$\text{UAC} = 50{,}000(A/P, 12\%, 10) + 35{,}000 - 5000(A/F, 12\%, 10)$$

$$= 50{,}000(0.17698) + 35{,}0000 - 5000(0.05698)$$

$$= 43{,}564$$

For the automatic machine,

$$\begin{aligned} \text{UAC} &= 100{,}000(A/P, 12\%, 7) + 10{,}000 \\ &= 100{,}000(0.21912) + 10{,}000 \\ &= 31{,}912 \end{aligned}$$

The automatic machine has a lower equivalent uniform annual cost and should therefore be selected.

The situation in this problem is typical of automation projects: A larger initial investment must be made for the sake of lower annual operating costs. Less labor is required to run the automatic machine.

Notice that in this problem, no revenues are given. Therefore, we must assume that the two alternatives are equivalent in terms of the work they can perform. The selection between the alternatives must be based on minimum UAC: Which alternative has the lowest equivalent uniform annual cost?

## 3.3 COSTS IN MANUFACTURING

### Fixed and variable costs

Manufacturing costs can be divided into two major categories, fixed costs and variable costs. The difference between the two is based on whether the expense varies in relation to the level of output.

A *fixed cost* is one that is constant for any level of production output. Examples of fixed costs include cost of the factory building, insurance, property taxes, and the cost of production equipment. All of these fixed costs can be expressed as annual costs. Those items which are capital investments (e.g., factory building and production equipment) can be converted to their equivalent uniform annual costs by the methods of the previous section.

A *variable cost* is one that increases as the level of production increases. Direct labor costs (plus fringe benefits), raw materials, and electrical power to operate the production machines are examples of variable costs. The ideal concept of variable cost is that it is directly proportional to output level. When fixed and variable costs are combined, we get the total cost of manufacturing as a function of output. A general plot of the relationship is shown in Figure 3.1.

### Overhead costs

Classification of costs as either fixed or variable is not always convenient for accountants and finance people. Fixed costs and variable costs are valid concepts, but the financial specialists of a manufacturing firm usually prefer to think in terms of direct labor cost, material cost, and overhead costs. The *direct labor cost* is the sum of the wages paid to the people who operate the

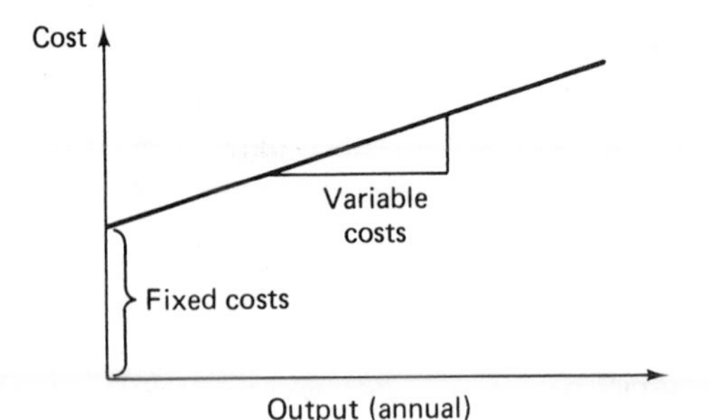


FIGURE 3.1 Plot of fixed and variable costs as a function of production output.

production machines and perform the processing and assembly operations. The *material cost* is the cost of all the raw materials that are used to produce the finished product of the firm. In terms of fixed and variable costs, direct labor and material costs must be considered as variable.

*Overhead costs* are all the other costs associated with running a manufacturing firm. Overhead can be divided into two categories: factory overhead (sometimes called factory expense) and corporate overhead. *Factory overhead* includes the costs of operating the factory other than direct labor and materials. The types of expenses included in this category are listed in Table 3.1. It can be seen that some of these costs are variable whereas others are fixed. The *corporate overhead cost* is the cost of running the company other than its manufacturing activities. A list of many of the expenses included under corporate overhead is presented in Table 3.2. Many manufacturing firms operate more than one plant, and this is one of the reasons for dividing overhead into factory and corporate categories.

TABLE 3.1 Typical Factory Overhead Costs

| | |
|---|---|
| Plant supervision | Applicable taxes |
| Line foremen | Insurance |
| Maintenance crew | Heat |
| Custodial services | Light |
| Security personnel | Power for machines |
| Tool crib attendant | Factory cost |
| Materials handling crew | Equipment cost |
| Shipping and receiving | Fringe benefits |

FACTORY OVERHEAD. The overhead costs of a firm can amount to several times the cost of direct labor. The overhead can be allocated according to a number of different bases, including direct labor cost, direct labor hours, space, material cost, and so on. We will use direct labor cost to illustrate how factory overhead rates are determined. Suppose that the total cost of operating a plant amounts to $900,000 per year. Of this total, $400,000 is direct labor cost. This means that $500,000 is indirect or overhead expense:

TABLE 3.2 Typical Corporate Overhead Expenses

| | |
|---|---|
| Corporate executives | Applicable taxes |
| Sales personnel | Cost of office space |
| Accounting department | Security personnel |
| Finance department | Heat |
| Legal counsel | Light |
| Research and development | Air conditioning |
| Design and engineering | Insurance |
| Other support personnel | Fringe benefits |

plant supervision, line foremen, annual cost of equipment, energy, maintenance personnel, and so on. The factory overhead rate for this plant would be figured as

$$\text{factory overhead rate} = \frac{\$500,000}{\$400,000} = 1.25$$

Overhead rates are often expressed as percentages, so this equals 125%. This rate could be applied to a particular production job, as illustrated in the following example.

### EXAMPLE 3.5

A batch of 100 parts is to be processed through the factory. Each part requires $\frac{1}{2}$ h of direct labor, and the cost of direct labor is \$6.00/h. If the factory overhead rate is 125%, compute the total cost of the job, including overhead.

The direct labor cost is

$$(100 \text{ parts})(0.5 \text{ h/part})(\$6.00/\text{h}) = \$300$$

The allocated overhead charge would be determined by multiplying \$300 by 125%:

$$\text{factory overhead} = \$300(125\%) = \$375$$

The allocated factory cost of the job = \$300 + \$375 = \$675.

CORPORATE OVERHEAD. The corporate overhead rate can be determined in a manner similar to that used for factory overhead. We will use an oversimplified example to illustrate. Suppose that the firm operates two plants with direct labor and factory overhead expenses as follows:

| | *Plant 1* | *Plant 2* | *Total* |
|---|---|---|---|
| Direct labor | \$400,000 | \$200,000 | \$600,000 |
| Factory expense | \$500,000 | \$300,000 | \$800,000 |
| Total cost | \$900,000 | \$500,000 | \$1,400,000 |

In addition, the cost of management, sales staff, engineering, accounting, and so on, amounts to $960,000. The corporate overhead rate would be based on the total direct labor of the two plants:

$$\text{corporate overhead rate} = \frac{\$960{,}000}{\$600{,}000}(100\%) = 160\%$$

Overhead rates, both factory and corporate, are simply a means for allocating expenses that are not directly associated with production. The principal concern in this book will be with determining the appropriate allocation of factory expenses, not corporate overhead.

## Cost of equipment usage

The trouble with overhead rates as we have developed them is that they are based on direct labor cost alone. A machine operator who runs an old, small engine lathe will be costed at the same overhead rate as the operator who runs a modern NC machining center representing a $250,000 investment. Obviously, the time on the automated machine should be valued at a higher rate. If differences between rates of different production machines are not recognized, manufacturing costs will not be accurately measured by the overhead rate structure.

To overcome this difficulty, it is appropriate to divide production costs (excluding raw materials) into two components: direct labor and machine cost. Associated with each will be the applicable factory overhead.[1] These cost components will apply not to the aggregate factory operations but to individual production work centers. A work center would typically be one worker–machine system or a small group of machines plus the labor to operate them.

The direct labor cost would consist of the wages paid to operate the work center. The applicable factory overhead allocated to direct labor might include fringe benefits and line supervision. These are factory expense items which are appropriately charged as direct labor overhead.

The machine cost is the capital cost of the machine apportioned over the life of the asset at the appropriate rate of return used by the firm. This provides an annual cost that may be expressed as an hourly rate (or any other time unit) by dividing the annual cost by the number of hours of use per year. The machine overhead rate is based on those factory expenses which are directly applicable to the machine. These would include power for the machine, floor space, maintenance and repair expenses, and so on. In

[1]It should be mentioned that not all of the overhead items listed in Table 3.1 are applicable in engineering economy calculations. Only those overhead costs that directly affect investment alternatives should be included in the computations. On the other hand, the firm must recover all its overhead costs, both factory and corporate (Tables 3.1 and 3.2), in pricing decisions.

separating the applicable factory overhead items of Table 3.1 between direct labor and machine, some arbitrary judgment must be used.

**EXAMPLE 3.6**

The determination of an hourly rate for a given work center can best be illustrated by means of an example. Given the following:

Direct labor rate = \$7.00/h.
Applicable labor factory overhead rate = 60%.
Capital investment in machine = \$100,000.
Service life = 8 years.
Salvage value = zero.
Applicable machine factory overhead rate = 50%.
Rate of return used = 10%.

The machine is operated one 8-h shift per day, 250 days per year. Determine the appropriate hourly cost for this worker–machine system.

The labor cost per hour is \$7.00(1 + 60%) = \$11.20/h. The machine cost must first be annualized:

$$\begin{aligned} \text{UAC} &= 100{,}000(A/P, 10\%, 8) \\ &= 100{,}000(0.18744) = \$18{,}744/\text{yr} \end{aligned}$$

The number of hours per year is $8 \times 250 = 2000$h/yr. Dividing the \$18,744 by 2000 gives \$9.37/h. Applying the 50% overhead rate, the machine cost per hour is \$9.37(1 + 50%) = \$14.06/h. So the

$$\begin{aligned} \text{total work center rate} &= \$11.20 + \$14.06 \\ &= \$25.26/\text{h} \end{aligned}$$

In subsequent chapters, there will be problems in which an hourly rate must be applied to a particular automated production system. Example 3.6 illustrates the general method by which this hourly rate is determined.

## 3.4 BREAK-EVEN ANALYSIS

*Break-even analysis* is a method of assessing the effect of changes in production output on costs, revenues, and profits. It is most commonly conceptualized in the form of a break-even chart. To construct the break-even chart the manufacturing costs are divided into fixed costs and variable costs. The sum of these costs is plotted as a function of production output. To plot the total cost, the variable cost per unit change in output must be determined. Revenues can also be plotted on a break-even chart as a function of production output.

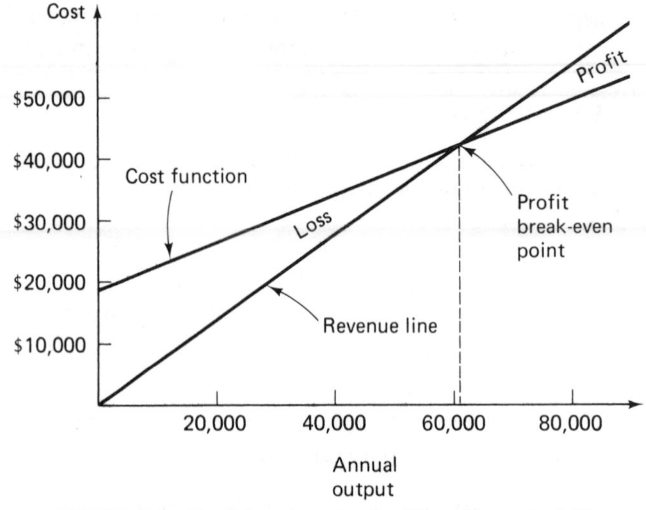


FIGURE 3.2 Profit break-even chart (see Example 3.7).

Break-even analysis can be used for either of two main purposes:

1. *Profit analysis*. In this case the break-even chart shows the effect of changes in output on costs and revenues. This gives a picture of how profits (or losses) will vary for different output levels. The break-even point is the output level at which total costs equal revenues and the profit is zero. An example of a break-even chart used for profit analysis is shown in Figure 3.2.

2. *Production method cost comparison*. In this case the break-even chart shows the effect of changes in output level on the costs of two (or more)

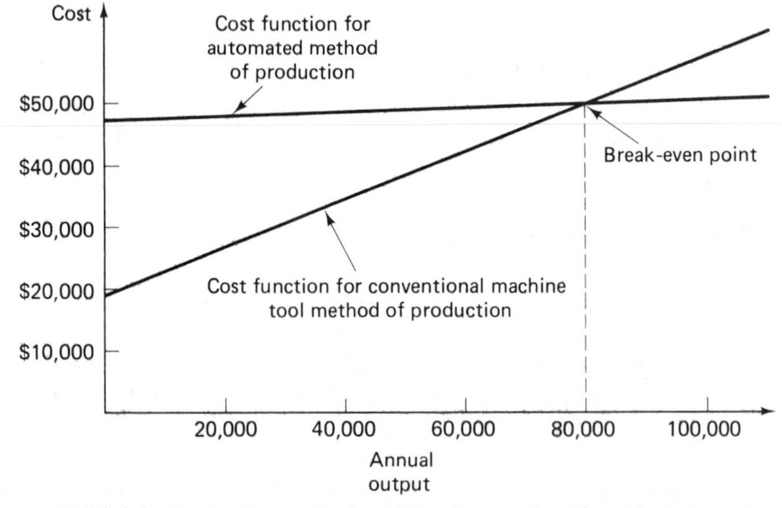


FIGURE 3.3 Production method cost break-even chart (see Example 3.8).

different methods of production. The break-even point for this chart is the output level at which the costs for the two production methods are equal. (When more than two production methods are plotted on the same chart, there will be a break-even point for each pair of production methods.) Figure 3.3 shows a break-even chart used for production method cost comparison.

We will illustrate the two types of break-even analysis by means of two examples.

## EXAMPLE 3.7

This example will demonstrate the break-even chart for profit analysis.

A conventional machine tool will cost \$70,000. It will have a service life of 8 years and an anticipated salvage value of \$5000 at the end of the life. The annual maintenance and repair cost for the machine is estimated at \$2000. A rate of return of 12% is expected from the investment. The machine will be used to produce one type of part and the production rate for this part will be 20 units/h. The labor rate to operate this machine is \$6.00/h. Applicable overhead rates are 20% on the machine and 30% on labor.

If the value added per piece for this operation is 70 cents, determine the annual production output at which break-even will occur.

Let $Q$ be the annual output level. The variable cost would be the labor rate divided by the production rate. The appropriate overhead rate would then be added to this. Hence, the variable cost per unit is

$$\frac{\$6.00/\text{h}}{20\text{ pieces/h}}(1+30\%)=\$0.39/\text{piece}$$

The corresponding annual variable cost as a function of $Q$ would be $\$0.39Q$.

The annual fixed cost is figured on the machine investment. First, ignoring overhead,

$$\begin{aligned}\text{UAC}&=70{,}000(A/P,12\%,8)+2000-5000(A/F,12\%,8)\\&=70{,}000(0.2013)+2000-5000(0.0813)\\&=15{,}684.50\end{aligned}$$

Adding the 20% overhead rate to this, the annual allocation of fixed cost would be $15{,}684.50(1+20\%)=\$18{,}821.40$.

The sum of the fixed and variable costs would be

$$\text{total cost}=\$18{,}821.40+0.39Q$$

This function is plotted in Figure 3.2.

Revenues as a function of $Q$ would be value added times quantity, or $\$0.70Q$. This is plotted in Figure 3.2. The break-even point occurs where the two lines intersect. To calculate this break-even point, the total cost function is set equal to the revenue function.

$$\begin{aligned}0.70Q&=18{,}821.40+0.39Q\\0.70Q-0.39Q=0.31Q&=18{,}821.40\\Q&=60{,}714\text{ pieces/yr}\end{aligned}$$

At a production rate of 20 pieces/h, this corresponds to 60,714/20 = 3035.7 h of operation per year. Above this level of production, the firm will make a profit on the investment. Below this output level, a loss will be suffered.

The next example will illustrate the use of break-even charts for comparison of production methods.

## EXAMPLE 3.8

An alternative to the conventional machine tool of Example 3.7 has been proposed. The alternative is an automated machine costing $125,000. Its expected service life is 5 years with an anticipated salvage value of zero. Annual maintenance costs are estimated at $5000. The same 12% rate of return and 20% overhead rates will be used. The production rate for this machine will be 60 units per hour, and one-fourth of one operator's time at $6.00 per hour will be needed to tend the machine. Assume an overhead rate for labor of 30%.

Variable cost would be

$$\frac{0.25(\$6.00/\text{h})}{60 \text{ pieces/h}}(1+30\%)Q = 0.0325Q$$

Fixed cost would be

$$[\$125{,}000(A/P, 12\%, 5) + 5000](1+20\%) = \$47{,}611.50$$

The total cost is obtained by summing the fixed and variable costs:

$$\text{total cost} = \$47{,}611.50 + 0.0325Q$$

The total cost functions for the two production methods are plotted in Figure 3.3. The break-even point is indicated by the intersection of the two lines. To calculate the break-even point, the two functions are set equal to each other.

$$18821.40 + 0.39Q = 47{,}611.50 + 0.0325Q$$

$$0.39Q - 0.0325Q = 47{,}611.50 - 18{,}821.40$$

$$0.3575Q = 28{,}790.10$$

$$Q = 80{,}532$$

At a production rate of 60 pieces/h on the automated machine, this would require 1342.2 h/yr, but on the conventional machine at 20 pieces per hour, 4026.6 h/yr are needed.

For annual production levels above 80,532 units, the automated machine would be advantageous, as shown in Figure 3.3.

## REFERENCES


[1] GRANT, E. L., IRESON, W. G., and LEAVENWORTH, R. S., *Principles of Engineering Economy*, 6th ed., John Wiley & Sons, Inc., New York, 1976.

[2] TARQUIN, A. J., and BLANK, L. T., *Engineering Economy*, McGraw-Hill Book Company, New York, 1976.

[3] THUESEN, G. J., FABRYCKY, W., and THUESEN, H. G., *Engineering Economy*, 5th ed., Prentice-Hall, Inc., Englewood Cliffs, N.J., 1977.


## PROBLEMS

**3.1.** What sum of money must be invested now if its value 5 years from now is to be $10,000 at an interest rate of 12%?

**3.2.** If $100 is deposited into a 10% interest account every year for 7 years starting at the end of the first year, how much money will have accumulated at the end of year 7?

**3.3.** Same as Problem 3.2 except that an initial payment of $100 is also made now.

**3.4.** A new automated production system has a first cost of $75,000, a service life of 5 years, and an anticipated salvage value at the end of this period of $10,000. Annual maintenance cost = $3000. The machine will be operated 2000 h/yr at a labor rate of $8.00/h. Compute the equivalent present worth of the system if the expected rate of return is 15%.

**3.5.** For the data of Problem 3.4, compute the equivalent uniform annual cost.

**3.6.** Two alternative production methods have been proposed, one manual method, the other an automatic machine. Data are given in the table.

| | *Manual* | *Automatic* |
|---|---:|---:|
| First cost | $15,000 | $95,000 |
| Annual operating cost | $30,000 | $10,000 |
| Salvage value | 0 | $15,000 |
| Service life (years) | 10 | 7 |

Use a rate of return of 10% to select the more economical alternative if the two alternatives are equivalent in terms of capability.

**3.7.** Solve Problem 3.6 using a rate of return of 20%. Why is the selection of method different from that of Problem 3.6?

**3.8.** A proposed automatic machine is to be used exclusively to produce one type of workpart. The machine has a first cost of $50,000 and its expected service life is 3 years with a salvage value of $20,000 at the end of the 3 years. The machine will be operated 4000 hours per year (two shifts) at $8.00/h (labor, power, maintenance, etc.). Its production rate is 10 units/h. Excluding raw material costs, compute the production cost per unit using a rate of return of 15%.

**3.9.** The following data apply to the operation of a particular automated manufacturing system:

Direct labor rate = \$6.00/h.
Number of operators required = 1.
Applicable labor factory overhead rate = 60%.
Capital investment in system = \$300,000.
Service life = 10 years.
Salvage value = \$30,000.
Applicable machine factory overhead rate = 40%.
The system is operated one shift (2000 h/yr).

Use a rate of return of 12% to determine the appropriate hourly rate for this worker–machine system.

**3.10.** Solve Problem 3.9 except using three-shift operation (6000 hours per year). Note the effect of increased machine utilization on the hourly rate for the system as compared to the results of Problem 3.9.

**3.11.** In Example 3.7, determine the profit (or loss) that would be made on the following production levels: (a) 50,000 pieces/yr; (b) 60,000 pieces/yr; (c) 70,000 pieces/yr; (d) 80,000 pieces/yr.

**3.12.** In Example 3.8, for levels of production of 80,000 and 90,000 pieces/yr, determine the following:
(a) The cost of production per year for each of the two production methods.
(b) The numbers of hours of production per year required for each of the two methods.
(c) The profit (or loss) for each of the production methods. Use a value added of \$0.70/piece to compute profit.

**3.13.** In Example 3.8, determine the profit break-even point for the automated production machine.

HW ✓

**3.14.** A piece of automated production equipment has a first cost of \$100,000. Service life = 6 years, anticipated salvage value is \$10,000, and annual maintenance costs are \$3000. The equipment will produce at the rate of 10 units/h, each unit worth \$2.00 in added revenue. One operator is required full time to tend the machine at a rate of \$6.00/hr. Assume that no overhead rates are applicable. Raw material costs equal \$0.20/unit. Use a rate of return of 10%.
(a) Compute the profit break-even point.
(b) How many hours of operation are required to produce the number of units indicated by the break-even point?
(c) How much profit (or loss) will be made if 25,000 units per year are produced?

**3.15.** In Problem 3.14, recompute the break-even point if the applicable overhead rates are 20% for the machine and 40% for labor.

**3.16.** The break-even point is to be determined for two methods of production, a manual method and an automated method. The manual method requires two operators at \$7.00/h each. Together, they produce at a rate of 35 units per hour. The automated method has an initial cost of $125,000, an 8-year service life, no salvage value, and annual maintenance costs of $3000. No labor (except for maintenance) is required to operate the machine, but the machine consumes energy at the rate of 50 kW/h when running. Cost of electricity is $0.03/kWh. If the production rate for this automated machine is 100 units/h, determine the break-even point for the two methods if a 15% rate of return is required.

## APPENDIX: TABLES OF INTEREST FACTORS

Tables of interest factors for several interest rates are shown on the following pages:

| | |
|---|---|
| Table A3.1 | 10% interest rate |
| Table A3.2 | 12% interest rate |
| Table A3.3 | 15% interest rate |
| Table A3.4 | 20% interest rate |

Tables are reprinted by permission from G. J. Thuesen, W. Fabrycky, and H. G. Thuesen, *Engineering Economy*, 5th ed., Prentice-Hall, Inc., Englewood Cliffs, N.J., 1977.

**TABLE A3.1 10% Interest Factors for Annual Compounding Interest**

| $n$ | Single Payment | | Equal Payment Series | | | |
|---|---|---|---|---|---|---|
| | Compound-amount factor | Present-worth factor | Compound-amount factor | Sinking-fund factor | Present-worth factor | Capital-recovery factor |
| | To find $F$ Given $P$ $F/P$ $i,n$ | To find $P$ Given $F$ $P/F$ $i,n$ | To find $F$ Given $A$ $F/A$ $i,n$ | To find $A$ Given $F$ $A/F$ $i,n$ | To find $P$ Given $A$ $P/A$ $i,n$ | To find $A$ Given $P$ $A/P$ $i,n$ |
| 1 | 1.100 | 0.9091 | 1.000 | 1.0000 | 0.9091 | 1.1000 |
| 2 | 1.210 | 0.8265 | 2.100 | 0.4762 | 1.7355 | 0.5762 |
| 3 | 1.331 | 0.7513 | 3.310 | 0.3021 | 2.4869 | 0.4021 |
| 4 | 1.464 | 0.6830 | 4.641 | 0.2155 | 3.1699 | 0.3155 |
| 5 | 1.611 | 0.6209 | 6.105 | 0.1638 | 3.7908 | 0.2638 |
| 6 | 1.772 | 0.5645 | 7.716 | 0.1296 | 4.3553 | 0.2296 |
| 7 | 1.949 | 0.5132 | 9.487 | 0.1054 | 4.8684 | 0.2054 |
| 8 | 2.144 | 0.4665 | 11.436 | 0.0875 | 5.3349 | 0.1875 |
| 9 | 2.358 | 0.4241 | 13.579 | 0.0737 | 5.7590 | 0.1737 |
| 10 | 2.594 | 0.3856 | 15.937 | 0.0628 | 6.1446 | 0.1628 |
| 11 | 2.853 | 0.3505 | 18.531 | 0.0540 | 6.4951 | 0.1540 |
| 12 | 3.138 | 0.3186 | 21.384 | 0.0468 | 6.8137 | 0.1468 |
| 13 | 3.452 | 0.2897 | 24.523 | 0.0408 | 7.1034 | 0.1408 |
| 14 | 3.798 | 0.2633 | 27.975 | 0.0358 | 7.3667 | 0.1358 |
| 15 | 4.177 | 0.2394 | 31.772 | 0.0315 | 7.6061 | 0.1315 |
| 16 | 4.595 | 0.2176 | 35.950 | 0.0278 | 7.8237 | 0.1278 |
| 17 | 5.054 | 0.1979 | 40.545 | 0.0247 | 8.0216 | 0.1247 |
| 18 | 5.560 | 0.1799 | 45.599 | 0.0219 | 8.2014 | 0.1219 |
| 19 | 6.116 | 0.1635 | 51.159 | 0.0196 | 8.3649 | 0.1196 |
| 20 | 6.728 | 0.1487 | 57.275 | 0.0175 | 8.5136 | 0.1175 |
| 21 | 7.400 | 0.1351 | 64.003 | 0.0156 | 8.6487 | 0.1156 |
| 22 | 8.140 | 0.1229 | 71.403 | 0.0140 | 8.7716 | 0.1140 |
| 23 | 8.954 | 0.1117 | 79.543 | 0.0126 | 8.8832 | 0.1126 |
| 24 | 9.850 | 0.1015 | 88.497 | 0.0113 | 8.9848 | 0.1113 |
| 25 | 10.835 | 0.0923 | 98.347 | 0.0102 | 9.0771 | 0.1102 |
| 26 | 11.918 | 0.0839 | 109.182 | 0.0092 | 9.1610 | 0.1092 |
| 27 | 13.110 | 0.0763 | 121.100 | 0.0083 | 9.2372 | 0.1083 |
| 28 | 14.421 | 0.0694 | 134.210 | 0.0075 | 9.3066 | 0.1075 |
| 29 | 15.863 | 0.0630 | 148.631 | 0.0067 | 9.3696 | 0.1067 |
| 30 | 17.449 | 0.0573 | 164.494 | 0.0061 | 9.4269 | 0.1061 |
| 31 | 19.194 | 0.0521 | 181.943 | 0.0055 | 9.4790 | 0.1055 |
| 32 | 21.114 | 0.0474 | 201.138 | 0.0050 | 9.5264 | 0.1050 |
| 33 | 23.225 | 0.0431 | 222.252 | 0.0045 | 9.5694 | 0.1045 |
| 34 | 25.548 | 0.0392 | 245.477 | 0.0041 | 9.6086 | 0.1041 |
| 35 | 28.102 | 0.0356 | 271.024 | 0.0037 | 9.6442 | 0.1037 |
| 40 | 45.259 | 0.0221 | 442.593 | 0.0023 | 9.7791 | 0.1023 |
| 45 | 72.890 | 0.0137 | 718.905 | 0.0014 | 9.8628 | 0.1014 |
| 50 | 117.391 | 0.0085 | 1163.909 | 0.0009 | 9.9148 | 0.1009 |
| 55 | 189.059 | 0.0053 | 1880.591 | 0.0005 | 9.9471 | 0.1005 |
| 60 | 304.482 | 0.0033 | 3034.816 | 0.0003 | 9.9672 | 0.1003 |
| 65 | 490.371 | 0.0020 | 4893.707 | 0.0002 | 9.9796 | 0.1002 |
| 70 | 789.747 | 0.0013 | 7887.470 | 0.0001 | 9.9873 | 0.1001 |
| 75 | 1271.895 | 0.0008 | 12708.954 | 0.0001 | 9.9921 | 0.1001 |
| 80 | 2048.400 | 0.0005 | 20474.002 | 0.0001 | 9.9951 | 0.1001 |
| 85 | 3298.969 | 0.0003 | 32979.690 | 0.0000 | 9.9970 | 0.1000 |
| 90 | 5313.023 | 0.0002 | 53120.226 | 0.0000 | 9.9981 | 0.1000 |
| 95 | 8556.676 | 0.0001 | 85556.760 | 0.0000 | 9.9988 | 0.1000 |
| 100 | 13780.612 | 0.0001 | 137796.123 | 0.0000 | 9.9993 | 0.1000 |

TABLE A3.2 12% Interest Factors for Annual Compounding Interest

| | Single Payment | | Equal Payment Series | | | |
|---|---|---|---|---|---|---|
| *n* | Compound-amount factor | Present-worth factor | Compound-amount factor | Sinking-fund factor | Present-worth factor | Capital-recovery factor |
| | To find *F* Given *P* *F*/*P* *i*, *n* | To find *P* Given *F* *P*/*F* *i*, *n* | To find *F* Given *A* *F*/*A* *i*, *n* | To find *A* Given *F* *A*/*F* *i*, *n* | To find *P* Given *A* *P*/*A* *i*, *n* | To find *A* Given *P* *A*/*P* *i*, *n* |
| 1 | 1.120 | 0.8929 | 1.000 | 1.0000 | 0.8929 | 1.1200 |
| 2 | 1.254 | 0.7972 | 2.120 | 0.4717 | 1.6901 | 0.5917 |
| 3 | 1.405 | 0.7118 | 3.374 | 0.2964 | 2.4018 | 0.4164 |
| 4 | 1.574 | 0.6355 | 4.779 | 0.2092 | 3.0374 | 0.3292 |
| 5 | 1.762 | 0.5674 | 6.353 | 0.1574 | 3.6048 | 0.2774 |
| 6 | 1.974 | 0.5066 | 8.115 | 0.1232 | 4.1114 | 0.2432 |
| 7 | 2.211 | 0.4524 | 10.089 | 0.0991 | 4.5638 | 0.2191 |
| 8 | 2.476 | 0.4039 | 12.300 | 0.0813 | 4.9676 | 0.2013 |
| 9 | 2.773 | 0.3606 | 14.776 | 0.0677 | 5.3283 | 0.1877 |
| 10 | 3.106 | 0.3220 | 17.549 | 0.0570 | 5.6502 | 0.1770 |
| 11 | 3.479 | 0.2875 | 20.655 | 0.0484 | 5.9377 | 0.1684 |
| 12 | 3.896 | 0.2567 | 24.133 | 0.0414 | 6.1944 | 0.1614 |
| 13 | 4.364 | 0.2292 | 28.029 | 0.0357 | 6.4236 | 0.1557 |
| 14 | 4.887 | 0.2046 | 32.393 | 0.0309 | 6.6282 | 0.1509 |
| 15 | 5.474 | 0.1827 | 37.280 | 0.0268 | 6.8109 | 0.1468 |
| 16 | 6.130 | 0.1631 | 42.753 | 0.0234 | 6.9740 | 0.1434 |
| 17 | 6.866 | 0.1457 | 48.884 | 0.0205 | 7.1196 | 0.1405 |
| 18 | 7.690 | 0.1300 | 55.750 | 0.0179 | 7.2497 | 0.1379 |
| 19 | 8.613 | 0.1161 | 63.440 | 0.0158 | 7.3658 | 0.1358 |
| 20 | 9.646 | 0.1037 | 72.052 | 0.0139 | 7.4695 | 0.1339 |
| 21 | 10.804 | 0.0926 | 81.699 | 0.0123 | 7.5620 | 0.1323 |
| 22 | 12.100 | 0.0827 | 92.503 | 0.0108 | 7.6447 | 0.1308 |
| 23 | 13.552 | 0.0738 | 104.603 | 0.0096 | 7.7184 | 0.1296 |
| 24 | 15.179 | 0.0659 | 118.155 | 0.0085 | 7.7843 | 0.1285 |
| 25 | 17.000 | 0.0588 | 133.334 | 0.0075 | 7.8431 | 0.1275 |
| 26 | 19.040 | 0.0525 | 150.334 | 0.0067 | 7.8957 | 0.1267 |
| 27 | 21.325 | 0.0469 | 169.374 | 0.0059 | 7.9426 | 0.1259 |
| 28 | 23.884 | 0.0419 | 190.699 | 0.0053 | 7.9844 | 0.1253 |
| 29 | 26.750 | 0.0374 | 214.583 | 0.0047 | 8.0218 | 0.1247 |
| 30 | 29.960 | 0.0334 | 241.333 | 0.0042 | 8.0552 | 0.1242 |
| 31 | 33.555 | 0.0298 | 271.293 | 0.0037 | 8.0850 | 0.1237 |
| 32 | 37.582 | 0.0266 | 304.848 | 0.0033 | 8.1116 | 0.1233 |
| 33 | 42.092 | 0.0238 | 342.429 | 0.0029 | 8.1354 | 0.1229 |
| 34 | 47.143 | 0.0212 | 384.521 | 0.0026 | 8.1566 | 0.1226 |
| 35 | 52.800 | 0.0189 | 431.664 | 0.0023 | 8.1755 | 0.1223 |
| 40 | 93.051 | 0.0108 | 767.091 | 0.0013 | 8.2438 | 0.1213 |
| 45 | 163.988 | 0.0061 | 1358.230 | 0.0007 | 8.2825 | 0.1207 |
| 50 | 289.002 | 0.0035 | 2400.018 | 0.0004 | 8.3045 | 0.1204 |

TABLE A3.3 15% Interest Factors for Annual Compounding Interest

| $n$ | Single Payment | | Equal Payment Series | | | |
|---|---|---|---|---|---|---|
| | Compound-amount factor | Present-worth factor | Compound-amount factor | Sinking-fund factor | Present-worth factor | Capital-recovery factor |
| | To find $F$ Given $P$ $F/P$ $i, n$ | To find $P$ Given $F$ $P/F$ $i, n$ | To find $F$ Given $A$ $F/A$ $i, n$ | To find $A$ Given $F$ $A/F$ $i, n$ | To find $P$ Given $A$ $P/A$ $i, n$ | To find $A$ Given $P$ $A/P$ $i, n$ |
| 1 | 1.150 | 0.8696 | 1.000 | 1.0000 | 0.8696 | 1.1500 |
| 2 | 1.323 | 0.7562 | 2.150 | 0.4651 | 1.6257 | 0.6151 |
| 3 | 1.521 | 0.6575 | 3.473 | 0.2880 | 2.2832 | 0.4380 |
| 4 | 1.749 | 0.5718 | 4.993 | 0.2003 | 2.8550 | 0.3503 |
| 5 | 2.011 | 0.4972 | 6.742 | 0.1483 | 3.3522 | 0.2983 |
| 6 | 2.313 | 0.4323 | 8.754 | 0.1142 | 3.7845 | 0.2642 |
| 7 | 2.660 | 0.3759 | 11.067 | 0.0904 | 4.1604 | 0.2404 |
| 8 | 3.059 | 0.3269 | 13.727 | 0.0729 | 4.4873 | 0.2229 |
| 9 | 3.518 | 0.2843 | 16.786 | 0.0596 | 4.7716 | 0.2096 |
| 10 | 4.046 | 0.2472 | 20.304 | 0.0493 | 5.0188 | 0.1993 |
| 11 | 4.652 | 0.2150 | 24.349 | 0.0411 | 5.2337 | 0.1911 |
| 12 | 5.350 | 0.1869 | 29.002 | 0.0345 | 5.4206 | 0.1845 |
| 13 | 6.153 | 0.1625 | 34.352 | 0.0291 | 5.5832 | 0.1791 |
| 14 | 7.076 | 0.1413 | 40.505 | 0.0247 | 5.7245 | 0.1747 |
| 15 | 8.137 | 0.1229 | 47.580 | 0.0210 | 5.8474 | 0.1710 |
| 16 | 9.358 | 0.1069 | 55.717 | 0.0180 | 5.9542 | 0.1680 |
| 17 | 10.761 | 0.0929 | 65.075 | 0.0154 | 6.0472 | 0.1654 |
| 18 | 12.375 | 0.0808 | 75.836 | 0.0132 | 6.1280 | 0.1632 |
| 19 | 14.232 | 0.0703 | 88.212 | 0.0113 | 6.1982 | 0.1613 |
| 20 | 16.367 | 0.0611 | 102.444 | 0.0098 | 6.2593 | 0.1598 |
| 21 | 18.822 | 0.0531 | 118.810 | 0.0084 | 6.3125 | 0.1584 |
| 22 | 21.645 | 0.0462 | 137.632 | 0.0073 | 6.3587 | 0.1573 |
| 23 | 24.891 | 0.0402 | 159.276 | 0.0063 | 6.3988 | 0.1563 |
| 24 | 28.625 | 0.0349 | 184.168 | 0.0054 | 6.4338 | 0.1554 |
| 25 | 32.919 | 0.0304 | 212.793 | 0.0047 | 6.4642 | 0.1547 |
| 26 | 37.857 | 0.0264 | 245.712 | 0.0041 | 6.4906 | 0.1541 |
| 27 | 43.535 | 0.0230 | 283.569 | 0.0035 | 6.5135 | 0.1535 |
| 28 | 50.066 | 0.0200 | 327.104 | 0.0031 | 6.5335 | 0.1531 |
| 29 | 57.575 | 0.0174 | 377.170 | 0.0027 | 6.5509 | 0.1527 |
| 30 | 66.212 | 0.0151 | 434.745 | 0.0023 | 6.5660 | 0.1523 |
| 31 | 76.144 | 0.0131 | 500.957 | 0.0020 | 6.5791 | 0.1520 |
| 32 | 87.565 | 0.0114 | 577.100 | 0.0017 | 6.5905 | 0.1517 |
| 33 | 100.700 | 0.0099 | 664.666 | 0.0015 | 6.6005 | 0.1515 |
| 34 | 115.805 | 0.0086 | 765.365 | 0.0013 | 6.6091 | 0.1513 |
| 35 | 133.176 | 0.0075 | 881.170 | 0.0011 | 6.6166 | 0.1511 |
| 40 | 267.864 | 0.0037 | 1779.090 | 0.0006 | 6.6418 | 0.1506 |
| 45 | 538.769 | 0.0019 | 3585.128 | 0.0003 | 6.6543 | 0.1503 |
| 50 | 1083.657 | 0.0009 | 7217.716 | 0.0002 | 6.6605 | 0.1501 |

**TABLE A3.4 20% Interest Factors for Annual Compounding Interest**

| | Single Payment | | Equal Payment Series | | | |
|---|---|---|---|---|---|---|
| $n$ | Compound-amount factor | Present-worth factor | Compound-amount factor | Sinking-fund factor | Present-worth factor | Capital-recovery factor |
| | To find $F$ Given $P$ $F/P$ $i, n$ | To find $P$ Given $F$ $P/F$ $i, n$ | To find $F$ Given $A$ $F/A$ $i, n$ | To find $A$ Given $F$ $A/F$ $i, n$ | To find $P$ Given $A$ $P/A$ $i, n$ | To find $A$ Given $P$ $A/P$ $i, n$ |
| 1 | 1.200 | 0.8333 | 1.000 | 1.0000 | 0.8333 | 1.2000 |
| 2 | 1.440 | 0.6945 | 2.200 | 0.4546 | 1.5278 | 0.6546 |
| 3 | 1.728 | 0.5787 | 3.640 | 0.2747 | 2.1065 | 0.4747 |
| 4 | 2.074 | 0.4823 | 5.368 | 0.1863 | 2.5887 | 0.3863 |
| 5 | 2.488 | 0.4019 | 7.442 | 0.1344 | 2.9906 | 0.3344 |
| 6 | 2.986 | 0.3349 | 9.930 | 0.1007 | 3.3255 | 0.3007 |
| 7 | 3.583 | 0.2791 | 12.916 | 0.0774 | 3.6046 | 0.2774 |
| 8 | 4.300 | 0.2326 | 16.499 | 0.0606 | 3.8372 | 0.2606 |
| 9 | 5.160 | 0.1938 | 20.799 | 0.0481 | 4.0310 | 0.2481 |
| 10 | 6.192 | 0.1615 | 25.959 | 0.0385 | 4.1925 | 0.2385 |
| 11 | 7.430 | 0.1346 | 32.150 | 0.0311 | 4.3271 | 0.2311 |
| 12 | 8.916 | 0.1122 | 39.581 | 0.0253 | 4.4392 | 0.2253 |
| 13 | 10.699 | 0.0935 | 48.497 | 0.0206 | 4.5327 | 0.2206 |
| 14 | 12.839 | 0.0779 | 59.196 | 0.0169 | 4.6106 | 0.2169 |
| 15 | 15.407 | 0.0649 | 72.035 | 0.0139 | 4.6755 | 0.2139 |
| 16 | 18.488 | 0.0541 | 87.442 | 0.0114 | 4.7296 | 0.2114 |
| 17 | 22.186 | 0.0451 | 105.931 | 0.0095 | 4.7746 | 0.2095 |
| 18 | 26.623 | 0.0376 | 128.117 | 0.0078 | 4.8122 | 0.2078 |
| 19 | 31.948 | 0.0313 | 154.740 | 0.0065 | 4.8435 | 0.2065 |
| 20 | 38.338 | 0.0261 | 186.688 | 0.0054 | 4.8696 | 0.2054 |
| 21 | 46.005 | 0.0217 | 225.026 | 0.0045 | 4.8913 | 0.2045 |
| 22 | 55.206 | 0.0181 | 271.031 | 0.0037 | 4.9094 | 0.2037 |
| 23 | 66.247 | 0.0151 | 326.237 | 0.0031 | 4.9245 | 0.2031 |
| 24 | 79.497 | 0.0126 | 392.484 | 0.0026 | 4.9371 | 0.2026 |
| 25 | 95.396 | 0.0105 | 471.981 | 0.0021 | 4.9476 | 0.2021 |
| 26 | 114.475 | 0.0087 | 567.377 | 0.0018 | 4.9563 | 0.2018 |
| 27 | 137.371 | 0.0073 | 681.853 | 0.0015 | 4.9636 | 0.2015 |
| 28 | 164.845 | 0.0061 | 819.223 | 0.0012 | 4.9697 | 0.2012 |
| 29 | 197.814 | 0.0051 | 984.068 | 0.0010 | 4.9747 | 0.2010 |
| 30 | 237.376 | 0.0042 | 1181.882 | 0.0009 | 4.9789 | 0.2009 |
| 31 | 284.852 | 0.0035 | 1419.258 | 0.0007 | 4.9825 | 0.2007 |
| 32 | 341.822 | 0.0029 | 1704.109 | 0.0006 | 4.9854 | 0.2006 |
| 33 | 410.186 | 0.0024 | 2045.931 | 0.0005 | 4.9878 | 0.2005 |
| 34 | 492.224 | 0.0020 | 2456.118 | 0.0004 | 4.9899 | 0.2004 |
| 35 | 590.668 | 0.0017 | 2948.341 | 0.0003 | 4.9915 | 0.2003 |
| 40 | 1469.772 | 0.0007 | 7343.858 | 0.0002 | 4.9966 | 0.2001 |
| 45 | 3657.262 | 0.0003 | 18281.310 | 0.0001 | 4.9986 | 0.2001 |
| 50 | 9100.438 | 0.0001 | 45497.191 | 0.0000 | 4.9995 | 0.2000 |

# part II

# High-Volume Discrete-Parts Production Systems

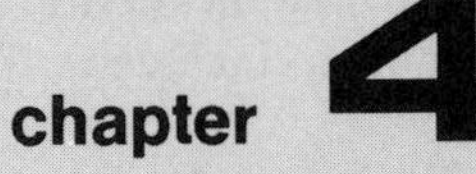


# Detroit-Type Automation

## 4.1 INTRODUCTION

The traditional symbol of automation is the mechanized flow line. Chronologically, this was the first example of automated production to appear. Its origins can be traced largely to the work of Henry Ford in the manufacture of automobiles. To be sure, Mr. Ford did not invent the automated flow line or its predecessor, the moving assembly line. However, in his efforts to improve the methods of automobile manufacture, he achieved such significant advances in assembly line mass-production techniques that the feasibility and potential of these methods were demonstrated. In turn, this led to the subsequent development of the fully automated transfer line. Because of the contributions of Henry Ford and others in the automotive industry, this type of mechanized production is often referred to as *Detroit automation.*

This chapter will consider the automated equipment used in the processing and assembly of discrete products in large volumes. Chapter 5 will concern itself with the quantitative analysis of automated flow lines. We will find that the performance of such lines can be estimated for a variety of different production line designs. Chapter 6 will deal with the problem of line

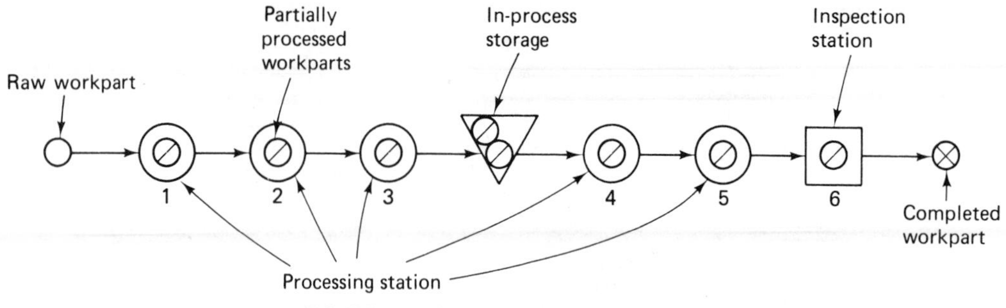


FIGURE 4.1 Schematic diagram of automated flow line.

balancing. Although this topic finds its principal application in the design of manual assembly lines, it is an appropriate topic to include in this discussion of Detroit automation.

An automated flow line consists of several machines or workstations which are linked together by work handling devices that transfer parts between the stations. The transfer of workparts occurs automatically and the workstations carry out their specialized functions automatically. The flow line can be symbolized as shown in Figure 4.1. A raw workpart enters one end of the line and the processing or assembly steps are performed sequentially as the part moves from one station to the next. It is possible to incorporate buffer storage zones into the flow line, either at a single location as pictured in Figure 4.1, or between every workstation. It is also possible to include inspection stations in the line to automatically perform intermediate checks on the quality of the workparts. Manual stations might also be located along the flow line to perform certain operations which are difficult or uneconomical to automate. These various features of mechanized flow lines will be discussed in subsequent sections.

Automated flow lines are generally the most appropriate means of production in cases of relatively stable product life; high product demand, which requires high rates of production; and where the alternative method of manufacture would involve a large labor content. The objectives of the use of flow line automation are, therefore:

To reduce labor costs.

To increase production rates.

To reduce work in progress with associated reductions in inventory costs.

To minimize distances moved between operations.

Specialization of operations.

Integration of operations.

## 4.2 PHYSICAL CONFIGURATION OF THE EQUIPMENT

Although Figure 4.1 shows the flow pattern of operations in a straight line, there are actually two general forms that the work flow can take. These two configurations are in-line and rotary.

### In-line type

The *in-line* configuration consists of a sequence of workstations in a more-or-less straight-line arrangement. The flow of work can take a few 90° turns, either for workpiece reorientation, factory layout limitations, or other reasons, and still qualify as a straight-line configuration. A common pattern of work flow, for example, is a rectangular shape, which would allow the same operator to load the starting workpieces and unload the finished workpieces. An example of an in-line transfer machine used for metal cutting operations is illustrated in Figure 4.2.

### Rotary type

In the *rotary* configuration, the workparts are indexed around a circular table or dial. The workstations are stationary and usually located around the outside periphery of the dial. The parts ride on the rotating table and are registered or positioned, in turn, at each station for its processing or assembly operation. This type of equipment is often referred to as an *indexing machine* or *dial index machine* and an example is shown in Figure 4.3.

### Selection

The choice between the two types depends on the application. The rotary type is commonly limited to smaller workpieces and to fewer stations. There is generally not as much flexibility in the design of the rotary configuration. For example, the dial-type design does not lend itself to providing for buffer storage capacity. On the other hand, the rotary configuration usually involves a lower-cost piece of equipment and typically requires less factory floor space.

The in-line design is preferable for larger workpieces and can accommodate a larger number of workstations. The number of stations on the dial index machine is more limited due to the size of the dial. In-line machines can be fabricated with a built-in storage capability to smooth out the effect of work stoppages at individual stations and other irregularities.

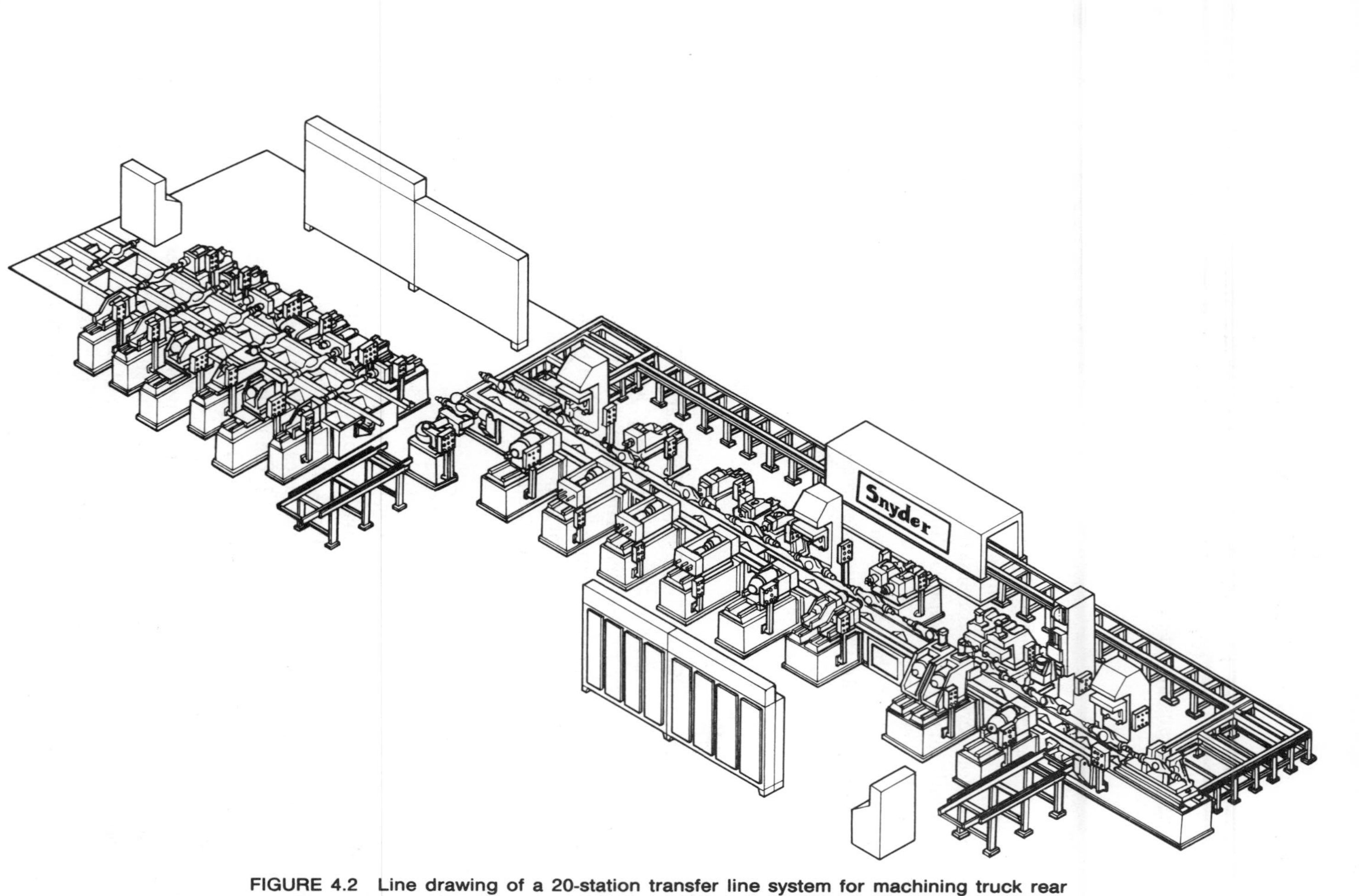


FIGURE 4.2 Line drawing of a 20-station transfer line system for machining truck rear axle housings. Line consists of two sections: a 7-station free transfer section and a 12-station palletized section. The station between the two sections reorients the workparts. Note the return loop for bringing pallets back to starting point. (Courtesy Snyder Corp.)

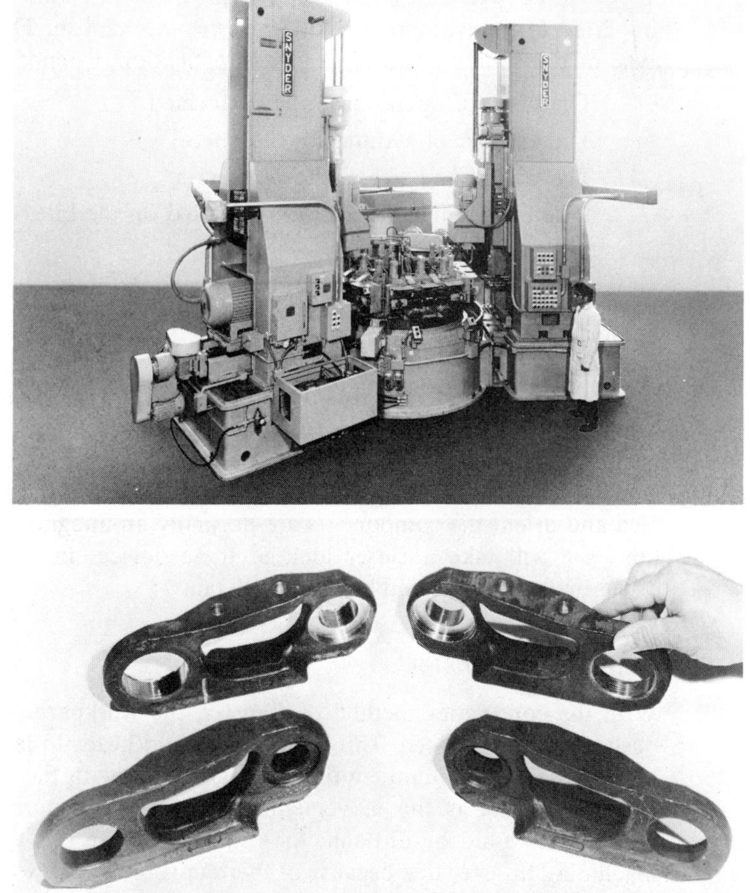


FIGURE 4.3 (Top view) Five-station dial index machine showing vertical and horizontal machining stations around periphery of rotary table. (Bottom view) Rough forgings and finished parts processed on the dial index machine. (Courtesy Snyder Corp.)

## 4.3 METHODS OF WORKPART TRANSPORT

The transfer mechanism of the automated flow line must not only move the partially completed workparts or assemblies between adjacent stations; it must also orient and locate the parts in the correct position for processing at each station. The general methods of transporting workpieces on flow lines can be classified into the following three categories:

Continuous transfer.

Intermittent or synchronous transfer.

Nonsynchronous or power-and-free transfer.

These three categories are distinguished by the kind of motion that is imparted to the workpiece by the transfer mechanism. The most appropriate type of transport system for a given application depends on such factors as:

The types of operation to be performed.

The number of stations on the line.

The weight and size of the workparts.

Whether manual stations are included on the line.

Production rate requirements.

Balancing the various process times on the line.

Before discussing the three types of work transport system, we should try to clarify a possible source of confusion. These transfer systems are used for both processing and assembly operations. In the case of automatic assembly machines, we are referring to the mechanisms that transport the partially completed assemblies between stations, not the feed mechanisms that present new components to the assemblies at a particular station. The devices that feed and orient the components are normally an integral part of the workstation. We will take a closer look at these devices in Section 4.7 when we discuss automatic assembly in more detail.

### Continuous transfer

With the continuous method of transfer, the workparts are moved continuously at constant speed. This requires the workheads to move during processing in order to maintain continuous registration with the workpart. For some types of operations, this movement of the workheads during processing is not feasible. It would be difficult, for example, to use this type of system on a machining transfer line because of inertia problems due to the size and weight of the workheads. In other cases, continuous transfer would be very practical. Examples of its use are in beverage bottling operations, packaging, manual assembly operations where the human operator can move with the moving flow line, and relatively simple automatic assembly tasks. In some bottling operations, for instance, the bottles are transported around a continuously rotating drum. Beverage is discharged into the moving bottles by spouts located at the drum's periphery. The advantage of this application is that the liquid beverage is kept moving at a steady speed and hence there are no inertia problems.

Continuous transfer systems are relatively easy to design and fabricate and can achieve a high rate of production.

### Intermittent transfer

As the name suggests, in this method the workpieces are transported with an intermittent or discontinuous motion. The workstations are fixed in position and the parts are moved between stations and then registered at the proper locations for processing. All workparts are transported at the same time and, for this reason, the term "synchronous transfer system" is also used to describe this method of workpart transport. Examples of applications of the intermittent transfer of workparts can be found in machining operations, pressworking operations or progressive dies, and mechanized assembly. Most of the transfer mechanisms reviewed in Section 4.4 provide the intermittent or synchronous type of workpart transport.

### Nonsynchronous transfer

This system of transfer, also referred to as a "power-and-free system," allows each workpart to move to the next station when processing at the current station has been completed. Each part moves independently of other parts. Hence, some parts are being processed on the line at the same time that others are being transported between stations.

Nonsynchronous transfer systems offer the opportunity for greater flexibility than do the other two systems, and this flexibility can be a great advantage in certain circumstances. In-process storage of workparts can be incorporated into the nonsynchronous systems with relative ease. Power-and-free systems can also compensate for line balancing problems where there are significant differences in process times between stations. Parallel stations or several series stations can be used for the longer operations and single stations can be used for the shorter operations. Therefore, the average production rates can be approximately equalized. Nonsynchronous lines are often used where there are one or more manually operated stations and cycle-time variations would be a problem on either the continuous or synchronous transport systems. Larger workparts can be handled on the nonsynchronous systems. A disadvantage of the power-and-free systems is that the cycle rates are generally slower than for the other types.

## 4.4 TRANSFER MECHANISMS

There are various types of transfer mechanisms used to move parts between stations. These mechanisms can be grouped into two types: those used to provide linear travel for in-line machines, and those used to provide rotary motion for dial indexing machines.

### Linear transfer mechanisms

We will explain the operation of three of the typical mechanisms: the walking beam transfer bar system, the powered roller conveyor system, and the chain-drive conveyor system. This is not a complete listing of all types, but it is a representative sample.

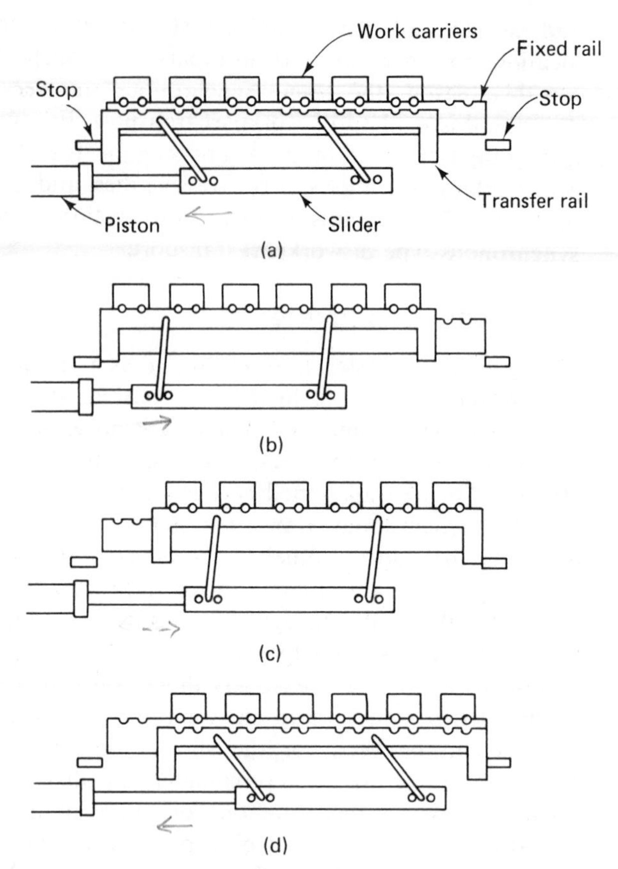


FIGURE 4.4 Walking beam transfer system, showing various stages during parts transfer cycle. (Reprinted from Boothroyd and Redford [2].)

Walking Beam Systems. With the walking beam transfer mechanism, the workparts are lifted up from their workstation locations by a transfer bar and moved one position ahead, to the next station. The transfer bar then lowers the parts into nests which more accurately position them for processing. This type of transfer device is illustrated in Figure 4.4.

Powered Roller Conveyor System. This type of system is used in general stock handling systems as well as in automated flow lines. The conveyor can be used to move parts or pallets possessing flat riding surfaces. The rollers can be powered by either of two mechanisms. The first is a belt drive, in which a flat moving belt beneath the rollers provides the rotation of the rollers by friction. A chain drive is the second common mechanism used to power the rollers. Powered roller conveyors are versatile transfer systems because they can be used to divert work pallets into workstations or alternate tracks (see Figure 4.5).

FIGURE 4.5 Roller conveyor system for transporting palletized workparts on "flexible manufacturing system" to be discussed in Chapter 19. (Courtesy White-Sundstrand Co.)

Chain-Drive Conveyor System. Figure 4.6 illustrates this type of transfer system. Either a chain or a flexible steel belt is used to transport the work carriers. The chain is driven by pulleys in either an "over-and-under" configuration, in which the pulleys turn about a horizontal axis, or an "around-the-corner" configuration, in which the pulleys rotate about a vertical axis.

This general type of transfer system can be used for continuous, intermittent, or nonsynchronous movement of workparts. In the nonsynchronous motion, the workparts are pulled by friction or ride on an oil film along a track with the chain or belt providing the movement. It is necessary to provide some sort of final location for the workparts when they arrive at their respective stations.

### Rotary transfer mechanisms

There are several methods used to index a circular table or dial at various equal angular positions corresponding to workstation locations. Those described below are meant to be a representative rather that a complete listing.

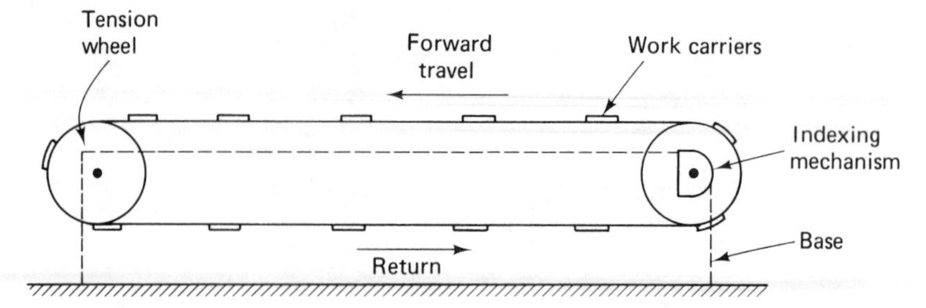


FIGURE 4.6 Chain-driven conveyor system, "over-and-under" type.

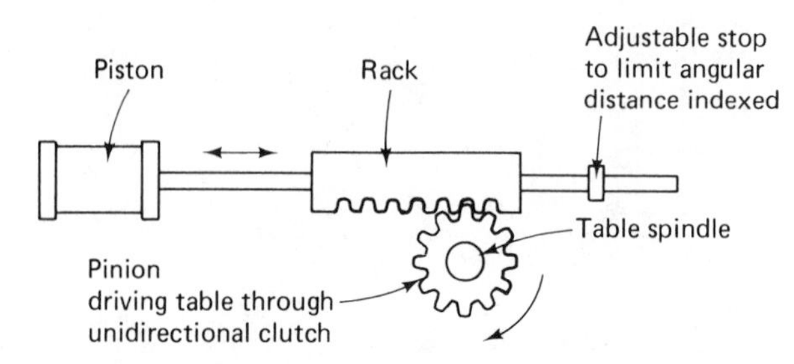


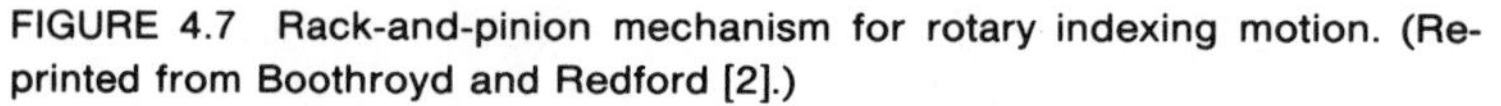

FIGURE 4.7 Rack-and-pinion mechanism for rotary indexing motion. (Reprinted from Boothroyd and Redford [2].)

**Rack and Pinion.** This mechanism is simple but is not considered especially suited to the high-speed operation often associated with indexing machines. The device is pictured in Figure 4.7 and uses a piston to drive the rack, which causes the pinion gear and attached indexing table to rotate. A clutch or other device is used to provide rotation in the desired direction.

**Ratchet and Pawl.** This drive mechanism is shown in Figure 4.8. Its operation is simple but somewhat unreliable, owing to wear and sticking of several of the components.

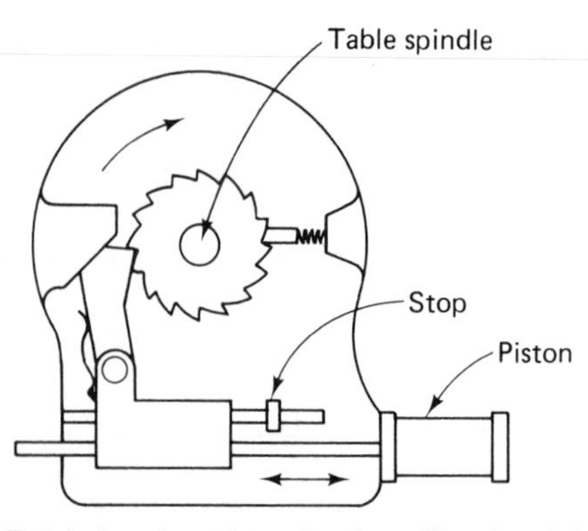


FIGURE 4.8 Ratchet-and-pawl mechanism. (Reprinted from Boothroyd and Redford [2].)

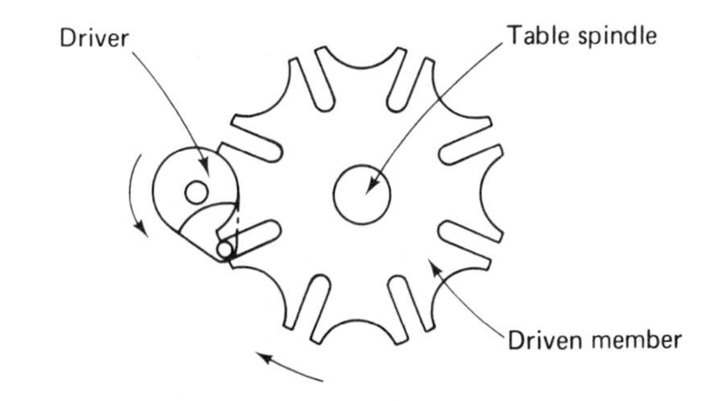


FIGURE 4.9 Geneva mechanism. (Reprinted from Boothroyd and Redford [2].)

GENEVA MECHANISM. The two previous mechanisms convert a linear motion into a rotational motion. The Geneva mechanism uses a continuously rotating driver to index the table as pictured in Figure 4.9. If the driven member has six slots for a six-station dial-indexing machine, each turn of the driver will cause the table to advance one-sixth of a turn. The driver only causes movement of the table through a portion of its rotation. For a six-slotted driven member, 120° of a complete rotation of the driver is used to index the table. The other 240° is dwell. For a four-slotted driven member, the ratio would be 90° for index and 270° for dwell. The usual number of indexings per revolution of the table is four, five, six, and eight.

CAM MECHANISMS. Various forms of cam mechanism, an example of which is illustrated in Figure 4.10, provide probably the most accurate and reliable method of indexing the dial. They are in widespread use in industry despite the fact that the cost is relatively high compared to alternative mechanisms. The cam can be designed to give a variety of velocity and dwell characteristics.

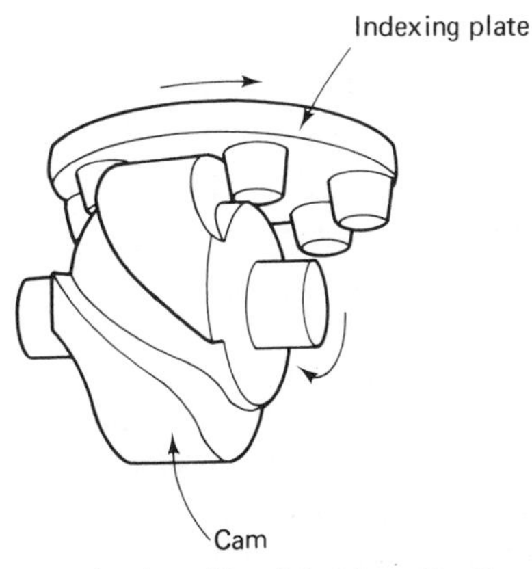


FIGURE 4.10 Cam mechanism. (Reprinted from Boothroyd and Redford [2].)

## 4.5 OPTIONAL FEATURES AT ADDITIONAL COST

Automated flow lines are often equipped with additional features beyond the basic transfer mechanisms and workstations. For example, the idea of using a buffer storage capacity between stations was introduced in Section 4.1. It is also possible to include various monitoring and control features in the line. All of these features add to the cost of the flow line, but their benefits often compensate for their added cost.

### Buffer storage

It is not uncommon for production flow lines to include storage zones for collecting banks of workparts along the line. One example of the use of storage zones would be two intermittent transfer systems, each without any storage capacity, linked together with a workpart inventory area. It is possible to connect three, four, or even more lines in this manner. Another example of workpart storage on flow lines is the nonsynchronous transfer line. With this system, it it is possible to provide a bank of workparts for every station on the line.

There are two principal reasons for the use of buffer storage zones. The first is to reduce the effect of individual station breakdowns on the line operation. The continuous or intermittent transfer system acts as a single integrated machine. When breakdowns occur at the individual stations or when preventive maintenance is applied to the machine, production must be halted. In many cases, the proportion of time the line spends out of operation can be significant, perhaps reaching 50% or more. Some of the common reasons for line stoppages are:

Tool failures or tool adjustments at individual processing stations.

Scheduled tool changes.

Defective workparts or components at assembly stations, which require that the feed mechanism be cleared.

Feed hopper needs to be replenished at an assembly station.

Limit switch or other electrical malfunction.

Mechanical failure of transfer system or workstation.

When a breakdown occurs on an automated flow line, the purpose of the buffer storage zone is to allow a portion of the line to continue operating while the remaining portion is stopped and under repair. For example, assume that a 20-station line is divided into two sections and connected by a parts storage zone which automatically collects parts from the first section and feeds them to the second section. If a station jam were to cause the first section of the line to stop, the second section could continue to operate as long as the supply of parts in the buffer zone lasts. Similarly, if the second section were to shut down, the first section could continue to operate as long as there is room in the buffer zone to store parts. Hopefully, the average production rate on the first section would be about equal to that of the second section. By dividing the line and using the storage area, the average production rate would be improved over the original 20-station flow line. A quantitative analysis of the effect of adding buffer inventory zones will be presented in Chapter 5. Figure 4.11 illustrates the case of two assembly lines separated by a storage buffer.

The second reason for using storage on flow lines is to smooth out the effects of variations in cycle times. These variations occur either between stations or, in the case of flow lines with one or more manual stations, they can occur from cycle to cycle at the same station. To illustrate the second case, suppose that we are considering an assembly line on which all the stations are mechanized except one. The manual station requires the operator to perform an alignment of two components and the time required tends to vary from cycle to cycle. For the transfer system in this line, we must choose between a synchronous system with no parts storage capacity and a nonsyn-

FIGURE 4.11 Two assembly machines connected by spiral-shaped parts storage buffer. The storage buffer allows a 1/2-h float between the two machines. The system assembles 16 parts of windshield washer pumps. (Courtesy Bodine Corp.)

chronous system which allows a "float" of workparts ahead of each station. We must select the transfer system that is more appropriate for our situation.

Assume that we have collected data on the operation and found the following distribution of operation times for a total of 100 cycles: 7 s: two occurrences, or 2%; 8 s: 10 %; 9 s: 18%; 10 s: 38%; 11 s: 20% and 12 s: 12%. This gives an average of 10 s. If this manual operation were used on the synchronous machine, the line would have to be set up with cycle time of 12 s to allow the operator time to finish all assemblies. This would give a production rate of 300 units/h from the line. If the cycle time were adjusted to 11 s, the cycle rate would increase to 327 per hour, but the operator would be unable to complete 12% of the assemblies. Thus, the actual production rate of completed assemblies would be only 288 units/h. If the cycle time were decreased to 10 s,the cycle rate would increase to 360 per hour. However, the operator would be unable to complete the assemblies requiring 11 s and 12 s. The actual production rate would suffer a decrease to 245 units/h.

With the nonsynchronous transfer system, the line could be arranged to collect a bank of workparts immediately before and after the manual station. Thus, the operator would be allowed a range of times to complete the alignment process. As long as the operator's average time were compatible with the cycle time of the transfer system, the flow line would run smoothly. The line cycle time could be set at 10 s and the production rate would be 360 good assemblies per hour.

The disadvantages of buffer storage on flow lines are increased factory floor space, higher in-process inventory, more materials handling equipment, and greater complexity of the overall flow line system. The benefits of buffer storage are often great enough to more than compensate for these disadvantages.

### Pallet fixtures

The transfer system is sometimes designed to accommodate some sort of pallet fixture. Workparts are attached to the pallet fixtures and the pallets are transferred between stations, carrying the part through its sequence of operations. The pallet fixture is designed so that it can be conveniently moved, located, and clamped in position at successive stations. Since the part is accurately located in the fixture, it is therefore correctly positioned for each operation. In addition to the obvious advantage of convenient transport and location of workparts, another advantage of pallet fixtures is that they can be designed to be used for a variety of similar parts.

The other method of workpart location and fixturing does not use pallets.[1] With this method, the workparts themselves are indexed from station

[1]The term "free transfer system" is sometimes applied to this type of line. However, the terminology is not entirely consistent, and a free transfer system may also refer to what we have called a nonsynchronous transfer system.

to station. When a part arrives at a station, it is automatically clamped in position for the operation.

The obvious benefit of this transfer method is that it avoids the cost of pallet fixtures.

### Control functions

Controlling an automated flow line is a complex problem, owing to the sheer number of sequential steps that must be carried out. There are three main functions that are utilized to control the operation of an automatic transfer system. The first of these is an operational requirement, the second is a safety requirement, and the third is a convenience.

1. *Sequence control.* The purpose of this function is to coordinate the sequence of actions of the transfer system and its workstations. The various activities of the automated flow line must be carried out with split-second timing and accuracy. On a metal machining transfer line, for example, the workparts must be transported, located, and clamped in place before the workheads can begin to feed. Sequence control is basic to the operation of the flow line.

2. *Hazard monitoring.* This function ensures that the transfer system does not operate in an unsafe or hazardous condition. Sensing devices may be added to make certain that the cutting tool status is satisfactory to continue to process the workpart in the case of a machining-type transfer line. Other checks might include monitoring certain critical steps in the sequence control function to make sure that these steps have all been performed and in the correct order. Hydraulic or air pressures might also be checked if these are crucial to the operation of automated flow lines.

3. *Quality monitoring.* The third control function is to monitor certain quality attributes of the workpart. Its purpose is to identify and possibly reject defective workparts and assemblies. The inspection devices required to perform quality monitoring are sometimes incorporated into existing processing stations. In other cases, separate stations are included in the line for the sole purpose of inspecting the workpart.

### Control systems

The traditional means of controlling the sequence of steps on the transfer system has been to use electromechanical relays. Relays are employed to maintain the proper order of activating the workheads, transfer mechanism, and other peripheral devices on the line. However, owing to their comparatively large size and relative unreliability, relays have lost ground to other control devices, such as programmable controllers and minicomputers. These more modern components offer opportunities for a higher level of control over the flow line, particularly in the areas of hazard monitoring and quality monitoring.

Conventional thinking on the control of the line has been to stop operation when a malfunction occurred. While there are certain malfunctions representing unsafe conditions that demand shutdown of the line, there are other situations where stoppage of the line is not required and perhaps not even desirable. For example, take the case of a feed mechanism on an automatic assembly machine that fails to feed its component. Assuming that the failures are random and infrequent, it may be better to continue to operate the machine and lock out the affected assembly from further operations. If the assembly machine were stopped, production would be lost at all other stations while the machine is down. Deciding whether it is better to stay in operation or stop the line must be based on the probabilities and economics of the particular case. The point is that there are two alternative control strategies to choose between, instantaneous control and memory control.

INSTANTANEOUS CONTROL. This mode of control stops the operation of the flow line immediately when a malfunction is detected. It is relatively simple, inexpensive, and trouble-free. Diagnostic features are often added to the system to aid in identifying the location and cause of the trouble to the operator so that repairs can be quickly made. However, stopping the machine results in loss of production from the entire line, and this is the system's biggest drawback.

MEMORY CONTROL. In contrast to instantaneous control, the memory system is designed to keep the machine operating. It works to control quality and/or protect the machine by preventing subsequent stations from processing the particular workpart and by segregating the part as defective at the end of the line. The premise upon which memory-type control is based is that the failures which occur at the stations will be random and infrequent. If, however, the station failures result from cause (a workhead that has gone out of alignment, for example) and tend to repeat, the memory system will not improve production but, rather, degrade it. The flow line will continue to operate, with the consequence that bad parts will continue to be produced. For this reason, a counter is sometimes used so that if a failure occurs at the same station for two or three consecutive cycles, the memory logic will cause the machine to stop for repairs.

In Section 5.4, we present an example problem that analyzes the difference between the instantaneous control and the memory control.

## 4.6 AUTOMATION FOR MACHINING OPERATIONS

Transfer systems have been designed to perform a great variety of different metal-cutting processes. In fact, it is difficult to think of machining operations that must be excluded from the list. Typical applications include operations such as milling, boring, drilling, reaming, and tapping. However, it is also feasible to carry out operations such as turning and grinding on transfer-type machines.

There is a well-developed classification scheme for mechanized and automated machines which perform a sequence of operations simultaneously on different workparts. These include dial indexing machines, trunnion machines, and transfer lines. To consider these machines in approximately the order of increasing complexity, we must begin with one that really does not belong in the list at all, the single-station machine.

### Single-station machine

These mechanized production machines perform several operations on a single workpart which is fixtured in one position throughout the cycle. The operations are performed on several different surfaces by workheads located around the piece. The available space surrounding a stationary workpiece limits the number of machining heads that can be used. This limit on the number of operations is the principal disadvantage of the single-station machine. Production rates are usually low to medium.

### Rotary indexing machine

To achieve higher rates of production, the rotary indexing machine performs a sequence of machining operations on several workparts simultaneously. Parts are fixtured on a horizontal circular table or dial, and indexed between successive stations. Six stations is about the maximum practical number that can be located around the table. An example of a dial indexing machine is shown in Figure 4.3.

### Trunnion machine

This machine, shown in Figure 4.12, uses a vertical drum mounted on a horizontal axis, so it is actually a variation of the dial indexing machine. The vertical drum is called a trunnion. Mounted on it are several fixtures which hold the workparts during processing.

Trunnion machines are most suitable for small workpieces. The configuration of the machine, with a vertical rather than horizontal indexing dial, provides the opportunity to perform operations on opposite sides of the workpart. Additional stations can be located on the outside periphery of the trunnion if this is required. The trunnion-type machine is appropriate for workparts in the medium production range.

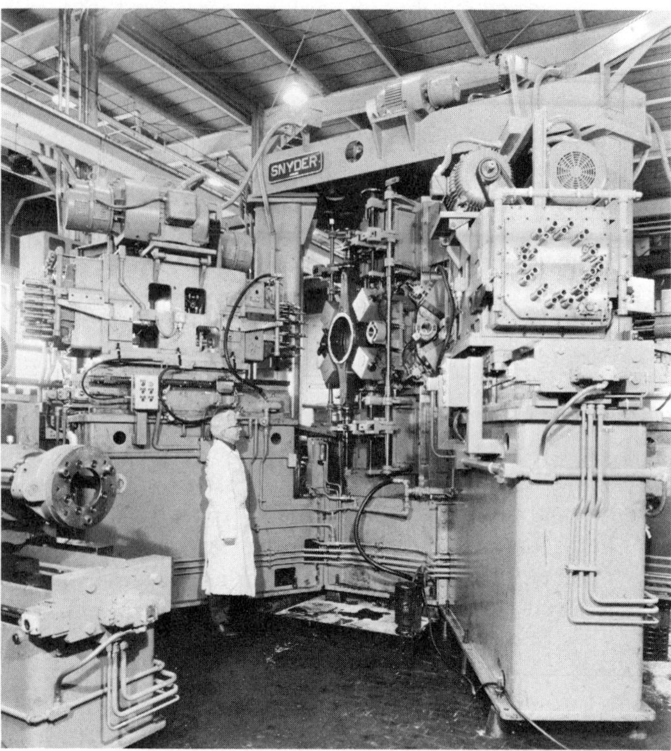

FIGURE 4.12 Six-station trunnion machine. (Courtesy Snyder Corp.)

## Center column machine

Another version of the dial indexing arrangement is the center column type, pictured in Figure 4.13. In addition to the radial machining heads located around the periphery of the horizontal table, vertical units are mounted on the center column of the machine. This increases the number of machining operations that can be performed as compared to the regular dial indexing type. The center column machine is considered to be a high-production machine which makes efficient use of floor space.

## Transfer machine

The most highly automated and versatile of the machines is the transfer line, illustrated in Figures 4.2 and 4.14. The workstations are arranged in a straight-line flow pattern and parts are transferred automatically from station to station. The transfer system can be synchronous or nonsynchronous, workparts can be transported with or without pallet fixtures, buffer storage

FIGURE 4.13 Ten-station center column machine that can be tooled to machine 34 different pump housing components. (Courtesy Snyder Corp.)

can be incorporated into the line operation if desired, and a variety of different monitoring and control features can be used to manage the line. Hence, the transfer machine offers the greatest flexibility of any of the machines discussed. The transfer line can accommodate larger workpieces than do the rotary-type indexing systems. Also, the number of stations, and therefore the number of operations, which can be included on the line is greater than for the circular arrangement. The transfer line has traditionally been used for machining a single product in high quantities over long production runs. More recently, transfer machines have been designed for ease of changeover to allow several different but similar workparts to be produced on the same line. These attempts to introduce flexibility into transfer line design can do nothing but add to the appeal of these high-production machines.

FIGURE 4.14 In-line transfer machine with 33 stations to process cast iron cylinder heads in foreground. (Courtesy F. Jos. Lamb Co.)

## 4.7 AUTOMATIC ASSEMBLY

*Automatic assembly* is a term that refers to the use of mechanized and automated devices to replace manual assembly operations. An automated assembly machine consists typically of the following elements:

1. Transfer system for transporting the partially completed assembly from workstation to workstation.
2. Automatic workstations to perform the various assembly steps. Included at a typical workstation is a part storage facility for holding the components to be added at that station, and a parts orientation and feeding mechanism to present the components to the workhead in the correct position for assembly. A most common example of a storage and feed mechanism is the vibratory bowl feeder, pictured in Figure 4.15.
3. Manual assembly stations where human operators carry out steps that are not easily mechanized.

The machine may also include automatic inspection devices, either at separate stations or included as part of the regular assembly stations, to check on quality. Automatic assembly machines can be of either in-line or rotary configuration, and can be designed with a variety of different workpart

FIGURE 4.15 Vibratory bowl feeder used to orient and feed metal caps for assembly. (Courtesy FMC Corp., Material Handling Equipment Div.)

FIGURE 4.16 Dial-type assembly machine. (Courtesy Bodine Corp.)

transport systems and other features discussed earlier in this chapter. Two in-line machines connected by a spiral-shaped buffer storage rack are shown in Figure 4.11. A dial-type assembly machine is pictured in Figure 4.16.

A great variety of different assembly operations can be performed on mechanized assembly machines. These include screw driving, nut running, press fitting, welding, soldering, staking, riveting, swaging, and adhesive bonding. In addition, other nonassembly operations have been performed on automatic assembly machines. Examples of these operations are sheetmetal forming, plastic molding, casting of metal around mating components, drilling and tapping operations, printing, and a variety of inspection processes.

One of the problems in the design of automated assembly machines is in orienting and feeding of components at the various workstations. Although this is a problem with any automated flow line, it is often a more difficult problem in assembly because of the variety of different component geometries. With a machining transfer line, the same workpart is handled throughout, so virtually the same materials handling solution can be used throughout. With assembly, a new solution must be developed for each different component in the assembly. The component to be added must be positioned at the workstation with high accuracy of location, and it must be done consistently on every cycle and usually at high rates of speed. Several different devices for orienting and feeding assembly components are illustrated in Figures 4.15 and 4.17.

Another problem area encountered in automatic assembly is that of defective components. Manufacturing variations are introduced into the production of the components that go into an assembly. These manufacturing variations lead to a certain fraction rate of defective components. Defects will generally lead to one of two situations during assembly, neither of which is desirable:

1. The defect will cause a jam in the feed mechanism or assembly workhead. This will cause a stoppage at the workstation, and perhaps in the entire line, with associated loss of production.
2. The defect will not cause a jam but will instead be added to the assembly, possibly rendering the entire assembly defective or causing a stoppage at a subsequent station.

The effect of these defective components on overall line performance can be analyzed and this will be taken up in Section 5.4.

Because of the various problems encountered in automating a particular assembly operation, the decision is sometimes made to use a manual station rather than an automatic one. Some assembly operations that would be relatively easy to perform manually would be quite difficult to automate.

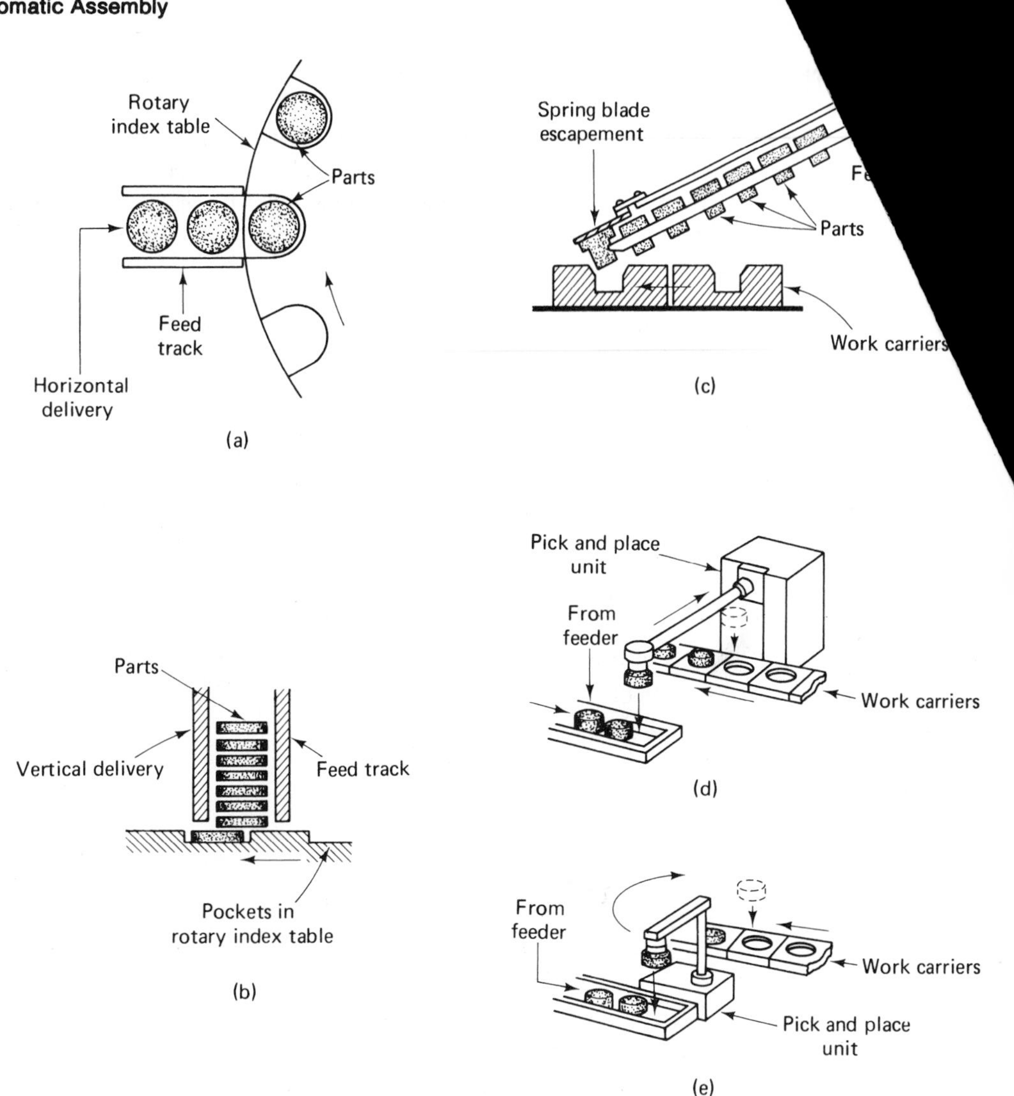


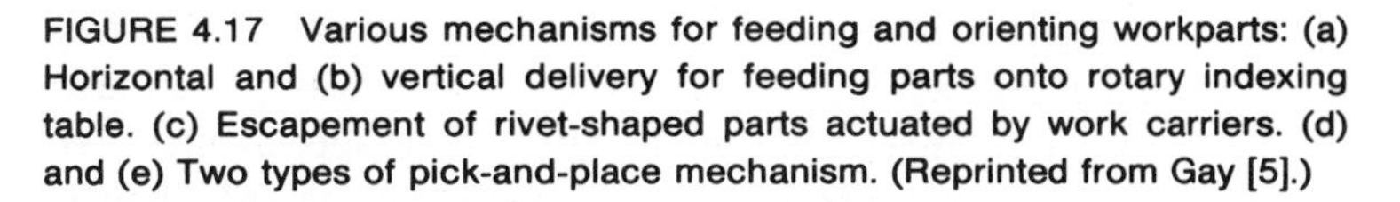
FIGURE 4.17 Various mechanisms for feeding and orienting workparts: (a) Horizontal and (b) vertical delivery for feeding parts onto rotary indexing table. (c) Escapement of rivet-shaped parts actuated by work carriers. (d) and (e) Two types of pick-and-place mechanism. (Reprinted from Gay [5].)

Examples would be steps that require a close visual alignment of mating parts, or that need the operator to apply his or her sense of touch or feel. Many inspection operations also fall into this category. A human operator may easily be able to detect a variety of defective attributes of a particular component, whereas it would be extremely difficult to fabricate an inspection device that could emulate this skill.

## ...TION CONSIDERATIONS

87

...firm decides that some form of automated flow line ...d of producing a particular workpart or assembly, ... specifications that must be decided. In designing ...mated flow line, some of the details to consider are the

...ed track

...her the flow line is to be engineered in-house or by a machine tool ...ilder.
Size, weight, geometry, and material if a processed workpart.
Size, weights, and number of components if an assembly.
Tolerance requirements.
Type and sequence of operations.
Production rate requirements.
Type of transfer system.
Methods of fixturing and locating workparts.
Methods of orienting and feeding components in the case of assemblies.
Reliability of individual stations and transfer mechanisms, as well as overall reliability of the line.
Buffer storage capability.
Ease of maintenance.
Control features desired.
Floor space available.
Flexibility of line in terms of possible future changes in product design.
Flexibility of line to accommodate more than a single workpart.
Initial cost of the line.
Operational and tooling cost for the line.

In developing the concept for a mechanized flow line, there are two general approaches that can be considered. The first is to use standard machine tools and other pieces of processing equipment at the workstations and to connect them with standard or special material handling equipment. The material handling hardware serves as the transfer system and moves, feeds, and ejects the work between the standard machines. The line of machines is referred to as an "integrated flow line" or *link line.* The individual machines must either be capable of operating on an automatic cycle or they must be manually operated. There may also be fixturing and location problems at the stations which are difficult to solve without some form of human assistance during the cycle. A firm will often prefer the integrated type of flow line because it can be made up from machine tools that are already in the plant, these machine tools can be reused when the production run is finished, and there is less debugging and maintenance. These flow lines can also be engineered by personnel within the firm, perhaps with the aid of material handling experts. The limitation of integrated flow lines is that they favor simpler workpiece shapes and smaller sizes since the work handling equipment is less sophisticated and more general-purpose. Greater future use of industrial robots as material handling devices will increase the attractiveness of the integrated flow line.

Integrated flow lines have found applications in such processes as plating and finishing operations, pressworking, rolling mill operations, gear manufacturing, and a variety of machining operations. An example is illustrated in Figure 4.18.

The alternative approach to building an automated line is to turn the problem over to a machine tool builder specializing in the fabrication of transfer lines, assembly machines, or other flow line equipment. Using the customer's blueprints and specifications, the builder will submit a proposal for the line. Typically, several machine tool builders will be asked to make proposals. Each proposed design will be based on the machine components comprising the builder's product line as well as the ingenuity and experience of the engineer preparing the proposal.

Once a particular proposal is accepted, the machine tool builder will proceed with the final detailed design. The resulting machine will utilize a "building-block" principle. That is, the specialized flow line, designed to produce the customer's particular product, will be constructed out of standard components. These standard components consist of the base or transfer system, and workheads for performing the various processing or assembly operations. These standard components will be fabricated into the special configuration required for the customer's product. Several examples of these standard transfer line components are pictured in Figures 4.19 and 4.20. For metal-cutting transfer lines, the workheads consist of the feed mechanism, spindle, and power source. The workheads must then be fitted with special

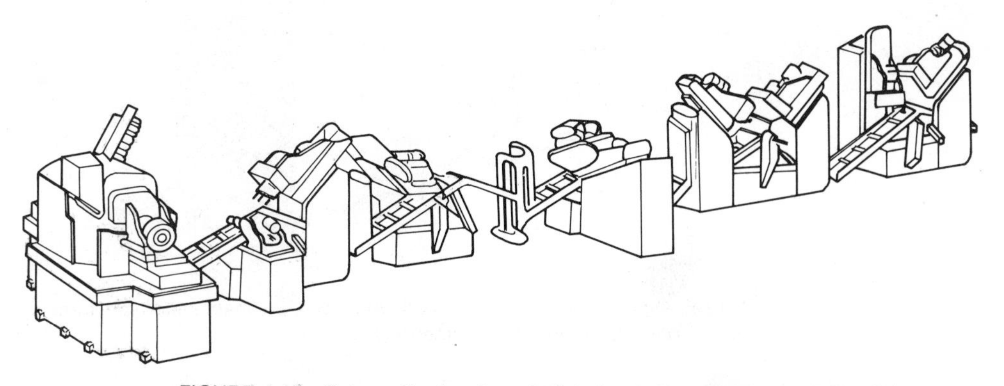

FIGURE 4.18 Schematic drawing of "link line." (Reprinted from Dallas [3].)

FIGURE 4.19 Standard rotary table component used on dial index machines. (Courtesy Ferguson Machine Co.)

FIGURE 4.20 Standard power feed unit used on machining-type transfer lines. (Courtesy Ferguson Machine Co.)

tools to carry out the particular process. For assembly machines, the workheads consist of the part storage and feed device (e.g., vibratory bowl feeder) and the mechanism that is to perform the particular assembly operation. As mentioned in Section 4.7, the workheads must be designed to handle the particular assembly component. Whether for machining or assembly operations, these workheads do not have a frame or worktable. Instead, they are attached to the transfer system frame, which has been specially adapted for the workpart.

When a flow line has been fabricated using this building-block principle, it is sometimes referred to as a *unitized flow line.* The standard machine tool components all go together and act as a single mechanical unit.

Higher production rates are generally possible with unitized construction compared to integrated flow lines. Also, less floor space is required because the unitized lines are typically much more compact. The higher cost of unitized equipment makes it suitable only for long production runs and on products not subject to frequent design changes. Equipment obsolescence becomes a danger if these two requirements are not met. Applications of this type of flow line construction are found in transfer lines for machining automotive engine parts and in assembly machines for pens, small hardware items, electrical assemblies, and so on. Figures 4.2, 4.3, 4.11, 4.12, 4.13, 4.14, and 4.16 illustrate this type of flow line construction.

## REFERENCES


[1] BERTHA, R. W., "Will Your Assembly System Be a Winner," *Automation,* September, 1973.

[2] BOOTHROYD, G., and REDFORD A. H. *Mechanized Assembly,* McGraw-Hill Publishing Co., London, 1968.

[3] DALLAS, D. B., *Tool and Manufacturing Engineers Handbook,* 3rd ed., Society of Manufacturing Engineers, Dearborn, Mich. (McGraw-Hill Book Company., New York), 1976, Chapter 12.

[4] EARY, D., and JOHNSON, G. E., *Process Engineering for Manufacturing,* Prentice-Hall, Englewood Cliffs, N.J., 1962, Chapter 13.

[5] GAY, D. S., "Ways to Place and Transport Parts," *Automation*, June, 1973.

[6] LAWSON, E. A., BLOCK, F. E., LONG, F. J., and COONFER, D.C., "Transfer Machines Today," *Automation,* June, 1971.

[7] LINDBERG, R. A., *Processes and Materials of Manufacture,* Allyn and Bacon, Inc., Boston, 1977, Chapter 7.

[8] RILEY, F. J., "Selecting Controls for Automatic Assembly," *Manufacturing Engineering and Management,* Part I, October, 1974; Part II, November, 1974.

[9] STEVENS, D. S., "The Current Technology of Assembly Machines," *Paper AD75-153,* Society of Manufacturing Engineers, Dearborn, Mich., 1975.

## PROBLEMS

**4.1.** For the eight-slotted driven member of the Geneva mechanism in Figure 4.9, determine the proportion of each complete revolution of the driver that motion is being imparted to the driven member. Express this in degrees. How many degrees of rotation represent dwell?

**4.2.** A Geneva mechanism with a six-slotted driven member is used in a dial-type assembly machine. The longest assembly operation takes exactly 1 s to complete, so the driven member must be in a stopped (dwell) position for this length of time.

(a) At what rotational speed must the driver be turned to accomplish this 1-s dwell?

(b) How much time will be required to index the dial to the next position?

(c) Determine the ideal production rate of the assembly machine if each index of the dial produces a completed workpart.

**4.3.** A certain transfer line performs a sequence of machining and assembly operations. One of the assembly operations near the end of the line is performed manually. The distribution of operation times for the manual station is as given in the table.

| *Operation time (s)* | *Frequency of occurrence (%)* |
|---|---|
| 15 | 2.7 |
| 16 | 6.1 |
| 17 | 12.1 |
| 18 | 25.9 |
| 19 | 32.1 |
| 20 | 10.9 |
| 21 | 6.9 |
| 22 | 3.3 |
| | 100.0 |

The slowest automatic station has a cycle time of 18 s.

(a) If the line uses a synchronous (intermittent) transfer system, determine the cycle rate and actual production rate per hour if the transfer system were set to index parts at each of the following intervals: once every 22 s, 21 s, 20 s, 19 s, 18 s. Time taken to move parts is negligible.

(b) If a nonsynchronous parts transfer system were used on this line to accumulate a float of parts before the manual station, at what cycle rate should the system be operated? Determine the corresponding production rate.

# Analysis of Automated Flow Lines

## 5.1 INTRODUCTION

In analyzing the performance of automated flow lines, two general problem areas must be distinguished. The first is related to the production processes used on the line. For example, consider a transfer line that performs a series of machining operations. There is an extensive body of knowledge related to the theory and practice of metal machining. This technology includes the proper specification and use of cutting tools, the metallurgy and machinability of the work material, chip control, machining economics, machine tool vibrations, and a host of other topics. Many of the problems encountered in the operation of a metal cutting transfer line are directly related to and can often be solved by the application of good machining principles. The same is true for other production processes. In each area of production, a technology has developed after many years of research and practical experience in the area. By making the best use of the given process technology, each individual workstation on the line can be made to operate at or near its maximum productive capability. However, even if it were possible to operate each station in an optimal way, this does not guarantee that the overall flow line will be optimized.

It is with this viewpoint of the overall flow line operation that we identify the second general problem area of flow line performance. This second area is concerned with the systems aspects of designing and running the line. Normally associated with the operation of an automated flow line is the problem of reliability. Since the line often operates as a single mechanism, failure of one component of the mechanism will often result in stoppage of the entire line. There are approaches to this problem that transcend the manufacturing processes at individual stations. What is the effect of the number of workstations on the performance of the line? How much improvement can be obtained by using one or more buffer storage zones? What is the effect of component quality on the operation of an automated assembly machine? How will the use of manual workstations affect the line? These are questions that can be analyzed using a systems approach.

In addition to the reliability problem, another systems design problem is one of properly balancing the line. In line balancing, the objective is to spread the total work load as evenly as possible among the stations in the line. The problem is normally associated with the design of manual assembly lines. It is also a consideration in automated flow lines, but the reliability problem usually predominates. We will consider line balancing in Chapter 6. The reliability of the line will be the principal concern of the present chapter.

## 5.2 GENERAL TERMINOLOGY AND ANALYSIS[1]

Flow line performance can be analyzed by means of three basic measures: average production rate, proportion of time the line is operating (line efficiency), and cost per item produced on the line. We will concern ourselves initially with flow lines that possess no internal buffer storage capacity.

To begin the analysis, we must assume certain basic characteristics about the operation of the line. A synchronous transfer system is assumed. Parts are introduced into the first workstation and are processed and transported at regular intervals to succeeding stations. This interval defines the ideal or theoretical cycle time $T_c$ of the flow line. $T_c$ is equal to the time required for parts to transfer plus the processing time at the longest workstation. The processing times at different stations will not be the same. Long holes take more time to drill than short holes. A milling operation may take longer than a tapping operation. The stations which require less time than the longest station will have a certain amount of idle time.

Because of breakdowns on the line, the actual average production time $T_p$ will be longer than the ideal cycle time. When a breakdown occurs at any one station, we assume the entire line is shut down. Let $T_d$ represent the

[1] Sections 5.2 and 5.3 are based largely on Groover [9].

average downtime to diagnose the problem and make repairs when a breakdown occurs. Since there may be more than one reason why a line is down, it is sometimes convenient to distinguish between the different reasons by subscripting the term as $T_{dj}$. The subscript $j$ is used to identify the reason for the breakdown (e.g., tool failure, part jam, feed mechanism, etc.). The frequency with which line stops per cycle occur for reason $j$ is denoted by $F_j$. Multiplying the frequency $F_j$ by the average downtime per stop $T_{dj}$ gives the mean time per cycle the machine will be down for reason $j$. If there is only one reason why the transfer machine may be down, the average production time $T_p$ is given by

$$T_p = T_c + FT_d \tag{5.1}$$

If there are several reasons why the line is down and we wish to distinguish among them, the average production time becomes

$$T_p = T_c + \sum_j F_j T_{dj} \tag{5.2}$$

It is possible that there are interactions between some of the causes of line stops. For example, if we were to introduce a scheduled maintenance or tool change program, this would presume to favorably influence the frequency of unplanned station breakdowns.

One of the important measures of performance for a transfer line or assembly machine is the average production rate $R_p$. We must differentiate between this production rate, which represents the actual output of the machine, and the theoretical production rate. The actual average production rate is based on the average production time $T_p$:

$$R_p = \frac{1}{T_p} \tag{5.3}$$

If the flow line produces less than a 100% yield, this production rate must be adjusted for the yield. For example, if 2% of the workparts are scrapped during processing, the production rate would be 98% of that calculated by Eq. (5.3). The theoretical production rate of the flow line, rarely achieved in practice, is computed as

$$R_c = \frac{1}{T_c} \tag{5.4}$$

The machine tool builder will use this value in describing the flow line, and will speak of it as the production rate at 100% efficiency. Unfortunately, the machine will not operate at 100% efficiency.

The line efficiency $E$ is simply the proportion of time the line is up and operating. We can compute it as follows:

$$E = \frac{T_c}{T_p} = \frac{T_c}{T_c + FT_d} \tag{5.5}$$

An alternative measure of performance is the proportion of downtime $D$ on the line:

$$D = \frac{FT_d}{T_p} = \frac{FT_d}{T_c + FT_d} \tag{5.6}$$

Certainly, the downtime proportion and the uptime proportion must add to 1:

$$E + D = 1 \tag{5.7}$$

The third measure of flow line performance is the cost per item produced. Let $C_m$ equal the cost of raw materials per product, where the product refers to the unit of output from the line. The raw materials will include the basic workpart plus any additional components assembled to it. Let $C_L$ represent the cost per minute to operate the line. This would include labor, overhead, maintenance, and the allocation of the capital cost of the equipment over its expected service life. The last item, capital cost, will likely be the largest portion of $C_L$. The cost of any disposable tooling should be computed on a per workpiece basis and is symbolized by $C_t$. Using these components, the general formula for calculating the cost per workpiece $C_{pc}$ is

$$C_{pc} = C_m + C_L T_p + C_t \tag{5.8}$$

This equation does not account for such things as yield or scrap rates, inspection costs associated with identifying defective items produced, or repair cost associated with fixing the defective items. However, these factors can be incorporated into Eq. (5.8) in a fairly straightforward manner.

We will let $n$ represent the number of workstations on the flow line and $Q$ will designate the quantity of workparts produced off the line. $Q$ may represent a batch size or it may mean the number of parts produced over a certain time period. We will use it in whatever way we find convenient. However, one note of caution is this: $Q$ may include a certain number of defects if the flow line has a habit of not producing 100% good product.

Let us consider an example to illustrate the terminology of flow line performance.

### EXAMPLE 5.1

Suppose that a 10-station transfer machine is under consideration to produce a component used in a pump. The item is currently produced by more conventional means, but demand for the item cannot be met. The manufacturing engineering department has estimated that the ideal cycle time will be $T_c = 1.0$ min. From similar transfer lines, it is estimated that breakdowns of all types will occur with a frequency, $F = 0.10$ breakdown/cycle, and that the average downtime per line stop will be 6.0 min. The scrap rate for the current conventional processing method is 5% and this is considered a good estimate for the transfer line. The starting casting for the component costs $1.50 each and it will cost $60.00/h or $1.00/min to operate the transfer line. Cutting tools are estimated to cost $0.15/workpart.

Using the foregoing data, it is desired to compute the following measures of line performance:

(a) Production rate.
(b) The number of hours required to meet a demand of 1500 units/week.
(c) Line efficiency.
(d) Cost per unit produced.

The average production time per piece can be calculated from Eq. (5.1).

$$T_p = 1.0 + 0.10(6.0) = 1.6 \text{ min}$$

The average production rate for the line would be determined by Eq. (5.3).

$$R_p = \frac{1}{1.6} = 0.625 \text{ piece/min or } 37.5 \text{ pieces/h}$$

However, correcting for the scrap rate of 5%, the actual production rate of good products is

$$R_p = 0.95(37.5) = 35.625 \text{ pieces/h}$$

To compute the number of hours required to produce 1500 units/week (we assume that this means 1500 good units plus scrap rather than 1500 with 5% scrap included), we divide the production rate of 35.625 units/h into the 1500-unit requirement:

$$\text{hours} = \frac{1500}{35.625} = 42.1 \text{ h}$$

The line efficiency is found by taking the ratio of the ideal cycle time to the average production time, according to Eq. (5.5).

$$E = \frac{1.0}{1.6} = 0.625$$

The proportion downtime is determined by Eq. (5.6).

$$D = \frac{0.10(6)}{1.6} = 0.375$$

The cost per product can be computed from Eq. (5.8) except that we must account for the scrap rate. This is accomplished by dividing the cost determined by Eq. (5.8) by the yield of good parts. In our example the yield is 0.95.

$$C_{pc} = \frac{1}{0.95}(1.50 + 1.00 \times 1.60 + 0.15)$$
$$= \$3.42/\text{good unit}$$

The \$3.42 represents the average cost per acceptable product under the assumption that we are discarding the 5% bad units at no salvage value and no disposal cost. Suppose that we could repair these units at a cost of \$5.00 per unit. Would it be economical to do so? To compute the cost per piece, the repair cost would have to be added to the other components of Eq. (5.8). Also, since repair of the defects means that our yield will be 100%, the 0.95 used above to obtain a cost of \$3.42 can be ignored.

$$C_{pc} = 1.50 + 1.00 \times 1.60 + 0.15 + 0.05(5.00)$$
$$= \$3.50/\text{unit}$$

The lower cost per unit is associated with the policy of scrapping the 5% defects rather than repairing them. Unless the extra units are needed to meet demand, the scrap policy seems preferable.

## 5.3 ANALYSIS OF TRANSFER LINES WITHOUT STORAGE

In this section we will consider the analysis of continuous and intermittent transfer machines without internal storage capacity. We will supplement the results of Section 5.2 by considering what happens at a workstation when it breaks down. There are two possibilities and we refer to their analyses as the upper-bound approach and the lower-bound approach. In practical terms, the difference between the two approaches is simply this: With the upper-bound approach, we assume that the workpart is not removed from the station when a breakdown occurs at that station. With the lower-bound approach, the workpart is taken out of the station when the station breaks down. The circumstances under which each approach is appropriate will be discussed in the following subsections.

### Upper-bound approach

The upper-bound approach provides an estimate of the upper limit on the frequency of line stops per cycle. We assume here that a breakdown at a station does not cause the part to be removed from that station. In this case, it is possible, perhaps likely, that there will be more than one line stop associated with a particular workpart. An example of this situation might be a hydraulic failure at a workstation which prevents the feed mechanism from working. Another possibility is that the cutting tool has nearly worn out and needs to be changed. Or, the workpart is close to being out of tolerance and a tool adjustment is required to correct the condition. With each of these examples, there is no reason for the part to be removed from the transfer machine.

Let $p_i$ represent the probability that a part will jam at a particular station $i$, where $i = 1, 2, \ldots, n$. Since the parts are not removed when a station jam occurs, it is possible (although not probable) that the part will jam at every station. The expected number of line stops per part passing through the line can be obtained by merely summing up the probabilities $p_i$ over the $n$ stations. Since each of the $n$ stations is processing a part each cycle, this expected number of line stops per part passing through the line is equal to the frequency of line stops per cycle. Thus,

$$F = \sum_{i=1}^{n} p_i \tag{5.9}$$

If the probabilities $p_i$ are all equal, $p_1 = p_2 = \ldots = p_n = p$, then

$$F = np \tag{5.10}$$

**EXAMPLE 5.2**

In a 10-station transfer line, the probability that a station breakdown will occur for a given workpart is equal to 0.01. This probability is the same for all 10 stations. Determine the frequency of line stops per cycle on this flow line using the upper-bound approach.

The value of $F$ can be calculated from Eq. (5.10).

$$F = 10(0.01) = 0.10$$

This is the value of $F$ used in Example 5.1.

## Lower-bound approach

The lower-bound approach gives an estimate of the lower limit on the expected number of line stops per cycle. In this approach, we assume that the station breakdown results in the destruction or damage of the workpiece. For example, a drill or tap breaks off in the part during processing. The broken tool must be replaced at the workstation and the workpart must be removed from the line for subsequent rework or scrap. Accordingly, the part cannot proceed to the next stations for further processing.

Again, let $p_i$ = the probability that a part will jam at a particular station $i$. Then, considering a given workpart as it proceeds through the line, $p_1$ = the probability that the part will jam at station 1, and $(1-p_1)$ = the probability that the part will not jam at station 1 and thus be available for subsequent processing. The quantity $p_2(1-p_1)$ = the probability that this given part will jam at station 2. Generalizing, the quantity

$$p_i(1-p_{i-1})(1-p_{i-2})\dots(1-p_2)(1-p_1) \qquad i=1,2,\dots,n$$

is the probability that the given part will jam at any station $i$. Summing all these probabilities from $i=1$ through $n$ would give the probability or what has previously been called the frequency of line stops per cycle. There is an easier way to determine this frequency.

The probability that a given part will pass through all $n$ stations without a line stop associated with its processing is given by

$$\prod_{i=1}^{n}(1-p_i)$$

Therefore, the frequency of line stops per cycle is provided by

$$F = 1 - \prod_{i=1}^{n}(1-p_i) \tag{5.11}$$

If the probabilities $p_i$ that a part will jam at a particular station are all equal, $p_1 = p_2 = \dots = p_n = p$, then

$$F = 1-(1-p)^n \tag{5.12}$$

We might be tempted to view this frequency, $F$, as the probability of a line stop per cycle, except that it is possible, in the upper-bound approach, for the frequency of line stops per cycle to exceed unity. A probability greater than 1 cannot be interpreted.

With the lower-bound approach, the number of workparts coming off the line will be less than the number starting. If the parts are removed from the line when a breakdown occurs, they are not available to be counted as part of the output of the line. Therefore, the production-rate formula given by Eq. (5.3) must be amended to reflect this reduction in output. Using the lower-bound approach, the production-rate formula becomes

$$R_p = \frac{1-F}{T_p}$$

where $F$ not only stands for the frequency of line stops but also the frequency of parts removal. If no rework is performed, $F$ is the scrap rate. Therefore, the term $(1-F)$ represents the yield of the transfer machine. $T_p$ is interpreted to mean the average cycle time of the machine.

### EXAMPLE 5.3

Compute the value of $F$ using the lower-bound approach for the data of Example 5.2. Also compute the production rate.

From Eq. (5.12) the value of $F$ is

$$F=1-(1-0.01)^{10}=0.0956$$

Although the value of $F$ is smaller as calculated by the lower-bound approach, the difference is small. This difference grows as the value of $p$ and the number of workstations increase.

To compute the production rate we must adjust Eq. (5.3), as indicated in the previous discussion, by the value of $F$. Production time, $T_p$, was calculated in the example of Section 5.2 as 1.6 min. Therefore, production rate would equal

$$R_p=\frac{1-0.0956}{1.6}=0.565 \text{ piece/min or } 33.9 \text{ pieces/h}$$

This is somewhat below the 35.6 pieces/h obtained in Example 5.1 where the scrap rate was 0.05 rather than the 0.0956 computed here. We can reason that the 5% scrap rate probably represents a mixture of the two cases assumed by the upper- and lower-bound approaches. When breakdowns occur, the workparts are sometimes removed from the line and other times they are not. The upper- and lower-bound approaches, as their names imply, provide upper and lower limits on the frequency of downtime occurrences. However, these two approaches also generate upper and lower limits on the production rate, assuming that station breakdowns are the sole cause of scrap. Of course, the 5% scrap figure may also include other conditions, such as poor quality of the workparts.

## Some comments and observations

Determining whether the upper- or the lower-bound approach is more appropriate for a particular transfer line requires knowledge about the operation of the line. The operator may be required to use judgment in each breakdown to determine whether the workpart should be removed. If parts are sometimes removed and sometimes not, the actual frequency of breakdowns will fall somewhere between the upper and lower bounds. Of the two approaches, the upper-bound approach is preferred, certainly for convenience of calculation and probably for accuracy also.

There are other reasons why line stops occur which are not directly related to workstations (e.g., transfer mechanism failure, scheduled tool changes for all stations, preventive maintenance, product changeover, etc.). These other factors must be taken into consideration when determining line performance.

The biggest difficulty in using Eqs. (5.9) through (5.12) lies in determining the values of $p_i$ for the various stations. Perhaps the best approach is to base the values of $p_i$ on previous historical data and experience with similar workstations.

There are a number of general truths about the operation of transfer lines which are revealed by the equations of Sections 5.2 and 5.3. First, the line efficiency decreases substantially as the number of stations increases. It is not uncommon for large transfer lines consisting of up to 100 stations to be down nearly 50% of the time. It is doubtful that such lines achieve the return on investment which their owners anticipated from them. The influence of the number of stations on line efficiency is dramatically displayed in Figure 5.1 for several assumed values of station breakdown probabilities.

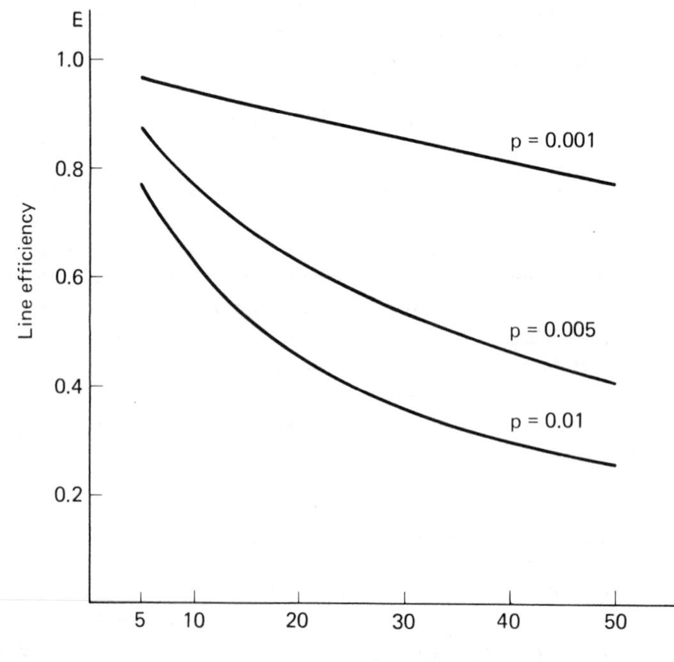


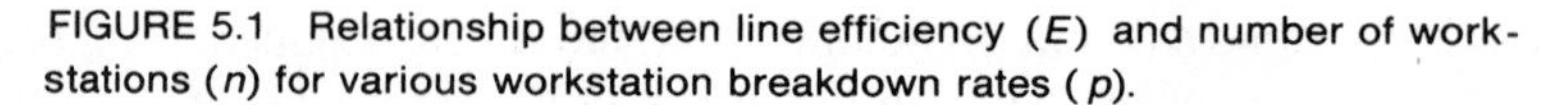
FIGURE 5.1 Relationship between line efficiency ($E$) and number of workstations ($n$) for various workstation breakdown rates ($p$).

This figure was calculated using the upper-bound approach. In comparing the upper- and lower-bound approaches, we find that the upper-bound calculations lead to a lower value of line efficiency but a higher value for the production rate. The reason for this apparent anomaly is this: using the lower-bound approach with its assumption of parts removal, the removed parts are not available at subsequent workstations to cause line stoppages.

Hence, the proportion of uptime on the line is greater. However, if the parts are removed from the line, the production rate of the line is reduced.

## 5.4 ANALYSIS OF ASSEMBLY MACHINES WITHOUT STORAGE

The analysis of automatic assembly machines shares much in common with the upper-bound approach used for metal-cutting transfer lines. Indeed, as we pointed out in Chapter 4, some flow lines are designed to perform both assembly and machining processes. With machines performing pure assembly operations, some modifications must be made in the analysis to account for the fact that components are being added at the various workstations. In developing equations that govern the operation of an assembly machine, we shall follow the general approach suggested by Boothroyd and Redford [1].

We assume that the typical operation occurring at a workstation of an assembly machine is one in which a component is added or joined in some fashion to an existing assembly. The existing assembly consists of a base part plus the components assembled to it at previous stations. The base part is launched onto the line either at or before the first workstation. The components that are added must be clean, uniform in size and shape, of high quality, and consistently oriented. When the feed mechanism and assembly workhead attempt to join a component that does not meet these specifications, the station can jam. When this occurs, it can result in the shutdown of the entire machine until the fault is corrected. Thus, in addition to the other mechanical and electrical failures that can interrupt the operation of a flow line, the problem of defective components is one that specifically plagues the operation of an automatic assembly machine. This is the problem we propose to deal with in this section.

### The assembly machine as a game of chance

Defective parts are a fact of manufacturing life. Defects occur with a certain fraction defective rate, $q$ ($q$ for queer). In the operation of an assembly workstation, $q$ can be considered as the probability that the next component is defective. When an attempt is made to feed and assemble a defective component, the defect might or might not cause the station to jam. Let $m$ equal the probability that a defect will result in the malfunction and stoppage of the workstation. Since the values of $q$ and $m$ may very well be different for different stations, we will subscript these terms as $q_i$ and $m_i$, where $i = 1, 2, \ldots, n$, the number of stations on the assembly machine.

Considering what happens at a particular workstation, station $i$, there are three possible events that might occur when the feed mechanism attempts to feed the component and the assembly device attempts to join it to the existing assembly:

1. The component is defective and causes a station jam.
2. The component is defective but does not cause a station jam.
3. The component is not defective.

The probability of the first event is the product of the fraction defective rate for the station, $q_i$, multiplied by the probability that a defect will cause the station to stop, $m_i$. This product is the same as the term $p_i$ of Section 5.3, the probability that a part will jam at station $i$. For an assembly machine,

$$p_i = m_i q_i \tag{5.13}$$

When the values of $m_i$ and $q_i$ are, respectively, equal for all workstations ($m_i = m$, and $q_i = q$ for all $i$), then

$$p = mq \tag{5.14}$$

In this case where the component is defective and causes a station jam, the defective component will be cleared and the next compoment will be allowed to feed and be assembled. We assume that this next component is not defective—the probability of two consecutive defects is very small, equal to $q^2$.

The second possible event, when the component is defective but does not cause a station jam, has a probability given by

$$(1 - m_i)q_i$$

With this outcome, a bad part is joined to the existing assembly, perhaps rendering the entire assembly defective.

The third possibility is obviously the most desirable. The probability that the component is not defective is equal to the proportion of good parts:

$$1 - q_i$$

The probabilities of the three possible events must sum to unity.

$$m_i q_i + (1 - m_i)q_i + (1 - q_i) = 1 \tag{5.15}$$

For the special case where $m_i = m$ and $q_i = q$ for all $i$, Eq. (5.15) reduces to

$$mq + (1 - m)q + (1 - q) = 1 \tag{5.16}$$

To determine the complete distribution of possible outcomes that can occur on an $n$-station assembly machine, we can multiply the terms of Eq. (5.15) together for all $n$ stations:

$$\prod_{i=1}^{n} \left[ m_i q_i + (1 - m_i) q_i + (1 - q_i) \right] = 1 \tag{5.17}$$

In the special case of Eq. (5.16), where all $m_i$ are equal and all $q_i$ are equal, Eq. (5.17) becomes the multinomial expansion

$$\left[ mq + (1 - m) q + (1 - q) \right]^n = 1 \tag{5.18}$$

Expansion of Eq. (5.17) or (5.18) will reveal the probabilities for all possible sequences of events which can take place on the assembly machine. Unfortunately, the number of terms in the expansion of Eq. (5.17) becomes very large for a machine with more than two stations. The exact number of terms is equal to $3^n$, where $n$ is the number of stations. For an eight-station line, the number of terms is equal to 6561, each term representing the probability of one of the 6561 possible sequences on the assembly machine.

## Measures of performance

Fortunately, we are not required to calculate every term to make use of the concept of assembly machine operation provided by Eqs. (5.15) and (5.17). One of the characteristics of performance that we might want to know is the proportion of assemblies that contain one or more defective components. Two of the three terms in Eq. (5.15) represent events that result in the addition of good components at a given station. The first term is $m_i q_i$, which indicates a line stop but also means that a defective component has not been added to the assembly. The other term is $(1 - q_i)$, which means that a good component has been added at the station. The sum of these two terms represents the probability that a defective component will not be added at station $i$. Multiplying these probabilities for all stations, we get the proportion of acceptable product coming off the line, $P_{ap}$.

$$P_{ap} = \prod_{i=1}^{n} (1 - q_i + m_i q_i) \tag{5.19}$$

If this is the proportion of assemblies with no defective components, the proportion of assemblies that contain at least one defect is given by

$$P_{qp} = 1 - \prod_{i-1}^{n} (1 - q_i + m_i q_i) \tag{5.20}$$

In the case of equal $m_i$ and equal $q_i$, these two equations become, respectively:

$$P_{ap} = (1 - q + mq)^n \tag{5.21}$$

$$P_{qp} = 1 - (1 - q + mq)^n \tag{5.22}$$

The proportions provided by Eqs. (5.19) or (5.21) give the "yield" of the assembly machine, certainly one important measure of the machine's performance. The proportions of assemblies with one or more defective components, given by Eqs. (5.20) or (5.22), must be considered a liability of the machine's operation. Either these assemblies must be identified through an inspection process and possibly repaired, or they will become mixed in with the good assemblies. The latter possibility would lead to undesirable consequences when the assemblies are placed in service.

In addition to the proportions of good and bad assemblies as measures of performance for an assembly machine, we are also interested in the machine's production rate, proportions of uptime and downtime, and average cost per unit produced.

To calculate production rate we must first determine the frequency of downtime occurrences per cycle, $F$. If each station jam results in a machine downtime occurrence, $F$ can be found by taking the expected number of station jams per cycle. Making use of Eq. (5.13), this value is given by

$$F = \sum_{i=1}^{n} p_i = \sum_{i=1}^{n} m_i q_i \tag{5.23}$$

If all $p_i$ are equal and all $m_i$ are equal, Eq. (5.23) becomes

$$F = nmq \tag{5.24}$$

The average production time per assembly is therefore given by

$$T_p = T_c + \sum_{i=1}^{n} m_i q_i T_d \tag{5.25}$$

where $T_c$ = ideal cycle time
$T_d$ = average downtime per occurrence

For the case of equal $m_i$ and equal $q_i$,

$$T_p = T_c + nmqT_d \tag{5.26}$$

From the average production time per assembly, we obtain the production rate from Eq. (5.3), repeated here for convenience:

$$R_p = \frac{1}{T_p}$$

However, it must be remembered that unless $m_i = 1$ for all stations, the production of assemblies will include some units with one or more defective components. Accordingly, the production rate should be corrected to give the rate of output of assemblies that contain no defects. We shall call this $R_{ap}$, the rate of production of acceptable product. By combining Eqs. (5.3) and (5.19), we obtain

$$R_{ap} = \frac{\prod_{i=1}^{n} (1 - q_i + m_i q_i)}{T_p} = \frac{P_{ap}}{T_p} \tag{5.27}$$

Using Eq. (5.21) rather than Eq. (5.22), we get the corresponding rate of production when $m_i$ are all equal and $q_i$ are all equal:

$$R_{ap} = \frac{(1 - q + mq)^n}{T_p} \tag{5.28}$$

Hence, Eq. (5.3) provides the average production cycle rate for an assembly machine, which includes production of both good and bad product. Equations (5.27) and (5.28) give values of average production rate for good product only.

The line efficiency is calculated as the ratio of ideal cycle time to average production time. This is the same as Eq. (5.5), except that average production time, $T_p$, is given by Eq. (5.25) or (5.26). Using Eq. (5.26) as an illustration,

$$E = \frac{T_c}{T_p} = \frac{T_c}{T_c + nmqT_d} \tag{5.29}$$

The proportion of downtime, $D$, is the average downtime per cycle divided by the average production time. For example, for the case of equal $m_i$ and equal $q_i$, the value of $D$ would be given by

$$D = \frac{nmqT_d}{T_p} = \frac{nmqT_d}{T_c + nmqT_d} \tag{5.30}$$

No attempt has been made to correct either the line efficiency $E$, or the proportion downtime $D$, for the yield of good assemblies. We are treating the assembly machine efficiency and the quality of units produced as separate issues in the computation of $E$ and $D$.

The cost per assembly produced, on the other hand, must take account of the output quality. Therefore, the general cost formula, given by Eq. (5.8), must be amended to include a correction for yield plus any additional costs such as inspection to identify defective assemblies. As an illustration, the correction for yield of good product would be incorporated into the cost formula as follows:

$$C_{pc} = \frac{C_m + C_L T_p + C_t}{(1 - q + mq)^n} = \frac{C_m + C_L T_p + C_t}{P_{ap}} \tag{5.31}$$

where the cost terms in the numerator were defined in Section 5.2. The denominator would tend to increase the cost of the assembly. Thus, as the quality of individual components deteriorates, this would result in an increased average cost per assembly produced.

It is appropriate to conclude this listing of equations on assembly machine performance with some examples. In addition to the traditional ways of indicating line performance (e.g., production rate, proportion uptime, cost per unit), we see an additional dimension of performance in the form of the yield. While the yield of good product is an important issue in any automated production line, we see that it can be explicitly included in the formulas for assembly machine performance by means of $q$ and $m$. We will demonstrate the influence of these factors with two example problems.

### EXAMPLE 5.4

In this example we want to examine the effects of $q$ and $m$ on line performance. We will assume that the values of $q$ and $m$ are the same for all stations. The performance measures are the yield of good product, production rate, and line efficiency. Given the following data:

$$n = 10 \text{ stations}$$
$$T_c = 6 \text{ s}$$
$$T_d = 60 \text{ s}$$

we will use two levels of parts quality, $q = 0.01$ and $0.02$, and three levels of station jam probability, $m = 0$, $0.5$, and $1.0$. The results are displayed in Figures 5.2, 5.3, and 5.4.

The effect of component parts quality, as indicated in the value of $q$, is obvious. As the fraction defective rate is increased, meaning that quality is deteriorating, all three measures of assembly machine performance suffer. Two exceptions should be noted. Considering line efficiency $E$ in Figure 5.4, when the value of $m$ is zero, parts quality has no effect on the proportion of uptime because the machine never stops for defective components. Line

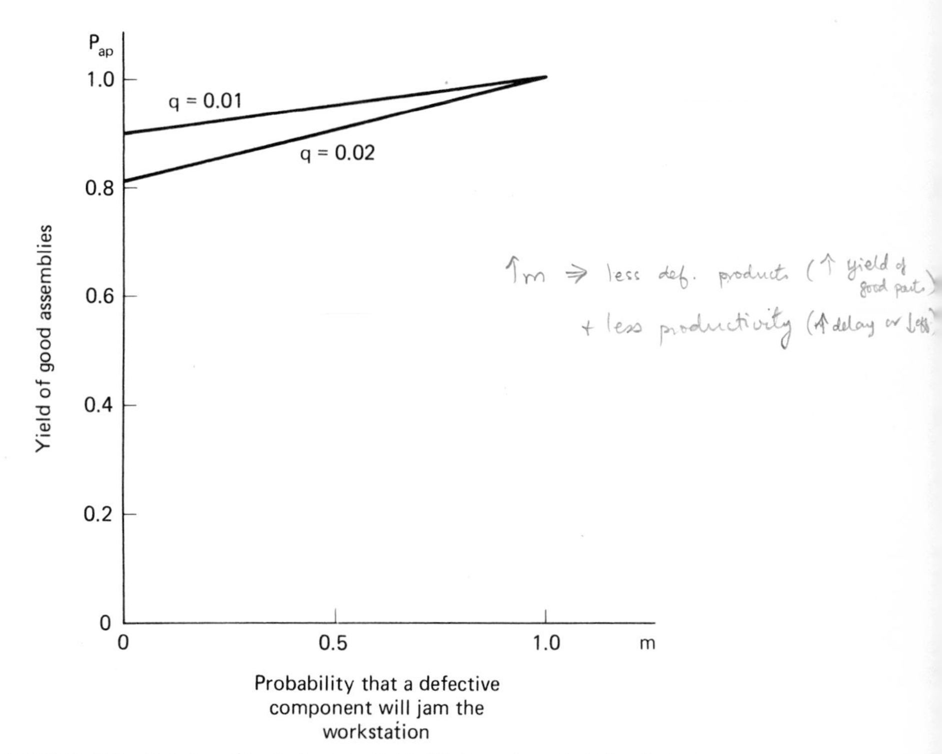


FIGURE 5.2 Relationship between yield ($P_{ap}$) and probability ($m$) that a defective component will jam the workstation. Two levels of parts quality ($q$) shown.

efficiency is 100% for both values of $q$. The other exception concerns the yield of good product, $P_{ap}$. As shown in Figure 5.2, when the value of $m = 1.0$, the yield is 100% for any value of $q$ because the machine always stops to prevent a defect from being assembled. Thus the assemblies produced consist of all nondefective components, no matter what the value of $q$.

Let us consider the additional effects of $m$, the probability that a defective component will cause a station to jam. As $m$ increases, the yield increases. As indicated above, if every defective component stops the line, the yield is 100%. Although this might seem to be beneficial, it produces side effects that are not beneficial.

The production rate of acceptable assemblies and the proportion uptime are adversely affected by increases in the value of $m$. This is shown in Figures 5.3 and 5.4. If every defective component causes a line stoppage, as represented by the case of $m = 1.0$, the production rate and line efficiency are significantly reduced. On the other hand, if the defective components are allowed to pass through the workhead and become part of the assembly, as represented by $m = 0$, the production rate is increased. This would suggest that it is more efficient for the machine to be designed to continue operating

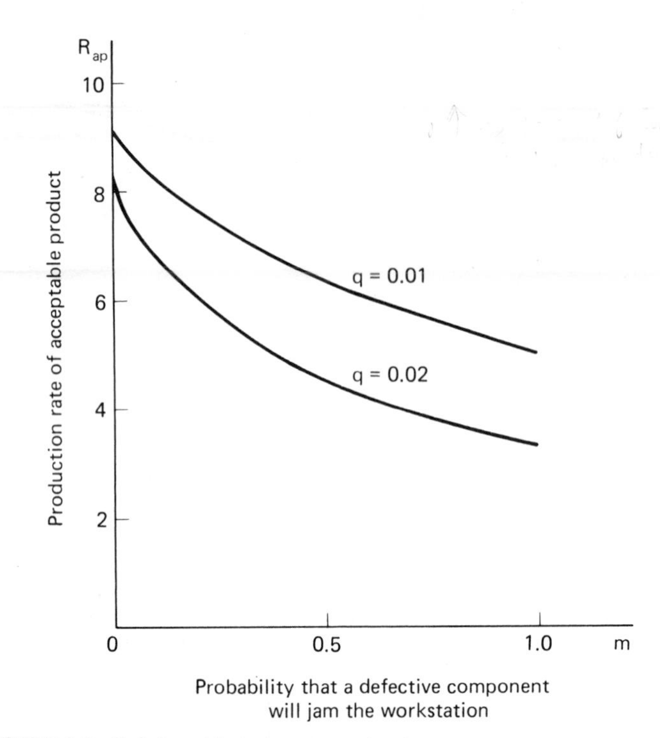


FIGURE 5.3 Relationship between production rate ($R_{ap}$) and probability ($m$) that a defective component will jam the workstation. Two levels of parts quality ($q$) shown.

rather than to stop when a defective part is encountered. In Section 4.5, we discussed two different types of controls, instantaneous control and memory control. The instantaneous type of control stops the machine when a defect (or any other machine malfunction) occurs. With memory control, the assembly machine is provided with a logic system that identifies when a defective component is encountered but does not stop the machine. Instead, it remembers the particular assembly containing the defect, perhaps locking it out from additional processing, and rejects the assembly at the last station. With the introduction of the variable $m$, we are now in a better position to compare the two types of controls. We will proceed to do this in the next example.

### EXAMPLE 5.5

The two types of controls are to be compared on a six-station indexing machine which performs an assembly operation at each station. The ideal cycle time of the assembly machine is 6 s, and the average time the machine spends down per line stop is 1 min. The fraction defect rate is 2% and this is considered equal for all components. Cost of the components averages \$0.20/assembly, which includes an allowance for the fraction of defective components. However, it does not include any allowance for

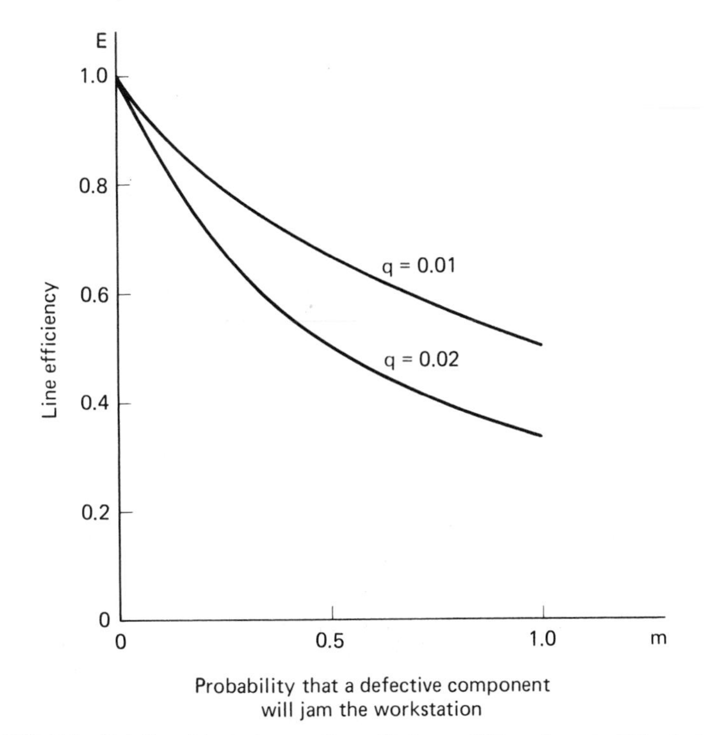


**FIGURE 5.4** Relationship between line efficiency ($E$) and probability ($m$) that a defective component will jam the workstation. Two levels of parts quality ($q$) shown.

assemblies that are spoiled during the assembly process. In other words, it does not consider the yield of good product from the assembly machine.

The cost to operate the machine using instantaneous control is \$60.00/h or \$1.00/min. Because of its greater complexity, the cost to operate the assembly machine using memory control is \$1.10/min. Special tooling costs are negligible.

Under ideal circumstances, the instantaneous control will result in a value of $m=1.0$, meaning that every defective component will cause the machine to stop. Likewise, the memory control will result in a theoretical value of $m=0$. In practice, neither of these ideals will be achieved. With instantaneous control, a portion of defects slips through undetected, so that the actual value of $m$ will be less than unity. With memory control, there will be cases of line stops resulting from defective components whose poor quality prevents them from being fed or assembled. Hence, the actual value of $m$ under memory control will be greater than the theoretical value of zero. However, we will ignore these complications and use the two extreme values of $m=1.0$ and $m=0$.

Under instantaneous control, from Eq. (5.26),

$$T_p = 0.1 + 6(1.0)(0.02)(1.0) = 0.22 \text{ min}$$

The proportion of good product would be, according to Eq. (5.21),

$$P_{ap} = [1 - 0.02 + 1.0(0.02)]^6 = 1.0$$

Hence, the cost per unit produced can be calculated from Eq. (5.31):

$$C_{pc} = 0.20 + \$1.00(0.22) = \$0.42/\text{unit}$$

Under memory control, similar computations proceed as follows:

$$T_p = 0.1 + 6(0)(0.02)(1.0) = 0.10 \text{ min}$$

$$P_{ap} = [1 - 0.02 + 0(0.02)]^6 = 0.8858$$

$$C_{pc} = \frac{0.20 + \$1.10(0.10)}{0.8858} = \$0.35/\text{unit}$$

The winner is memory control. (based on unit cost)

## 5.5 PARTIAL AUTOMATION

There are many examples of flow lines in industry that contain both automated and manually operated workstations. These cases of partially automated production lines occur for two main reasons. First, mechanization of a manually operated flow line is often introduced gradually. Automation of the simpler manual operations is carried out initially, and the transition toward a fully automated line is accomplished progressively. A long period of time may elapse before the transformation has been completed. Meanwhile, the line operates as a partially automated system. The second reason for partial automation is based strictly on economics. Some manual operations are difficult to automate and it may be uneconomical to do so. Therefore, when the sequence of workstations is planned for the flow line, certain stations are designed to be manually operated while the rest are designed to be automatic. Examples of operations that are often difficult to automate are processes requiring alignment or special human skills to carry out. Many assembly operations fall into this category. Also, inspection operations often create problems when automation is being considered to substitute for a human operator. Defects in a workpart that can be easily perceived by a human inspector are sometimes extremely difficult to identify by an automatic inspection device. Another problem is that the automatic device can only check the defects for which it was designed, whereas a human operator is capable of sensing a variety of unforeseen imperfections in the part.

To analyze the performance of a partially automated flow line, we will build on the developments of the previous sections. Our analysis will be confined to the operation of a system without buffer storage. Later in this section, we will speculate on the improvements that would result if in-process inventory banks were used for the manual workstations. The ideal cycle time, $T_c$, will be determined by the slowest station in the line, which generally will be one of the manual stations. We assume that breakdowns occur only at the automatic stations and that the reasons for breakdowns are as varied as they are for a completely automated system (e.g., tool failures, defective components, electrical and mechanical malfunctions, etc.). Breakdowns do not occur at the manual stations because the human operators are flexible enough, we assume, to adapt to the kinds of variations and disruptions that cause an automatic workhead to stop. This, of course, is not always true, but our analysis will be based on this assumption. Let $p$ equal the probability of a station breakdown for an automatic workhead. The values of $p$ may be different for different stations, but we will leave to the reader the task of generalizing upon our particular case of equal $p$ for all stations.

Since $n$ is the total number of workstations on the line, let $n_a$ equal the number of automatic stations and $n_o$ equal the number of manual operator stations. The sum of $n_a$ and $n_o$ must be $n$. In our previous discussion of costs in Section 5.2, $C_L$ was used to represent the cost to operate the line, including labor, capital, and overhead. For a partially automated system, we must separate this term into several components. Let $C_o$ equal the operator cost per manual station. This is a labor cost in \$/minute which applies to each manually operated workstation. The cost per automatic workstation in \$/minute will be denoted by $C_{as}$. This cost will often vary for different stations, depending on the level of sophistication of the mechanism. We can allow for these differences in $C_{as}$ with little difficulty, as will be shown in a subsequent example problem. The last cost component of $C_L$ is the cost per minute of the automatic transfer mechanism which will be used for all stations, manual and automatic, to transport the workparts. Let this cost be symbolized by $C_{at}$. It is not a cost per station, but rather a cost that includes all $n$ stations. Combining these costs, the total line cost $C_L$ is given by

$$C_L = n_o C_o + n_a C_{as} + C_{at} \tag{5.32}$$

To figure the average production time, the ideal cycle time is added to the average downtime per cycle as follows:

$$\begin{aligned} T_p &= T_c + FT_d \\ &= T_c + n_a p T_d \end{aligned} \tag{5.33}$$

By using $n_a$ (the number of automatic workstations), we are consistent with our assumption that breakdowns do not occur at manual stations. Equation (5.33) is based on the upper-bound approach of Section 5.3.

By substituting Eqs. (5.32) and (5.33) into Eq. (5.8), we obtain the cost per unit produced:

$$C_{pc} = C_m + (n_o C_o + n_a C_{as} + C_{at})(T_c + n_a p T_d) + C_t \tag{5.34}$$

### EXAMPLE 5.6

A proposal has been made to replace one of the current manual stations with an automatic workhead on a 10-station assembly machine. The current system has six automatic workheads and four manual stations. The current cycle time is 30 s. The bottleneck station is the manual station that is the candidate for replacement. The proposed automatic station would allow the cycle time to be reduced to 24 s. The new station is costed at \$0.25/min. Other cost data for the existing assembly machine:

$$C_o = \$0.15/\text{min}$$
$$C_{as} = \$0.10/\text{min}$$
$$C_{at} = \$0.10/\text{min}$$

Breakdowns occur at each of the six automatic workstations with a probability $p = 0.01$. The average downtime per breakdown is 3 min. It is estimated that the value of $p$ for the new automatic station would be $p = 0.02$. The average downtime for the line would be unaffected.

Component parts of the product cost \$0.50/assembly. Tooling costs can be neglected ($C_t = 0$).

It is desired to compare the challenger (the new automated station) with the defender (the current manual station) on the basis of cost per unit.

For the current machine, the production time is

$$T_p = 0.5 + 6(0.01)(3) = 0.68 \text{ min/unit}$$
$$C_L = 4(0.15) + 6(0.10) + 0.10 = \$1.30/\text{min}$$
$$C_{pc} = 0.50 + 1.30(0.68) = \$1.384/\text{unit}$$

For the assembly machine with the new workhead replacing the current manual station, similar computations are as follows:

$$T_p = 0.4 + (6 \times 0.01 + 0.02)(3) = 0.64 \text{ min/unit}$$
$$C_L = 3\,(0.15) + 6(0.10) + 0.25 + 0.10 = \$1.40/\text{min}$$
$$C_{pc} = 0.50 + 1.40(0.64) = \$1.396/\text{unit}$$

The conclusion is that the improved performance of the automatic workhead does not justify its greater cost. It should be noted that the reduced reliability of the automatic workheads figures prominently in the cost calculations. If the value of $p$ for the new automatic workhead had been equal to 0.01 instead of 0.02, the conclusion would have been reversed.

## Buffer storage

The preceding analysis assumes no buffer storage between stations on the assembly machine. Therefore, when the automated portion of the line breaks down, the manual stations must also stop working for lack of workparts. It would be beneficial to the line operation to build up an inventory of parts before and after each manual station. In this manner, these stations, which usually operate at a slower pace than the automatics, could continue to produce while the automated portion of the line is down. This would tend to help solve the line balancing problem that often occurs on partially automated machines. To illustrate, consider the previous example of the current assembly system whose ideal cycle time is 0.50 min. Under the current method of operation, both manual and automatic stations are out of operation when a breakdown occurs. The 0.50-min cycle time is caused by one of the manual stations. Suppose that the ideal cycle time on the automatic portion of the machine could be set at 0.32 minute. The resulting average production time on the automatic stations would then be

$$T_p = 0.32 + 6(0.01)(3) = 0.50 \text{ min}$$

If a bank of workparts could be provided for the human operators to work on during the downtime occurrences, the average production time of the entire assembly system would be 0.50 min rather than 0.68 min as computed in Example 5.6. The resulting cost per assembly, ignoring any added cost as a result of the buffer storage capacity, would be

$$C_{pc} = 0.50 + 1.30(0.50) = \$1.15/\text{unit}$$

which is a substantial reduction from the \$1.384/unit previously calculated for the current line.

We will treat the topic of buffer storage in more detail in the next section.

## 5.6 AUTOMATED FLOW LINES WITH INTERNAL PARTS STORAGE

One of the methods by which flow lines can be made to operate more efficiently is to add one or more parts storage buffers between workstations along the line. The buffer zones divide the line into stages. If one storage buffer is used, the line is divided into two stages. If two storage buffers are used at two different locations along the line, a three-stage line results. The upper limit on the number of storage zones is to have in-process inventory capacity between every workstation. The number of stages will then be equal to the number of stations. For an $n$-stage line, there will be $n-1$ storage buffers. This, of course, does not include the inventory of raw workparts which are available for feeding onto the front of the flow line. Nor does it include the inventory of finished parts that accumulate at the end of the line.

In a flow line without internal workpart storage, the workstations are interdependent. When one station breaks down, all the other stations will be affected, either immediately or by the end of a few cycles of operation. The other workstations will be forced to stop production for one of two reasons:

1. *Starving of stations.* If a workstation cannot continue to operate because it has no parts to work on, the station is said to be *starved* of parts. When a breakdown occurs at a given station in the line, the stations following the broken-down station will become starved.

2. *Blocking of stations.* This occurs when a station is prevented from passing its workpart along to the following station. When a station breakdown occurs, the preceding or upstream stations are said to be *blocked*, because they are unable to transfer workparts to the station that is down.

The terms "starving" and "blocking" are often used in reference to manual flow lines. However, they are useful for explaining how buffer storage zones can be used to improve the efficiency of automated flow lines. When an automated flow line is divided into stages by storage buffers, the overall efficiency and production rate of the line are improved. When one of the stages in the line breaks down, the storage buffers prevent the other stages from being immediately affected. Consider a two-stage transfer line with one storage buffer between the stages. If the first stage breaks down, the inventory of workparts that has accumulated in the storage buffer will allow the second stage to continue operating until the inventory has been exhausted. Similarly, if the second stage breaks down, the storage zone will accept the output of the first stage. The first stage can continue to operate until the capacity of the storage buffer is reached.

The purpose served by the storage buffer can be extended to flow lines with more than two stages. The presence of these in-process inventories allows each stage to operate somewhat independently. It is clear that the extent of this independence between one stage and the next depends on the storage capacity of the buffer zone separating the two stages.

## Limits of storage buffer effectiveness

The two extreme cases of storage buffer effectiveness can be identified as:

1. No buffer storage capacity at all.
2. Storage buffers with infinite capacity.

NO BUFFER STORAGE CAPACITY. In this case, the flow line acts as one machine. When a station breakdown occurs, the entire line is forced down. The efficiency of the line was previously given by Eq. (5.5). We will rewrite this equation as

$$E_0 = \frac{T_c}{T_c + FT_d} \tag{5.35}$$

The subscript 0 will identify this as the efficiency of the line with no buffer storage capacity. It represents a starting point from which the line efficiency can be improved by adding provision for in-process storage.

INFINITE-CAPACITY STORAGE BUFFERS. The opposite extreme is the case where buffer zones of infinite capacity are installed between each stage. If we assume that each of these buffer zones is half full (in other words, each buffer zone has an infinite supply of parts as well as the capability to accept an infinite number of additional parts), then each stage is independent of the rest. Infinite storage buffers means that no stage will ever be blocked or starved because of the breakdown of some other stage.

Of course, a flow line with infinite-capacity buffers cannot be realized in practice. However, if such a line could be approximated in real life, the overall line efficiency would be determined by the bottleneck stage. We would have to limit the production on the other stages to be consistent with that of the bottleneck stage. Otherwise, the in-process inventory upstream from the bottleneck would grow indefinitely, eventually reaching some practical maximum, and the in-process inventory in the downstream buffers would decline to zero. As a practical matter, therefore, the upper limit on the efficiency of such a line is defined by the efficiency of the bottleneck stage. If we assume the cycle time $T_c$ to be the same for all stages, the efficiency of any stage is given by

$$E_k = \frac{T_c}{T_c + F_k T_{dk}} \tag{5.36}$$

where the subscript $k$ is used to identify the stage.

The overall line efficiency would be given by

$$E_\infty = \min_k E_k \tag{5.37}$$

where the subscript $\infty$ is used to indicate the infinite storage buffers between stages.

BUFFER STOCK EFFECTIVENESS IN PRACTICE. By providing one or more buffer zones on a flow line, we expect to improve the line efficiency above $E_0$ but we cannot expect to achieve $E_\infty$, simply because buffer zones of infinite capacity are not possible. Hence, the actual value of $E$ will lie somewhere between these extremes:

$$E_0 < E < E_\infty \tag{5.38}$$

Before embarking on the problem of evaluating $E$ for realistic levels of buffer capacity, it seems appropriate to comment on the practical implications of Eqs. (5.35) through (5.38):

1. Equation (5.38) indicates that if $E_0$ and $E_\infty$ are nearly equal in value, relatively little advantage will be gained from the addition of storage buffers to the line. If $E_\infty$ is significantly larger than $E_0$, storage buffers will provide a pronounced improvement in the line's efficiency.

2. The workstations along the line should be divided into stages so as to make the efficiencies of all stages as equal as possible. In this way the maximum difference between $E_0$ and $E_\infty$ will be achieved, and no single stage will stand out as a significant bottleneck.

3. The efficiency of an automated flow line with buffer storage can be maximized under the following conditions:

   a. By setting the number of stages equal to the number of workstations. That is, all adjacent stations will be separated by storage buffers.
   b. By providing that all workstations have an equal probability of breakdown.
   c. By designing the storage buffers to be of large capacity. The actual capacity would be determined by the average downtime. If the average downtime (in particular, the ratio $T_d/T_c$) is large, more buffer capacity must be provided to ensure adequate insulation between workstations.

4. Although it is not obvious from Eqs. (5.35) to (5.38), the "law of diminishing returns" operates as the number of stages increases. The biggest improvement in line efficiency comes from adding the first storage buffer to the line. As more storage buffers are added, the efficiency improves, but at an ever-slower rate. This will be demonstrated in Example 5.8 later in this section.

### Analysis of a two-stage line

Much of the preceding discussion was based on the work of Buzacott [3], who has pioneered the analytical research on flow lines with buffer stocks. Several of his publications on the subject are listed in the references at the end of this chapter ([3], [4], [5], [6], [7]). Other researchers have also studied the problem of flow lines with storage buffers, especially the two-stage line, and the interested reader is referred to their work ([12], [15], [16]).

Our presentation of the two-stage flow line problem will closely follow Buzacott's analysis developed in reference [3]. The two-stage line is separated by a storage buffer of capacity $b$. The capacity is expressed in terms of the number of workparts the storage zone can hold. Let $F_1$ and $F_2$ represent, respectively, the breakdown rates of stages 1 and 2. We will use the term $r$ to define the ratio of breakdown rates as follows:

$$r=\frac{F_2}{F_1} \tag{5.39}$$

The ideal cycle time, $T_c$, is the same for both stages. We assume that the downtime distributions of all stations within a stage are the same, and that the average downtimes of stages 1 and 2 are $T_{d1}$ and $T_{d2}$.

Over the long run, both stages must have equal efficiencies. For example, if the efficiency of stage 1 would tend to be greater than the stage 2 efficiency, the inventory in the storage buffer would tend to build up until its capacity $b$ is reached. Thereafter, stage 1 would be blocked whenever it tried to outproduce stage 2. Likewise, if the efficiency of stage 2 is greater, the buffer inventory would become depleted, thus starving stage 2. Accordingly, the efficiencies of the two stages would tend to equalize over time.

The overall line efficiency for the two stages can be expressed as

$$E=E_0+D_1'h(b) \tag{5.40}$$

where $E_0$ is the line efficiency without buffer storage. The value of $E_0$ was given by Eq. (5.35), but we will rewrite it below to explicitly define the two-stage system efficiency when $b=0$:

$$E_0=\frac{T_c}{T_c+F_1T_{d1}+F_2T_{d2}} \tag{5.41}$$

The term $D_1'h(b)$ appearing in Eq. (5.40) represents the improvement in efficiency that results from adding buffer capacity ($b>0$).

$D_1'$ is the proportion of total time that stage 1 is down. However, Buzacott defines $D_1'$ as

$$D_1'=\frac{F_1T_{d1}}{T_c+F_1T_{d1}+F_2T_{d2}} \tag{5.42}$$

Note that the form of this equation is different from the case given by Eq. (5.6) when we were considering the flow line to be an independent and integral mechanism, uncomplicated by the presence of buffer stocks.

The term $h(b)$ is the ideal proportion of the downtime $D_1'$ (when stage 1 is down) that stage 2 could be up and operating within the limits of buffer capacity $b$. Buzacott [3] presents equations for evaluating $h(b)$ using a Markov chain analysis. The equations cover several different downtime distributions and are based on the assumption that the probability of both stages being down at the same time is negligible. Four of these equations are presented in Table 5.1.

TABLE 5.1 Formulas for Computing $h(b)$ for a Two-stage Flow Line under Several Downtime Situations

---

The following definitions and assumptions apply to Eqs. (5.44) through (5.47) used to compute $h(b)$: Assume that the two stages have equal repair times and equal cycle times. That is,

$$T_{d_1} = T_{d_2} = T_d$$

and

$$T_{c_1} = T_{c_2} = T_c$$

Let

$$b = B\frac{T_d}{T_c} + L \tag{5.43}$$

where $B$ is the largest integer satisfying the relation

$$b\frac{T_c}{T_d} \geqslant B$$

and $L$ represents the leftover units, the number by which $b$ exceeds $BT_d/T_c$.

Finally, let $r = F_2/F_1$, as given by Eq. (5.39).

With these definitions and assumptions, we can express the relationships for two theoretical downtime distributions as developed by Buzacott [3]:

1. *Constant repair distribution.* Each downtime occurrence is assumed to require a constant repair time, $T_d$.

$$r \neq 1: \quad h(b) = r\frac{1-r^B}{1-r^{B+1}} + L\frac{T_c}{T_d}\frac{r^{B+1}(1-r)^2}{(1-r^{B+1})(1-r^{B+2})} \tag{5.44}$$

$$r = 1: \quad h(b) = \frac{B}{B+1} + L\frac{T_c}{T_d}\frac{1}{(B+1)(B+2)} \tag{5.45}$$

2. *Geometric repair distribution.* This downtime distribution assumes that the probability that repairs are completed during any cycle is independent of the time since repairs began. Define the parameter $K$:

$$K = \frac{1 + r - T_c/T_d}{1 + r - rT_c/T_d}$$

Then for the two cases

$$r \neq 1: \quad h(b) = \frac{r(1-K^b)}{1-rK^b} \tag{5.46}$$

$$r = 1: \quad h(b) = \frac{bT_c/T_d}{2 + (b-1)T_c/T_d} \tag{5.47}$$

---

In sample calculations, the author has found that Eq. (5.40) tends to overestimate line efficiency. This is because of the assumption implicit in the computation of $h(b)$ that both stages will not be broken down at the same time. Another way of stating this assumption is this: During the downtime of stage 1, stage 2 is assumed to always be operating. However, it is more realistic to believe that during the downtime of stage 1, stage 2 will be down a certain portion of the time, this downtime determined by the efficiency of stage 2. Hence, it seems more accurate to express the overall line efficiency of a two-stage line as follows, rather than by Eq. (5.40):

$$E = E_0 + D_1' h(b) E_2 \tag{5.48}$$

where $E_2$ represents the efficiency of stage 2 by Eq. (5.36).

In work subsequent to the research documented in reference [3], Buzacott develops a more exact solution to the two-stage problem than that provided by Eq. (5.40). This is reported in reference [5] and seems to agree closely with the results given by Eq. (5.48).

## EXAMPLE 5.7

This example will illustrate the use of Eq. (5.48) and Table 5.1 to calculate line efficiency for a transfer line with one storage buffer. The line has 10 workstations, each with a probability of breakdown of 0.02. The cycle time of the line is 1 min, and each time a breakdown occurs, it takes exactly 5 min to make repairs. The line is to be divided into two stages by a storage bank so that each stage will consist of five stations. We want to compute the efficiency of the two-stage line for various buffer capacities.

First, let us compute the efficiency of the line with no buffer capacity.

$$F = np = 10(0.02) = 0.2$$

By Eq. (5.35),

$$E_0 = \frac{1.0}{1.0 + 0.2(5)} = 0.50$$

Next, dividing the line into two equal stages by a buffer zone of infinite capacity, each stage would have an efficiency given by Eq. (5.36).

$$F_1 = F_2 = 5(0.02) = 0.1$$

$$E_1 = E_2 = \frac{1.0}{1.0 + 0.1(5)} = 0.6667$$

By Eq. (5.37), $E_\infty = 0.6667$ represents the maximum possible efficiency that could be achieved by using a storage buffer of infinite capacity.

Now, we will see how Eq. (5.48) and the formulas in Table 5.1 are used. Let us investigate the following buffer capacities: $b = 1, 10, 100$, and $\infty$.

When $b=1$, according to Eq. (5.43) in Table 5.1, the buffer capacity is converted to $B=0$, and $L=1$. We are dealing with a constant repair-time distribution and the breakdown rates are the same for both stages, so $r=1$. We will therefore use Eq. (5.45) to compute the value of $h(1)$:

$$h(1)=0+\frac{1(1.0)}{5.0}\frac{1}{(0+1)(0+2)}=0.10$$

Using Eq. (5.42) to get the proportion of total time that stage 1 is down,

$$D_1'=\frac{0.1(5.0)}{1.0+0.2(5.0)}=0.25$$

Now the line efficiency can be computed from Eq. (5.48):

$$E=0.50+0.25(0.10)(0.6667)=0.5167$$

Only a very modest improvement results from the use of a storage buffer with capacity of one workpart.

When $b=10, \mathrm{B}=2$ and $L=0$. The value of $h(10)$ is again computed from Eq. (5.45):

$$h(10)=\frac{2}{2+1}+0=0.6667$$

The resulting line efficiency is

$$E=0.50+0.25(0.6667)(0.6667)=0.6111$$

We see a 22% increase in line efficiency over $E_0$ from using a buffer capacity of 10 work units.

When $b=100$, $B=20$, $L=0$, and

$$h(100)=\frac{20}{21}+0=0.952$$

Therefore,

$$E=0.50+0.25(0.952)(0.6667)=.6587$$

A 32% increase in line efficiency results when the buffer capacity equals 100. Comparing this to the case when $b=10$, we can see the law of diminishing returns operating.

When the storage capacity is infinite ($b=\infty$),

$$h(\infty)=1.0$$

and

$$E=0.50+0.25(1.0)(0.6667)=0.6667$$

according to Eq. (5.48), which equals the result given by Eq. (5.37).

It is of instructional value to compare this last result for $b = \infty$ given by Eq. (5.48), with the result given by Eq. (5.40), which neglects the possibility of overlapping stage downtime occurrences. According to Eq. (5.40), the line efficiency would be calculated as

$$E = 0.50 + 0.25(1.0) = 0.75$$

which, of course, exceeds the maximum possible value given by Eq. (5.37). The corrected formula, Eq. (5.48), provides better calculated results, since it accounts for the occurrence of overlapping breakdowns for the two stages.

## Flow lines with more than two stages

We will not consider exact formulas for computing efficiencies for flow lines consisting of more than two stages. It is left to the interested reader to consult some of the references, in particular [1], [3], and [5]. However, the previous discussion in this section should provide a general guide for deciding on the configuration of multistage lines. Let us consider some of these decisions in the context of an example.

### EXAMPLE 5.8

Suppose that the flow line under consideration here has 16 stations with cycle time of 15 s (assume that all stations have roughly equal process times). When station breakdowns occur, the average downtime is 2 min. The breakdown frequencies for each station are presented in the following table:

| *Station* | $p_i$ | *Station* | $p_i$ |
|---|---|---|---|
| 1 | 0.01 | 9 | 0.03 |
| 2 | 0.02 | 10 | 0.01 |
| 3 | 0.01 | 11 | 0.02 |
| 4 | 0.03 | 12 | 0.02 |
| 5 | 0.02 | 13 | 0.02 |
| 6 | 0.04 | 14 | 0.01 |
| 7 | 0.01 | 15 | 0.03 |
| 8 | 0.01 | 16 | 0.01 |

We want to consider the relative performances when the line is separated into two, three, or four stages.

First, the performance of the line can be determined for the single-stage case (no storage buffer). From the table, the frequency of downtime occurrences is given by Eq. (5.9):

$$F = \sum_{i=1}^{16} p_i = 0.30$$

The line efficiency is, by Eq. (5.35),

$$E_0 = \frac{0.25}{0.25 + 0.30(2)} = 0.2941$$

To divide the line into stages, we must first decide the optimum locations for the storage buffers. The stations should be grouped into stages so as to make the efficiencies of the stages as close as possible. Then, to assess relative performances for the two-, three-, and four-stage lines, we will base the comparison on the use of storage buffers with infinite capacity.

For the two-stage line, the breakdown frequency of $F = 0.30$ should be shared evenly between the two stages. From the table of $p_i$ values on page 123, it can be determined that the storage buffer should be located between stations 8 and 9. This will yield equal $F$ values for the two stages.

$$F_1 = \sum_{i=1}^{8} p_i = 0.15$$

$$F_2 = \sum_{i=9}^{16} p_i = 0.15$$

The resulting stage efficiencies are

$$E_1 = E_2 = \frac{0.25}{0.25 + 0.15(2)} = 0.4545$$

From Eq. (5.37), $E_\infty = 0.4545$.

Similarly, for the three-stage configuration, the frequency of breakdowns should be divided as equally as possible among the three stages. Accordingly, the line should be divided as follows:

| *Stage* | *Stations* | $F_k$ | $E_k$ |
|---|---|---|---|
| 1 | 1–5 | 0.09 | 0.5814 |
| 2 | 6–10 | 0.10 | 0.5556 |
| 3 | 11–16 | 0.11 | 0.5319 |

The third stage, with the lowest efficiency, determines the overall three-stage line efficiency: $E_\infty = 0.5319$.

Finally, for four stages the division would be the following:

| *Stage* | *Stations* | $F_k$ | $E_k$ |
|---|---|---|---|
| 1 | 1–4 | 0.07 | 0.6410 |
| 2 | 5–8 | 0.08 | 0.6098 |
| 3 | 9–12 | 0.08 | 0.6098 |
| 4 | 13–16 | 0.07 | 0.6410 |

The resulting line efficiency for the four-stage configuration would be $E_\infty = 0.6098$.

The example shows the proper approach for dividing the line into stages. It also shows how the line efficiency continues to increase as the number of storage banks is increased. However, it can be seen that the rate of improvement in efficiency drops off as more storage buffers are added. In the foregoing calculations we assumed an infinite buffer capacity. For realistic capacities, the efficiency would be less as was illustrated in Example 5.7.

The maximum possible efficiency would be achieved by using an infinite storage bank between every workstation. Station 6 would be the bottleneck stage. We leave it for the reader to figure out the line efficiency for the 16-stage case.

## 5.7 COMPUTER SIMULATION OF AUTOMATED FLOW LINES

A number of studies have been concerned with simulating the operation of flow lines, both automated and nonautomated. Some of these studies have been performed by large industrial concerns and the models developed as well as the results obtained are often considered proprietary. References [8], [10], [11], [13], and [14] are a sampling of the reports of these studies.

*Production* magazine [8] reports some of the results provided by a computer simulation model developed by Ingersoll Manufacturing consultants. However, very little information is given about the details of the model itself. Phillips and Slovick [13] report on an analysis of an actual continuous production line at one of the Western Electric plants. The line is manually operated with seven work stations. A GERTS-type simulation language was used to develop the model of the line. Although the study was limited in scope to the particular assembly and test line, the investigators found the GERTS simulation approach to be quite valuable for production control purposes. Phillips and Pritsker [14] explore the possibilities of using GERTS on a number of production system situations.

Perhaps the most comprehensive automated flow line simulation study reported in the literature was performed by Hanifin ([10],[11]). Hanifin used as the object of his investigation several actual transfer lines at the Kokomo Works of Chrysler Corporation. His computer model, using GPSS as the simulation language, was based on the operation of these machining flow lines. The model was developed to deal with several somewhat specific problem areas at Chrysler. Yet the results of his study can be generalized to some extent. His investigation considered the effect of adding up to three storage buffer areas at three specific locations along the line. He also investigated the effect of different average tool change times.

## REFERENCES


[1] Boothroyd, G., and Redford, A. H., *Mechanized Assembly*, McGraw-Hill Publishing Co., London, 1968, Chapters 7 and 8.

[2] Buzacott, J. A., "Prediction of the Efficiency of Production Systems without Internal Storage," *International Journal of Production Research*, Vol. 6, No. 3, 1968, pp. 173–188.

[3] Buzacott, J. A., "Automatic Transfer Lines with Buffer Stocks," *International Journal of Production Research*, Vol. 5, No. 3, 1967, pp. 183–200.

[4] Buzacott, J. A., "Methods of Reliability Analysis of Production Systems Subject to Breakdowns," *Operations Research and Reliability*, ed. by D. Grouchko, Gordon and Breach, New York, 1971, pp. 211–232.

[5] Buzacott, J. A., "The Role of Inventory Banks in Flow-Line Production Systems," *International Journal of Production Research*, Vol. 9, No. 4, 1971, pp. 425–436.

[6] Buzacott, J. A., and Hanifin, L. E., "Models of Automatic Transfer Lines with Inventory Banks—A Review and Comparison," *AIIE Transactions*, Vol. 10, No. 2, 1978, pp. 197–207.

[7] Buzacott, J. A., and Hanifin, L. E., "Transfer Line Design and Analysis—An Overview," *Proceedings*, 1978 Fall Industrial Engineering Conference of AIIE, December, 1978.

[8] "Computer Techniques Isolate Production Snags," *Production*, February, 1977.

[9] Groover, M. P., "Analyzing Automatic Transfer Machines," *Industrial Engineering*, Vol. 7, No. 11, 1975, pp. 26–31.

[10] Hanifin, L. E., "Increased Transfer Line Productivity Utilizing Systems Simulation," *D. Eng. dissertation*, University of Detroit, 1975.

[11] Hanifin, L. E., Liberty, S. G., and Taraman, K., "Improved Transfer Line Efficiency Utilizing Systems Simulation," *Tech. Paper MR*75–169, Society of Manufacturing Engineers, Dearborn, Mich., 1975.

[12] Okamura, K., and Yamashina, H., "Analysis of the Effect of Buffer Storage Capacity in Transfer Line Systems," *AIIE Transactions*, Vol. 9, No. 2, 1977, pp. 127–135.

[13] Phillips, D. T., and Slovick, R. F., "A GERTS III Q Application to a Production Line," *Proceedings*, 1974 Spring Annual Conference of AIIE, May, 1974.

[14] Phillips, D. T., and Pritsker, A. A. B., "GERT Network Analysis of Complex Production Systems," *International Journal of Production Research*, Vol. 13, No. 3, 1975, pp. 223–237.

[15] Rao, N. P., "Two-Stage Production Systems with Intermediate Storage," *AIIE Transactions*, Vol. 7, No. 4, 1975, pp. 414–421.

[16] Sheskin, T. J., "Allocation of Interstage Storage along an Automatic Production Line," *AIIE Transactions*, Vol. 8, No. 1, 1976, pp. 146–152.

## PROBLEMS

**5.1.** An eight-station rotary indexing machine operates with an ideal cycle time of 20 s. The frequency of line stop occurrences is 0.06 stop/cycle on the average. When a stop occurs, it takes an average of 3 min to make repairs. Determine the following:

(a) Average production time, $T_p$.
(b) Average production rate, $R_p$.
(c) Line efficiency, $E$.
(d) Proportion of downtime, $D$.

**5.2.** Assume that the frequency of line stop occurrences in Problem 5.1 is due to random mechanical and electrical failures on the line. Suppose, in addition to the foregoing reasons for line stops, that the workstation tools must be reset and/or changed at regular intervals. It takes an average of 4 min to inspect and adjust or change the tools at all eight stations. This procedure is performed every 200 cycles. Recompute $T_p$, $R_p$, $E$, and $D$ with these additional data for the indexing machine.

**5.3.** Component costs associated with the operation of the indexing machine of Problem 5.2 are as follows:

Cost of raw workpart = \$0.35/workpiece.
Cost to operate the line = \$0.50/min.
Cost of disposable tooling = \$0.02/workpiece.

Compute the average cost per workpiece produced off the rotary indexing machine.

**5.4.** Recalculate the average cost per workpiece in Problem 5.3 if it is known that the indexing machine produces the parts at a 7% defective rate.

**5.5.** In the operation of a certain 15 station transfer line, the ideal cycle time is 0.58 min. Breakdowns occur at a rate of once every 10 cycles, and the average downtime per breakdown ranges between 2 min and 9 min, with an average of 4.2 min. The plant in which the transfer line is located works an 8-h day, 5 days/week. How many parts will the line be capable of producing during an average week?

**5.6.** An automatic assembly machine has 10 workheads and a theoretical cycle time of 10 s. Jams occur with a frequency of 0.04 stop/cycle and the average downtime is 2.0 min. The last workstation is an automatic inspection device which performs a quality check on the completed assembly. If the check reveals a defect, the defect is separated from the good-quality assemblies. Past records indicate that the scrap rate averages 3%. How many hours must the assembly machine be operated to produce the production quota per week of 10,000 assemblies?

**5.7.** In the assembly machine operation of Problem 5.6, the cost of the components used per assembly is \$0.25. Cost to operate the machine during regular time is

\$0.90/min and \$1.10/min for overtime. Overtime would be paid for time above the 40 h/week. Determine the following:

(a) Cost per week to produce the required 10,000 assemblies.
(b) Average cost per assembly.
(c) If the defective assemblies could be repaired at an average cost of \$0.20/defect, would this repair work pay for itself in reduced overtime cost?

**5.8.** The following data apply to a 12-station in-line transfer machine:

$$p = 0.01 \text{ (all stations have an equal probability of failure)}$$

$$T_c = 0.3 \text{ min}$$

$$T_d = 3.0 \text{ min}$$

Using the upper-bound approach, compute the following for the transfer machine:

(a) $F$, the frequency of line stops.
(b) $R_p$, the average production rate.
(c) $E$, the line efficiency.

**5.9.** Solve Problem 5.8 using the lower-bound approach. What proportion of workparts are removed from the transfer line?

**5.10.** A circular indexing machine performs 10 assembly operations at 10 separate stations. The total cycle time, including transfer time between stations, is 10 s. Stations break down with a probability $p = 0.007$, which can be considered equal for all 10 stations. When these work stoppages occur, it takes an average of 2 minutes to correct the fault. Parts are not normally removed from the machine when these stops occur. Compute the proportion of downtime, the efficiency, and the production rate of this circular indexing machine.

**5.11.** A transfer machine has six stations as follows:

| *Station* | *Operation* | $p_i$ | *Process time (min)* |
|---|---|---|---|
| 1 | Load part | 0 | 0.78 |
| 2 | Drill three holes | 0.02 | 1.25 |
| 3 | Ream two holes | 0.01 | 0.90 |
| 4 | Tap two holes | 0.04 | 1.42 |
| 5 | Mill flats | 0.01 | 1.42 |
| 6 | Unload part | 0 | 0.45 |

The time to transfer between stations = 0.28 min. If the part stops due to a jam in the line, it is removed as defective. It takes an average of 8 min to determine the fault and correct the problem and remove the part. Also, there is a scheduled tool change every 40 parts which takes 6 min to complete. There are 20,000 parts to be started onto the transfer machine.

(a) How many defective parts will be removed from the line?
(b) How many total hours will be consumed in the manufacturing process?
(c) Find the proportion of downtime.
(d) Find the rate of production of acceptable parts.

**5.12.** In Problem 5.11, the cost of operating the transfer machine is \$60/h, as estimated by the accounting department of the company. It is proposed that a minicomputer and sensors be installed to aid in diagnosing breakdowns when they occur. The anticipated savings are 2 min off the 8 min to identify and correct the fault when it occurs. The computer will have no effect on tool changes. The estimated cost of installing the monitoring system is \$11,000. Will the system pay for itself?

**5.13.** A 23-station transfer line has been logged for 5 days (a total of 2400 min). During this time, there were a total of 158 downtime occurrences on the line. The accompanying table identifies the type of downtime occurrence, how many occurrences, and how much total time for the type of occurrence.

| *Type of occurrence* | *Number of occurrences* | *Total minutes lost* |
|---|---|---|
| Associated with stations | 132 | 793 |
| Tool-related causes at stations | 104 | 520 |
| Mechanical failures at stations | 21 | 189 |
| Other miscellaneous station failures | 7 | 84 |
| Associated with transfer mechanism | 26 | 78 |

The transfer line performs a sequence of machining operations, the longest of which takes 0.42 min. The transfer mechanism takes 0.08 min to index a part from one station to the next position. Assuming no parts removal when the line jams, determine the following characteristics for this line for the 5-day period:

(a) How many parts are produced.
(b) Downtime proportion.
(c) Production rate and production time.
(d) Determine the frequency rate $p$ associated with the transfer mechanism breakdowns.

**5.14.** A six-station automatic assembly machine has an ideal cycle time of 12 s. Downtime occurs for two reasons. First, mechanical and electrical failures of the workheads occur with a frequency of once every 100 cycles. Average downtime per occurrence for this reason is 3 minutes. Second, defective components also result in downtime occurrences. The fraction defect rate of each of the six components added to the base part at the six stations is $q=2\%$. The probability that a defective part will cause a station jam is $m=0.5$ for all stations. Downtime per occurrence for defective parts is 2.0 min. Determine the following:

(a) Average production rate $R_p$.
(b) Line efficiency $E$.
(c) Yield of assemblies that are free of defective components.
(d) The proportion of assemblies that contain at least one defective component.
(e) The proportion of downtime that is due to mechanical and electrical failures of the workheads.

(f) The proportion of downtime that is due to defective component jams at workstations.

**5.15.** A partially automated flow line has a mixture of mechanized and manual workstations. There are a total of six stations and the overall theoretical cycle time is 1.0 min. This includes a transfer time of 6 s. The six stations possess characteristics as follows:

| *Station* | *Type* | *Process time (s)* | $p_i$ |
|---|---|---|---|
| 1 | Manual | 30 | 0 |
| 2 | Automatic | 15 | 0.01 |
| 3 | Automatic | 20 | 0.02 |
| 4 | Automatic | 25 | 0.01 |
| 5 | Manual | 54 | 0 |
| 6 | Manual | 30 | 0 |

Cost of the transfer mechanism is $0.10/min to operate. Cost to run each of the automatic workheads is figured at $0.12/min for each of the three automatic stations. Labor cost to operate each of the manual stations is $0.15/min for each of stations 1, 5, and 6. It has been proposed to substitute an automatic workhead for the current manual station 5. The cost of this workhead will be 0.25/min and its breakdown rate $p$ will be 0.02, but its process time will be only 30 s compared to 54 s for the current line. The average downtime per breakdown of the current and proposed configuration is 3.0 min. Should the proposal be accepted?

**5.16.** A 16-station automatic transfer line has the following operating characteristics:

$$T_c = 0.75 \text{ min}$$
$$T_d = 3.0 \text{ min}$$
$$p = 0.01 \text{ for all workstations}$$

A proposal has been submitted to place a storage buffer between stations 8 and 9 to improve the overall efficiency of the transfer line. What is the current line efficiency (use the upper-bound approach) and what is the maximum possible efficiency that would result from use of a storage buffer?

**5.17.** If the capacity of the storage buffer in Problem 5.16 is to be 10 workparts, calculate the line efficiency and the average production rate of the two-stage transfer line. Assume downtime ($T_d = 3.0$ min) is a constant.

**5.18.** Solve Problem 5.17 assuming that downtime is geometrically distributed.

**5.19.** In the transfer line of Problem 5.18, suppose that it is more technically feasible to place the storage buffer between stations 6 and 7. Using a storage buffer capacity of 10 workparts, calculate the line efficiency and the average production rate. Assume that downtime is constant.

**5.20.** Solve Problem 5.19 except assume that downtime is geometrically distributed.

# Flow Line Balancing

## 6.1 INTRODUCTION

One of the problems in the design of production flow lines is to allocate the total work performed on the line to its various stations. This is called the *line balancing problem* because the objective is to assign an equal share of the work to each station. The goal is to balance the workload on the line.

In this chapter we present the line balancing problem and several methods for its solution. Achieving an even distribution of work among workstations is highly desirable in an automated flow line. However, it is with manual assembly lines and partially mechanized flow lines that the line balancing problem is more commonly associated. The reason for this is that the total work content performed on a manual line can usually be divided into many individual small tasks or work elements. Then these individual work elements are arranged into various possible groups, each group constituting the amount of work to be done at a particular station. Since the number of arrangements of work elements into groups can be quite large, the problem of finding the optimum arrangement is a significant one. By contrast, the work elements performed on automated flow lines are often restricted by fixed machine cycles. The flexibility to subdivide these operations into smaller tasks is not available as it is with manual assembly work. Consider a transfer machine as an illustration. The drilling of a hole or milling of a surface of the workpart cannot conveniently be subdivided into smaller work elements. Therefore, each of these tasks would be completed at a single workstation. Manual assembly work tends to be more flexible, and it is this flexibility to allocate the work to stations in so many ways that gives rise to the need for solving the line balancing problem.

As an introduction to our discussion of line balancing, let us consider the various ways in which a conventional manual flow line can be operated. There are two basic classes of manual flow lines:

1. *Nonmechanical lines.* In this arrangement, no belt or conveyor is used to move the parts between operator workstations. Instead, the parts are passed from station to station by hand. Several problems result from this mode of operation:

Starving at stations, where the operator has completed his or her work but must wait for parts from the preceding station.

Blocking of stations, where the operator has completed his or her work but must wait for the next operator to finish the task before passing along the part.

As a result of these problems, the flow of work on a nonmechanical line is usually uneven. The cycle times will vary, and this will contribute to the overall irregularity. Buffer stocks of parts between workstations are often used to smooth out the production flow.

2. *Moving belt lines.* These lines use a continuously moving belt or conveyor to move parts between stations. Some of the potential problems here are the following:

Starving can occur as with nonmechanical lines.

Incomplete items are sometimes produced when the operator is unable to finish the current part and the next part travels right by on the conveyor. Blocking does not occur.

Again, buffer stocks are sometimes used to overcome these problems. Also, station overlaps can sometimes be allowed, where the worker is permitted to travel beyond the normal boundaries of the station in order to complete work.

In the moving belt line, it is possible to achieve a higher level of control over the production rate of the line. This is accomplished by means of the feed rate, which refers to the time interval between workparts on the moving belt. Let $f_b$ denote this feed rate. It is measured in workpieces per time and depends on two factors: the speed with which the belt moves, and the spacing of workparts along the belt. Let $V_b$ equal the belt speed (feet per minute or meters per second) and $s_p$ equal the spacing between parts on the moving belt (feet or meters per workpiece). Then the feed rate is determined by

$$f_b = \frac{V_b}{s_p} \tag{6.1}$$

To control the feed rate of the line, raw workparts would be launched onto the line at regular intervals. As the parts flow along the line, the operator has a certain time period during which he or she must begin work on each piece. Otherwise, the part will flow past the station. This time period is called the tolerance time $T_t$. It is determined by the belt speed and the length of the workstation. This length we will symbolize by $L_s$, and it is determined largely by the operator's reach at the workstation. The tolerance time is therefore defined by

$$T_t = \frac{L_s}{V_b} \tag{6.2}$$

For example, suppose that the desired production rate on a manual flow line with moving conveyor belt was 60 units/h. This would necessitate a feed rate of 1 part/min. This could be achieved by a belt speed of 0.5 m/min and a part spacing of 0.5 m. (Other combinations of $V_b$ and $s_p$ would also provide the same feed rate.) If the length of each workstation were 1.5 m, the tolerance time available to the operators for each workpiece would be 3 min. It is generally desirable to make the tolerance time large to compensate for worker process time variability.

In both the nonmechanical lines and the moving belt lines it is highly desirable to assign work to the stations so as to equalize the process or assembly times at the workstations. The problem is sometimes complicated by the fact that the same production line may be called upon to process more than one type of product. This complication gives rise to the identification of three flow line cases (and therefore three different types of line balancing problems).

The three production situations on flow lines are defined according to the product or products to be made on the line. Will the flow line be used exclusively to produce one particular model? Or, will it be used to produce several different models, and if so, how will they be scheduled on the line? There are three cases which can be defined in response to these questions:

1. *Single-model line.* This is a specialized line dedicated to the production of a single model or product. The demand rate for the product is

great enough that the line is devoted 100% of the time to the production of that product.

2. *Batch-model or multimodel line.* This line is used for the production of two or more models. Each model is produced in batches on the line. The models or products are usually similar in the sense of requiring a similar sequence of processing or assembly operations. It is for this reason that the same line can be used to produce the various models.

3. *Mixed-model lines.* This line is also used for the production of two or more models, but the various models are intermixed on the line so that several different models are being produced simultaneously rather than in batches. Automobile and truck assembly lines are examples of this case.

To gain a better perspective of the three cases, the reader might consider the following. In the case of the batch-model line, if the batch sizes are very large, the batch-model line approaches the case of the single model line. If the batch sizes become very small (approaching a batch size of one), the batch-model line approximates to the case of the mixed-model line.

In principle, the three cases can be applied in both manual flow lines and automated flow lines. However, in practice, the flexibility of human operators makes the latter two cases more feasible on the manual assembly line. It is anticipated that future automated lines will incorporate quick changeover and programming capabilities within their designs to permit the batch-model, and eventually the mixed-model, concepts to become practicable.

Achieving a balanced allocation of workload among the stations of the line is a problem in all three cases. The problem is least formidable for the single-model case. For the batch-model line, the balancing problem becomes more difficult; and for the mixed-model case, the problem of line balancing becomes quite complicated.

In this chapter, we will consider only the single-model line balancing problem, although the same concepts and similar terminology and methodology apply for the batch- and mixed-model cases.

## 6.2 THE LINE BALANCING PROBLEM

In flow line production there are many separate and distinct processing and assembly operations to be performed on the product. Invariably, the sequence of processing or assembly steps is restricted, at least to some extent, in terms of the order in which the operations can be carried out. For example, a threaded hole must be drilled before it can be tapped. In mechanical fastening, the washer must be placed over the bolt before the nut can be turned and tightened. These restrictions are sometimes called *precedence constraints* in the language of line balancing. It is generally the case that the product must be manufactured at some specified production rate in order to satisfy demand for the product. Whether we are concerned with performing these processes and assembly operations on automatic machines or manual flow lines, it is desirable to design the line so as to satisfy all of the foregoing specifications as efficiently as possible.

The line balancing problem is to arrange the individual processing and assembly tasks at the workstations so that the total time required at each workstation is approximately the same. If the work elements can be grouped so that all the station times are exactly equal, we have perfect balance on the line and we can expect the production to flow smoothly. In most practical situations it is very difficult to achieve perfect balance. When the workstation times are unequal, the slowest station determines the overall production rate of the line.

In order to discuss the terminology and relationships in line balancing, we shall refer to the following example problem. Later, when discussing the various solution techniques, we shall apply the techniques to this problem.

## EXAMPLE 6.1

A new small electrical appliance is to be assembled on a production flow line. The total job of assembling the product has been divided into minimum rational work elements. The industrial engineering department has developed time standards based on previous similar jobs. This information is given in Table 6.1. In the right-hand column are the immediate predecessors for each element as determined by precedence requirements. Production demand will be 120,000 units/yr. At 50 weeks/yr and 40 h/week, this reduces to an output from the line of 60 units/h or 1 unit/min.

TABLE 6.1 Table of Work Elements

| *No.* | *Element description* | $T_{ej}$ | *Must be preceded by* |
|---|---|---|---|
| 1 | Place frame on workholder and clamp | 0.2 | — |
| 2 | Assemble plug, grommet to power cord | 0.4 | — |
| 3 | Assemble brackets to frame | 0.7 | 1 |
| 4 | Wire power cord to motor | 0.1 | 1, 2 |
| 5 | Wire power cord to switch | 0.3 | 2 |
| 6 | Assemble mechanism plate to bracket | 0.11 | 3 |
| 7 | Assemble blade to bracket | 0.32 | 3 |
| 8 | Assemble motor to brackets | 0.6 | 3, 4 |
| 9 | Align blade and attach to motor | 0.27 | 6, 7, 8 |
| 10 | Assemble switch to motor bracket | 0.38 | 5, 8 |
| 11 | Attach cover, inspect, and test | 0.5 | 9, 10 |
| 12 | Place in tote pan for packing | 0.12 | 11 |

## Terminology

Let us define the following terms in line balancing. Some of the terms should be familiar to the reader from Chapters 4 and 5, but we want to give very specific meanings to these terms for our purposes in this chapter.

MINIMUM RATIONAL WORK ELEMENT. In order to spread the job to be done on the line among its stations, the job must be divided into its component tasks. The minimum rational work elements are the smallest practical indivisible tasks into which the job can be divided. These work elements cannot be subdivided further. For example, the drilling of a hole would normally be considered as a minimum rational work element. In manual assembly, when two components are fastened together with a screw and nut, it would be reasonable for these activities to be taken together. Hence, this assembly task would constitute a minimum rational work element. We can symbolize the time required to carry out this minimum rational work element as $T_{ej}$, where $j$ is used to identify the element out of the $n_e$ elements that make up the total work or job. For instance, the element time, $T_e$, for element 1 in Example 6.1 is 0.2 min.

The time $T_{ej}$ of a work element is considered a constant rather than a variable. An automatic workhead most closely fits this assumption, although the processing time could probably be altered by making adjustments in the station. In a manual operation, the time required to perform a work element will, in fact, vary from cycle to cycle.

Another assumption implicit in the use of $T_e$ values is that they are additive. The time to perform two work elements is the sum of the times of the individual elements. In practice, this might not be true. It might be that some economy of motion could be achieved by combining two work elements at one station, thus violating the additivity assumption.

TOTAL WORK CONTENT. This is the aggregate of all the work elements to be done on the line. Let $T_{wc}$ be the time required for the total work content. Hence,

$$T_{wc} = \sum_{j=1}^{n_e} T_{ej} \tag{6.3}$$

For the example, $T_{wc} = 4.00$ min.

WORKSTATION PROCESS TIME. A workstation is a location along the flow line where work is performed, either manually or by some automatic device. The work performed at the station consists of one or more of the individual work elements and the time required is the sum of the times of the work elements done at the station. We will use $T_{si}$ to indicate the process time at station $i$ of an $n$-station line. It should be clear that the sum of the station process times should equal the sum of the work element times.

$$\sum_{i=1}^{n} T_{si} = \sum_{j=1}^{n_e} T_{ej} \tag{6.4}$$

Cycle Time. This is the ideal or theoretical cycle time of the flow line, which is the time interval between parts coming off the line. The design value of $T_c$ would be specified according to the required production rate to be achieved by the flow line. Allowing for downtime on the line, the value of $T_c$ must meet the following requirement:

$$T_c \leqslant \frac{E}{R_p} \tag{6.5}$$

where $E$ is the line efficiency as defined in Chapter 5, and $R_p$ the required production rate. As we observed in Chapter 5, the line efficiency of an automated line will be somewhat less than 100%. For a manual line, where mechanical malfunctions are less likely, the efficiency will be closer to 100%.

In Example 6.1, the required production rate is 60 units/h or 1 unit/min. At a line efficiency of 100%, the value of $T_c$ would be 1.0 min. At efficiencies less than 100%, the ideal cycle time must be reduced (or what is the same thing, the ideal production rate $R_c$ must be increased) to compensate for the downtime.

The minimum possible value of $T_c$ is established by the bottleneck station, the one with the largest value of $T_s$. That is,

$$T_c \geqslant \max T_{si} \tag{6.6}$$

If $T_c = \max T_{si}$, there will be idle time at all stations whose $T_s$ values are less than $T_c$.

Finally, since the station times are comprised of element times,

$$T_c \geqslant T_{ej} \qquad (\text{for all } j = 1, 2, \ldots, n_e) \tag{6.7}$$

This equation states the obvious: that the cycle time must be greater than any of the element times.

In Chapter 5, we defined the ideal cycle time to include the transfer time. In Eqs. (6.5) to (6.7), and in the remainder of this chapter, we assume that the transfer time is negligible. If this is not true, a correction must be made in the value of $T_c$ used in Eqs. (6.6) and (6.7) to allow for parts transfer time.

Precedence Constraints. These are also referred to as "technological sequencing requirements." The order in which the work elements can be accomplished is limited, at least to some extent. In Example 6.1, the switch must be mounted onto the motor bracket before the cover of the appliance can be attached. The right-hand column in Table 6.1 gives a complete listing of the precedence constraints for assembling the hypothetical electrical appliance. In nearly every processing or assembly job, there are precedence requirements that restrict the sequence in which the job can be accomplished.

In addition to the precedence constraints described above, there may be other types of constraints on the line balancing solution. These concern the restrictions on the arrangement of the stations rather than the sequence of work elements. The first is called a *zoning constraint*. A zoning constraint may be either a positive constraint or a negative constraint. A *positive* zoning constraint means that certain work elements should be placed near each other, preferably at the same workstation. For example, all the spray-painting elements should be performed together since a special semienclosed workstation has to be utilized. A *negative* zoning constraint indicates that work elements might interfere with one another and should therefore not be located in close proximity. As an illustration, a work element requiring fine adjustments or delicate coordination should not be located near a station characterized by loud noises and heavy vibrations.

Another constraint on the arrangement of workstations is called a *position constraint*. This would be encountered in the assembly of large products such as automobiles or major appliances. The product is too large for one worker to perform work on both sides. Therefore, for the sake of facilitating the work, operators are located on both sides of the flow line. This type of situation is referred to as a position constraint.

In the example problem there are no zoning constraints or position constraints given. The line balancing methods presented in Section 6.3 are not equipped to deal with these constraints conveniently. However, in real-life situations, they may constitute a significant consideration in the design of the flow line.

PRECEDENCE DIAGRAM. This is a graphical representation of the sequence of work elements as defined by the precedence constraints. It is customary to use nodes to symbolize the work elements, with arrows connecting the nodes to indicate the order in which the elements must be performed. Elements that must be done first appear as nodes at the left of the diagram. Then the sequence of processing and/or assembly progresses to the right. The precedence diagram for Example 6.1 is illustrated in Figure 6.1. The element times are recorded above each node for convenience.

BALANCE DELAY. Sometimes also called balancing loss, this is a measure of the line inefficiency which results from idle time due to imperfect allocation of work among stations. It is symbolized as $d$ and can be computed

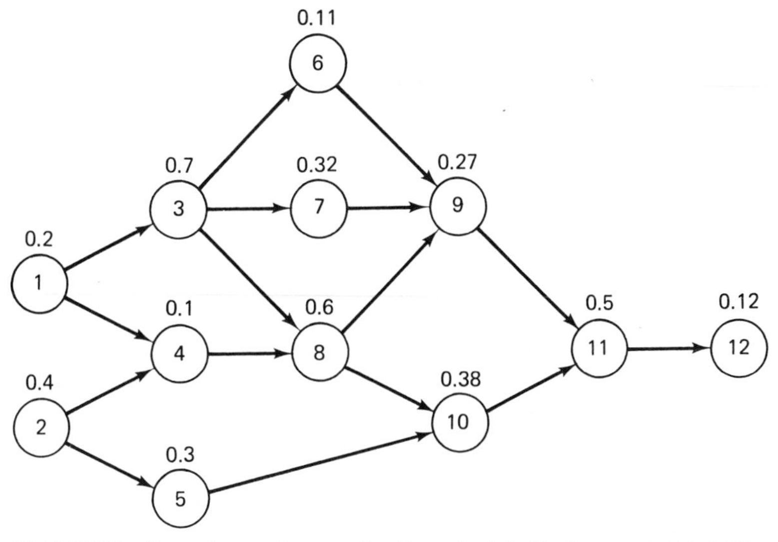


FIGURE 6.1 Precedence diagram for Example 6.1. Nodes represent work elements. The times ($T_e$) for each work element are shown above each node. Arrows indicate the sequence in which the elements must be accomplished.

for the flow line as follows:

$$d = \frac{nT_c - T_{wc}}{nT_c} \tag{6.8}$$

The balance delay is often expressed as a percent rather than as a decimal fraction in Eq. (6.8).

The balance delay should not be confused with the proportion downtime, $D$, of an automated flow line, as defined by Eq. (5.6). $D$ is a measure of the inefficiency that results from line stops. The balance delay measures the inefficiency from imperfect line balancing.

Considering the data given in Example 6.1, the total work content $T_{wc} = 4.00$ min. We shall assume that $T_c = 1.0$ min. If it were possible to achieve perfect balance with $n = 4$ workstations, the balance delay would be, according to Eq. (6.8):

$$d = \frac{4(1.0) - 4.0}{4(1.0)} = 0$$

If the line could only be balanced with $n = 5$ stations for the 1.0 min cycle, the balance delay would be

$$d = \frac{5(1.0) - 4.0}{5(1.0)} = 0.20 \quad \text{or} \quad 20\%$$

Both of these solutions provide the same theoretical production rate. However, the second solution is less efficient because an additional workstation, and therefore an additional assembly operator, is required. One possible way to improve the efficiency of the five-station line is to decrease the cycle time $T_c$. To illustrate, suppose that the line could be balanced at a cycle time of $T_c = 0.80$ min. The corresponding measure of inefficiency would be

$$d = \frac{5(0.80) - 4.0}{5(0.80)} = 0$$

This solution (if it were possible) would yield a perfect balance. Although five workstations are required, the theoretical production rate would be $R_c = 1.25$ units/min, an increase over the production rate capability of the four-station line. The reader can readily perceive that there are many combinations of $n$ and $T_c$ that will produce a theoretically perfect balance. Each combination will give a different production rate. In general, the balance delay $d$ will be zero for any values of $n$ and $T_c$ that satisfy the relationship

$$nT_c = T_{wc} \tag{6.9}$$

Unfortunately, because of precedence constraints and because the particular values of $T_{ej}$ usually do not permit it, perfect balance might not be achievable for every $nT_c$ combination that equals the total work content time. In other words, the satisfaction of Eq. (6.9) is a necessary condition for perfect balance, but not a sufficient condition.

As indicated by Eq. (6.5), the desired maximum value of $T_c$ is specified by the production rate required of the flow line. Therefore, Eq. (6.9) can be cast in a different form to determine the theoretical minimum number of workstations required to optimize the balance delay for a specified $T_c$. Since $n$ must be an integer, we can state:

$$\text{minimum } n \text{ is the smallest integer} \geqslant \frac{T_{wc}}{T_c} \tag{6.10}$$

Applying this rule to our example problem, with $T_{wc} = 4.0$ min and $T_c = 1.0$ min, the minimum $n = 4$ stations.

In the next section, we will examine several methods that attempt (but do not guarantee) to provide line balancing solutions with minimum balance delay for a given $T_c$.

## 6.3 MANUAL METHODS OF LINE BALANCING

In this section we will consider several methods for solving the line balancing problem by hand, using Example 6.1 for purposes of illustration. These methods are heuristic approaches, meaning that they are based on logic and common sense rather than on mathematical proof. None of the methods guarantees an optimal solution, but they are likely to result in good solutions which approach the true optimum. The manual methods to be presented are:

1. Largest-candidate rule.
2. Kilbridge and Wester's method.
3. Ranked positional weights method.

In Section 6.4, we will consider some computer procedures for solving the line balancing problem.

### Largest-candidate rule

This is the easiest method to understand. The work elements are selected for assignment to stations simply on the basis of the size of their $T_e$ values. The steps used in solving the line balancing problem are listed below, followed by Example 6.2, which is the application of these steps to Example 6.1.

PROCEDURE

*Step 1.* List all elements in descending order of $T_e$ value, largest $T_e$ at the top of the list.

*Step 2.* To assign elements to the first workstation, start at the top of the list and work down, selecting the first feasible element for placement at the station. A feasible element is one that satisfies the precedence requirements and does not cause the sum of the $T_e$ values at the station to exceed the cycle time $T_c$.

*Step 3.* Continue the process of assigning work elements to the station as in step 2 until no further elements can be added without exceeding $T_c$.

*Step 4.* Repeat steps 2 and 3 for the other stations in the line until all the elements have been assigned.

One comment should be made which applies not only to the largest-candidate rule but to the other methods as well. Starting with a given $T_c$ value, it is not usually clear how many stations will be required on the flow line. Of course, the most desirable number of stations is that which satisfies Eq. (6.10). However, the practical realities of the line balancing problem may not permit the realization of this number.

### EXAMPLE 6.2

The work elements of Example 6.1 are listed in Table 6.2 in the manner prescribed by step 1. Also listed are the immediate predecessors for each element. This is of value in determining feasibility of elements that are candidates for assignment to a given station.

Following step 2, we start at the top of the list and search for feasible work elements. Element 3 is not feasible because its immediate predecessor is element 1, which has not yet been assigned. The first feasible element encountered is element 2. It is therefore assigned to station 1. We then start the search over again from the top of the list. Steps 2 and 3 result in the assignment of elements 2, 5, 1, and 4 to station 1. The total of their element times is 1.00 min. Hence, $T_{s1} = 1.0$, which equals $T_c$, and station 1 is filled. Continuing the procedure for the remaining stations results in the allocation shown in Table 6.3. There are five stations, and the balance delay for this assignment is

$$d = \frac{5(1.0) - 4.0}{5(1.0)} = 0.20 = 20\%$$

The solution is illustrated in Figure 6.2. The largest-candidate rule provides an approach that is appropriate for only simpler line balancing problems. More sophisticated techniques are required for more complex problems.

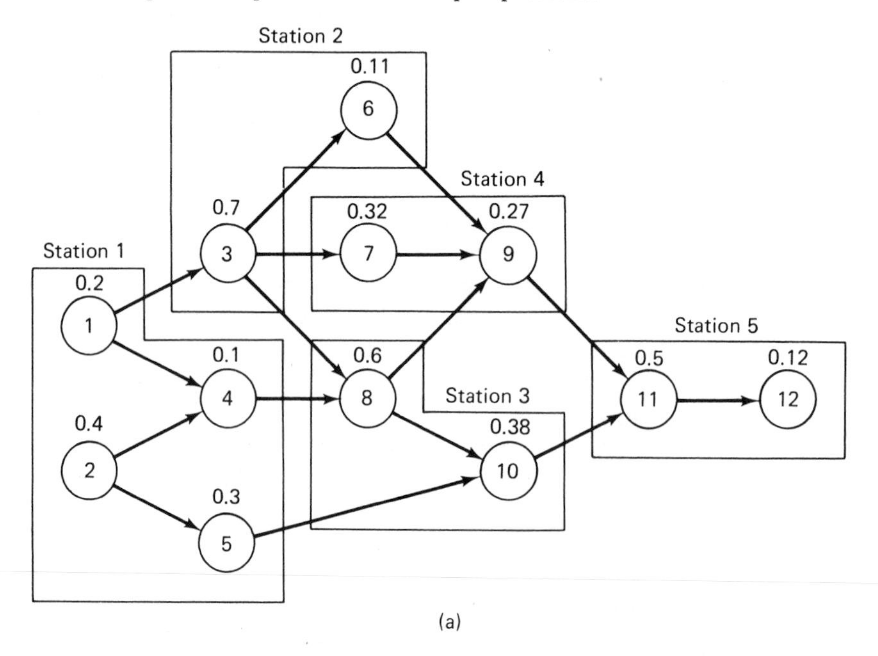


(a)

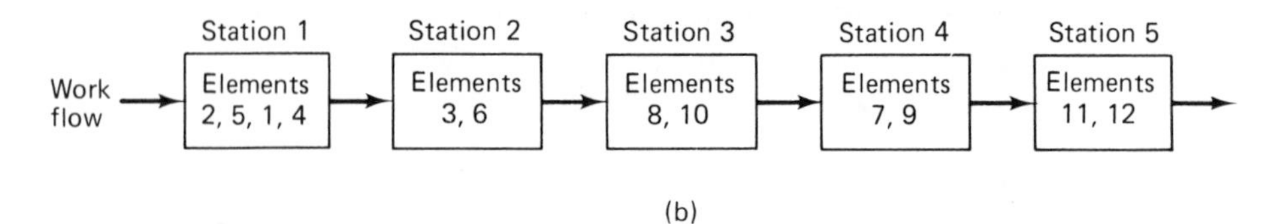


(b)

FIGURE 6.2 Solution of sample problem (Example 6.2) using largest candidate rule: (a) grouping of elements at stations in precedence diagram; (b) schematic diagram of workstation layout identifying elements assigned to each station.

TABLE 6.2 Work Elements Arranged According to $T_e$ Value for the Largest-Candidate Rule

| *Work element* | $T_e$ | *Immediate predecessors* |
|---|---|---|
| 3 | 0.7 | 1 |
| 8 | 0.6 | 3,4 |
| 11 | 0.5 | 9,10 |
| 2 | 0.4 | — |
| 10 | 0.38 | 5,8 |
| 7 | 0.32 | 3 |
| 5 | 0.3 | 2 |
| 9 | 0.27 | 6,7,8 |
| 1 | 0.2 | — |
| 12 | 0.12 | 11 |
| 6 | 0.11 | 3 |
| 4 | 0.1 | 1,2 |

TABLE 6.3 Work Elements Assigned to Stations According to the Largest-Candidate Rule

| *Station* | *Element* | $T_e$ | $\Sigma T_e$ *at station* |
|---|---|---|---|
| 1 | 2 | 0.4 | |
| | 5 | 0.3 | |
| | 1 | 0.2 | |
| | 4 | 0.1 | 1.00 |
| 2 | 3 | 0.7 | |
| | 6 | 0.11 | 0.81 |
| 3 | 8 | 0.6 | |
| | 10 | 0.38 | 0.98 |
| 4 | 7 | 0.32 | |
| | 9 | 0.27 | 0.59 |
| 5 | 11 | 0.5 | |
| | 12 | 0.11 | 0.62 |

## Kilbridge and Wester's method

This technique has received a good deal of attention in the literature since its introduction in 1961 [6]. The technique has been applied to several rather complicated line balancing problems with apparently good success [10]. It is a heuristic procedure which selects work elements for assignment to stations according to their position in the precedence diagram. The elements at the front of the diagram are selected first for entry into the solution. This overcomes one of the difficulties with the largest candidate rule, with which elements at the end of the precedence diagram might be the first candidates to be considered, simply because their $T_e$ values are large.

We will demonstrate Kilbridge and Wester's method on our sample problem. However, our problem is elementary enough that many of the difficulties which the procedure is designed to solve are missing. The interested reader is invited to consult the references, especially [2], [6], or [10], which apply the Kilbridge and Wester procedure to several more realistic problems.

## PROCEDURE AND EXAMPLE 6.3

It will be convenient to discuss the method with reference to our sample problem.

*Step 1.* Construct the precedence diagram so that nodes representing work elements of identical precedence are arranged vertically in columns. This is illustrated in Figure 6.3. Elements 1 and 2 appear in column I, elements 3, 4, and 5 are in column II, and so forth. Note that element 5 could be located in either column II or III without disrupting precedence constraints.

*Step 2.* List the elements in order of their columns, column I at the top of the list. If an element can be located in more than one column, list all the columns by the element to show the transferability of the element. This step is presented for the problem in Table 6.4. The table also shows the $T_e$ value for each element and the sum of the $T_e$ values for each column.

*Step 3.* To assign elements to workstations, start with the column I elements. Continue the assignment procedure in order of column number until the cycle time is reached. $T_c$ in our sample problem is 1.0 minute. The sum of the $T_e$ values for the columns is helpful because we can see how much of the cycle time is contained in each column. The total time of the elements in column I is 0.6 min, so all of the

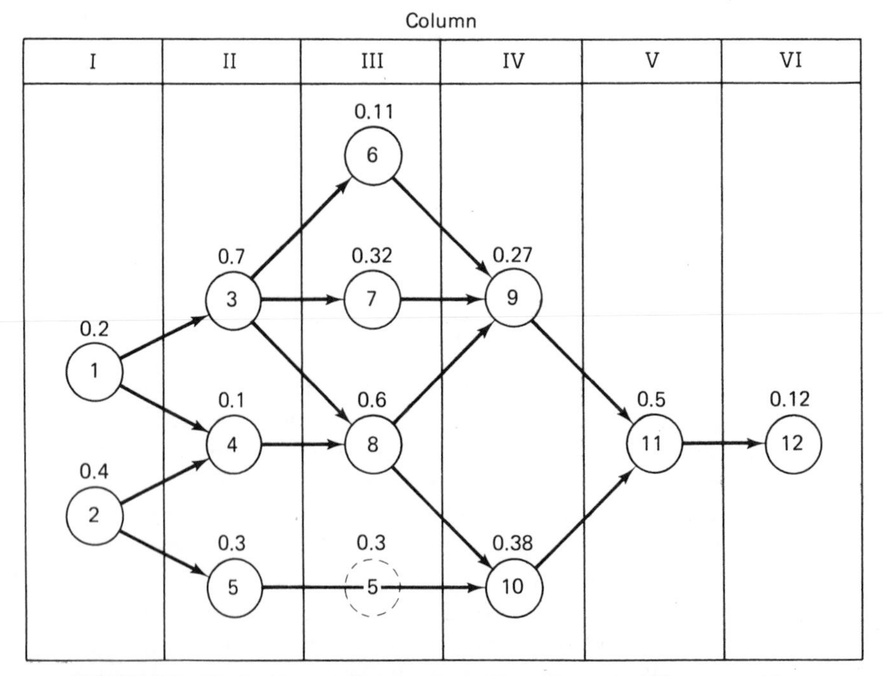


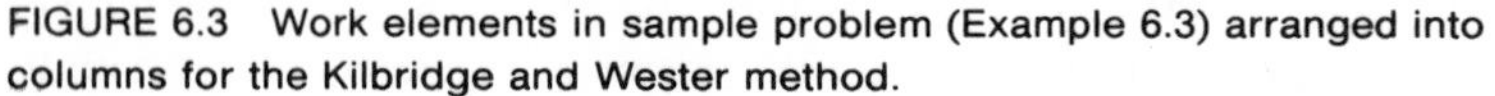
FIGURE 6.3 Work elements in sample problem (Example 6.3) arranged into columns for the Kilbridge and Wester method.

TABLE 6.4 Work Elements Arranged According to Columns from Figure 6.3 in the Kilbridge and Wester Method

| *Work element* | *Column* | $T_e$ | *Sum of column* $T_e$*s* |
|---|---|---|---|
| 1 | I | 0.2 | |
| 2 | I | 0.4 | 0.6 |
| 3 | II | 0.7 | |
| 4 | II | 0.1 | |
| 5 | II, III | 0.3 | 1.1 |
| 6 | III | 0.11 | |
| 7 | III | 0.32 | |
| 8 | III | 0.6 | 1.03 |
| 9 | IV | 0.27 | |
| 10 | IV | 0.38 | 0.65 |
| 11 | V | 0.5 | 0.5 |
| 12 | VI | 0.12 | 0.12 |

first-column elements can be entered at station 1. We immediately see that the column II elements cannot all fit at station 1. To select which elements from column II to assign, we must choose those which can still be entered without exceeding $T_c$. Immediately, element 3 is discarded from consideration since $T_{e3}=0.7$ min. When added to the column I elements, $T_s$ would exceed 1.0 min. Accordingly, elements 4 and 5 are added to station 1 to make the total process time at that station equal to $T_c$. Unlike the largest-candidate rule, we need not concern ourselves with precedence requirements, since this is automatically taken care of by ordering the elements according to columns.

To begin on the second station, element 3 from column II would be entered first. The column III elements would be considered next. Element 6 is the only one that can be entered.

The assignment process continues in this fashion until all elements have been allocated. Table 6.5 shows the line balancing solution yielded by the Kilbridge and

TABLE 6.5 Work Elements Assigned to Stations According to Kilbridge and Wester's Method

| *Station* | *Element* | $T_e$ | $\Sigma T_e$ *at station* |
|---|---|---|---|
| 1 | 1 | 0.2 | |
| | 2 | 0.4 | |
| | 4 | 0.1 | |
| | 5 | 0.3 | 1.00 |
| 2 | 3 | 0.7 | |
| | 6 | 0.11 | 0.81 |
| 3 | 7 | 0.32 | |
| | 8 | 0.6 | 0.92 |
| 4 | 9 | 0.27 | |
| | 10 | 0.38 | 0.65 |
| 5 | 11 | 0.5 | |
| | 12 | 0.12 | 0.62 |

Wester method. Since five stations are required, the balance delay is again equal to 20%, the same as that provided by the largest-candidate rule. However, note that the work elements which make up the five stations are not the same as those in Table 6.3. Also, for stations that do have the same elements, the sequence in which the elements are assigned is not necessarily identical. Station 1 illustrates this difference.

In general, the Kilbridge and Wester method will provide a superior line balancing solution when compared with the largest-candidate rule. However, this is not always true, as demonstrated by our sample problem.

## Ranked positional weights method

The ranked positional weights procedure was introduced by Helgeson and Birnie in 1961 [5]. In a sense, it combines the strategies of the largest-candidate rule and Kilbridge and Wester's method. A ranked positional weight value (call it the RPW for short) is computed for each element. The RPW takes account of both the $T_e$ value of the element and its position in the precedence diagram. Then, the elements are assigned to work stations in the general order of their RPW values.

PROCEDURE

*Step 1.* Calculate the RPW for each element by summing the element's $T_e$ together with the $T_e$ values for all the elements that follow it in the arrow chain of the precedence diagram.

*Step 2.* List the elements in the order of their RPW, largest RPW at the top of the list. For convenience, include the $T_e$ value and immediate predecessors for each element.

*Step 3.* Assign elements to stations according to RPW, avoiding precedence constraint and time-cycle violations.

### EXAMPLE 6.4

Applying the RPW method to our example problem, we first compute a ranked positional weight value for each element. For element 1, the elements that follow it in the arrow chain (see Figure 6.1) are 3, 4, 6, 7, 8, 9, 10, 11, and 12. The RPW for element 1 would be the sum of the $T_e$'s for all these elements, plus $T_e$ for element 1. This RPW value is 3.30. The reader can see that the trend will be toward lower values of RPW as we get closer to the end of the precedence diagram.

Table 6.6 lists the work elements according to RPW. We begin the assignment process by considering elements at the top of the list and working downward. Each time an element is entered into solution, we go back to the top of the list. The reader should follow through the solution in Table 6.7 to verify the order in which the work elements are assigned.

In the RPW line balance, the number of stations required is five, as before, but the maximum station process time is 0.92 min at number 3. Accordingly, the line could be operated at a cycle time of $T_e = 0.92$ rather than 1.0 min. This would, of course, be beneficial, since the production rate would be increased to $R_c = 1.075$ units/min. The corresponding balance delay is

$$d = \frac{5(0.92) - 4.0}{5(0.92)} = 0.13 = 13\%$$

The RPW solution represents a more efficient assignment of work elements to stations than either of the two preceding solutions. However, it should be noted that we have accepted a cycle time different from that which was originally specified for the problem. If the problem were reworked with $T_c = 0.92$ min using the largest-candidate rule or Kilbridge and Wester's method, it might be possible to duplicate the efficiency provided by the RPW method.

TABLE 6.6 Work Elements Arranged in Order of RPW Value in the Ranked Positional Weights Method

| *Element* | *RPW* | $T_e$ | *Immediate predecessor* |
|---|---|---|---|
| 1 | 3.30 | 0.2 | — |
| 3 | 3.00 | 0.7 | 1 |
| 2 | 2.67 | 0.4 | — |
| 4 | 1.97 | 0.1 | 1,2 |
| 8 | 1.87 | 0.6 | 3,4 |
| 5 | 1.30 | 0.3 | 2 |
| 7 | 1.21 | 0.32 | 3 |
| 6 | 1.00 | 0.11 | 3 |
| 10 | 1.00 | 0.38 | 5,8 |
| 9 | 0.89 | 0.27 | 6,7,8 |
| 11 | 0.62 | 0.5 | 9,10 |
| 12 | 0.12 | 0.12 | 11 |

TABLE 6.7 Work Elements Assigned to Stations According to the Ranked Positional Weights Method

| *Station* | *Element* | $T_e$ | $\Sigma T_e$ *at station* |
|---|---|---|---|
| 1 | 1 | 0.2 | |
| | 3 | 0.7 | 0.9 |
| 2 | 2 | 0.4 | |
| | 4 | 0.1 | |
| | 5 | 0.3 | |
| | 6 | 0.11 | 0.91 |
| 3 | 8 | 0.6 | |
| | 7 | 0.32 | 0.92 |
| 4 | 10 | 0.38 | |
| | 9 | 0.27 | 0.65 |
| 5 | 11 | 0.5 | |
| | 12 | 0.12 | 0.62 |

For large balancing problems, involving perhaps several hundred work elements, these manual methods of solution become awkward. A number of computer programs have been developed to deal with these larger assembly line cases. In the following section, we survey some of these computerized approaches.

## 6.4 COMPUTERIZED LINE BALANCING METHODS

The three methods described in the preceding section are generally carried out manually. This does not preclude their implementation on the digital computer. In fact, computer programs have been developed based on several of the heuristic approaches. However, the use of the computer allows a more complete enumeration of the possible solutions to a line balancing problem than is practical with a manual solution method. Accordingly, the computer line balancing algorithms are normally structured to explore a wide range of alternative allocations of elements to workstations. In this section, we will discuss some of the techniques for solving large-scale line balancing problems based on the use of the computer. The first of these, COMSOAL, is the only one for which we will detail the procedure.

### COMSOAL

This acronym stands for Computer Method of Sequencing Operations for Assembly Lines. It is a method developed at Chrysler Corporation and reported by Arcus in 1966 [1]. Although it was not the first computerized line balancing program to be developed, it seems to have attracted considerably more attention than those which preceded it. The procedure is to iterate through a sequence of alternative solutions and keep the best one. Let us outline the basic algorithm of COMSOAL and proceed to discuss it with regard to our sample problem.

**PROCEDURE AND EXAMPLE 6.5**

*Step 1.* Construct list A, showing all work elements in one column and the total number of elements that immediately precede each element in an adjacent column. This is illustrated in Table 6.8. Note that these types of data would be quite easy to compile and manipulate by the computer.

*Step 2.* Construct list B (Table 6.9), showing all elements from list A that have no immediate predecessors.

*Step 3.* Select at random one of the elements from list B. The computer would be programmed to perform this random selection process. The only constraint is that the element selected must not cause the cycle time $T_c$ to be exceeded.

*Step 4.* Eliminate the element selected in step 3 from lists A and B and update both lists, if necessary. Updating may be needed because the selected element was

TABLE 6.8 List A in COMSOAL at the Beginning of the Sample Problem

| *Element* | *Number of immediate predecessors* |
|---|---|
| 1 | 0 |
| 2 | 0 |
| 3 | 1 |
| 4 | 2 |
| 5 | 1 |
| 6 | 1 |
| 7 | 1 |
| 8 | 2 |
| 9 | 3 |
| 10 | 2 |
| 11 | 2 |
| 12 | 1 |

TABLE 6.9 List B in COMSOAL at the Beginning of the Sample Problem

| *Elements with no immediate predecessors* |
|---|
| 1 |
| 2 |

probably an immediate predecessor for some other element(s). Hence, there may be changes in the number of immediate predecessors for certain elements in list A; and there may now be some new elements having no immediate predecessors that should be added to list B.

To illustrate, suppose in step 3 that element 1 is chosen at random for entry into the first workstation. This would mean that element 3 no longer has any immediate predecessors. Tables 6.10 and 6.11 show the updated lists A and B, respectively.

*Step 5.* Again select one of the elements from list B which is feasible for cycle time.

TABLE 6.10 List A in COMSOAL after Step 3

| *Element* | *Number of immediate predecessors* |
|---|---|
| 2 | 0 |
| 3 | 0 |
| 4 | 1 |
| 5 | 1 |
| 6 | 1 |
| 7 | 1 |
| 8 | 2 |
| 9 | 3 |
| 10 | 2 |
| 11 | 2 |
| 12 | 1 |

TABLE 6.11 List B in COMSOAL after Step 3

| Elements with no immediate predecessors |
|---|
| 2 |
| 3 |

TABLE 6.12 One Possible Solution with COMSOAL

| Station | Element | $T_e$ | Site at station |
|---|---|---|---|
| 1 | 1 | 0.2 | |
| | 2 | 0.4 | |
| | 5 | 0.3 | |
| | 4 | 0.1 | 1.0 |
| 2 | 3 | 0.7 | |
| | 6 | 0.11 | 0.81 |
| 3 | 8 | 0.6 | |
| | 10 | 0.38 | 0.98 |
| 4 | 7 | 0.32 | |
| | 9 | 0.27 | 0.59 |
| 5 | 11 | 0.5 | |
| | 12 | 0.12 | 0.62 |

*Step 6.* Repeat steps 4 and 5 until all elements have been allocated to stations within the $T_c$ constraint.

One possible solution to the problem is shown in Table 6.12. The balance delay is again $d=20\%$, the same efficiency as obtained with the largest candidate rule and the Kilbridge and Wester method.

*Step 7.* Retain the current solution and repeat steps 1 through 6 to attempt to determine an improved solution. If an improved solution is obtained, it should be retained.

The steps involved in the COMSOAL algorithm represent an uncomplicated data manipulation procedure. It is therefore ideally suited to computer programming. Although there is much iteration in the algorithm, this is of minor consequence because of the speed with which the computer is capable of performing the iterations.

## CALB

The Advanced Manufacturing Methods Program (AMM) of the IIT Research Institute seems to be the nucleus for today's research in line balancing methodology [7]. In 1968, this group introduced a computer package called CALB (for Computer Assembly Line Balancing or Computer-Aided Line

Balancing), which has more or less become the industry standard. Its applications have included a variety of assembled products, including automobiles and trucks, electronic equipment, appliances, military hardware, and others.

CALB can be used for both single-model and mixed-model lines. For the single-model case, the data required to use the program include the identification of each work element $T_e$ for each element, the predecessors, and other constraints that may apply to the line. Also needed to balance the line is information on minimum and maximum allowable time per workstation (in other words, cycle time data). The CALB program starts by sorting the elements according to their $T_e$ and precedence requirements. Based on this sort, elements are assigned to stations so as to satisfy the minimum and maximum allowable station times. Line balancing solutions with less than 2% idle time have been described as common [10] and the cost to operate the computer program is modest. Typical costs of less than $10/run have been cited for problems with around 300 work elements [7].

To use CALB on mixed-model lines, additional data are required such as the production requirements per shift for each model to be run on the line, and a definition of relative element usage per model. The solutions obtained by CALB are described as near-optimum and are achieved with only a few seconds of time on the computer.

Another computer package developed by the AMM Program of IITRI is called GALS (for Generalized Assembly Line Simulator). It has been described in the following terms: "The GALS program was developed as a tool for evaluating the design of a new line or proposed changes to existing lines. It can provide an estimate of expected productivity of the line and pinpoint potential problem areas."[1]

## ALPACA

This computer system was developed by one of the major users of assembly flow lines, General Motors [11]. The acronym represents "Assembly Line Planning and Control Activity." It was first implemented in 1967, but improvements in computer hardware since that time have reduced the cost of using the package to 10% of the original usage cost. ALPACA is described as an interactive line balancing system in which the user can transfer work from one station to another along the flow line and immediately assess the relative efficiency of the change. One of the complex problems facing the automotive industry is the proliferation of car models and options. ALPACA was designed to cope with the complications on the assembly line that arise from this problem. The system user can quickly determine what changes in work element assignments should be made to maintain a reasonable line balance for the ever-changing product flow.

[1]E. L. Magad, "Cooperative Manufacturing Research," *Industrial Engineering*, Vol. 4, No. 1, January, 1972, pp. 36–40.

## 6.5 OTHER WAYS TO IMPROVE THE LINE BALANCE

The line balancing techniques described in Sections 6.3 and 6.4 represent strict and precise procedures for allocating work elements to stations according to a specified cycle time. For most flow line situations, these techniques result in allocations that possess a high degree of balance efficiency. However, the designer of a flow line, either manual or automatic, should not overlook other possible ways for improving the operation of the line. In this section, let us examine some of these possibilities.

### Dividing work elements

In Section 6.2, a minimum rational work element was defined as the smallest practical indivisible task, which cannot be subdivided further. In some instances, it may be perfectly reasonable to define certain tasks as minimum rational work elements even though it would be technically possible to further subdivide these elements. For example, it is reasonable to identify the drilling of a hole as a work element, and therefore to perform this work element all at one station. However, if the drilling of a deep hole at one station were to cause a bottleneck situation, it could be argued that the drilling operation should be separated into two steps. The advantage of this would be not only to eliminate the bottleneck but also to increase the tool lives of the drill bits.

This kind of situation is more prevalent on mechanized processing lines, where operations such as the drilling process described above are performed. In the initial definition of work elements to be done on the line, it may not seem to be as technologically feasible to subdivide such process operations as it is in the case of assembly operations.

### Changing workhead speeds at automatic stations

This again refers to automated (or semiautomated) lines such as machining transfer lines. It may be possible to effect a reduction in the process time of a bottleneck station by increasing its speed or feed rate. There will normally be a penalty associated with such changes in the form of a shorter tool life. This will result in more frequent line stops for tool replacement.

There is an opposite side to the coin. At a station where there is idle time, the workhead feed and speed should probably be reduced to prolong the tool life. This would tend to reduce the frequency of downtime occurrences on the line.

Through a process of increasing the speed/feed combinations at the stations with long process times, and reducing the speed/feed combinations at stations with idle time, it should be possible to improve the balance on the flow line.

## Parallel stations

One of the restrictions implicit in the previous heuristic methods is that the stations must be arranged sequentially. If this restriction can be disregarded, it allows the use of parallel stations where bottlenecks previously existed. To illustrate, suppose that a five-station line had station process times of 1 min at all but the fifth workstation. At this station the process time was 2 min. With the stations arranged in series, the output rate would be limited by the 2 min process time of the fifth station to $R_c = 30$ units/h. However, if two stations were arranged in parallel at the fifth station position, the output could be increased to $R_c = 60$ units/h. Each of the parallel stations would have a production rate of 0.5 unit/min, but since there are two of them, their effective output would be 1 unit/min. Most problems are not as easy as this.

### EXAMPLE 6.6

Consider the sample problem (Example 6.1) we have been using to illustrate the various line balancing techniques. The total work content time is $T_{wc} = 4.0$ min. Therefore, with a cycle time of $T_c = 1.0$ min, the theoretical minimum number of workstations needed should be four. However, none of the methods discussed in Sections 6.3 or 6.4 was able to achieve a solution with less than five stations.

By utilizing a configuration with parallel stations, it is possible to achieve a perfectly balanced solution with four stations. The allocation of elements is listed in Table 6.13. Stations 1 and 2 both have the same work elements: 1, 2, 3, 4, and 8, whose total process time is 2.0 min. Since there are two stations, however, the effective output rate of the parallel arrangement is 1 unit/min. Stations 3 and 4 each have process times of 1 min, so the desired line cycle time of $T_c = 1.0$ min is achieved. The solution is illustrated in Figure 6.4.

TABLE 6.13 Solution of Sample Problem Using Parallel Stations (Example 6.6)

| *Station* | *Elements* | $T_e$ | $\Sigma T_e$ *at stations* |
|---|---|---|---|
| 1 and 2 in parallel | 1 | 0.2 | |
| | 2 | 0.4 | |
| | 3 | 0.7 | |
| | 4 | 0.1 | |
| | 8 | 0.6 | 2.0 |
| 3 | 5 | 0.3 | |
| | 6 | 0.11 | |
| | 7 | 0.32 | |
| | 9 | 0.27 | 1.0 |
| 4 | 10 | 0.38 | |
| | 11 | 0.5 | |
| | 12 | 0.12 | 1.0 |

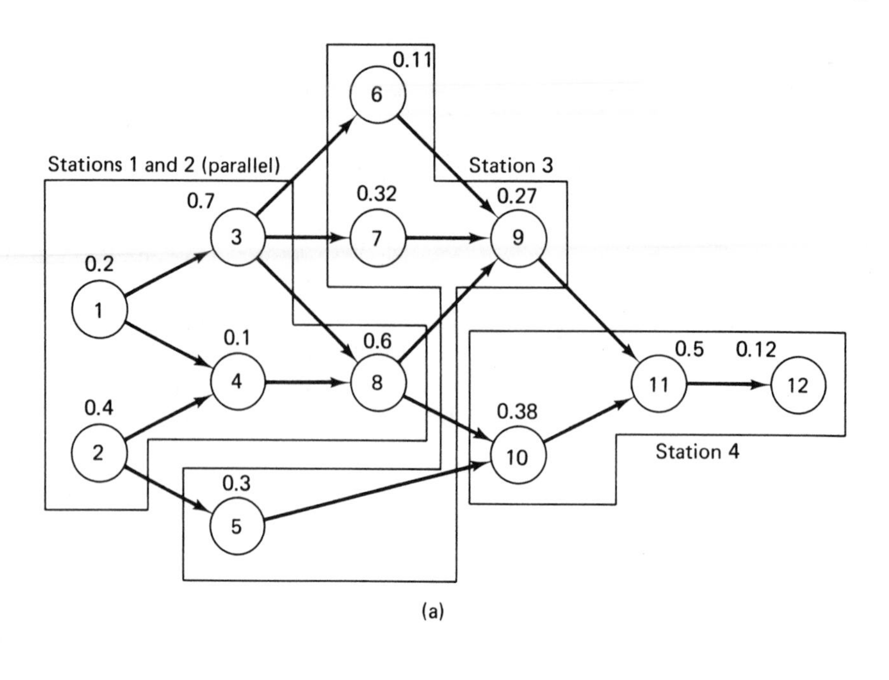


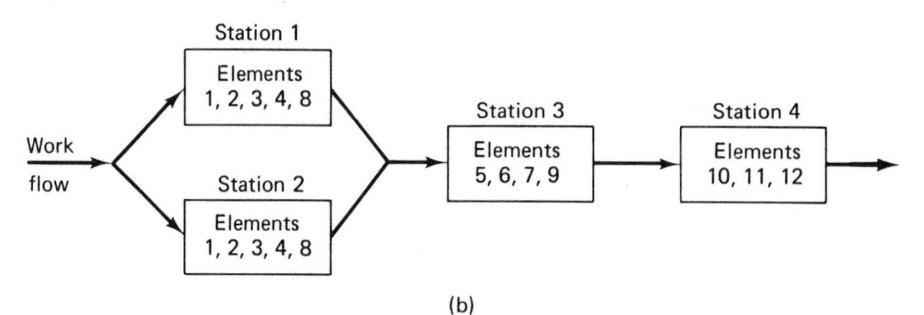


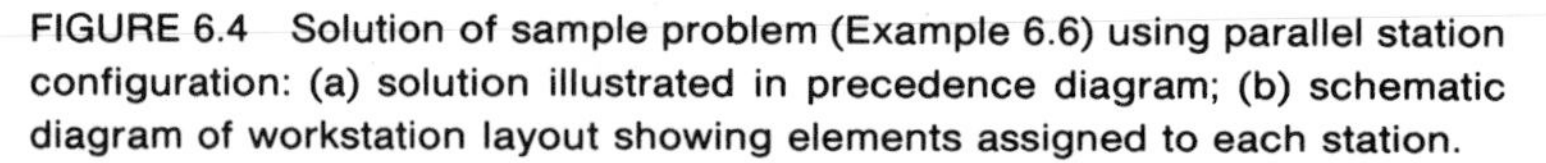
FIGURE 6.4 Solution of sample problem (Example 6.6) using parallel station configuration: (a) solution illustrated in precedence diagram; (b) schematic diagram of workstation layout showing elements assigned to each station.

There is no formalized procedure for developing a solution that utilizes parallel stations such as the above. Rather, a certain ingenuity and flexibility of mind are required in order for the analyst to perceive that the traditional approaches can be improved upon.

The use of parallel stations in a manual flow line is easily accomplished because of the inherent flexibility of the human operator. In an automated line it is necessary to incorporate a switching device in the workpart transfer mechanism which will alternate work units between the two parallel stations.

### Inventory buffers between stations

The justification for using storage buffers on automated flow lines was discussed in Section 5.6. On manual flow lines, storage buffers can also be of benefit. Their principal use is to smooth the flow of work, which might otherwise be disrupted by worker process time variability. Although the line balancing techniques assume that process times are constant, any human activity (and most other physical processes, for that matter) is characterized by random variations. These variations are manifested in process time differences from cycle to cycle. The buffer stocks between workstations help to level these differences.

### Methods analysis

The customary use of the term *methods analysis* implies the study of human work activity for possible improvements. Such an analysis seems an obvious requirement for a manual flow line job, since the work elements need to be defined for the job before any line balancing can be performed. In addition, methods analysis can also be used to increase the rate of production at those stations that turn out to be the bottlenecks. The methods analysis may result in better workplace layout, redesigned tooling and fixturing, or improved hand and body motions. All of these improvements are likely to yield a superior balance of work on the manual flow line.

Analysis of the operations on automated lines may also lead to improvements in work flow and balance. However, attempts to optimize the line balance on automated process lines are usually emphasized during the planning and design stages, since alterations of the finished line are difficult because of the fixed nature of the equipment.

## 6.6 A GUIDE TO THE LITERATURE

There is much available in the technical literature on line balancing, and the interested reader will be able to explore the subject in much more detail than space permits us to do here. Two excellent books on the subject of manual flow line design, with emphasis on the line balancing problem, are those by Wild [12] and Prenting and Thomopoulos [10]. The book by Wild, in addition to discussing the single-model flow line problem, also presents algorithms for solving the batch-model and mixed-model case. The book by Prenting and Thomopoulos is basically a collection of the best technical articles that have been written on the subject of assembly line systems. It is an excellent reference for anyone interested in pursuing this field of production.

In 1966, Mastor concluded his doctoral research at UCLA [8], which was concerned with an evaluation of the various line balancing techniques which had been developed to that time (which includes all of those presented in this chapter except the IITRI programs and ALPACA). Mastor subsequently contributed his findings to the general literature [9]. Although the scope of the comparison was limited to two problems, the reader may find this report of interest. A more recent comparative study was carried out by Dar-El [4].

For the practicing engineer or manager with a flow line balancing problem to solve, unless it is a relatively simple problem for which the techniques in this chapter will produce good results, we recommend that the computer programs developed by the Advanced Manufacturing Methods Group at IITRI (CALB and others) be considered.

## REFERENCES

[1] ARCUS, A. L., "COMSOAL—A Computer Method of Sequencing Operations for Assembly Lines," *International Journal of Production Research*, Vol. 4, No. 4, 1966, pp. 259–277.

[2] BUFFA, E. S., and TAUBERT, W. H., *Production-Inventory Systems: Planning and Control*, Richard D. Irwin, Inc., Homewood, Ill.,1972, Chapters 8, 9.

[3] BUXLEY, G. M., SLACK, H. D., and WILD, R., "Production Flow Line System Design—A Review" *AIIE Transactions*, Vol. 5, No. 1, 1973, pp. 37–48.

[4] DAR-EL, E. M., "Solving Large Single-Model Assembly Line Balancing Problems—A Comparative Study," *AIIE Transactions*, Vol. 7, No. 3, 1975, pp. 302–310.

[5] HELGESON, W. B., and BIRNIE, D. P., "Assembly Line Balancing Using Ranked Positional Weight Technique," *Journal of Industrial Engineering,* Vol. 12, No. 6, 1961, pp. 394–398.

[6] KILBRIDGE, M. D., and WESTER, L., "A Heuristic Method of Assembly Line Balancing," *Journal of Industrial Engineering*, Vol. 12, No. 4, 1961, pp. 292–298.

[7] MAGAD, E. L., "Cooperative Manufacturing Research," *Industrial Engineering*, Vol. 4, No. 1, 1972, pp. 36–40.

[8] MASTOR, A. A., "An Experimental Investigation and Comparative Evaluation of Production Line Balancing Techniques," Unpublished Ph.D. dissertation, UCLA, 1966.

[9] MASTOR, A. A., "An Experimental Investigation and Comparative Evaluation of Production Line Balancing Techniques," *Management Science*, Vol. 16, No. 11, 1970, pp. 728–746.

[10] PRENTING, T. O., and THOMOPOULOS, N. T., *Humanism and Technology in Assembly Line Systems*, Spartan Books, Hayden Book Co., Rochelle Park, N.J., 1974.

[11] SHARP, W. I., JR., "Assembly Line Balancing Techniques," Paper MS77–313, *Society of Maufacturing Engineers,* Dearborn, Mich., 1977.

[12] WILD, RAY, *Mass-Production Management*, John Wiley & Sons Ltd., London, 1972.

## PROBLEMS

**6.1.** A manual production flow line is arranged with six stations and a conveyor system is used to move parts along the line. The belt speed is 4 ft/min and the spacing of raw workparts along the line is one every 3 ft. The total line length is 30 ft, hence each station length equals 5 ft. Determine the following:
(a) Feed rate $f_b$.
(b) Tolerance time $T_t$.
(c) Theoretical cycle time $T_c$.

**6.2.** Given the physical flow line configuration of Problem 6.1, is it likely that the line could be utilized to produce a job whose total work content time = 5.0 min? What about a total work content time of 4.0 min? 3.0 min?

**6.3.** A manual assembly line is to be designed with a production rate of 100 completed assemblies per hour. The line will have eight stations and the length of each station is 1.0 m. The minimum allowable tolerance time is to be 2.0 min. If the line is figured to have an uptime efficiency of 97% (estimated from previous similar lines), determine the following parameters for the line:
(a) Ideal cycle time $T_c$.
(b) Belt speed $V_b$.
(c) Feed rate $f_b$.
(d) Part spacing $s_p$ along the belt.

**6.4.** The total work content time of a certain assembly job is 7.8 min. The estimated downtime of the line is $D = 5\%$, and the required production rate is $R_p = 80$ units/h.
(a) Determine the theoretical minimum number of workstations required to optimize the balance delay.
(b) For the number of stations determined in part (a), compute the balance delay $d$.
(c) What feed rate should be specified if a moving belt line is to be used?

**6.5.** In an automated transfer line, a series of 10 operations is performed at 10 machining stations. The average station process time is 21 s and the longest process time is 28 s. It takes 7 s to transfer the workparts from one station to the next. The proportion downtime on the line is measured at $D = 25\%$.
(a) What is the ideal cycle time $T_c$?
(b) Compute the balance delay $d$. How would you treat the transfer time of 7 s?
(c) What is the average production rate for the line?

(d) If balance delay and proportion downtime were combined into one efficiency factor to represent the total equipment utilization, what would be the value of that efficiency factor?

(e) What steps might be taken on this line to improve the balance delay?

**6.6.** The following list defines the precedence relationships and element times for a new model toy:

| *Element* | $T_c$ *(min)* | *Immediate predecessors* |
|---|---|---|
| 1 | 0.5 | — |
| 2 | 0.3 | 1 |
| 3 | 0.8 | 1 |
| 4 | 0.2 | 2 |
| 5 | 0.1 | 2 |
| 6 | 0.6 | 3 |
| 7 | 0.4 | 4, 5 |
| 8 | 0.5 | 3, 5 |
| 9 | 0.3 | 7, 8 |
| 10 | 0.6 | 6, 9 |

(a) Construct the precedence diagram for this job.

(b) If the ideal cycle time is to be 1.0 min, what is the theoretical minimum number of stations required to minimize the balance delay?

(c) Compute the balance delay for the answer found in part (b).

**6.7.** Determine the assignment of work elements to stations using the largest-candidate rule for Problem 6.6.

(a) How many stations are required?

(b) Compute the balance delay.

**6.8.** Solve Problem 6.6 using the Kilbridge and Wester method.

**6.9.** Solve Problem 6.6 using the ranked positional weights method.

**6.10.** Solve for one iteration of Problem 6.6 using COMSOAL.

**6.11.** A proposal has been submitted to replace a group of assembly workers, each working individually, with an assembly line. The accompanying table gives the individual work elements.

| *Element* | $T_e$ *(min)* | *Immediate predecessors* |
|---|---|---|
| 1 | 1.0 | — |
| 2 | 0.5 | — |
| 3 | 0.8 | 1, 2 |
| 4 | 0.3 | 2 |
| 5 | 1.2 | 3 |
| 6 | 0.2 | 3, 4 |
| 7 | 0.5 | 4 |
| 8 | 1.5 | 5, 6, 7 |

The demand rate for this job is 1600 units/week (assume 40 h/week) and the current number of operators required to meet this demand is eight using the individual manual workers.

(a) Construct the precedence diagram from the data provided on work elements.

(b) Use the largest-candidate rule to assign work elements to stations. What is the balance delay for the solution?

(c) The initial cost to install the assembly line is $20,000. If the hourly rate for workers is $5.00/h, will the assembly line be justified using a 3-year service life? Assume 50 weeks/yr. Use a rate of return = 10%.

**6.12.** Solve Problem 6.11(b) using the Kilbridge and Wester method.

**6.13.** Solve Problem 6.12(b) using the ranked positional weights method.

**6.14.** A new small electrical appliance for the home do-it-yourselfer is to be assembled manually on a production flow line. The total job of assembling the product has been divided into minimum rational work elements and these are described in Table 6.14. Also given in this table are tentative time standards as estimated by the industrial engineering department from similar jobs done

TABLE 6.14 List of Work Elements

| *No.* | *Element description* | $T_e$ *(min)* | *Immediate predecessor* |
|---|---|---|---|
| 1 | Place frame on workholder and clamp | 0.15 | — |
| 2 | Assemble fan to motor | 0.37 | — |
| 3 | Assemble bracket 1 to frame | 0.21 | 1 |
| 4 | Assemble bracket 2 to frame | 0.21 | 1 |
| 5 | Assemble motor to frame | 0.58 | 1,2 |
| 6 | Affix insulation to bracket 1 | 0.12 | 3 |
| 7 | Assemble angle plate to bracket 1 | 0.29 | 3 |
| 8 | Affix insulation to bracket 2 | 0.12 | 4 |
| 9 | Attach link bar to motor and bracket 2 | 0.30 | 4,5 |
| 10 | Assemble three wires to motor | 0.45 | 5 |
| 11 | Assemble nameplate to housing | 0.18 | — |
| 12 | Assemble light fixture to housing | 0.20 | 11 |
| 13 | Assemble blade mechanism to frame | 0.65 | 6,7,8,9 |
| 14 | Wire switch, motor, and light | 0.72 | 10,12 |
| 15 | Wire blade mechanism to switch | 0.25 | 13 |
| 16 | Attach housing over motor | 0.35 | 14 |
| 17 | Test blade mechanism, light, etc. | 0.16 | 15,16 |
| 18 | Affix instruction label to cover plate | 0.12 | — |
| 19 | Assemble grommet to power cord | 0.10 | — |
| 20 | Assemble cord and grommet to cover plate | 0.23 | 18,19 |
| 21 | Assemble power cord leads to switch | 0.40 | 17,20 |
| 22 | Assemble cover plate to frame | 0.33 | 21 |
| 23 | Final inspect and remove from workholder | 0.25 | 22 |
| 24 | Package | 1.75 | 23 |

previously. In the extreme right-hand column of the table are the immediate predecessors established by precedence requirements. The small appliance is to be assembled at the rate of one product per minute off the production line. You are to design the layout of stations along the line so as to meet this production requirement.

Use one of the methods of line balancing presented in Section 6.3 to balance the line as much as possible. How many stations are required? If the production rate is increased or decreased slightly (by not more than 20%), could the balance be improved? What is the percent balance delay? Make a sketch of the flow line layout, showing positions of stations and operators along the line.

part 

# Numerical Control Manufacturing Systems

# chapter 7

# Numerical Control

## 7.1 INTRODUCTION

In Chapter 1, we noted that there were two basic categories of automated manufacturing systems: fixed automation and programmable automation. The Detroit-type flow lines discussed and analyzed in Chapters 4 through 6 are prime examples of fixed automation. The configuration of the equipment is fixed to produce one type of part or product in large quantities. The advantage of this arrangement is that the products can be made at high rates of production and low cost per unit. The disadvantage is that this type of automated system is not at all flexible. It can only be used for the one product (or a limited range of similar products). When the product run is finished, the automated equipment is of no use unless a major changeover is performed on it.

Programmable automation, on the other hand, is designed to accommodate changes in product. Its principal applications are found in low to medium production runs of different parts and jobs. Probably the single most important example of programmable automation, certainly in discrete metal parts manufacture, is numerical control.

In the next several chapters we will provide a comprehensive discussion of numerical control, with particular emphasis on its principal application—the machining process. In the present chapter, we define the term and describe how numerical control works and where it should be used. In Chapter 8, we consider how to program in numerical control. We have described it as programmable automation. Well, then, how is it programmed? In Chapter 9, we discuss several extensions or developments that are directly related to numerical control.

## 7.2 WHAT IS NUMERICAL CONTROL?

*Numerical control* (often abbreviated NC) can be defined as a form of programmable automation in which the process is controlled by numbers, letters, and symbols. In NC, the numbers form a program of instructions designed for a particular workpart or job. When the job changes, the program of instructions is changed. This capability to change the program for each new job is what gives NC its flexibility. It is much easier to write new programs than to make major changes in the production equipment.

NC equipment is used in all areas of metal parts fabrication and comprises roughly 15% of the modern machine tools in industry today. Since numerically controlled machines are considerably more expensive than their conventional counterparts, the asset value of industrial NC machine tools is proportionally much larger than their numbers. Equipment utilizing numerical control has been designed to perform such diverse operations as drilling, milling, turning, grinding, sheetmetal pressworking, spot welding, arc welding, riveting, assembly, drafting, inspection, and parts handling. And this is by no means a complete list. Numerical control should be considered as a possible mode of controlling the operation for any production situation possessing the following characteristics:

Similar workparts in terms of raw material (e.g., metal stock for machining).

The workparts are produced in various sizes and geometries.

The workparts are produced in batches of small to medium-size quantities.

A sequence of similar processing steps is required to complete the operation on each workpiece.

Many machining jobs meet these conditions. The machined workparts are metal, they are specified in many different sizes and shapes, and most machined parts produced in industry today are made in small to medium-size lot sizes. To produce each part, a sequence of drilling operations may be required, or a series of turning or milling operations. The suitability of NC for these kinds of jobs is the reason for the tremendous growth of numerical control in the metalworking industry over the last 25 years.

### Basic components of an NC system

An operational numerical control system consists of the following three basic components:

1. Program of instructions.
2. Controller unit, also called machine control unit (MCU).
3. Machine tool or other controlled process.

The general relationship among the three components is illustrated in Figure 7.1. The program of instructions serves as the input to the controller unit, which in turn commands the machine tool or other process to be controlled.

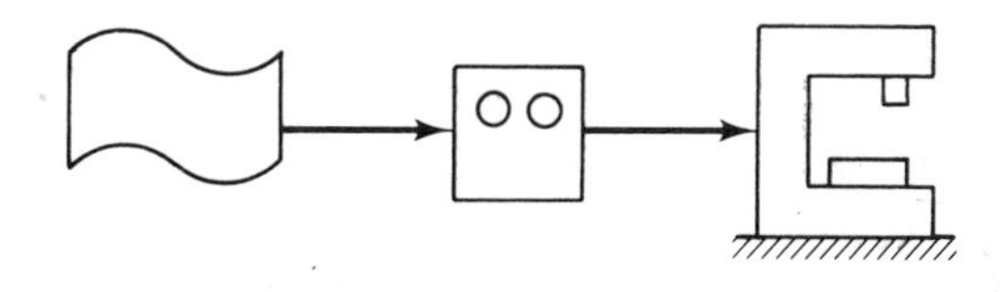


FIGURE 7.1 Three basic components of a numerical control system.

PROGRAM OF INSTRUCTIONS. The program of instructions is the detailed step-by-step set of directions which tell the machine tool what to do. It is coded in numerical or symbolic form on some type of input medium that can be interpreted by the controller unit. The most common input medium is 1-in.-wide punched tape. Over the years, other forms of input media have been used, including punched cards, magnetic tape, and even 35-mm motion picture film.

There are two other methods of input to the NC system which should be mentioned. The first is by manual entry of instructional data to the controller unit. This is time-consuming and is rarely used except as an auxiliary means of control or when only one or a very limited number of parts are to be made. The second method of input is by means of a direct link with a computer. This is called *direct numerical control,* or DNC, and we will discuss this later in Chapter 9.

The program of instructions is prepared by someone called a part programmer. The programmer's job is to provide a set of detailed instructions by which the sequence of processing steps is to be performed. For a machining operation, the processing steps involve the relative movement of the machine tool table and the cutting tool.

Controller Unit. The second basic component of the NC system is the controller unit. This consists of the electronics and hardware that read and interpret the program of instructions and convert it into mechanical actions of the machine tool. The typical elements of the controller unit include the tape reader, a data buffer, signal output channels to the machine tool, feedback channels from the machine tool, and the sequence controls to coordinate the overall operation of the foregoing elements.

The tape reader is an electrical-mechanical device for winding and reading the punched tape containing the program of instructions. The data contained on the tape are read into the data buffer. The purpose of this device is to store the input instructions in logical blocks of information. A block of information usually represents one complete step in the sequence of processing elements. For example, one block may be the data required to move the machine table to a certain position and drill a hole at that location.

The signal output channels are connected to the servomotors and other controls in the machine tool. Through these channels, the instructions are sent to the machine tool from the controller unit. To make certain that the instructions have been properly executed by the machine, feedback data are sent back to the controller via the feedback channels. The most important function of this return loop is to assure that the table and workpart have been properly located with respect to the tool. Most NC machine tools in use today are provided with position feedback controls for this purpose and are referred to as *closed-loop systems.* However, in recent years there has been a growth in the use of *open-loop systems,* which do not make use of feedback signals to the controller unit. The advocates of the open-loop concept claim that the reliability of the system is great enough that feedback controls are not needed and are an unnecessary extra cost. We shall consider the subject of feedback control systems in greater detail in Chapter 13.

Sequence controls coordinate the activities of the other elements of the controller unit. The tape reader is actuated to read data into the buffer from the tape, signals are sent to and from the machine tool, and so on. These types of operations must be synchronized and this is the function of the sequence controls.

Another element of the NC system, which may be physically part of the controller unit or part of the machine tool, is the control panel. The control panel or control console contains the dials and switches by which the machine operator runs the NC system. It may also contain data displays to provide information to the operator. Although the NC system is an automatic system, the human operator is still needed to turn the machine on and off, to change tools (some NC systems have automatic tool changers), to load and unload the machine, and to perform various other duties. To be able to discharge these duties, the operator must be able to control the system, and this is done through the control panel.

MACHINE TOOL. The third basic component of an NC system is the machine tool or other controlled process. It is the part of the NC system which performs useful work. In the most common example of an NC system, one designed to perform machining operations, the machine tool consists of the worktable and spindle as well as the motors and controls necessary to drive them. It also includes the cutting tools, work fixtures, and other auxiliary equipment needed in the machining operation. Figure 7.2 illustrates an NC machine tool.

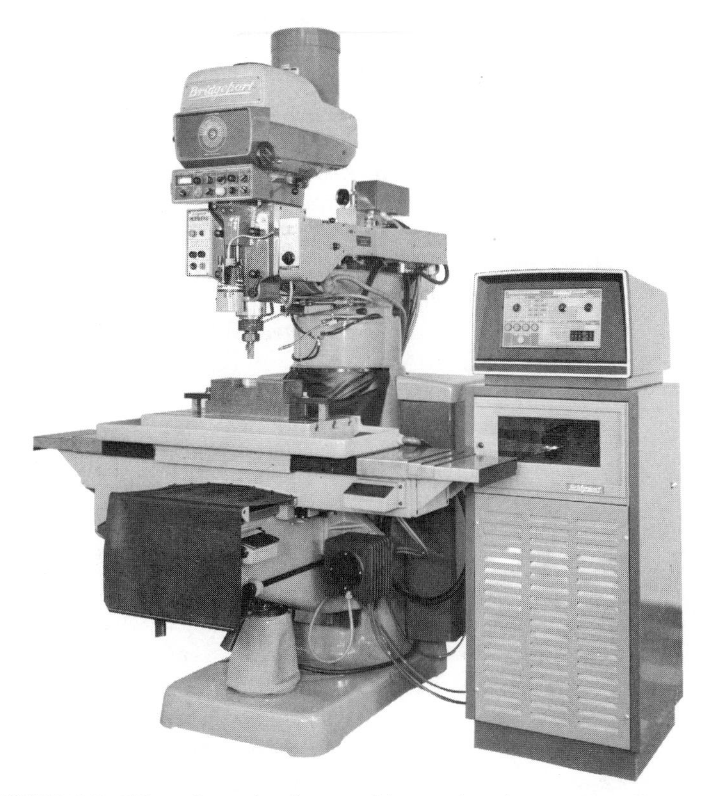


FIGURE 7.2 NC system showing machine tool and controller. (Courtesy Bridgeport Machines Division of Textron Inc.)

## The NC method

To utilize numerical control in manufacturing, the following steps must be accomplished:

1. The engineering drawing of the workpart must be interpreted in terms of manufacturing processes to be used. We assume that a portion of this processing is to be performed on an NC machine.

2. The part programmer plans the process for the portion of the job to be done on NC.

3. A punched tape is prepared from the part programmer's process plan.

4. The tape is checked or verified.

5. The job is produced on an NC machine under tape control.

We will refer to these steps as the NC method or the NC procedure. In today's industrial world, designers conceive of products and document their designs on engineering drawings. There are drawings of the assembled products and drawings of the component parts. A method to economically manufacture the products must then be planned by process engineers (step 1). Depending on the type of component, it may be most practical to perform some of the processing on an NC machine. This is where the part programmer enters the procedure (step 2).

The part programmer is knowledgeable about the machining process and has also been trained to program for numerical control. It is his responsibility to plan the sequence of machining steps to be performed by NC and to write these down in a special format. There are two ways to program for numerical control:

1. Manual part programming.
2. Computer-assisted part programming.

In manual programming, the machining instructions are prepared on a document called a part program manuscript. Basically, the manuscript is a listing of the relative cutting tool/workpiece positions which must be followed in order to machine the workpiece. A punched tape is then prepared (step 3) directly from the part program manuscript.

In computer-assisted part programming, much of the tedious computational work required in manual part programming is transferred to the computer. For complex workpart geometries or jobs with many machining steps, use of the computer results in significant savings in part programming time. The part programmer's work consists typically of two tasks. First, he must define the configuration of the workpart in terms of basic geometric elements: points, lines, planes, circles, and so on. Second, he must direct the cutting tool to perform the machining steps along these geometric elements: drill the holes, mill the flat surfaces, and so on. The part programmer accomplishes these tasks by using a special programming language. The computer then interprets the program and performs the various calculations necessary to prepare the punched tape (step 3) for the NC machine tool.

Chapter 8 will be devoted exclusively to the subject of part programming, both manual and computer-assisted. It seems understandable that the brief explanation of part programming given above may leave the reader with a few unanswered questions. We attempt to answer all those questions in Chapter 8.

After the punched tape has been prepared, a method is usually provided for checking the accuracy of the tape (step 4 in the NC procedure). Sometimes the tape is checked by running it through a computer program which plots the various tool movements (or table movements) on paper. In this way, major errors in the tape can be discovered. The "acid test" of the tape involves trying it out on the machine tool to make the part. Programming errors are not uncommon, and it may require about three attempts before the tape is correct and ready to use.

The final step in the NC procedure (step 5) is to use the NC tape in production. This involves ordering the raw workparts, specifying and preparing the tooling and any special fixturing that may be required, and setting up the NC machine tool for the job. The machine tool operator's function during production is to load the raw workpart in the machine and establish the starting position of the cutting tool relative to the workpiece. The NC system then takes over and machines the part according to the instructions on tape. When the part is completed, the operator removes it from the machine and loads the next part.

It is customary when manufacturing a batch of parts for the quality control inspector to thoroughly check the first part for conformance to the engineering drawing. After the accuracy of this initial part has been verified, production of the remaining workparts in the batch can proceed with relative assurance that the first part will be duplicated each cycle.

## Historical perspective

The development of numerical control owes much to the United States Air Force, which recognized the need to develop more efficient manufacturing methods for modern aircraft. Following World War II, the components used to fabricate jet aircraft became more complex and required more machining. Most of the machining involved milling operations, so the Air Force sponsored a research project at the Massachusetts Institute of Technology to develop a prototype NC milling machine. This prototype was produced by retrofitting a conventional tracer mill with numerical control servomechanisms for the three axes of the machine. In March 1952, the MIT Labs held the first demonstration of the NC machine. The machine tool builders gradually began initiating their own development projects to introduce commercial NC units. Also, certain user industries, especially airframe builders, worked to devise numerical control machines to satisfy their own particular production needs. The Air Force continued its encouragement of NC development by sponsoring additional research at MIT to design a part programming language that could be used for controlling the NC machines. This work resulted in the APT language, which stands for Automatically Programmed Tools. The objective of the APT research was to provide a means by which the part programmer could communicate the machining instructions to the machine tool in simple English-like statements. Although the APT language has been criticized as being too large for many computers, it nevertheless represents a major accomplishment. APT is still widely used in industry today, and most other modern part programming languages are based on APT concepts.

## 7.3 COORDINATE SYSTEM AND MACHINE MOTIONS

In order for the part programmer to plan the sequence of positions and movements of the cutting tool relative to the workpiece, it is necessary to establish a standard axis system by which the relative positions can be specified. Using an NC drill press as an example, the drill spindle is in a fixed vertical position, and the table is moved and controlled relative to the spindle. However, to make things easier for the programmer, we adopt the viewpoint that the workpiece is stationary while the drill bit is moved relative to it. (This is a viewpont that we will reinforce in Chapter 8.) Accordingly, the coordinate system of axes is established with respect to the machine table.

Two axes, $x$ and $y$, are defined in the plane of the table, as shown in Figure 7.3. The $z$-axis is perpendicular to this plane and movement in the $z$ direction is controlled by the vertical motion of the spindle. The positive and negative directions of motion of tool relative to table along these axes are as shown in Figure 7.3. NC drill presses are classified as either two-axis or three-axis machines, depending on whether or not they have the capability to control the $z$-axis.

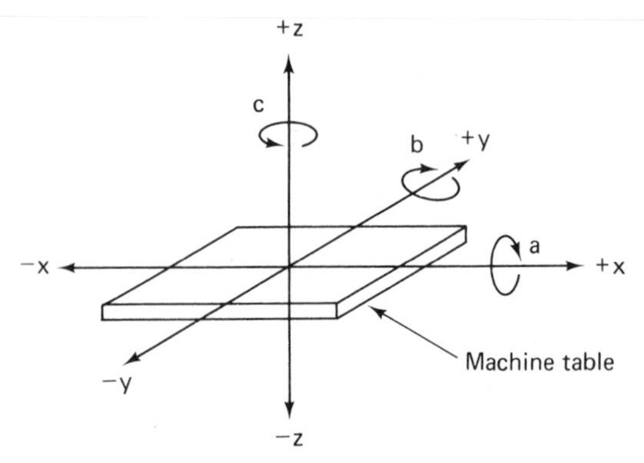


FIGURE 7.3 Machine tool axis system for NC.

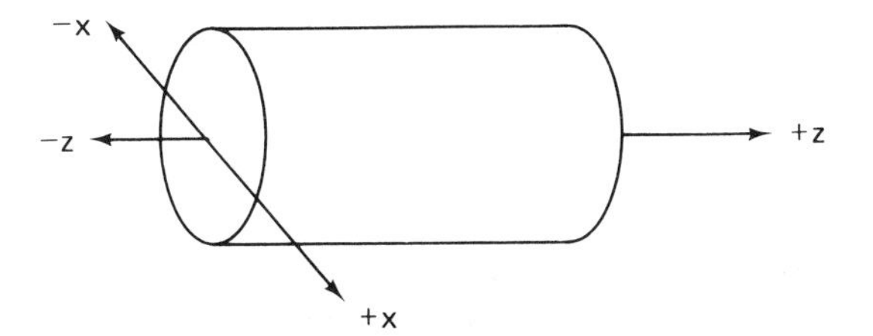


FIGURE 7.4 The *x*- and *z*-axes in NC turning.

A numerical control milling machine and similar machine tools (boring mill, for example) use an axis system similar to that of the drill press. However, in addition to the three linear axes, these machines may possess the capacity to control one or more rotational axes. Three rotational axes are defined in NC: the *a*, *b*, and *c* axes. These axes are used to specify angles about the *x*, *y*, and *z* axes, respectively. To distinguish positive from negative angular motions, the "right-hand rule" can be used. Using the right hand with the thumb pointing in the positive linear axis direction (*x*, *y*, or *z*), the fingers of the hand are curled to point the positive rotational direction. This is illustrated in Figure 7.3.

For turning operations, two axes are normally all that are required to command the movement of the tool relative to the rotating workpiece. The *z*-axis is the axis of rotation of the workpart, and *x*-axis defines the radial location of the cutting tool. This arrangement is illustrated in Figure 7.4.

## Other features of the location system

The purpose of the coordinate system is to provide a means of locating the tool in relation to the workpiece. Depending on the NC machine, the part programmer may have several different options available for specifying this location.

FIXED ZERO VERSUS FLOATING ZERO. The programmer must determine the position of the tool relative to the origin (zero point) of the coordinate system. NC machines have either of two methods for specifying the zero point. The first possibility is for the machine to have a *fixed zero.* In this case, the origin is always located at the same position on the machine table. Usually, that position is the southwest corner (lower left-hand corner) of the table and all tool locations will be defined by positive *x* and *y* coordinates.

The second and more common feature on modern NC machines allows the machine operator to set the zero point at any position on the machine table. This feature is called *floating zero.* The part programmer is the one who decides where the zero point should be located. The decision is based on part programming convenience. For example, the workpart may be symmetrical and the zero point should be established at the center of symmetry. The location of the zero point is communicated to the machine operator. At the beginning of the job, the operator moves the tool under manual control to some "target point" on the table. The target point is some convenient place on the workpiece or table for the operator to position the tool. For example, it might be a predrilled hole in the workpiece. The target point has been referenced to the zero point by the part programmer. In fact, the programmer may have selected the target point as the zero point for tool positioning. When the tool has been positioned at the target point, the machine operator presses a "zero" button on the machine tool console, which tells the machine where the origin is located for subsequent tool movements.

With fixed-zero systems, the part programmer and machine operator must reference the job to the machine's permanent zero point. This is a less convenient arrangement.

ABSOLUTE VERSUS INCREMENTAL POSITIONING. Another option sometimes available to the part programmer is to use either an absolute system of tool positioning or an incremental system. *Absolute positioning* means that the tool locations are always defined in relation to the zero point. If a hole is to be drilled at a spot that is 8 in. above the $x$-axis and 6 in. to the right of the $y$-axis, the coordinate location of the hole would be specified as $x = +6.000$ and $y = +8.000$. By contrast, *incremental positioning* means that the next tool location must be defined with reference to the previous tool location. If in the previous drilling example, the previous hole had been drilled at an absolute position of $x = +4.000$ and $y = +5.000$, the incremental position instructions would be specified as $x = +2.000$ and $y = +3.000$ in order to move the drill to the desired spot. Figure 7.5 illustrates the difference between absolute and incremental positioning.

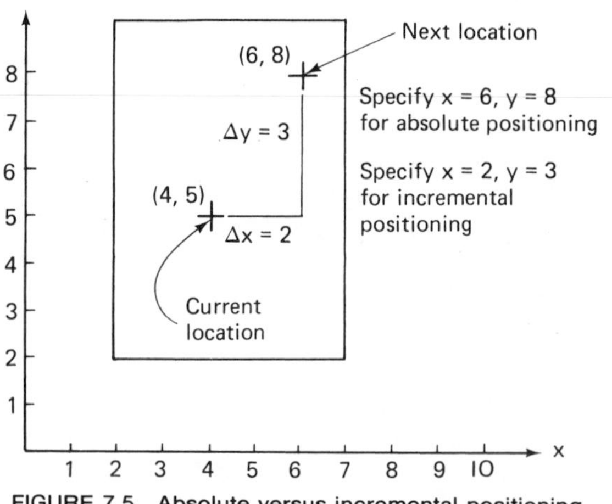


FIGURE 7.5 Absolute versus incremental positioning.

## 7.4 TYPES OF NC SYSTEMS

When classified according to the machine tool control system, there are three basic types of NC systems:

1. Point-to-point.
2. Straight cut.
3. Contouring.

This classification is concerned with the amount of control over the relative motion between the workpiece and cutting tool. The least control is exerted over the tool motion with point-to-point systems. Contouring represents the highest level of control.

### Point-to-point NC

*Point-to-point* (PTP) is also sometimes called a positioning system. In PTP, the objective of the machine tool control system is to move the cutting tool to a predefined location. The speed or path by which this movement is accomplished is not important in point-to-point NC. Once the tool reaches the desired location, the machining operation is performed at that position.

NC drill presses are a good example of PTP systems. The spindle must first be positioned at a particular location on the workpiece. This is done under PTP control. Then, the drilling of the hole is performed at that location, the tool is moved to the next hole location, and so forth. Since no cutting is performed between holes, there is no need for controlling the relative motion of the tool and workpiece between hole locations. On positioning systems, the speeds and feeds used by the machine tool are often controlled by the operator rather than by the NC tape. Figure 7.6 illustrates the point-to-point type of control.

Positioning systems are the simplest machine tool control systems and are therefore the least expensive of the three types. However, for certain processes such as drilling operations and spot welding, PTP is perfectly suited to the task and any higher level of control would be unnecessary.

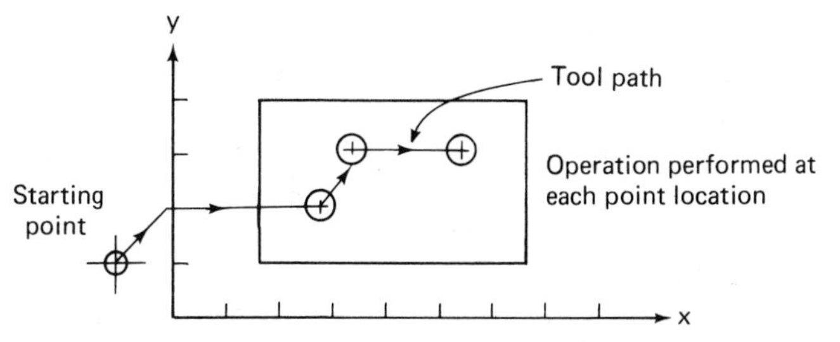


FIGURE 7.6 Point-to-point (positioning) NC operations.

## Straight-cut NC

*Straight-cut* control systems are capable of moving the cutting tool parallel to one of the major axes at a controlled rate suitable for machining. It is therefore appropriate for performing milling operations to fabricate workpieces of rectangular configurations. With this type of NC system it is not possible to combine movements in more than a single axis direction. Therefore, angular cuts on the workpiece would not be possible. An example of a straight-cut operation is shown in Figure 7.7.

An NC machine tool capable of performing straight-cut movements is also capable of point-to-point movements.

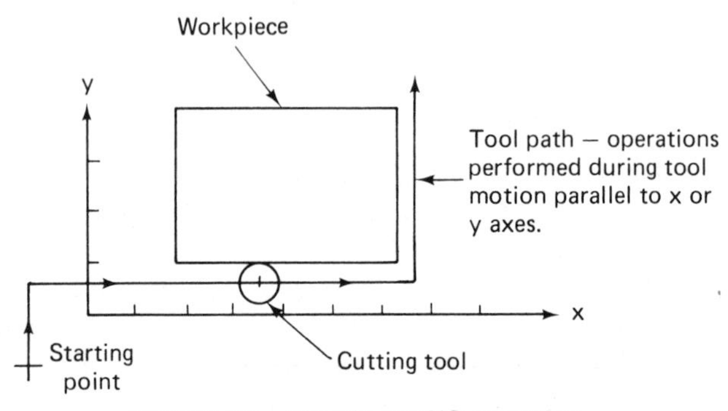


FIGURE 7.7 Straight-cut NC operation.

## Contouring NC

*Contouring* is the most complex, the most flexible, and the most expensive type of machine tool control. It is capable of performing both PTP and straight-cut operations. In addition, the distinguishing feature of contouring NC systems is their capacity for simultaneous control of more than one axis movement of the machine tool. The path of the cutter is continuously controlled to generate the desired geometry of the workpiece. For this reason, contouring systems are also called continuous-path NC systems. Straight or plane surfaces at any orientation, circular paths, conical shapes, or most any other mathematically definable form are possible under contouring control. Figures 7.8 and 7.9 illustrate the versatility of continuous-path NC. Milling and turning operations are common examples of the use of contouring control.

For the mathematically oriented reader, it might be useful to distinguish between PTP, straight-cut, and contouring in the following way. Consider a two-axis control system, where the table is moved in the $xy$ plane. With point-to-point systems, control is achieved over the $x$ and $y$ coordinates. With straight-cut systems, control is provided for either $dx/dt$ or $dy/dt$, but only one at a time. With contouring systems, both of the rates $dx/dt$ and $dy/dt$ can be controlled simultaneously. In order to cut a straight path at some

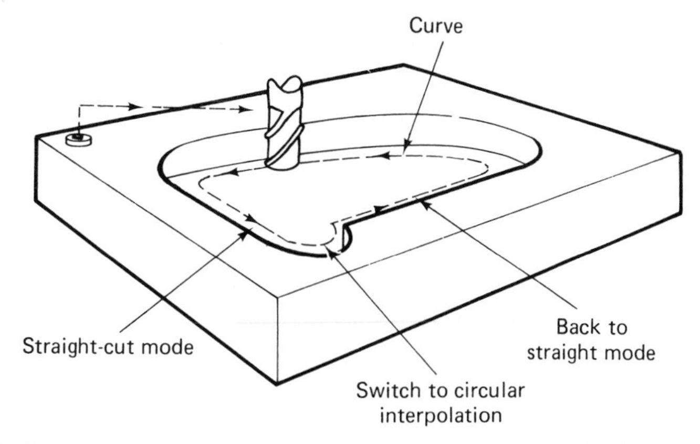


FIGURE 7.8 Two-dimensional NC contouring. (Reprinted from Olesten [6].)

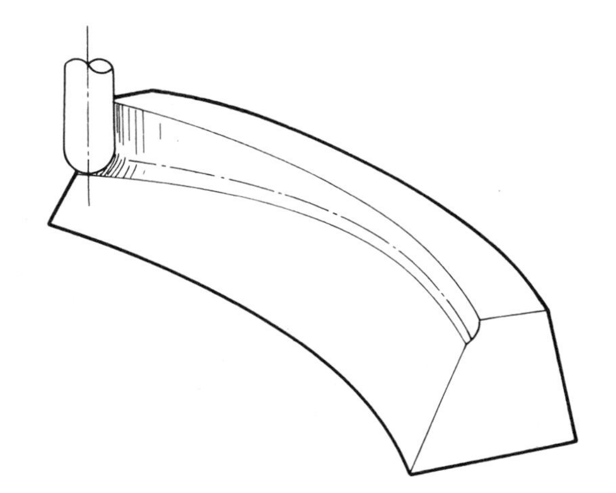

FIGURE 7.9 Three-dimensional NC contouring. (Reprinted from Olesten [6].)

angle, the relative values of $dy/dt$ and $dx/dt$ must be maintained in proportion to the tangent of the angle. In order to machine along a curved path, the values of $dy/dt$ and $dx/dt$ must continually be changed so as to follow the path. This is accomplished by breaking the curved path into very short straight-line segments that approximate the curve. Then, the tool is commanded to machine each segment in succession. What results is a machined outline that closely approaches the desired shape. The maximum error between the two can be controlled by the length of the individual line segments as illustrated in Figure 7.10.

The reader can easily imagine the complexity involved when more than two axes movements must be controlled by a contouring system. Some NC machine tools possess the capability to simultaneously control five axes to achieve the desired machined surface.

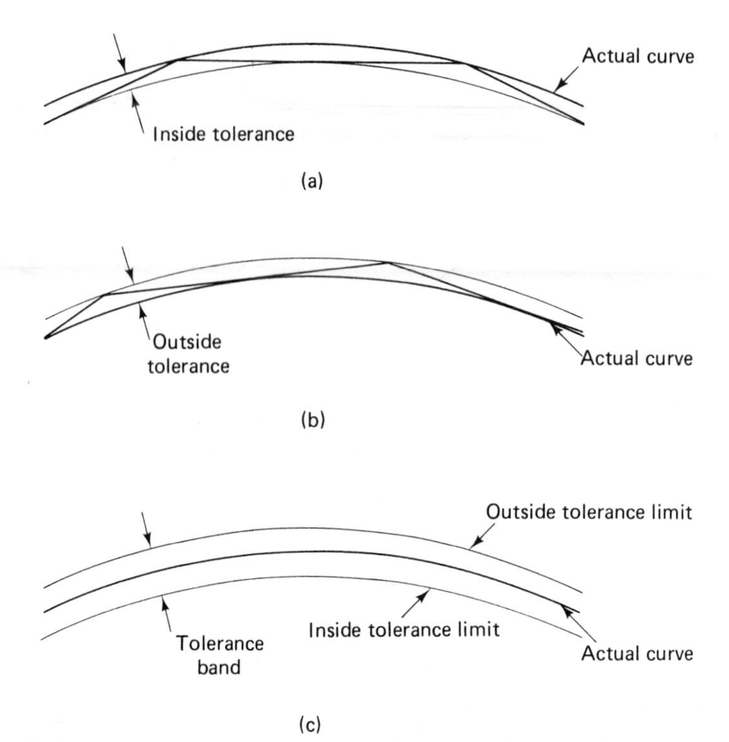


FIGURE 7.10 Approximation of a curved path in NC by a series of straight-line segments. The accuracy of the approximation is controlled by the "tolerance" between the actual curve and the maximum deviation of the straight-line segments. In (a), the tolerance is defined on the inside of the curve. It is also possible to define the tolerance on the outside of the curve, as in (b). Finally, the tolerance can be specified on both inside and outside, as shown in (c).

## 7.5 APPLICATIONS OF NUMERICAL CONTROL

As indicated in Section 7.2, NC has been used in a variety of applications. A complete listing of the applications is probably not possible, but such a list would have to include the following:

Metal cutting machine tools.
Pressworking machine tools.
Welding machines.
Inspection machines.
Automatic drafting.
Assembly machines.
Tube bending.

Flame cutting.
Industrial robots (not in all cases).
Automated knitting machines.
Cloth cutting.
Automatic riveting.

By far, the most common application of numerical control is for metal-cutting machine tools. Within this category alone, NC equipment has been built to perform virtually the entire range of material removal processes, including:

Milling.
Drilling and related operations.
Boring.
Turning.
Grinding.
Sawing.

The versatility that can be acquired by combining NC with metal-cutting machine tools is best represented by the *NC Machining Center*. The machining center incorporates several time-saving features into one machine tool to

FIGURE 7.11 NC machining center uses computer in control unit. (Courtesy Kearney & Trecker Corp.)

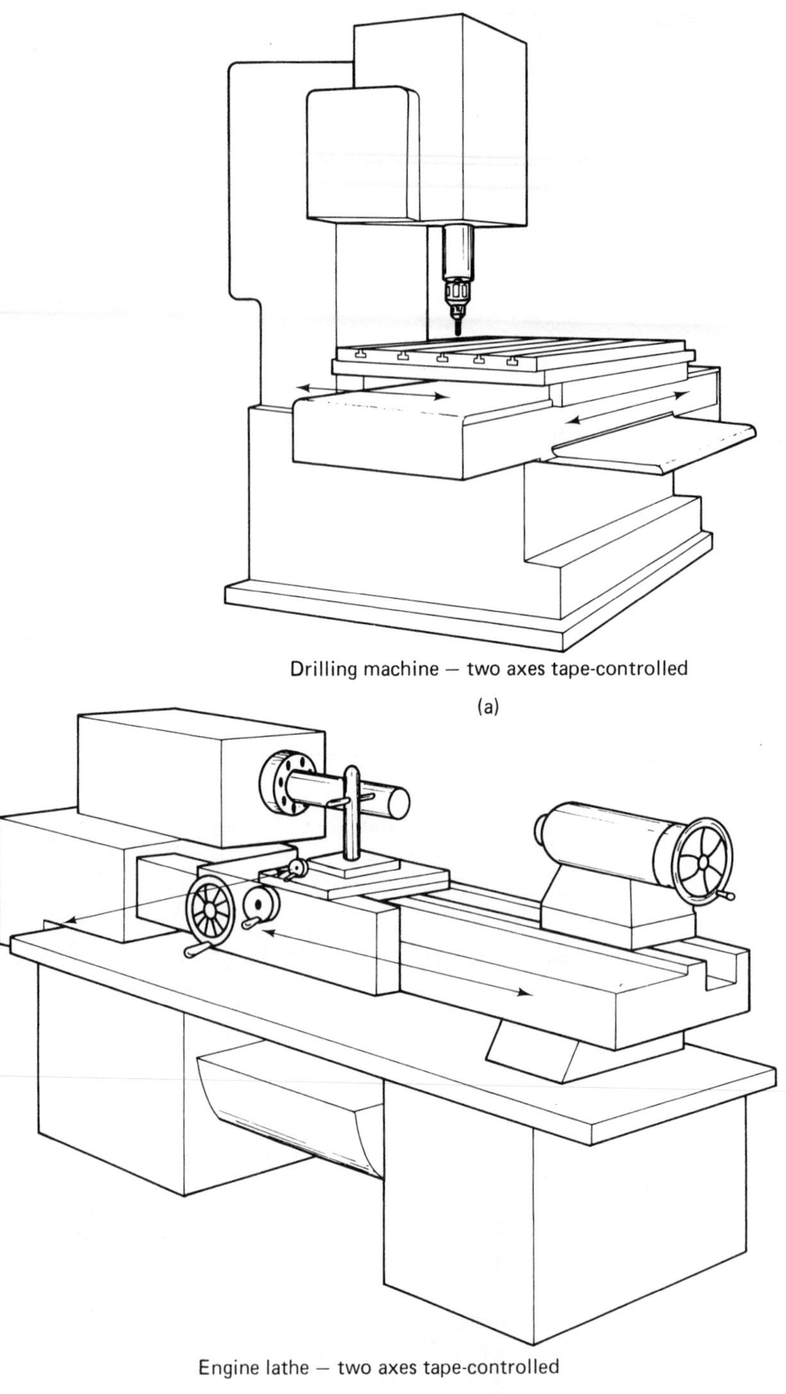

Drilling machine — two axes tape-controlled

(a)

Engine lathe — two axes tape-controlled

(b)

FIGURE 7.12 (a) NC drilling, (b) NC turning. (Reprinted from *NC Handbook*, 2nd ed., Bendix Corporation, Detroit, Mich., 1969.)

achieve a high level of automated yet flexible production. First, a machining center is capable of performing a variety of different operations: drilling, tapping, reaming, milling, and boring. Second, it has the capacity to change tools automatically under tape command. A variety of machining operations means that a variety of cutting tools is required. The tools are kept in a tool drum or other holding device. When the tape calls a particular tool, the drum rotates to position the tool for insertion into the spindle. The automatic tool changer then grasps the tool and places it into the spindle chuck. At the termination of the current operation, the tool changer makes an exchange with the tool for the next operation. A third capability of the NC machining center is workpiece positioning. The machine table can orient the job so that it can be machined on several surfaces, as required. Finally, the fourth feature of many machining centers is that it has virtually two tables on which the workpiece can be set up. While machining is being performed on one workpart, the operator can be unloading the previously completed piece, and loading the next one. This improves machine tool utilization because the machine does not have to stand idle during loading and unloading of the workparts. An example of an NC machining center is shown in Figure 7.11.

Other numerical control machines and applications are illustrated in Figures 7.12 through 7.14.

FIGURE 7.13 NC milling machine with computer control. (Courtesy Kearney & Trecker Corp.)

FIGURE 7.14 NC traveling column machining center shown machining a table base (left) and upright (right) for an NC machining center. This machine tool is, in effect, reproducing itself. (Courtesy Kearney & Trecker Corp.)

## 7.6 THE ECONOMICS OF NUMERICAL CONTROL

There are a number of reasons why NC systems are being adopted so widely by the metalworking industry. It has been estimated that 75% of manufacturing is carried out in lot sizes of 50 or less. As indicated above, these small lot sizes are the typical applications for NC. Following are the advantages of numerical control when it is utilized in these small production quantities:

1. *Reduced nonproduction time.* Numerical control cannot change the basic metal cutting process, but it can increase the proportion of time the machine is engaged in cutting metal. It accomplishes this decrease in nonproductive time by means of fewer setups, less setup time, reduced workpiece handling time, automatic tool changes on some machines, and so on.

2. *Reduced fixturing.* NC requires simpler fixtures because the positioning is done by the NC tape rather than the fixture or jig.

3. *Reduced lead time.* Jobs can be set up more quickly with NC.

4. *Greater manufacturing flexibility.* NC adapts better to changes in jobs, production schedules, and so on.

5. *It is easier to accommodate engineering design changes on the workpiece with NC.* Instead of making alterations in a complex fixture, the tape can be altered.

6. *Improved accuracy and reduced human error.* NC is ideal for complicated parts where the chances of human mistakes are high.

## Where is NC most appropriate?

It is clear from the advantages listed above that NC is appropriate only for certain parts, not all parts. The general characteristics of jobs for which NC is most appropriate are the following:

1. Parts are processed frequently and in small lot sizes.
2. Complex part geometry.
3. Close tolerances must be held on the workpart.
4. Many operations must be performed on the part in its processing.
5. Much metal needs to be removed.
6. Engineering design changes are likely.
7. It is an expensive part where mistakes in processing would be costly.
8. Parts requiring 100% inspection.

In order to justify that a job be processed by NC, it is not necessary that the job possess every one of these attributes. However, the more of these characteristics that are present, the more likely that the part is a good application for numerical control. Unfortunately, there are many situations in which it is not clear whether NC or conventional processing should be used.

## Deciding between NC and conventional machining[1]

Both numerical control and conventional machining have their relative advantages and disadvantages. However, it is not always obvious whether a particular part should be processed by one method or the other. To utilize the potential benefits of numerical control, only those parts that are appropriate for NC must be processed on it. Currently, there is no universally accepted procedure for deciding on parts to be processed by numerical control. The decision is usually based on the experience of the process planner and the facilities available within the machine shop. When the choice between NC and conventional machine tools is not clear, alternative process plans must be

[1]This section is based on Groover et al. [4].

developed for both methods. This requries time and expense and does not guarantee that a correct decision will be made. It would be highly desirable if the decision on NC could be made without the necessity of designing several alternative process plans.

A study was conducted in several of the machine shops of the Ingersoll Rand Co. The objective of this study was to develop a standard scoring system and decision table for determining whether to process a part on a numerical control or on conventional equipment. In the use of the scoring system, an analysis is made of the physical characteristics shown on the part print and of other known information about the part, such as lot size, lots/year, and so on. These factors are assigned weighting values. The assigned weighting factors are then summed up, and a decision is made whether to produce the part by NC or by conventional methods, depending on the sum of the factor weights. The procedure consists of two steps. The first step is concerned with a conditional decision. If numerical control were to be used, what type of NC machine would it be? The second step involves a comparison of the relative merits of using conventional processing methods against the particular NC machine selected in the first step. This comparison involves the use of 22 process planning factors that might influence the decision on processing method. Each factor is assigned a value which depends on the part characteristics. The values are added and if the sum exceeds a certain threshold value, NC should be used; otherwise, conventional machines should be used.

Figure 7.15 illustrates a sample worksheet that was developed to carry out the two-step decision procedure. The first step uses a decision table to select the appropriate type of NC machine tool for the part in question. There were three basic types available in the machine shop of Ingersoll Rand's Rock Drill Division, where the study was performed. These are: NC Turning Center, NC Turret Drill, and NC Machining Center. Depending on the combination of "yes" or "no" answers to the four conditions given in the top half of the decision table, the choice of equipment is made accordingly.

The second step (decision making) requires each of the 22 weighting factors to be evaluated. The range of possible values is illustrated in Figure 7.15 for each of the factors. The analysis required to obtain these values is not presented here but can be found in reference [4]. When all 22 factors have been assigned weighting values, they are summed. If the sum is greater than 110, this indicates that NC should be used for the particular workpart. A sum less than 110 indicates that conventional processing should be used.

At the time of this writing, the procedure is being employed on a trial basis in the Rock Drill Division. In evaluations of the procedure on a number of test cases with several different process planners using the procedure, the scoring system yielded the same decisions as were obtained using the longer traditional procedure. However, the decisions were reached with reduced time and effort.

Part name: ______________________ Part no. ______________________

Drg. no. ______________________ Date ____________ Analyst ____________

Step 1: NC equipment selection

Codes:

Yes = Y
No = N
* = Action
S = Total

| Conditions | 1 | 2 | 3 | 4 | 5 | 6 |
|---|---|---|---|---|---|---|
| Lathe operation(s) | Y | N | N | N | Y | Y |
| Required pallets | N | Y | N | N | Y | N |
| Milling operation | N | – | – | – | – | Y |
| Extensive tooling | – | – | N | Y | – | – |
| Lathe NC turning center | * | | | | | |
| NC turret drill | | | * | | | |
| NC machining center | | * | | * | | |
| Error | | | | | * | * |

Step 2: Decision-making

| Process criteria | | Weighting factors | | | | | | | | | | | Assigned weights |
|---|---|---|---|---|---|---|---|---|---|---|---|---|---|
| | | 0 | 1 | 2 | 3 | 4 | 5 | 6 | 7 | 8 | 9 | 10 | |
| Cycle index | | | | | | | | | | | | | |
| Fixture ratio | | | | | | | | | | | | | |
| Lot size | | | | | | | | | | | | | |
| Lots per year | | | | | | | | | | | | | |
| Complexity | no. of holes/bores/dia.'s | | | | | | | | | | | | |
| | angular holes/surfaces | | | | | | | | | | | | |
| | positioning system | | | | | | | | | | | | |
| | tolerance | | | | | | | | | | | | |
| | over 3-controlled axes | | | | | | | | | | | | |
| | math. defined surfaces | | | | | | | | | | | | |
| Machinability | contouring | | | | | | | | | | | | |
| | overall size | | | | | | | | | | | | |
| | overall weight | | | | | | | | | | | | |
| | surface finish | | | | | | | | | | | | |
| | expensive part | | | | | | | | | | | | |
| | no. of surfaces | | | | | | | | | | | | |
| Changes | engineering | | | | | | | | | | | | |
| | method | | | | | | | | | | | | |
| Family of parts | | | | | | | | | | | | | |
| 100% inspection | | | | | | | | | | | | | |
| Special skills required | | | | | | | | | | | | | |
| Lead time | | | | | | | | | | | | | |
| | | | | | | | | | | | | Total | |

Rule 1: $50 \leqslant S < 110$ . . . . Use conventional processing

Rule 2: $110 < S \leqslant 170$ . . . Use NC processing

Decision: . . . . . . . . . . . . . .

FIGURE 7.15 Part process analysis worksheet to decide between NC and conventional processing for a given workpart.

Proposed Equipment: ______________________________

I. Required Investment

| | |
|---|---|
| 1. Installed cost of proposed equipment | $ ________ |
| 2. Disposal value of any equipment retired | ________ |
| 3. Capital required in absence of proposed equipment | ________ |
| 4. Total investment released or avoided (2 + 3) | ________ |
| 5. Net investment required (1 − 4) | ________ |

II. Effect of Investment on Operating Costs

| | A. Savings | B. Losses |
|---|---|---|
| 6. Direct machine operator labor | ____ | ____ |
| 7. Programming costs | ____ | ____ |
| 8. Inspection costs | ____ | ____ |
| 9. Indirect labor | ____ | ____ |
| 10. Fringe benefits | ____ | ____ |
| 11. Tooling (expendable tools) | ____ | ____ |
| 12. Tool setting | ____ | ____ |
| 13. Fixtures | ____ | ____ |
| 14. Supplies (cutting fluids, etc.) | ____ | ____ |
| 15. Maintenance | ____ | ____ |
| 16. Scrap and rework | ____ | ____ |
| 17. Downtime (excluding maintenance) | ____ | ____ |
| 18. Power | ____ | ____ |
| 19. Floor space released or added | ____ | ____ |
| 20. Property taxes, insurance | ____ | ____ |
| 21. Workpart handling | ____ | ____ |
| 22. Inventory (storage, insurance, etc.) | ____ | ____ |
| 23. Safety – estimate if applicable | ____ | ____ |
| 24. Flexibility – estimate a value | ____ | ____ |
| 25. Other factors specific to proposal | ____ | ____ |
| 26. Totals (sum 6 through 25) | A. ____ | B. ____ |
| 27. Net gain (26A-26B) | ____ | $ ____ |

III. Effect of Investment on Revenues

| | A. Gains | B. Losses |
|---|---|---|
| 28. From change in product quality | ____ | ____ |
| 29. From change in volume of output | ____ | ____ |
| 30. Totals (28 + 29) | A. ____ | B. ____ |
| 31. Net gain (30A − 30B) | $ ____ | |

IV. Other Factors Related to Proposed Equipment

32. Estimated service life = ________ years
33. Estimated salvage value $ ________
34. Depreciation method: ______________________________
35. Rate of return = ________ %

FIGURE 7.16 Typical NC machine economic analysis form.

## Economic analysis of an NC machine investment

The previous discussion was concerned with the situation in which both conventional machines and NC machines are available in the factory to process a particular job. The analysis was intended to answer the question: Which type of machine is appropriate for the job?

Another question that must be answered is whether or not to invest in a particular piece of NC equipment. The investment in a numerical control machine tool must be justified using the same criteria as any other capital expenditure. The investment must satisfy the firm's rate-of-return objectives.

The generally accepted methods for performing an economic analysis of an investment proposal were considered in Chapter 3. Although individual companies have their own ways of evaluating a capital equipment proposal, the typical factors that most firms would include in an NC evaluation are shown in Figure 7.16. The "NC Machine Economic Analysis Form" illustrated in Figure 7.16 was compiled from references [2], [3], and [6]. Many of the entries in the analysis form would be difficult to estimate. If reliable estimates cannot be obtained, the entries should be considered as intangible factors. Subjective judgment would have to be used to assess the importance of the factor in deciding on the investment.

# REFERENCES


[1] Childs, J. J., *Numerical Control Part Programming,* Industrial Press, Inc., New York, 1973.

[2] Dallas, D. B. (Editor), *Tool and Manufacturing Engineers Handbook,* 3rd ed., Society of Manufacturing Engineers (McGraw-Hill Book Company, New York), 1976, Chapter 12.

[3] DeVries, M. F., "Two Case Studies of the Investment Justification for N/C Machinery Using the MAPI Method," *Educational Module,* Manufacturing Productivity Educational Committee, Purdue University, 1977.

[4] Groover, M. P., Onitiri, T., and Marshall, C., "A Point Scoring System for Decisions on Numerical Control Part Processing," *Paper MS77-295,* Society of Manufacturing Engineers, Dearborn, Mich., 1977.

[5] Howe, R. E. (Editor), *Introduction to Numerical Control in Manufacturing,* Society of Manufacturing Engineers, Dearborn, Mich., 1969.

[6] Olesten, N. O., *Numerical Control,* Wiley–Interscience, New York, 1970.

[7] Pressman, R. S., and Williams, J. E., *Numerical Control and Computer-Aided Manufacturing,* John Wiley & Sons, Inc., New York, 1977.

[8] "A Look at NC Today," *Manufacturing Engineering,* October, 1975, pp. 24–28.

# chapter 8 NC Part Programming

## 8.1 THE PART PROGRAMMING PROCESS

Part programming involves the planning and specification of the sequence of processing steps to be performed on the NC machine. It also involves, although less directly, the preparation of the punched tape (or other input medium) by which the processing instructions are communicated to the machine. As indicated in Section 7.2, there are two methods by which this programming is accomplished: manual part programming and computer-assisted part programming. In this chapter, we will consider both methods. However, there are really three issues:

1. How is the punched tape coded so that it can be read and interpreted by the NC system?
2. How is the punched tape prepared using manual part programming?
3. How is the punched tape prepared using computer-assisted part programming?

To appreciate the part programming process, the reader must possess an understanding of the first issue. The coding of the punched tape is concerned with the basic symbols used to communicate a complex set of instructions to the NC machine tool. In numerical control, the punched tape must be generated whether the part programming is done manually or with the assistance of some computer package. With either method of part programming, the tape is the net result of the programming effort. In Section 8.2, our attention will be focused on the punched tape and the structure of the basic language used by the NC system.

Once the reader has seen how the punched tape is coded in NC, we can begin on the principal topic of this chapter: Part programming. Section 8.3 covers manual programming, and Sections 8.4 and 8.5 deal with computer-assisted part programming. The scope and purpose of this book does not permit a complete treatment of the subject. Entire books have been written on part programming, and we direct the interested reader to references [1] and [4] for further study. Our purpose is to develop an appreciation and basic understanding of the part programming process.

## 8.2 PUNCHED TAPE AND TAPE FORMAT

The part program is converted into a sequence of machine tool actions by means of the input medium, which contains the program, and the controller unit, which interprets the input medium. The controller unit and the input medium must be compatible. That is, the input medium uses coded symbols which represent the part program, and the controller unit must be capable of reading those symbols. The most common input medium is the punched tape. The tape has been standardized so that tape punchers are manufactured to prepare the NC tapes, and tape readers (part of the controller unit) can be manufactured to read the tapes. Without this standardization, there would be a hopeless plethora of different systems of tape preparation and use. Surely, numerical control machines would not have received their wide acceptance of today unless the machine tool manufacturers of yesterday had agreed upon industry standards. Our discussion of input media for numerical control will be restricted to the punched tape.

### The NC tape

The punched tape used for NC is 1 in. wide. It is standardized as shown in Figure 8.1 by the Electronics Industries Association (EIA), which has been responsible for many of the important standards in the NC industry. The tape can be made out of several materials. Paper tape is common. Although its

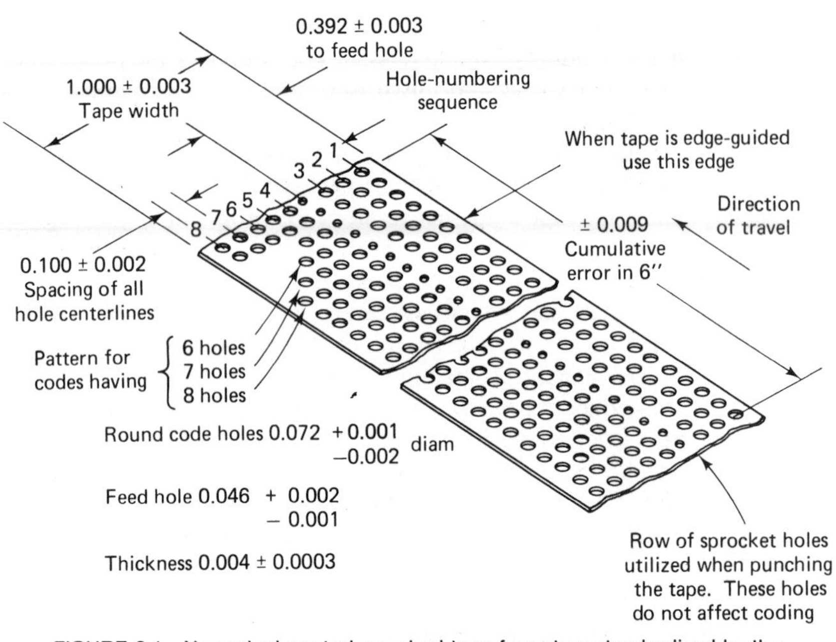


FIGURE 8.1 Numerical control punched tape format as standardized by the Electronics Industries Association (EIA). (Reprinted from Childs [1].)

cost is low, it is not durable and therefore not appropriate for repeated use. Stronger tape materials suitable for higher production use include Mylar-reinforced paper, Mylar-coated aluminum, and certain plastics. Paper is often used for the initial preparation and testing of the part program. Then, a production tape is duplicated out of one of the more durable materials for shop floor use. The punched paper tape is retained as the master copy.

In manual part programming, the tape is prepared on a typewriter-like machine, the most common example of which is the Friden Flexowriter. This device, in addition to its typewriter keyboard, also has a mechanism for perforating blank tapes and interpreting punched tapes. In the use of the Flexowriter, the operator types directly from the part programmer's manuscript. This produces both a typed version of the program and a punched tape.

In computer-assisted part programming, the punched tape is prepared from the computer output by a device called a "tape punch."

By either method of preparation, the tape is ready for use (assuming that testing has verified its accuracy). During production, the tape is fed through the tape reader once for each workpiece. It is advanced through the tape reader one instruction at a time. While the machine tool is performing one instruction, the next instruction is being read into the controller unit's data buffer. The metal-cutting function and the tape-reading function occur simultaneously. This makes the operation of the NC system more efficient. After the last instruction has been read into the controller, the tape is rewound back to the start of the program to be ready for the next workpart.

## Tape coding

As shown in Figure 8.1, there are eight regular columns of holes running in the lengthwise direction of the tape. There is also a ninth column of holes between the third and fourth regular columns. However, these are smaller and are used as sprocket holes for feeding the tape.

Figure 8.1 shows a hole present in nearly every position of the tape. However, the coding of the tape is provided by either the presence or absence of a hole in the various positions. Because there are two possible conditions for each position—either the presence or absence of a hole—this coding system is called the binary code. It uses the base 2 number system, which can represent any number in the more familiar base 10 or decimal system.

In the binary system, there are only two numbers, 0 and 1. The meaning of successive digits in the binary system is based on the number 2 raised to successive powers. The first digit is $2^0$, the second digit is $2^1$, the third is $2^2$, and so forth. The value of $2^0$ is 1, $2^1=2$, $2^2=4$, $2^3=8$, and so on. The two numbers, 0 or 1, in the successive digit positions indicate either the presence or absence of the value. Table 8.1 shows how the binary system is used to represent numbers in the decimal system.

For example, the decimal number 5 is represented in the binary system by 0101. The conversion from binary to decimal systems makes use of the following type of computation:

$$1\times2^0+0\times2^1+1\times2^2+0\times2^3=1\times1+0\times2+1\times4+0\times8=5$$

The reader can see from Table 8.1 that four digits are required in the binary system to represent any of the single-digit numbers in the decimal system. Yet there are eight regular columns of holes in the standard NC

TABLE 8.1 Comparison of Binary and Decimal Number Systems

| *Binary* | *Decimal* |
|---|---|
| 0000 | 0 |
| 0001 | 1 |
| 0010 | 2 |
| 0011 | 3 |
| 0100 | 4 |
| 0101 | 5 |
| 0110 | 6 |
| 0111 | 7 |
| 1000 | 8 |
| 1001 | 9 |

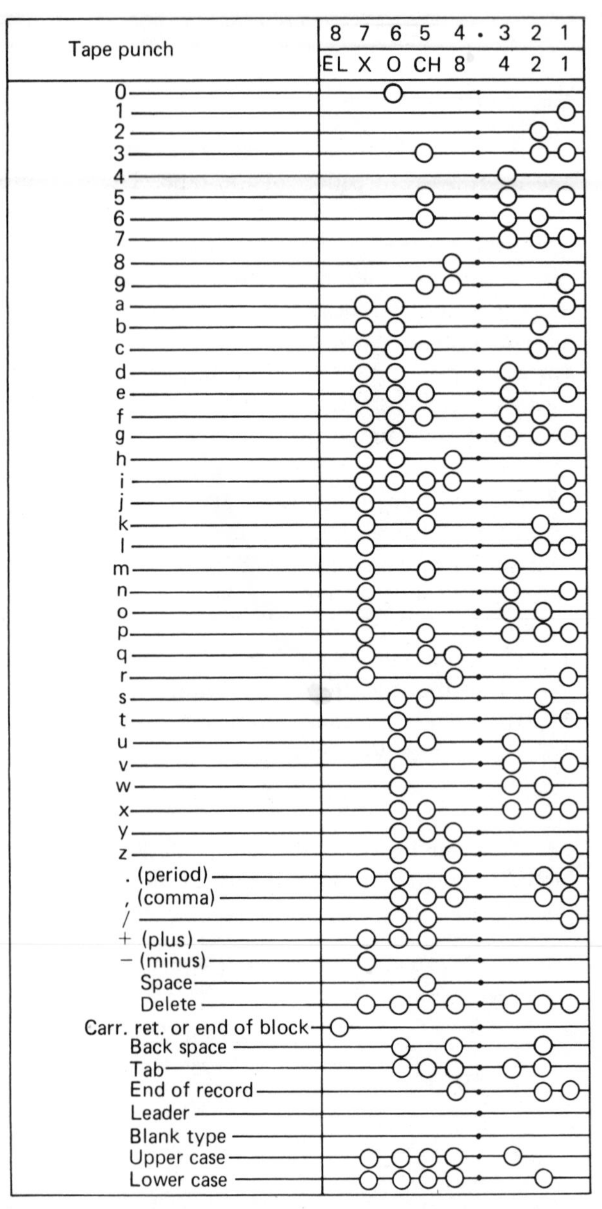


FIGURE 8.2 Standard EIA tape coding for NC. (Reprinted from Howe [3].)

punched tape. The reason that eight columns are needed on the tape is because there are other symbols that must be coded on the tape besides numbers. Alphabetical letters, plus and minus signs, and other symbols are also needed in NC tape coding. The standard EIA tape coding is shown in Figure 8.2. Eight columns provide more than enough binary digits to define any of the required symbols. In fact, the fifth-column position is used exclusively as a check, called *parity*, on the correctness of the tape. The way the parity check works is this: The NC tape reader is designed to read an odd number of holes across the width of the tape. Whenever the particular number or symbol being punched calls for an even number of holes, an extra hole is punched in column 5, hence making the total an odd number. For example, the decimal number 5 uses a punched hole in columns 1 and 3, an even number of holes. Therefore, a parity hole would be added. The decimal number 7 already uses an odd number of holes (columns 1, 2, and 3), so a parity hole is not needed. The parity check helps to assure that the tape punch mechanism on the Flexowriter has perforated a complete hole in all required positions. If the tape reader counts an even number of holes in the tape, it would signal the operator that a parity error had occurred.

## How instructions are formed

A binary digit is called a *bit*. It has a value of 0 or 1 depending on the absence or presence of a hole in a certain row and column position on the tape. (Columns of hole positions run lengthwise along the tape. Row positions run across the tape.) Out of a row of bits, a *character* is made. A character is a combination of bits, which represents a letter, number, or other symbol. A *word* is a collection of characters used to form part of an instruction. Typical NC words are $x$-position, $y$-position, cutting speed, and so on. Out of a collection of words, a *block* is formed. A block of words is a complete NC instruction. Using an NC drilling operation as an example, a block might contain information on the $x$ and $y$ coordinates of the hole location, the speed and feed at which the cut should be run, and perhaps even a specification of the cutting tool.

To separate blocks, an end-of-block (EOB) symbol is used (in the EIA standard, this is a hole in column 8). The tape reader feeds the data from the tape into the buffer in blocks. That is, it reads in a complete instruction at a time.

## NC words

Following is a list of the different types of words used in the formation of a block. Not every NC machine uses all the words. Also, the manner in which the words are expressed will differ between machines. By convention, the words in a block are given in the order below:

SEQUENCE NUMBER (n-WORDS). This is used to identify the block.

PREPARATORY WORK (g-WORDS). This word is used to prepare the controller for instructions that are to follow. For example, the word g02 is used to prepare the NC controller unit for circular interpolation along an arc in the clockwise direction. The preparatory word is needed so that the controller can correctly interpret the data that follow it in the block. Some typical examples of g-words are given in Table 8.2.

TABLE 8.2 Some Common g-Words

| *Code* | *Preparatory function* |
|---|---|
| g00 | Used with contouring systems to prepare for a point-to-point operation. |
| g01 | Linear interpolation in contouring systems. |
| g02 | Circular interpolation, clockwise. |
| g03 | Circular interpolation, counterclockwise |
| g04 | Dwell. |
| g05 | Hold until operator restarts. |
| g08 | Acceleration code, causes machine to accelerate smoothly. |
| g09 | Deceleration code, causes machine to decelerate smoothly. |
| etc. | |

COORDINATES. (x-, y-, AND z-WORDS). These give the coordinate positions of the tool. In a two-axis system, only two of the words would be used. In a four- or five-axis machine, additional a-words and/or b-words would specify the angular positions.

Although different NC systems use different formats for expressing a coordinate, we will adopt the convention of expressing it in the familiar decimal form: For example, x-7.235 or y-0.500. Some formats do not use the decimal point in writing the coordinate. The + sign to define a positive coordinate location is optional. The negative sign is, of course, mandatory.

FEED RATE (f-WORD). This specifies the feed in a machining operation. Units are inches per minute (ipm) by convention.

CUTTING SPEED (s-WORD). This specifies the cutting speed of the process, the rate at which the spindle rotates. Units are revolutions per minute (rpm). In a machining operation it is usually desirable for the tool engineer to specify the speed in terms of the relative surface speed of the tool and work. The units would be feet per minute (fpm). It is therefore necessary for the part programmer to make the conversion from fpm to rpm. The conversion formula is

$$S = \frac{12V}{\pi D} \tag{8.1}$$

where $S$ = spindle speed, rpm
$V$ = surface speed, fpm
$D$ = drill or milling cutter diameter, in.; for turning, $D$ would be the diameter of the workpiece

For example, if the tool engineer called for a $\frac{1}{2}$-in.-diameter high-speed steel drill to be operated at 100 fpm, the corresponding spindle speed would be $S = 764$ rpm. Similarly, a feed rate expressed in inches per revolution (ipr) would have to be converted into inches per minute. The conversion formula is

$$f = Sf_r \tag{8.2}$$

where $f$ = feed rate, ipm
$S$ = spindle speed, rpm
$f_r$ = feed, ipr

TOOL SELECTION (t-WORD). This word would only be needed for machines with a tool turret or automatic tool changer. The t-word specifies which tool is to be used in the operation. For example, t05 might be the designation of a $\frac{1}{2}$-in. drill bit in turret position 5 on an NC turret drill.

MISCELLANEOUS FUNCTION (m-WORD). The m-word is used to specify certain miscellaneous or auxiliary functions which may be available on the machine tool. Of course, the machine must possess the function that is being called. A partial but representative list of miscellaneous functions is given in Table 8.3.

The miscellaneous function is the last word in the block. To identify the end of the instruction, an end-of-block (EOB) symbol is punched on the tape.

TABLE 8.3 Some Typical m-Words

| *Code* | *Miscellaneous function* |
|---|---|
| m00 | Stops the machine; operator must restart. |
| m03 | Start spindle in clockwise direction. |
| m04 | Start spindle in counterclockwise direction. |
| m05 | Stop spindle. |
| m06 | Execute tool change, either automatically or manually. Does not include selection of tool, which is done by t-word or by the operator. If operator changes tool, he must restart machine. |
| m07 | Turn coolant on. |
| m09 | Turn coolant off. |
| m13 | Start spindle in clockwise direction and turn coolant on. |
| m14 | Start spindle in counterclockwise direction and turn coolant on. |
| m30 | End-of-tape command, which tells tape reader to rewind the tape. |

## Tape formats

The organization of words within blocks is called the *tape format*. Three tape formats seem to enjoy the most widespread use:

1. Word address format.
2. Tab sequential format.
3. Fixed block format.

The tape format refers to the method of writing the words in a block of instruction. Within each format there are variations because of differences in machining process, type of machine, features of the machine tool, and so on.

WORD ADDRESS FORMAT. In this format, a letter precedes each word and is used to identify the word type and to address the data to a particular location in the controller unit. The x-prefix identifies an $x$-coordinate word, an s-prefix identifies spindle speed, and so forth. The standard sequence of words for a two-axis NC system would be:

n-word
g-word
x-word
y-word
f-word
s-word
t-word
m-word
EOB

However, since the type of word is designated by the prefix letter, the words can be presented in any sequence. Also, if a word remains unchanged from the previous block or is not needed, it can be deleted from the block.

The word address system seems most common with contouring systems. However, in Example 8.1 we will demonstrate this type format using point-to-point data.

TAB SEQUENTIAL FORMAT. This tape format derives its name from the fact that words are listed in a fixed sequence and separated by depressing the tab key (TAB) when typing the manuscript on a Flexowriter. The TAB symbol in the EIA standard is coded as 01111100 (holes in columns 2 through 6) on the tape. Since the words are written in a set order, no address letter is required. The order of words within the block follows the previously mentioned standard. If a word remains the same as in the previous block, it need not be retyped. However, the TAB code is required to maintain the sequence of the words.

Example 8.1 will illustrate the tab sequential format for point-to-point data. PTP tape preparation most commonly uses this tape format.

Fixed Block Format. This is the least flexible and probably the least desirable of the three formats. Not only must the words in each block be in identical sequence, but the characters within each word must be the same length and format. If a word remains the same from block to block, it must nevertheless be repeated in each block.

### EXAMPLE 8.1

In an NC drilling operation, two holes must be drilled in sequence at the following coordinate locations:

Hole 1: $x=2.000$ $y=2.500$

Hole 2: $x=4.000$ $y=2.500$

No prepatory or miscellaneous words are required. Tooling is changed manually, so no t-word is required. The holes are to be drilled to $\frac{1}{2}$-in. diameter at 75 surface feet per minute and 0.005 ipr. Write the two instruction blocks in each of the three tape formats.

First we must convert the surface speed to spindle rotational speed in rpm using Eq. (8.1):

$$S=\frac{75\text{ fpm}}{0.5\pi\text{ ipr}}(12\text{ in./ft})=573\text{ rpm}$$

Now to convert 0.005 ipr into rpm, multiply by spindle speed as given in Eq. (8.2):

$$f=(0.005\text{ ipr})(573\text{ rpm})=2.87\text{ ipm}$$

There are five words to be coded on the NC tape for each hole:

| | *Hole 1* | *Hole 2* |
|---|---|---|
| n-word | 001 | 002 |
| x-word | 2.000 | 4.000 |
| y-word | 2.500 | 2.500 |
| f-word | 2.87 | 2.87 |
| s-word | 573 | 573 |

In the word address format, the two statements would read:

n001 x2.000 y2.500 f2.87 s573 EOB

n002 x4.000 EOB

In the tab sequential format, the two blocks would be:

001 TAB 2.00 TAB 2.50 TAB 2.87 TAB 573 EOB

002 TAB 4.00 TAB TAB TAB EOB

We are using TAB and EOB to denote the codes for the tab key and end-of-block (carriage return) on the Flexowriter. In the fixed block format, the two blocks would be:

001 +02.000 +02.500 2.87 573 EOB
002 +04.000 +02.500 2.87 573 EOB

Now that we have considered the manner in which the NC tape is coded and the type of data that must be provided to the numerical control system, let us next examine the part programmer's place in the procedure.

## 8.3 MANUAL PART PROGRAMMING

In manual programming, the part programmer specifies the machining instructions on a form called a *manuscript*. Manuscripts come in various forms, depending on the machine tool and tape format to be used. For example, the manuscript form for a two-axis point-to-point drilling machine would be different than one for a three-axis contouring machine. Three representative manuscript forms are illustrated in Figure 8.3.

As mentioned in Chapter 7, the manuscript is a listing of the relative tool and workpiece locations. It also includes other data, such as preparatory commands, miscellaneous instructions, and speed/feed specifications, all of which are needed to operate the machine under tape control. The manuscript is designed so that the NC tape can be typed directly from it on a Flexowriter.

We shall divide manual programming jobs into two categories: point-to-point jobs and contouring jobs. Except for complex workparts with many holes to be drilled, manual programming is ideally suited for point-to-point applications. On the other hand, except for the simplest milling and turning jobs, manual programming can become quite time-consuming for applications requiring continuous-path control of the tool. Accordingly, we shall only concern ourselves with manual part programming for point-to-point operations in this chapter. Manual contour programming requires such tedious and detailed calculations that the space needed for the topic would be more than is warranted by the basic purpose of this book, which is to survey the field of automated manufacturing systems. Contouring is much more appropriate for computer-assisted part programming (Sections 8.4 and 8.5).

The basic method of manual part programming for a point-to-point application is best demonstrated by means of an example.

NC Part Programming Manuscript

Two-Axis PTP or Contouring Machine
Word Address Format

Part No. ________________ Date ________________

Part Name ________________ Prepared by ________________

| n-WORD | g-WORD | x-WORD | y-WORD | f-WORD | s-WORD | t-WORD | m-WORD | EOB | COMMENTS |
|---|---|---|---|---|---|---|---|---|---|
| | | | | | | | | | |

NC Part Programming Manuscript

Two-Axis Point-to-Point Machine
Tab Sequential Format

Part No. ________________ Date ________________

Part Name ________________ Prepared by ________________

| SEQUENCE NO. | TAB/EOB | x-COORD | TAB/EOB | y-COORD | TAB/EOB | m-WORD | TAB/EOB | COMMENTS |
|---|---|---|---|---|---|---|---|---|
| | | | | | | | | |

NC Part Programming Manuscript

Two-Axis Point-to-Point Machine
Fixed Block Format

Part No. ________________ Date ________________

Part Name ________________ Prepared by ________________

| SEQUENCE NO. | x-COORD | y-COORD | m-WORD | COMMENTS |
|---|---|---|---|---|
| | | | | |

FIGURE 8.3 Three part programming manuscript forms: word address format, tab sequential format, and fixed block format.

## EXAMPLE 8.2

Suppose that the part to be programmed is a drilling job. The engineering drawing for the part is presented in Figure 8.4. Three holes are to be drilled at a diameter of 31/64 in. The close hole size tolerance requires reaming to 0.500 in. diameter. Recommended speeds and feeds are given[1] as follows:

| | *Speed (fpm)* | *Feed (ipr)* |
|---|---|---|
| 0.484-in.-diameter drill | 75 | 0.006 |
| 0.500-in.-diameter reamer | 50 | 0.010 |

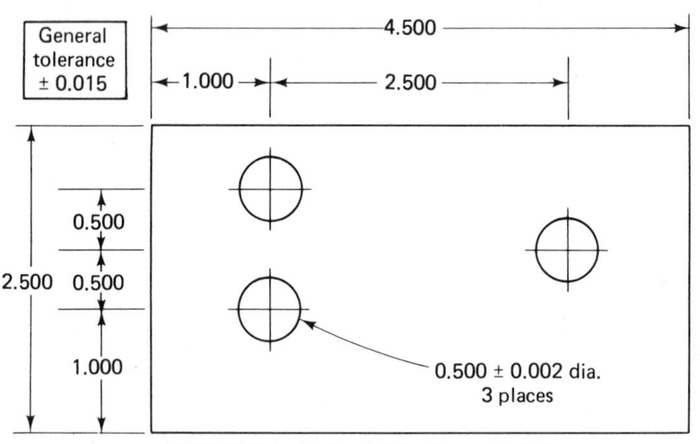


FIGURE 8.4 Part print for Example 8.2.

The NC drill press uses the tab sequential tape format. Drill bits are manually changed by the machine operator, but speeds and feeds must be programmed on the tape. The machine has the floating-zero feature and absolute positioning.

The first step in preparing the part program is to define the axis coordinates in relation to the workpart. We assume that the outline of the part has already been machined before the drilling operation. Therefore, the operator can use one of the corners of the part as the target point. Let us define the lower left-hand corner as the target point and the origin of our axis system. The coordinates are shown in Figure 8.5 for the example part. The $x$ and $y$ locations of each hole can be seen in the figure.

The machine settings for speed and feed must next be determined. For the drill, the spindle speed would be computed from Eq. (8.1):

$$S = \frac{12(75)}{0.484\pi} = 592 \text{ rpm}$$

[1] Recommended cutting speeds and feeds could be obtained from machinability data handbooks.

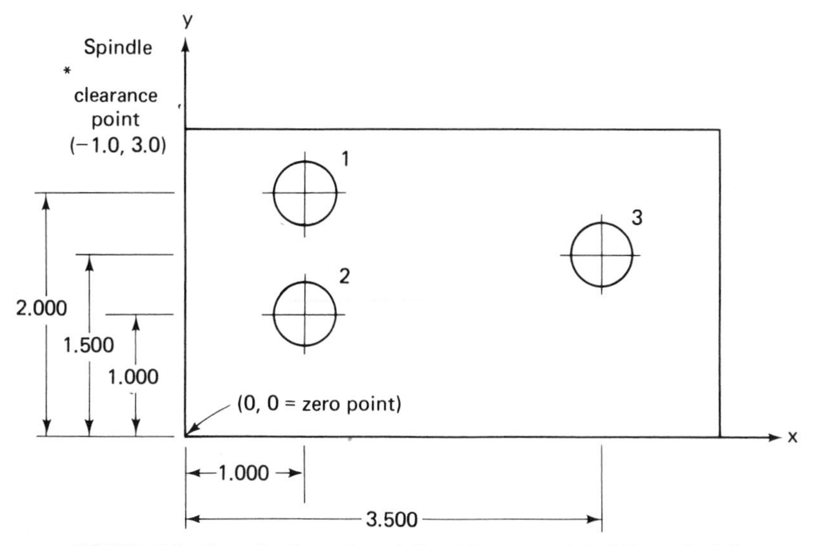

FIGURE 8.5 Coordinate system defined for part print of Example 8.2.

The feed rate is determined from Eq. (8.2):

$$f = 592(0.006) = 3.55 \text{ ipm}$$

Similarly, for the reaming operation,

$$S = 382 \text{ rpm}$$

$$f = 3.82 \text{ ipm}$$

The completed manuscript would appear as in Figure 8.6. The first line shows the $x$ and $y$ coordinates at the zero point. The machine operator would insert the tape and read this first block into the system. (A block of instruction corresponds generally to one line on the manuscript form.) The tool would then be positioned over the target point on the machine table. The operator would then press the zero buttons to set the machine.

The next line on the manuscript is RWS, which stands for rewind–stop. This signal is coded into the tape as holes in columns 1, 2, and 4. The symbol stops the tape after it has been rewound. The last line on the tape contains the m30 word, causing the tape to be rewound at the end of the machining cycle.

Other m-words used in the program are m06, which stops the machine for an operator tool change, and m13, which turns spindle and coolant on. Note in the last line that the tool has been removed from the work area to allow for changing the workpiece.

NC part programming manuscript
Two axis point-to-point drill press
tab sequential format

Part No. EXAMPLE 8.2 Date 4/4/79

Part Name HOLE PLATE Prepared by MPG

| Seq. No. | Tab EOB | x-COORD | Tab EOB | y-COORD | Tab EOB | Feed | Tab EOB | Speed | Tab EOB | m-WORD | Tab EOB | Comments |
|---|---|---|---|---|---|---|---|---|---|---|---|---|
| 00 | TAB | 0.0 | TAB | 0.0 | EOB | | | | | | | ZERO |
| RWS | | | | | | | | | | | | |
| 01 | TAB | 1.0 | TAB | 2.0 | TAB | 3.55 | TAB | 592 | TAB | 13 | EOB | DRILL 1 |
| 02 | TAB | | TAB | 1.0 | EOB | | | | | | | DRILL 2 |
| 03 | TAB | 3.5 | TAB | 1.5 | EOB | | | | | | | DRILL 3 |
| 04 | TAB | -1.0 | TAB | 3.0 | TAB | | TAB | | TAB | 06 | EOB | TOOL CHG |
| 05 | TAB | 3.5 | TAB | 1.5 | TAB | 3.82 | TAB | 382 | TAB | 13 | EOB | REAM 3 |
| 06 | TAB | 1.0 | TAB | 1.0 | EOB | | | | | | | REAM 2 |
| 07 | TAB | | TAB | 2.0 | EOB | | | | | | | REAM 1 |
| 08 | TAB | -1.0 | TAB | 3.0 | TAB | | TAB | | TAB | 06 | EOB | TOOL CHG |
| 09 | TAB | | TAB | | TAB | | TAB | | TAB | 30 | EOB | REWIND & CHG PART |

FIGURE 8.6 Part program manuscript for Example 8.2.

## 8.4 COMPUTER-ASSISTED PART PROGRAMMING

The workpart of Example 8.2 was relatively simple. It was a suitable application for manual programming. Most parts machined on NC systems are considerably more complex. In the more complicated point-to-point jobs and in contouring applications, manual part programming becomes an extremely tedious task and subject to error. In these instances it is much more appropriate to employ the high-speed digital computer to assist in the part programming process. Many part programming language systems have been developed to automatically perform most of the calculations which the programmer would otherwise be forced to do. This saves time and results in a more accurate and more efficient part program.

### The part programmer's job

In computer-assisted part programming, the NC procedure for preparing the tape from the engineering drawing is followed as described in Section 7.2. The difference in the part programmer's job between manual programming and computer-assisted programming is this: With manual programming, a manuscript is used which is formatted so the NC tape can be typed directly from it. With computer-assisted part programming, the machining instructions are written in English-like statements of the NC programming language, which are then processed by the computer to prepare the tape. The computer automatically punches the tape in the proper tape format for the particular NC machine.

When utilizing one of the NC programming languages, part programming can be summarized as consisting basically of two tasks:

1. Defining the geometry of the workpart.
2. Specifying the tool path and/or operation sequence.

Let us now consider these two tasks in computer-assisted part programming. Our frame of reference will be for a contouring application, but the concepts apply for a positioning application as well.

WORKPART GEOMETRY DEFINITION. No matter how complicated the workpart may appear, it is composed of basic geometric elements. Using a relatively simple workpart to illustrate, consider the component shown in Figure 8.7. Although somewhat irregular in overall appearance, the outline of the part consists of intersecting straight lines and a partial circle. The holes in the part can be expressed in terms of the center location and radius of the hole. Nearly any component that can be conceived by a designer can be described by points, straight lines, planes, circles, cylinders, and other

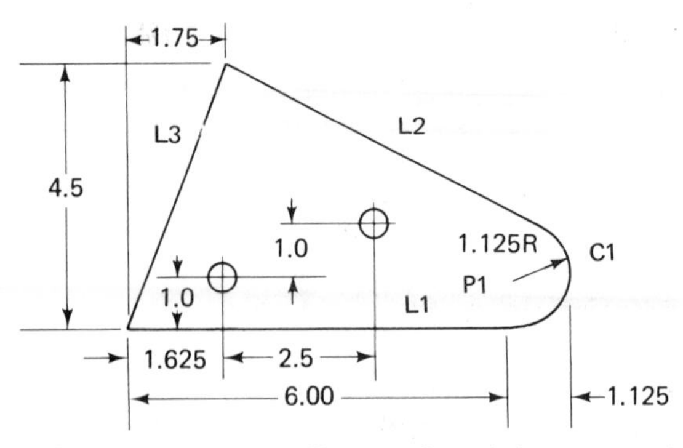


FIGURE 8.7 Sample workpart to illustrate how it is composed of basic geometric elements such as points, lines, and circles.

mathematically defined surfaces. It is the part programmer's task to enumerate the component elements out of which the workpart is formed. Each geometric element must be identified and the dimensions and location of the element explicitly defined. Using the APT programming language as an example, the following statement might be used to define a point:

P1 = POINT/6.0, 1.125, 0

The point is identified by the symbol P1 and is located at $x = 6.0$, $y = 1.125$, and $z = 0$.

Similarly, a circle in the *xy* plane might be defined by the APT statement

C1 = CIRCLE/CENTER, P1, RADIUS, 1.125

The center of circle C1 is P1 (previously defined) and the radius is 1.125.

The various geometric elements in the drawing of Figure 8.7 would be identified in a similar fashion by the part programmer.

TOOL PATH CONSTRUCTION. After defining the workpart geometry, the programmer must next construct the path that the cutter will follow to machine the part. This tool path specification involves a detailed step-by-step sequence of cutter moves. The moves are made along the geometry elements which have previously been defined. To illustrate, using Figure 8.7 and the APT language, the following statement could be used to command the tool to make a left turn from line L2 onto line L3:

GOLFT/L3, PAST, L1

This assumes the tool was previously located at the intersection of lines L2 and L3 and had just finished a cut along L2. The statement directs the tool to cut along L3 until it just passes line L1.

By using statements similar to the above, the tool can be directed to machine along the workpart surfaces, to go to point locations, to drill holes at those point locations, and so on. In addition to geometry definition and tool path specification, the part programmer also provides other commands to the NC system. However, let us await Section 8.5, where we will consider a wide range of possible APT statements.

## The computer's job

The computer's job in computer-assisted part programming consists of the following steps:

1. Input translation.
2. Arithmetic calculations.
3. Cutter offset computation.
4. Post processor.

The sequence of these steps and their relationships to the part programmer and the machine tool are illustrated in Figure 8.8.

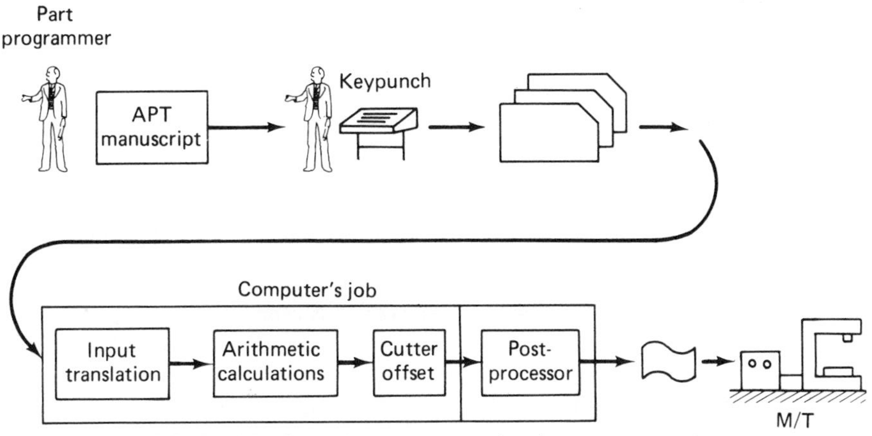


FIGURE 8.8 Steps in computer-assisted part programming.

Input Translation. The part programmer submits his program as a deck of computer cards (or alternative medium) written in the APT or other language. The input translation component converts the coded instructions contained in the program into computer-usable form, preparatory to further processing.

Arithmetic Calculations. The arithmetic calculations unit of the system consists of a comprehensive set of subroutines for solving the mathematics required to generate the part surface. These subroutines are called by the various part programming language statements. The arithmetic unit is really the fundamental element in the part programming package. This unit frees the programmer from the time-consuming geometry and trigonometry calculations to concentrate on the workpart processing.

CUTTER OFFSET COMPUTATION. When we described the second task of the part programmer as that of constructing the tool path, we ignored one basic factor: the size of the cutting tool. The actual tool path is different from the part outline. This is because the tool path is the path taken by the center of the cutter. It is at the periphery of the cutter that machining takes place.

The purpose of the cutter offset computation is to offset the tool path from the desired part surface by the radius of the cutter. This means that the part programmer can define the exact part outline in his geometry statements. Thanks to the cutter offset calculation provided by the programming system, he need not concern himself with this task. The cutter offset problem is illustrated in Figure 8.9.

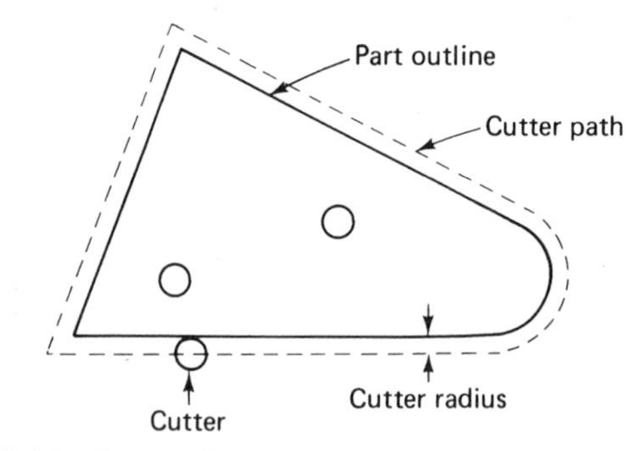


FIGURE 8.9 Cutter offset problem in contour part programming.

POSTPROCESSOR. As we have noted previously, NC machine tool systems are different. They have different features and capabilities. They use different NC tape formats. Nearly all of the part programming languages, including APT, are designed to be general-purpose languages, not limited to one or two machine tool types. Therefore, the final task of the computer in computer-assisted part programming is to take the general instructions and make them specific to a particular machine tool system. The unit that performs this task is called a *postprocessor*.

The postprocessor is really a separate computer program that has been written to prepare the punched tape for a specific machine tool. The input to the postprocessor is the output from the other three components: a series of cutter locations and other instructions. The output of the postprocessor is the NC tape written in the correct format for the machine on which it is to be used.

## NC part programming languages

Probably over 100 NC part programming languages have been developed since the initial MIT research on NC programming systems in 1956. Most of the languages were developed to serve particular needs and machines and have not survived the test of time. However, a good number of languages are still in use today. In this subsection we shall review some of those which are generally considered important.

APT (AUTOMATICALLY PROGRAMMED TOOLS). The APT language was the product of the MIT developmental work on NC programming systems. Its development began in June, 1956, and it was first used in production around 1959. Today it is the most widely used language in the United States. Although first intended as a contouring language, modern versions of APT can be used for both positioning and continuous-path programming in up to five axes.

AUTOSPOT (AUTOMATIC SYSTEM FOR POSITIONING TOOLS). This was developed by IBM and first introduced in 1962 for PTP programming. Today's version of AUTOSPOT can be used for contouring as well.

SPLIT (SUNDSTRAND PROCESSING LANGUAGE INTERNALLY TRANSLATED). This is a proprietary system intended for Sundstrand's machine tools. It can handle up to five axis positioning and possesses contouring capability as well. One of the unusual features of SPLIT is that the postprocessor is built into the program. Each machine tool uses its own SPLIT package, thus obviating the need for a special postprocessor.

COMPACT II. This is a package available from Manufacturing Data Systems, Inc. (MDSI), a firm based in Ann Arbor, Michigan. The NC language is similar to SPLIT in many of its features. MDSI leases the COMPACT II system to its users on a time-sharing basis. The part programmer uses a remote terminal to feed the program into one of the MDSI computers, which in turn produces the NC tape.

ADAPT (ADAPTATION OF APT). Several part programming languages are based directly on the APT program. One of these is ADAPT, which was developed by IBM under Air Force contract. It was intended to provide many of the features of APT but to utilize a significantly smaller computer. The full APT program requires a large computing system (256K). This precludes its use by many small and medium-size firms that do not have access to a large computer. ADAPT is not as powerful as APT, but can be used to program for both positioning and contouring jobs.

EXAPT (EXTENDED SUBSET OF APT). This was developed in Germany starting around 1964 and is based on the APT language. There are three versions: EXAPT I—designed for positioning (drilling and also straight-cut milling), EXAPT II—designed for turning, and EXAPT III—designed for limited contouring operations. One of the important features of EXAPT is that it attempts to compute optimum feeds and speeds automatically.

UNIAPT. The UNIAPT package represents another attempt to adapt the APT language to use on smaller computers. The name derives from the developer, the United Computing Corp. of Carson, California. Their efforts have provided a limited version of APT to be implemented on minicomputers, thus allowing many smaller shops to possess computer-assisted programming capacity.

As indicated previously, the most widely used part programming language is APT. APT also forms the basis for several other NC languages, including those mentioned above: ADAPT, EXAPT, and UNIAPT. Accordingly, if we are to present the fundamentals of any part programming language, it seems appropriate to direct our attention to the APT system.

## 8.5 THE APT LANGUAGE

In this section, we will present some of the fundamental principles of computer-assisted part programming using APT. The reader will certainly not become an expert part programmer after completing this introduction. Our objectives are much less ambitious. What we hope to accomplish is merely to demonstrate the English-like statements of APT and to show how they can be formulated to command the cutting tool through its sequence of machining operations.

APT is not only an NC language; it is also the computer program that performs the calculations to generate cutter positions based on APT statements. We will not concern ourselves with the internal workings of the computer program. Instead, we will concentrate on the language that the part programmer must use.

APT is a three-dimensional system that can be used to control up to five axes. We will limit our discussion to the more familiar three axes, $x$,$y$, and $z$, and exclude rotational coordinates. APT can be used to control a variety of different machining operations. We will cover only drilling and milling applications. There are over 400 words in the APT vocabulary. Only a small (but important) fraction will be covered here.

To program in APT, the workpart geometry must first be defined. Then the tool is directed to various point locations and along surfaces of the workpart to carry out the machining operations. The viewpoint of the part programmer is that the workpiece remains stationary and the tool is instructed to move relative to the part.

There are four types of statements in the APT language:

1. *Geometry statements.* These define the geometric elements that comprise the workpart. They are also sometimes called definition statements.
2. *Motion statements.* These are used to describe the path taken by the cutting tool.
3. *Postprocessor statements.* These apply to the specific machine tool and control system. They are used to specify feeds and speeds and to actuate other features of the machine.
4. *Auxiliary statements.* These are miscellaneous statements used to identify the part, tool, tolerances, and so on.

## Geometry statements

When the tool motions are specified, their description is in terms of points and surfaces. Therefore, the points and surfaces must be defined before tool motion commands can be given.

The general form of an APT geometry statement is this:

symbol = geometry type/descriptive data (8.3)

An example of such a statement is

P1 = POINT/5.0, 4.0, 0.0 (8.4)

The statement is made up of three sections. The first is the symbol used to identify the geometric element. A symbol can be any combination of six or fewer alphabetic and numeric characters. At least one of the six must be an alphabetic character. Also, although it may seem obvious, the symbol cannot be one of the APT vocabulary words. Some examples may help to show what is permissible as a symbol, and what is not permissible:

| | |
|---|---|
| PZL | Permissible |
| PABCDE | Permissible |
| PABCDEF | No; too many characters |
| 123789 | No; must have alphabetic character |
| POINT | No; APT vocabulary word |
| P1.2 | No; only alphabetic and numeric characters are allowed |

The second section of the geometry statement is an APT vocabulary word that identifies the type of geometry element. Besides POINT, other geometry elements in the APT vocabulary include LINE, PLANE, and CIRCLE.

The third section of the geometry statement is the descriptive data that define the element precisely, completely, and uniquely. These data may include quantitative dimensional and positional data, previously defined geometry elements, and other APT words.

The punctuation used in the APT geometry statement is illustrated in the example, Eq. (8.4). The statement is written as an equation, the symbol being equated to the surface type. A slash separates the surface type from the descriptive data. Commas are used to separate the words and numbers in the descriptive data.

There are a variety of ways to specify the different geometry elements. The Appendix at the end of this chapter presents a dictionary of APT vocabulary words as well as a sampling of statements for defining the geometry elements we will be using: points, lines, circles, and planes. The reader may benefit from a few examples below:

To Specify a Point. In addition to listing the $x$,$y$, and $z$ coordinates of the point, it can also be defined by the intersection of two lines:

P2=POINT/INTOF, L1, L2

In the descriptive data, INTOF stands for "intersection of." This is followed by the symbols for the two lines.

Other methods for defining a point are given in the Appendix under "POINT."

To Specify a Line. The easiest way to specify a line is by two points through which the line passes:

L3=LINE/P3, P4

The part programmer may find it convenient to define a new line parallel to another line which has previously been defined. One way of doing this would be:

L4=LINE/P5, PARLEL, L3

This states that the line L4 must pass through point P5 and be parallel (PARLEL) to line L3.

To Specify a Plane. A plane can be defined by specifying three points through which it passes:

PL1=PLANE/P1, P4, P5

It can also be defined as being parallel to another plane, similar to the previous line parallelism statement.

PL2 = PLANE/P2, PARLEL, PL1

Plane PL2 is parallel to plane PL1 and passes through point P2.

TO SPECIFY A CIRCLE. A circle can be specified by its center and its radius.

C1 = CIRCLE/CENTER, P1, RADIUS, 5.0

The two APT descriptive words are used to identify the center and radius. The orientation of the circle perhaps seems undefined. By convention, it is a circle located in the $x$-$y$ plane.

GROUND RULES. There are several rules that must be followed in formulating an APT geometry statement:

1. The coordinate data must be specified in the order $x,y,z$. For example, the statement

P1 = POINT/5.0, 4.0, 0.0

is interpreted by the APT program to mean a point at $x = 5.0$, $y = 4.0$, and $z = 0.0$.

2. Any symbols used as descriptive data must have been previously defined. For example, in the statement

P2 = POINT/INTOF, L1, L2

the two lines L1 and L2 must have been previously defined. In setting up the list of geometry statements, the APT programmer must be sure to define symbols before using them in subsequent statements.

3. A symbol can be used to define only one geometry element. The same symbol cannot be used to define two different elements. For example, the following sequence would be incorrect:

P1 = POINT/1.0, 1.0, 1.0
P1 = POINT/2.0, 3.0, 4.0

4. Only one symbol can be used to define any given element. For example, the following two statements in the same program would render the program incorrect:

P1 = POINT/1.0, 1.0, 1.0
P2 = POINT/1.0, 1.0, 1.0

5. Lines defined in APT are considered to be of infinite length in both directions. Similarly, planes extend indefinitely and circles defined in APT are complete circles.

## Motion statements

APT motion statements have a general format, just as the geometry statements do. The general form of a motion statement is

$$\text{motion command/descriptive data} \tag{8.5}$$

An example of a motion statement is

$$\text{GOTO/P1} \tag{8.6}$$

The statement consists of two sections separated by a slash. The first section is the basic motion command, which tells the tool what to do. The second section is comprised of descriptive data, which tell the tool where to go. In the example statement above, the tool is commanded to go to point P1, which has been defined in a preceding geometry statement.

At the beginning of the motion statements, the tool must be given a starting point. This point is likely to be the target point, the location where the operator has positioned the tool at the start of the job. The part programmer keys into this starting position with the following statement:

$$\text{FROM/TARG} \tag{8.7}$$

The FROM is an APT vocabulary word which indicates that this is the initial point from which others will be referenced. In the statement above, TARG is the symbol given to the starting point. Any other APT symbol could be used to define the target point. Another way to make this statement is

$$\text{FROM}/-2.0,-2.0,0.0$$

where the descriptive data in this case are the $x, y$, and $z$ coordinates of the target point. The FROM statement occurs only at the start of the motion sequence.

It is convenient to distinguish between PTP movements and contouring movements when discussing the APT motion statements.

Point-to-Point Motions. There are only two basic PTP motion commands: GOTO and GODLTA. The GOTO statement instructs the tool to go to a particular point location specified in the descriptive data. Two examples would be:

```
GOTO/P2
GOTO/2.0, 7.0, 0.0
```

In the first statement, P2 is the destination of the tool point. In the second statement, the tool has been instructed to go to the location whose coordinates are $x=2.0$, $y=7.0$, and $z=0$.

The GODLTA command specifies an incremental move for the tool. For example, the statement

```
GODLTA/2.0, 7.0, 0.0
```

instructs the tool to move from its present position 2 in. in the $x$-direction and 7 in. in the $y$-direction. The $z$-coordinate remains unchanged.

The GODLTA command is useful in drilling and related operations. The tool can be directed to a particular hole location with the GOTO statement. Then the GODLTA command would be used to drill the hole, as in the following sequence:

```
GOTO/P2
GODLTA/0, 0, -1.5
GODLTA/0, 0, +1.5
```

**EXAMPLE 8.3**

Example 8.2 was a PTP job which was programmed manually. Let us write the APT geometry and motion statements necessary to perform the drilling portion of this job. We will set the plane defined by $z=0$ about $\frac{1}{4}$ in. above the part surface. The part will be assumed to be $\frac{1}{2}$-in. thick.

```
P1 = POINT/1.0, 2.0, 0
P2 = POINT/1.0, 1.0, 0
P3 = POINT/3.5, 1.5, 0
P0 = POINT/-1.0, 3.0, +2.0
FROM/P0
GOTO/P1
GODLTA/0, 0, -1.0
GODLTA/0, 0, +1.0
GOTO/P2
GODLTA/0, 0, -1.0
GODLTA/0, 0, +1.0
GOTO/P3
GODLTA/0, 0, -1.0
GODLTA/0, 0, +1.0
GOTO/P0
```

This is not a complete APT program because it does not contain the necessary auxiliary and postprocessor statements. However, the statement sequence demonstrates how geometry and motion statements can be combined to command the tool through a series of machining steps.

CONTOURING MOTIONS. Contouring commands are somewhat more complicated because the tool's position must be continuously controlled throughout the move. To accomplish this control, the tool is directed along two intersecting surfaces as shown in Figure 8.10. These surfaces have very specific names in APT:

1. *Drive surface.* This is the surface (it is pictured as a plane in Figure 8.10) that guides the side of the cutter.

2. *Part surface.* This is the surface (again shown as a plane in Figure 8.10) on which the bottom of the cutter rides. The reader should note that the "part surface" may or may not be an actual surface of the workpart. The part programmer must define this plus the drive surface for the purpose of maintaining continuous path control of the tool.

There is one additional surface that must be defined for APT contouring motions:

3. *Check surface.* This is the surface that stops the movement of the tool in its current direction. In a sense, it checks the forward movement of the tool.

There are several ways in which the check surface can be used. This is determined by APT modifier words within the descriptive data of the motion statement. The three main modifier words are TO, ON, and PAST, and their use with regard to the check surface is shown in Figure 8.11. A fourth

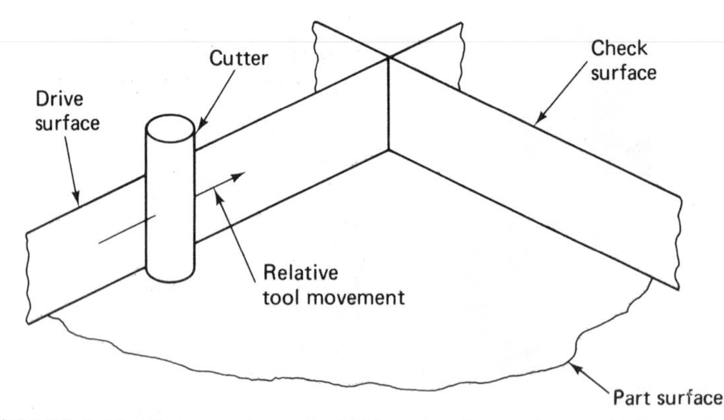


FIGURE 8.10 Three surfaces in APT contouring motions which guide the cutting tool.

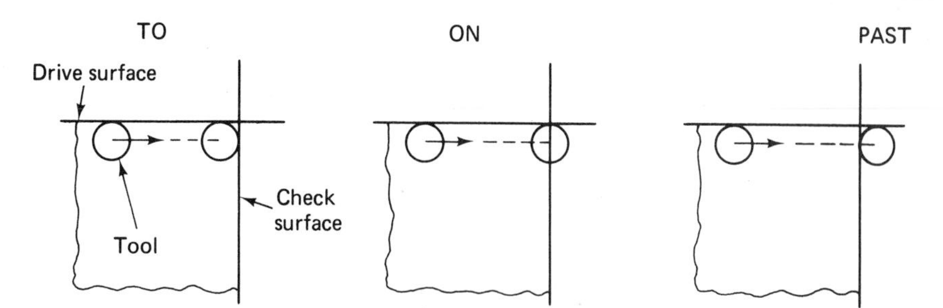


FIGURE 8.11 Use of APT modifier words in a motion statement: TO, ON, and PAST. TO moves tool into initial contact with check surface. ON moves tool until tool center is on check surface. PAST moves tool just beyond check surface.

modifier word is TANTO. This is used when the drive surface is tangent to a circular check surface, as illustrated in Figure 8.12. In this case the cutter can be brought to the point of tangency with the circle by use of the TANTO modifier word.

The APT contour motion statement commands the cutter to move along the drive and part surfaces and the movement ends when the tool is at the check surface. There are six motion command words:

| GOLFT | GOFWD | GOUP |
|---|---|---|
| GORGT | GOBACK | GODOWN |

Their interpretation is illustrated in Figure 8.13. In commanding the cutter, the programmer must keep in mind where it is coming from. As the tool reaches the new check surface, does the next movement involve a right turn or an upward turn or what? The tool is directed accordingly by one of the six motion words.

To begin the sequence of motion commands, the FROM statement, Eq. (8.7), is used in the same manner as for PTP moves. The statement following the FROM statement defines the initial drive surface, part surface, and check

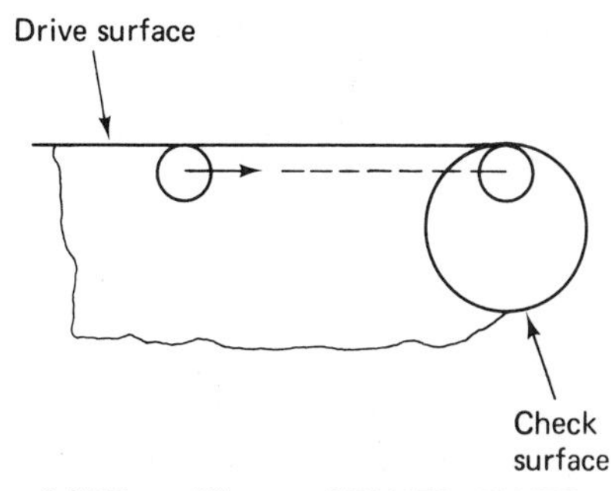


FIGURE 8.12 Use of APT modifier word TANTO. TANTO moves tool to point of tangency between two surfaces, at least one of which is circular.

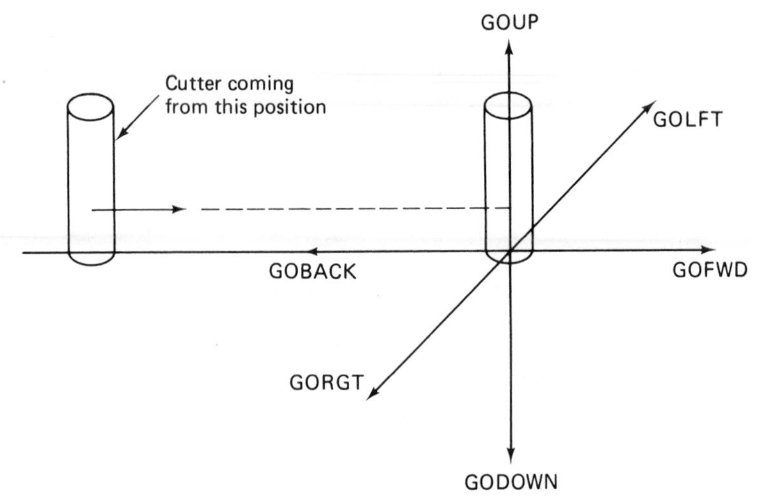


FIGURE 8.13 Use of APT motion commands.

surface. The sequence is of the following form:

FROM/TARG
GO/TO, PL1, TO, PL2, TO, PL3

The symbol TARG represents the target point where the operator has set up the tool. The GO command instructs the tool to move to the intersection of the drive surface (PL1), the part surface (PL2), and the check surface (PL3). The periphery of the cutter is tangent to PL1 and PL3, and the bottom of the cutter is on PL2. This cutter location is defined by use of the modifier word TO. The three surfaces included in the GO statement must be specified in the order: drive surface first, part surface second, and check surface last.

Note that the GO/TO command is different from the GOTO command. GOTO is used only for PTP motions. GO/TO is used to initialize the sequence of contouring motions.

After initialization, the tool is directed along its path by one of the six command words. It is not necessary to repeat the symbol of the part surface after it has been defined. For instance, consider Figure 8.14. The cutter has been directed from TARG to the intersection of surfaces PL1, PL2, and PL3. It is desired to move the tool along plane PL3. The following command would be used:

GORGT/PL3, PAST, PL4

This would direct the tool to move along PL3, using it as the drive surface. The tool would continue until past surface PL4, which is the new check surface. Although the part surface (PL2) may remain the same throughout the motion sequence, the drive surface and check surface are redefined in each new command.

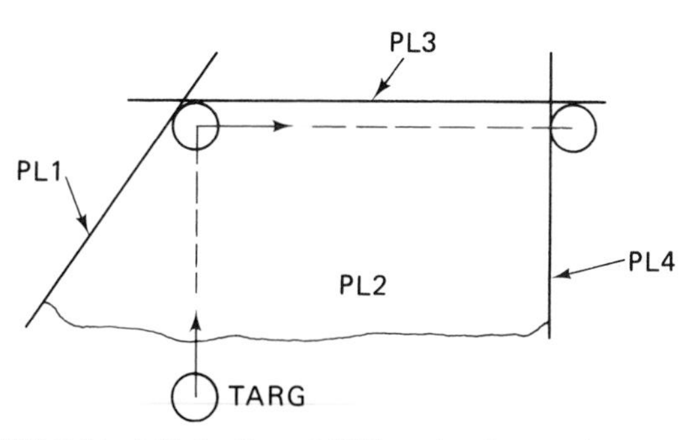


FIGURE 8.14 Initialization of APT contouring motion sequence.

Let us consider an alternative statement to the above which would accomplish the same motion but would lead to easier programming:

GORGT/L3, PAST, L4

We have substituted lines L3 and L4 for planes PL3 and PL4, respectively. When looking at a part drawing, such as Figure 8.7, the sides of the part appear as lines. On the actual part, they are three-dimensional surfaces, of course. However, it would be more convenient for the part programmer to define these surfaces as lines and circles rather than planes and cylinders. Fortunately, the APT computer program allows the geometry of the part to be defined in this way. Hence, the lines L3 and L4 in the foregoing motion statement are treated as the drive surface and check surface. This substitution can only be made when the part surfaces are perpendicular to the *xy*-plane.

**EXAMPLE 8.4**

We will write the APT geometry and motion statements for the workpart of Figure 8.7, which is repeated in Figure 8.15 with coordinate axes given.

```
P0=POINT/0, -1.0, 0
P1=POINT/6.0, 1.125, 0
P2=POINT/0, 0, 0
P3=POINT/6.0, 0, 0
P4=POINT/1.75, 4.5, 0
L1=LINE/P2, P3
C1=CIRCLE/CENTER, P1, RADIUS, 1.125
L2=LINE/P4, LEFT, TANTO, C1
L3=LINE/P2, P4
PL1=PLANE/P2, P3, P4
FROM/P0
GO/TO, L1, TO, PL1, TO, L3
GORGT/L1, TANTO, C1
GOFWD/C1, PAST, L2
GOFWD/L2, PAST, L3
GOLFT/L3, PAST, L1
GOTO/P0
```

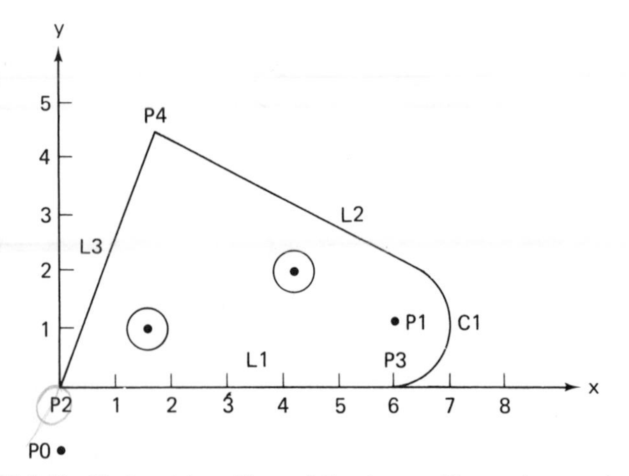


FIGURE 8.15 Workpart from Figure 8.7 redrawn with *x* and *y* axes included for Example 8.4.

The reader may have questioned the location of the part surface (PL1) in the APT sequence. For this machining job, the part surface must be defined below the bottom plane of the workpiece in order for the cutter to machine the entire thickness of the piece. Therefore, the part surface is really not a surface of the part at all. Example 8.4 raises several other questions: How is the cutter size accounted for in the APT program? How are feeds and speeds specified? These and other questions are answered by the postprocessor and auxiliary statements.

## Postprocessor statements

To write a complete part program, statements must be written that control the operation of the spindle, the feed, and other features of the machine tool. These are called *postprocessor statements*. Some of the common postprocessor statements that appear in the Appendix at the end of the chapter are:

| | |
|---|---|
| COOLNT/ | RAPID |
| END | SPINDL/ |
| FEDRAT/ | TURRET/ |
| MACHIN/ | |

The postprocessor statements, and the auxiliary statements in the following subsection, are of two forms: either with or without the slash(/). The statements without the slash are self-contained. No additional data are needed. The APT words with the slash require descriptive data after the slash. These descriptions are given for each word in the Appendix.

The FEDRAT/ statement should be explained. FEDRAT stands for feed rate and the interpretation of feed differs for different machining operations. In a drilling operation the feed is in the direction of the drill bit axis. The feed rate can be determined for a drilling operation from Eq. (8.2). However, in an end milling operation, typical for NC, the feed would be in a direction perpendicular to the axis of the cutter. To determine the feed rate in inches per minute (ipm), the desired chip load [inches per tooth (ipt)] must be specified. The following equation can be used to make the conversion:

$$f = S f_t n_t \tag{8.8}$$

where $f$ = feed rate, ipm
$S$ = spindle speed, rpm
$f_t$ = feed, ipt
$n_t$ = number of teeth in the cutter

Equation (8.8) is used to compute ipm for the FEDRAT/ command in milling. And Eq. (8.2) is used for FEDRAT/ in drilling.

### Auxiliary statements

The complete APT program must also contain various other statements, called *auxiliary statements*. These are used for cutter size definition, part identification, and so on. The following APT words used in auxiliary statements are defined in the Appendix to this chapter:

| | |
|---|---|
| CLPRNT | INTOL/ |
| CUTTER/ | OUTTOL/ |
| FINI | PARTNO |

The offset calculation of the tool path from the part outline is based on the CUTTER/ definition. For example, the statement

CUTTER/.500

would instruct the APT program that the cutter diameter is 0.500 in. Therefore, the tool path must be offset from the part outline by 0.250 in.

### EXAMPLE 8.5

We are now in a position to write a complete APT program. The workpart of Example 8.4 will be used to illustrate the format of the APT program.

We will assume that the workpiece is a plain low-carbon steel plate, cut out in the rough shape of the part outline. The tool is a two-flute, $\frac{1}{2}$-in.-diameter, HSS end-milling cutter. Typical cutting conditions might be recommended as follows: cutting speed = 75 fpm and feed = 0.002 ipt.

From Eq. (8.1), the spindle speed should be

$$S = \frac{12(75)}{\pi(0.5)} = 573 \text{ rpm}$$

```
Column
1     6 | 8 | 10                                            72
PARTNO  |   | EXAMPLE PART
        |   | MACHIN/MILL, 1
        |   | CLPRNT
        |   | INTOL/.001
        |   | OUTTOL/.001
        |   | CUTTER/.500
P0      | = | POINT/0, -1.0, 0
P1      | = | POINT/6.0, 1.125, 0
P2      | = | POINT/0, 0, 0
P3      | = | POINT/6.0, 0, 0
P4      | = | POINT/1.75, 4.5, 0
L1      | = | LINE/P2, P3
C1      | = | CIRCLE/CENTER, P1, RADIUS, 1.125
L2      | = | LINE/P4, LEFT, TANTO, C1
L3      | = | LINE/P2, P4
PL1     | = | PLANE/P2, P3, P4
        |   | SPINDL/573
        |   | FEDRAT/2.29
        |   | COOLNT/ON
        |   | FROM/P0
        |   | GO/TO, L1, TO, PL1, TO, L3
        |   | GORGT/L1, TANTO, C1
        |   | GOFWD/C1, PAST, L2
        |   | GOFWD/L2, PAST, L3
        |   | GOLFT/L3, PAST, L1
        |   | RAPID
        |   | GOTO/P0
        |   | COOLNT/OFF
        |   | FINI
```

FIGURE 8.16 APT program for Example 8.5.

The feed rate can be computed from Eq. (8.8):

$$f = 573(0.002)(2) = 2.29 \text{ ipm}$$

Figure 8.16 presents the program with correct character spacing identified at the top as if it were to be keypunched onto computer cards.

## Summary

We have attempted in this section to demonstrate the APT language for part programming. The complete APT package contains more than 400 words. It is obvious that we have omitted much in our presentation of the subject. However, enough of the APT language has been included to permit the reader to try some of the programming exercises at the end of the chapter. By working through these problems, the reader can acquire a firsthand experience in NC computer-assisted part programming.

A more complete coverage of part programming with emphasis on the APT language is contained in several of the references (e.g., [1] and [4]). We especially recommend the book by Childs [1].

## REFERENCES


[1] CHILDS, J. J., *Numerical Control Part Programming*, Industrial Press, Inc., New York, 1973.

[2] DALLAS, D. B. (Editor), *Tool and Manufacturing Engineers Handbook*, 3rd ed., Society of Manufacturing Engineers (McGraw-Hill Book Company, New York), 1976, Chapter 12.

[3] HOWE, R. E. (Editor), *Introduction to Numerical Control in Manufacturing*, Society of Manufacturing Engineers, Dearborn, Mich., 1969.

[4] Illinois Inst. of Technology Research Institute (IITRI), *APT Part Programming*, McGraw-Hill Book Company, New York, 1967.

[5] LESLIE, W. H. P., *Numerical Control Users Handbook*, McGraw-Hill Book Company, New York, 1970.

[6] OLESTEN, N. O., *Numerical Control*, Wiley–Interscience, New York, 1970.

[7] PRESSMAN, R. S., and WILLIAM, J. E., *Numerical Control and Computer-Aided Manufacturing*, John Wiley & Sons, Inc., New York, 1977.


## PROBLEMS

**8.1.** A drilling operation is to be carried out using a $\frac{1}{4}$-in.-diameter high-speed steel (HSS) drill. The work material is a machinable grade of aluminum. Recommended cutting speed is 400 surface feet per minute and recommended feed rate is 0.003 ipr. Convert these recommendations into machine tool speed and feed: rpm and ipm, respectively.

**8.2.** The same work material as in Problem 8.1 is to be milled with a $\frac{3}{4}$-in.-diameter four-flute end milling cutter. Recommendations are: cutting speed = 400 fpm and feed = 0.003 ipt. Convert these values to rpm and ipm.

**8.3.** A cast-iron workpiece is to be face-milled on an NC machine using cemented carbide inserts. The cutter has 16 teeth and is 5 in. in diameter. The recommended cutting speed is 250 fpm. Recommended feed is 0.002 ipt. Convert these to rpm and ipm for use by the machine tool.

**8.4.** A part program is to be written to drill the holes in the workpart of Figure P8.4 on page 220. The part is $\frac{3}{8}$ in. thick.

(a) Define the $x$ and $y$ axes for the job.

(b) Write the part program manuscript (manual part programming) using the word address format and an absolute position system. The words that

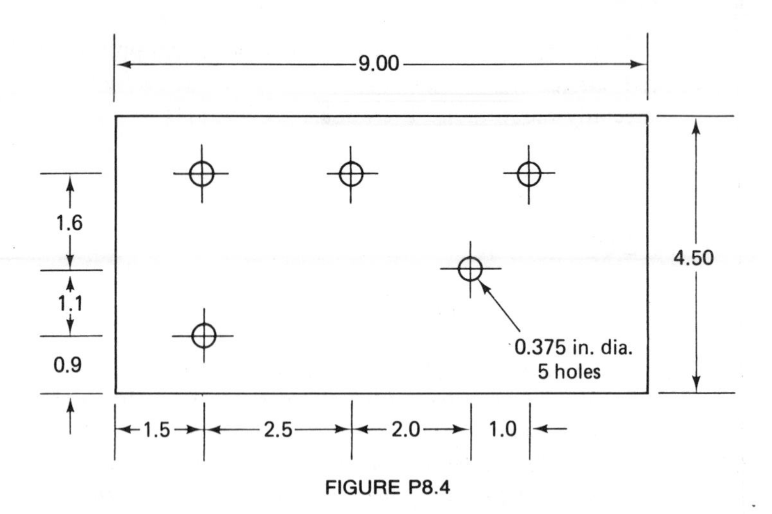


FIGURE P8.4

must be specified for the particular NC drill press are n-, x-, y-, and m-words. The speed and feed are manually set by the operator.

**8.5.** Solve Problem 8.4 except that speed and feed specifications must be included in the program (n-, x-, y-, f-, s-, and m-words must be given). Assume that the tool material is HSS and the work material is aluminum so that the cutting conditions would be as determined in Problem 8.1.

**8.6.** Solve Problem 8.4 except use the tab sequential format rather than word address format.

**8.7.** Solve Problem 8.5 except use the tab sequential format.

**8.8.** Solve Problem 8.4 except use the fixed block format.

**8.9.** The part in Figure P8.9 is to be drilled on a turret-type NC drill press. The $\frac{3}{8}$-in.-diameter holes (4) are to be drilled with a $\frac{23}{64}$-in.-diameter drill and reamed

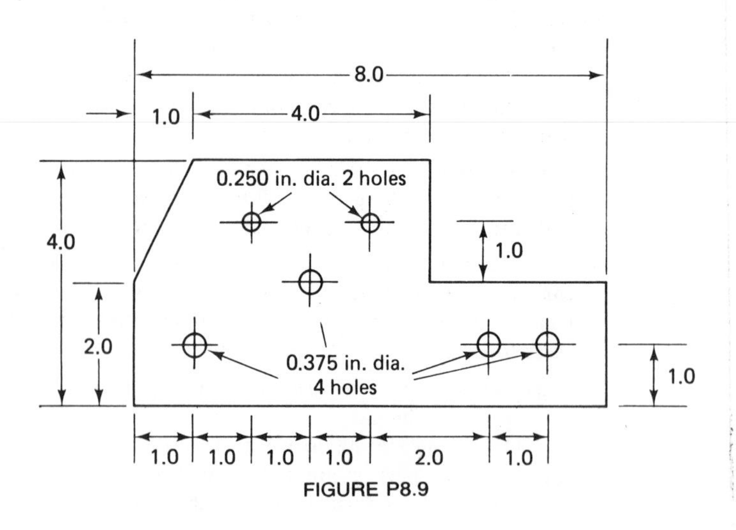


FIGURE P8.9

to final size. All tooling is HSS. The turret on the drill press has six positions, but only three are required for this operation sequence. The part is $\frac{3}{8}$ in. thick.

(a) Designate the three tools for turret positions t01, t02, and t03.

(b) Define the $xy$ coordinate system.

(c) Write the part program manuscript using the word address format and absolute positioning. The following words must be specified in the program for the particular drill press: n-, x-, y-, t-, and m-words. Cutting conditions are set by the operator.

**8.10.** Solve Problem 8.9 except that the f- and s-words must be included. Recommended cutting conditions are as follows:

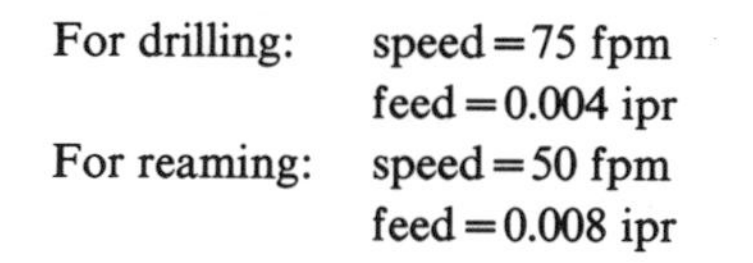

| | |
|---|---|
| For drilling: | speed = 75 fpm |
| | feed = 0.004 ipr |
| For reaming: | speed = 50 fpm |
| | feed = 0.008 ipr |

**8.11.** Write the APT part program to perform the drilling in Problem 8.5.

**8.12.** Write the APT part program to solve Problem 8.10.

**8.13.** The outline of the part in Figure P8.13 is to be machined in an end-milling operation. Write the APT geometry statements that define the part outline. Do not consider the two $\frac{3}{8}$-in. holes. They will be used for clamping the part during machining.

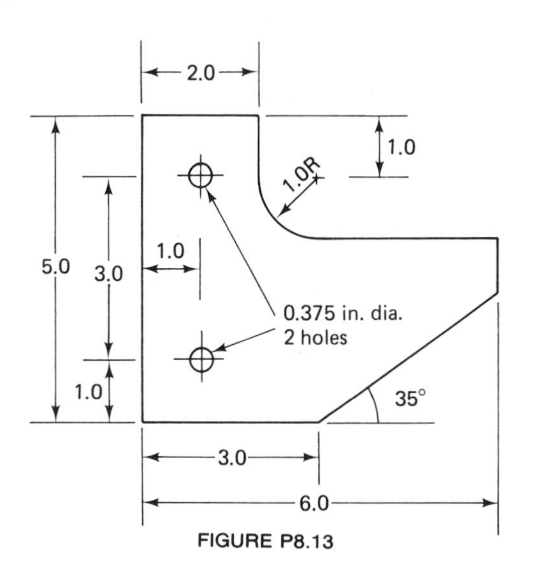


FIGURE P8.13

**8.14.** Write the complete APT program for the part of Problem 8.13. The postprocessor call statement is MACHIN/TURDRL, 02. The cut will be made with a $\frac{3}{4}$-in.-diameter end mill. Speed = 500 rpm; feed = 4.0 ipm. Inside tolerance on the circular approximation is 0.001 in. No outside tolerance is allowed.

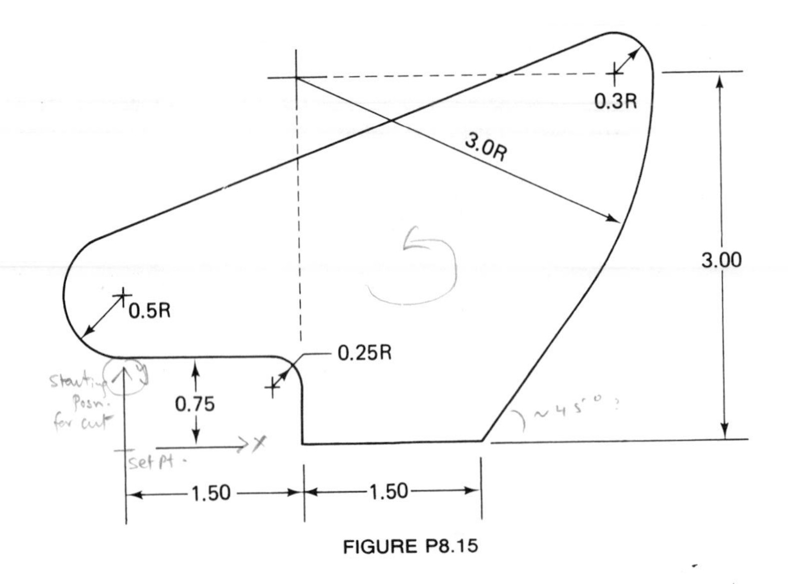


FIGURE P8.15

**8.15.** The outline of the cam shown in Figure P8.15 is to be milled using a two-flute, $\frac{1}{2}$-in.-diameter end mill.

(a) Write the geometry statements in APT to define the part outline.

(b) Write the motion statement sequence using the geometry elements defined in part (a).

(c) Write the complete APT program. Inside and outside tolerances should be 0.0005 in. Feed rate = 3 ipm; speed = 500 rpm. Postprocessor call statement is MACHIN/MILL, 01. Assume that the rough outline for the part has been obtained in a bandsaw operation. Ignore clamping problems with this part.

**8.16.** The part outline of Figure P8.16 is to be milled in two passes with the same tool. The tool is a 1-in.-diameter end mill. The first cut is to leave 0.050 in. of stock on the part outline. The second cut will take the part to size. Write the APT geometry and motion statements to perform the two passes.

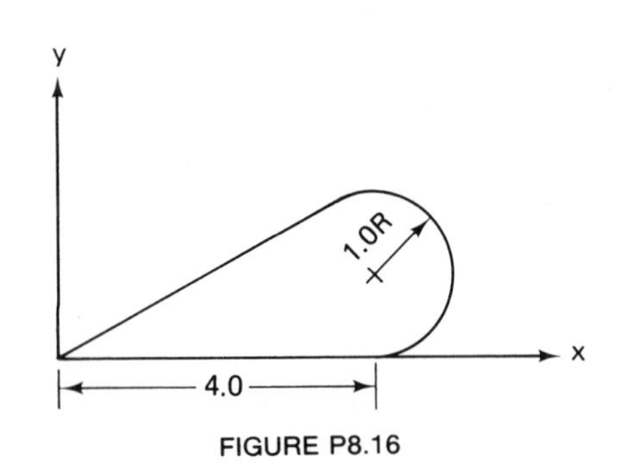


FIGURE P8.16

**8.17.** The top surface of a large cast iron plate is to be face-milled flat. The area to be machined is 15 in. wide and 27 in. long. The cutter to be used and the machining conditions are as described in Problem 8.3. Sketch the part surface in relation to an assumed set of axes. Write the APT program to complete the job. The postprocessor call statement is MACHIN/MILL, 05. Use a coolant for this job.

## APPENDIX: APT WORD DEFINITIONS

ATANGL: At angle (descriptive data). Indicates that the data that follow represent a specified angle. Angle is given in degrees. *See* LINE.

CENTER: Center (descriptive data). Used to indicate the center of circle. *See* CIRCLE.

CIRCLE: Circle (geometry type). Used to define a circle in the *xy*-plane. Methods of definition:

1. By the coordinates of the center and the radius. See Figure A8.1.

   C1 = CIRCLE/CENTER, 4.0, 3.0, 0.0, RADIUS, 2

   C1 = CIRCLE/4.0, 3.0, 0.0, 2.0

2. By the center point and the radius. See Figure A8.1.

   C1 = CIRCLE/CENTER, P1, RADIUS, 2.0

3. By the center point and tangent to a line. See Figure A8.1.

   C1 = CIRCLE/CENTER, P1, TANTO, L1

4. By three points on the circumference. See Figure A8.1.

   C1 = CIRCLE/P2, P3, P4

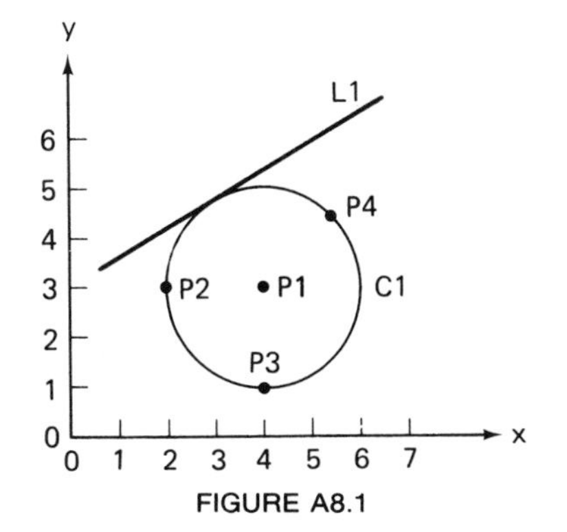


FIGURE A8.1

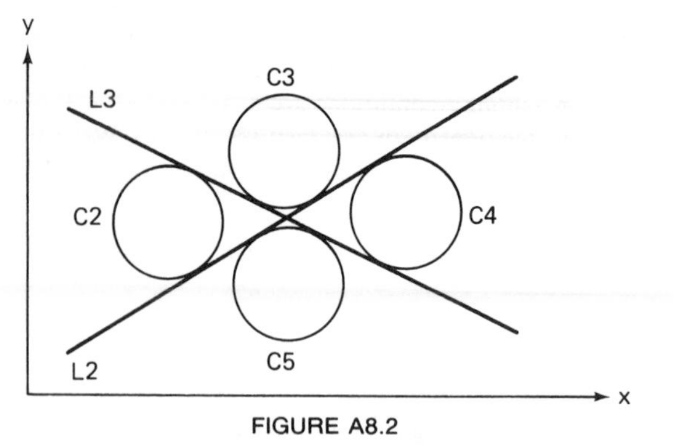


FIGURE A8.2

5. By two intersecting lines and the radius. See Figure A8.2.

C2=CIRCLE/XSMALL, L2, YSMALL, L3, RADIUS, .375
C3=CIRCLE/YLARGE, L2, YLARGE, L3, RADIUS, .375
C4=CIRCLE/XLARGE, L2, YLARGE, L3, RADIUS, .375
C5=CIRCLE/YSMALL, L2, YSMALL, L3, RADIUS, .375

CLPRNT: Cutter location print (auxiliary statement). Can be used to obtain a computer printout of the cutter location sequence on the NC tape.

COOLNT: Coolant (postprocessor statement). Turns coolant on, off, and actuates other coolant options that may be available. Examples:

COOLNT/ON    COOLNT/OFF
COOLNT/FLOOD    COOLNT/MIST

CUTTER: Cutter (auxiliary statement). Defines cutter diameter to be used in tool offset computations. The statement

CUTTER/1.0

defines a 1.0-in.-diameter milling cutter. Cutter path would be offset from part outline by one-half the diameter.

END: End (postprocessor statement). Used to stop the machine at the end of a section of the program. Can be used to change tools manually. Meaning may vary between machine tools. To continue program, a FROM statement should be used.

FEDRAT: Feed rate (postprocessor statement). Used to specify feet rate in inches per minute.

FEDRAT/6.0

FINI: Finish (auxiliary statement). Must be the last word in the APT program. Used to indicate the end of the complete program.

FROM: From the tool starting location (motion startup command). Used to specify the starting point of the cutter, from which other tool movements will be measured. The starting point is specified by the part programmer and set up by the machine operator. Methods of specification:

1. By a previously defined starting point (TARG).

FROM/TARG

2. By the coordinates of the starting point.

FROM/ − 1.0, − 1.0, 0.0

GO: Go (motion startup command in contouring). Used to bring the tool from the starting point against the drive surface, part surface, and check surface. In the statements:

GO/TO, L1, TO, PL1, TO, L2
GO/PAST, L1, TO, PL1, ON, L2

the initial drive surface is the line L1, the part surface is PL1, and the initial check surface is L2.

GODLTA: Go delta (PTP motion command). Instructs the tool to move in increments as specified from the current tool location. In the statement

GODLTA/2.0, 3.0, −4.0

the tool is instructed to move 2.0 in. in the $x$-direction, 3.0 in. in the $y$-direction, and −4.0 in. in the $z$-direction from its present position.

GOBACK: Go back (contour motion command). Instructs the tool to move back relative to its previous direction of movement. In the statement

GOBACK/PL5, TO, L1

the tool is instructed to move in the opposite general direction relative to its previous path. It moves on the drive surface PL5 until it reaches L1. The part surface has been specified in a previous GO statement.

In specifying the motion command the part programmer must pretend to be riding on top of the cutter and must give the next move (GOBACK, GOFWD, GOUP, GODOWN, GORGT, GOLFT) according to the tool's preceding motion. Also, the motion command should indicate the largest direction component. For example, if the next tool move was both forward and to the left, the motion command (GOFWD vs. GOLFT) would be determined by whichever direction component was larger. See Figure 8.13 in Section 8.5.

GODOWN: Go down (contour motion command). *See* GOBACK.

GOFWD: Go forward (contour motion command). *See* GOBACK.

GOLFT: Go left (contour motion command). *See* GOBACK.

GORGT: Go right (contour motion command). *See* GOBACK.

GOTO: Go to (PTP motion command). Used to move the tool center to a specified point location. Methods of specification:

1. By using a previously defined point. GOTO/P1
2. By defining the coordinates of the point.

GOTO/2.0, 5.0, 0.0

GOUP: Go up (contour motion command). *See* GOBACK.

INTOF: Intersection of (descriptive data). Indicates that the intersection of two geometry elements is the specified point. *See* POINT.

INTOL: Inside tolerance (auxiliary statement). Indicates the allowable tolerance between the inside of a curved surface and any straight-line segments used to approximate the curve. See Figure A8.3.

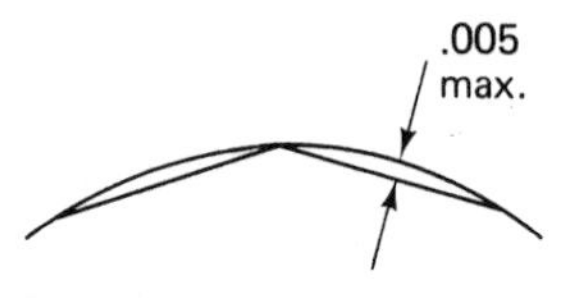


INTOL/.005

FIGURE A8.3

LEFT: Left (descriptive data). Used to indicate which of two alternatives, left or right, is desired. *See* LINE.

LINE: Line (geometry type). Used to define a line that is interpreted by APT as a plane perpendicular to the *xy* plane. Methods of definition:

1. By the coordinates of two points. See Figure A8.4.

L1=LINE/2,1,0,5,3,0

2. By two points. See Figure A8.4.

L1=LINE/P1, P2

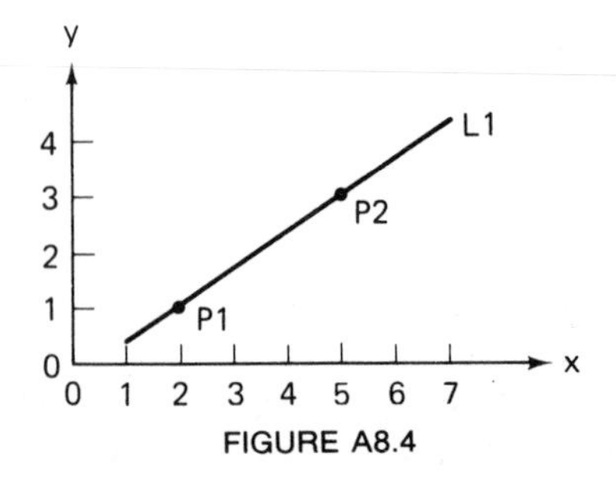


FIGURE A8.4

3. By a point and tangent to a circle. See Figure A8.5 on page 227.

L1=LINE/P1, LEFT, TANTO, C1
L2=LINE/P1,RIGHT,TANTO,C1

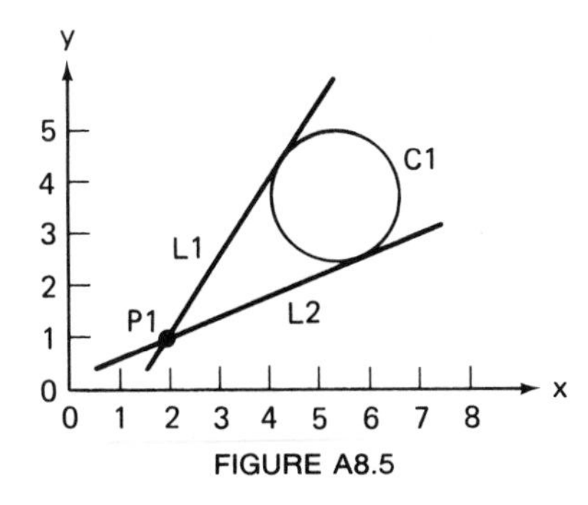

FIGURE A8.5

The descriptive words LEFT and RIGHT are used by looking from the point toward the circle.

4. By a point and the angle of the line to the $x$-axis or another line. See Figure A8.6.

```
L3=LINE/P1, ATANGL, 20
L4=LINE/P1, ATANGL, 30, L3
```

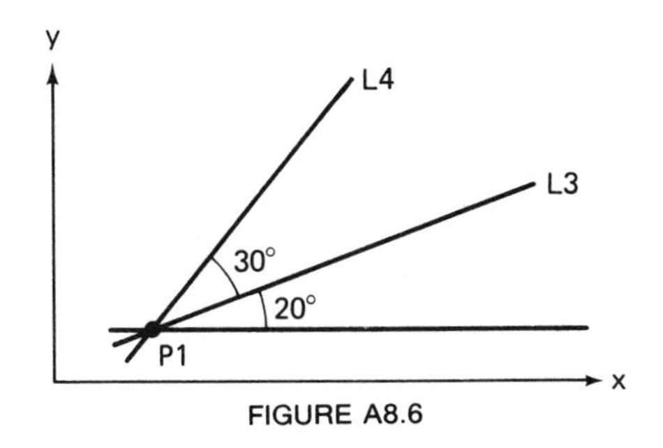

FIGURE A8.6

5. By a point and being parallel to or perpendicular to another line. See Figure A8.7.

```
L5=LINE/P2, PARLEL, L3
L6=LINE/P2, PERPTO, L3
```

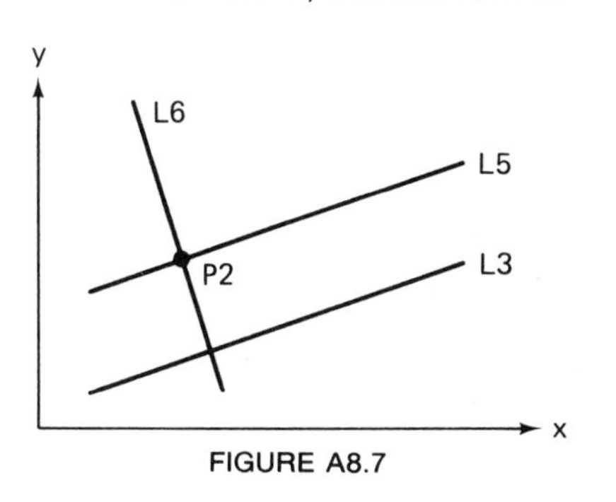

FIGURE A8.7

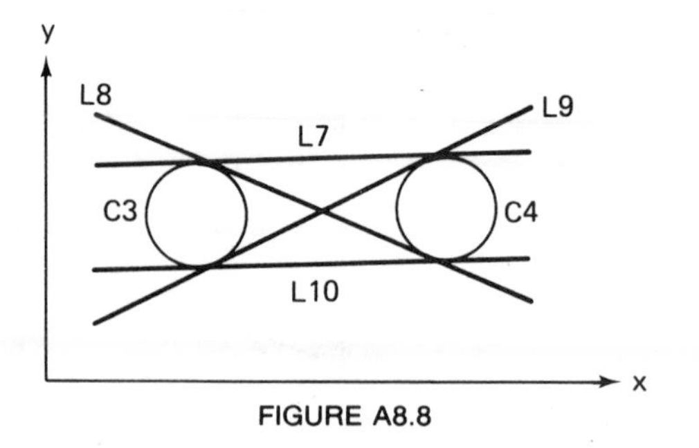


FIGURE A8.8

6. By being tangent to two circles. See Figure A8.8.

L7=LINE/LEFT, TANTO, C3, LEFT, TANTO, C4
L8=LINE/LEFT, TANTO, C3, RIGHT, TANTO, C4
L9=LINE/RIGHT, TANTO, C3, LEFT, TANTO, C4
L10=LINE/RIGHT, TANTO, C3, RIGHT, TANTO, C4

The descriptive words LEFT and RIGHT are used by looking from the first circle written toward the second circle. For example, another way to specify L7 would be

L7=LINE/RIGHT, TANTO, C4, RIGHT, TANTO, C3

MACHIN: Machine (postprocessor statement). Used to specify the machine tool and to call the postprocessor for that machine tool. In the statement

MACHIN/MILL, 1

the MILL identifies the machine tool type and 1 identifies the particular machine and postprocessor. The APT system then calls the specified postprocessor to prepare the NC tape for that machine.

ON: On (motion modifier word) used with three other motion modifier words—TO, PAST, and TANTO—to indicate the point on the check surface where the tool motion is to terminate. See Figure A8.9.

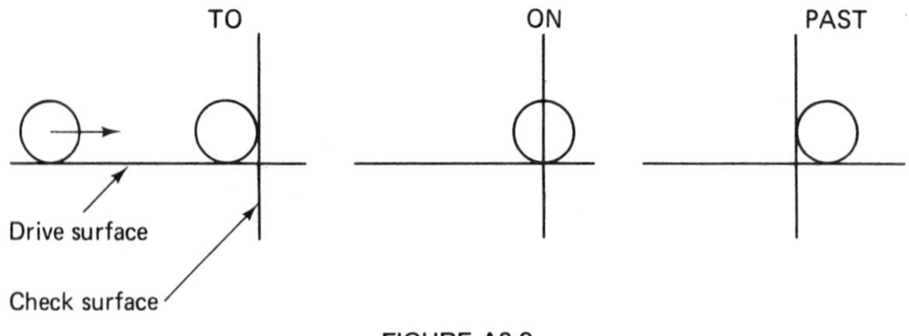


FIGURE A8.9

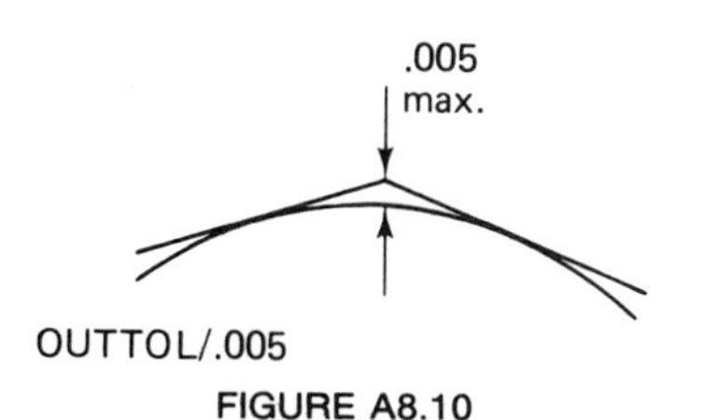


FIGURE A8.10

OUTTOL: Outside Tolerance (auxiliary statement). Indicates the allowable tolerance between the outside of a curved surface and any straight-line segments used to approximate the curve. See Figure A8.10.

*Note*: The INTOL and OUTTOL statements can be used together to indicate allowable tolerances on both inside and outside of the curved surface. See Figure A8.11.

INTOL/.0025

OUTTOL/.0025

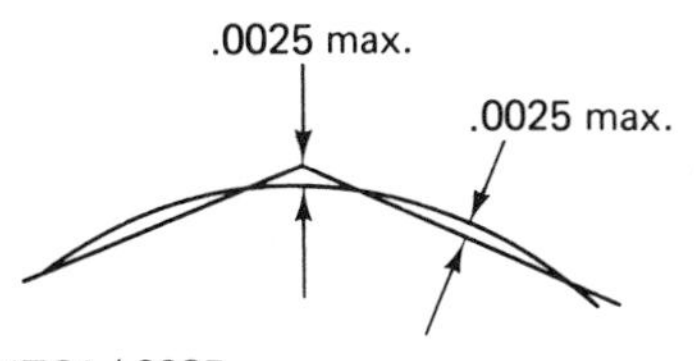


FIGURE A8.11

PARLEL: Parallel (descriptive data). Used to define a line or plane as being parallel to another line or plane. *See* LINE and PLANE.

PARTNO: Part number (auxiliary statement). Used at start of program to identify the part program. PARTNO must be typed in columns 1 through 6 of the first computer card in the deck.

PARTNO MECHANISM PLATE 47320

PAST: Past (motion modifier word). *See* ON.

PERPTO: Perpendicular to (descriptive data). Used to define a line or plane as being perpendicular to some other line or plane. *See* LINE and PLANE.

PLANE: Plane (geometry type). Used to define a plane.

Methods of definition:

1. By three points that do not lie on the same straight line. See Figure A8.12 on page 230.

PL1 = PLANE/P1, P2, P3

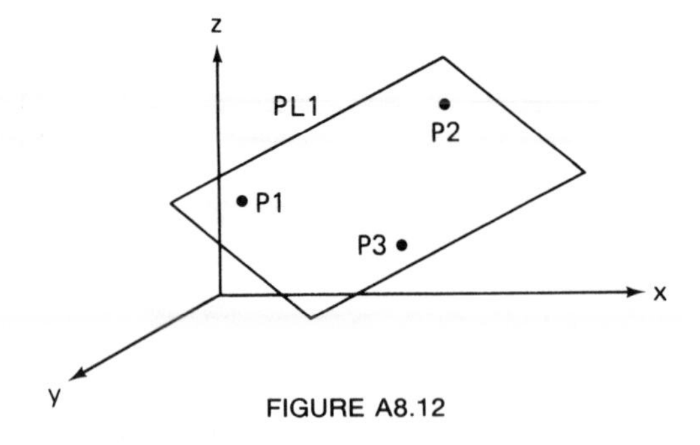

FIGURE A8.12

2. By a point and being parallel to another plane. See Figure A8.13.

PL2 = PLANE/P4, PARLEL, PL1

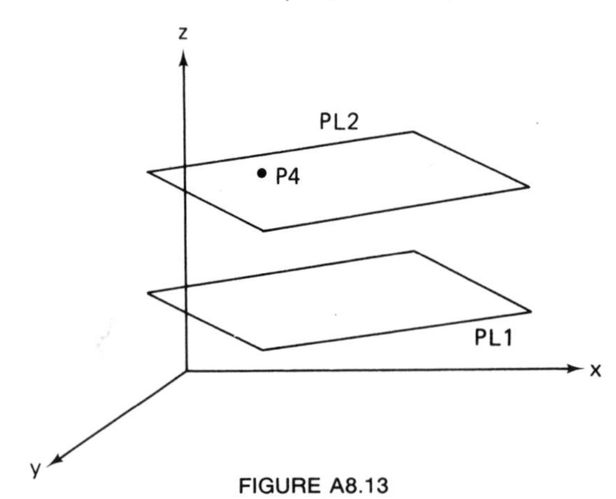

FIGURE A8.13

3. By two points and being perpendicular to another plane. See Figure A8.14.

PL3 = PLANE/PERPTO, PL1, P5, P6

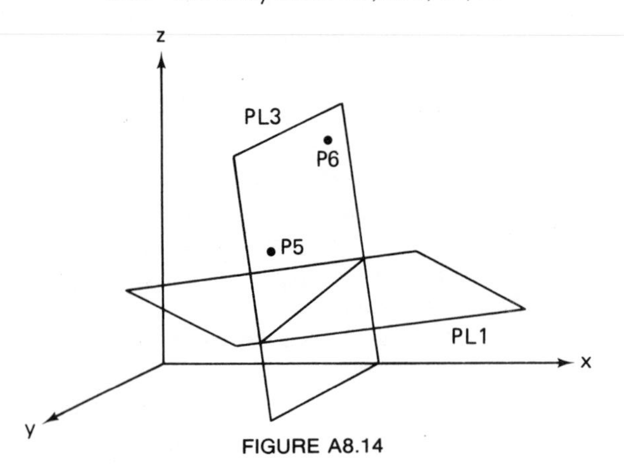

FIGURE A8.14

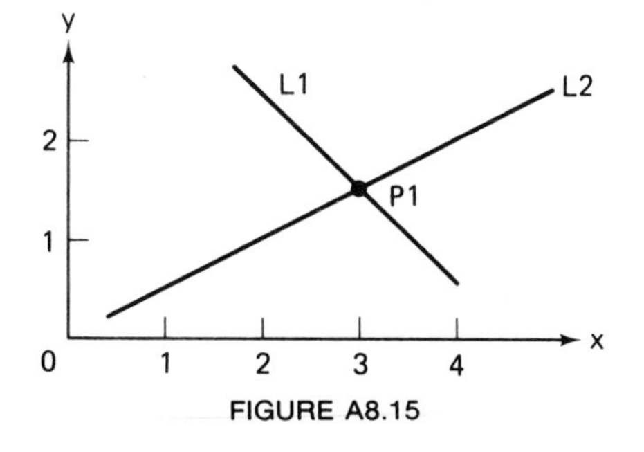


FIGURE A8.15

POINT: Point (geometry type). Used to define a point.

Methods of definition:

1. By the x, y, and z coordinates. See Figure A8.15.

P1 = POINT/3.0, 1.5, 0.0

2. By the intersection of two lines. See Figure A8.15.

P1 = POINT/L1, L2

3. By the intersection of a line and a circle. See Figure A8.16.

P2 = POINT/YLARGE, INTOF, L3, C1
P3 = POINT/XLARGE, INTOF, L3, C1

Any of the descriptive words—XLARGE, XSMALL, YLARGE, YSMALL—can be used to indicate the relative position of the point. For example, for point P2, YLARGE or XSMALL could be used. For point P3, YSMALL or XLARGE could be used.

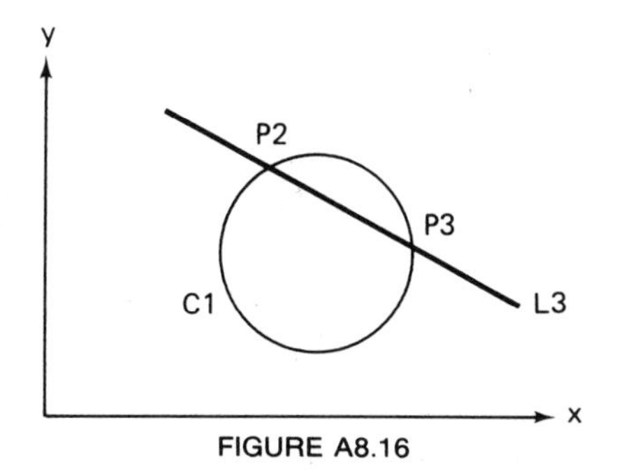


FIGURE A8.16

4. By two intersecting circles. See Figure A8.17.

P4 = POINT/YLARGE, INTOF, C1, C2
P5 = POINT/YSMALL, INTOF, C1, C2

5. By the center of a circle:

P6 = POINT/CENTER, C1

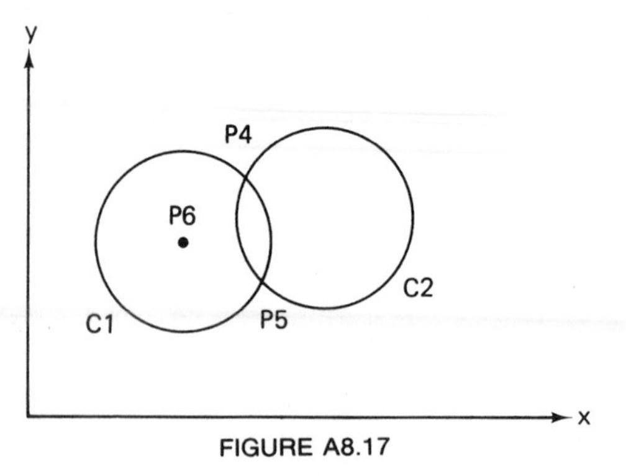


FIGURE A8.17

RADIUS: Radius (descriptive data). Used to indicate the radius of a circle. *See* CIRCLE.

RIGHT: Right (descriptive data). *See* LEFT and LINE.

TANTO: Tangent to (two uses: descriptive data and motion modifier word).

1. As descriptive data, it is used to indicate tangency of one geometry element to another. *See* CIRCLE and LINE.
2. As a motion modifier word, it is used to indicate that the tool motion is to terminate at the point of tangency between the drive surface and the check surface. See Figure A8.18.

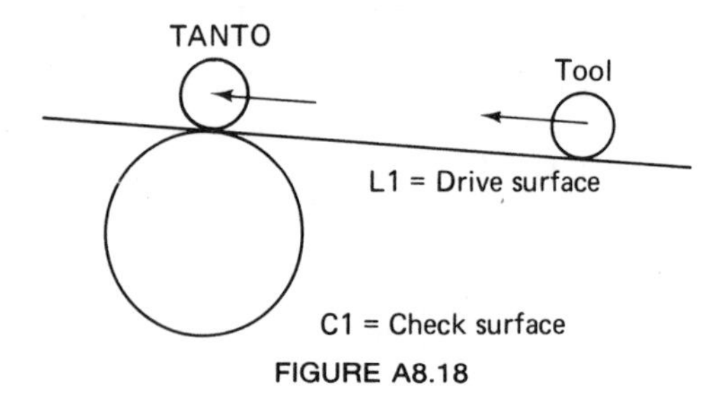


FIGURE A8.18

TO: To (motion modifier word). *See* ON.

TURRET: Turret (postprocessor statement). Used to specify a turret position on a turret lathe or drill or to call a specific tool from an automatic tool changer. Example: TURRET/T30.

XLARGE: In the positive $x$-direction (descriptive data). Used to indicate the relative position of one geometric element with respect to another when there are two possible alternatives. *See* CIRCLE and POINT.

XSMALL: In the negative $x$-direction (descriptive data). *See* XLARGE.

YLARGE: In the positive $y$-direction (descriptive data). *See* XLARGE.

YSMALL: In the negative $y$-direction (descriptive data). *See* XLARGE.

chapter 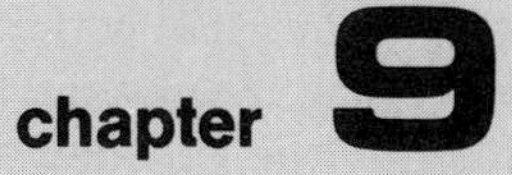

# Extensions of Numerical Control

## 9.1 INTRODUCTION

Numerical control has caused a virtual revolution in the discrete metal parts manufacturing industry. It seems to be an economic axiom that the success of any invention should breed further developments and improvements. The success of NC has led to a number of extensions of numerical control concepts and technology. Four of the more important developments are the following:

1. Direct numerical control (DNC).
2. Computer numerical control (CNC).
3. Adaptive control.
4. Industrial robots.

Both DNC and CNC involve a marriage between computer technology and numerical control technology. *Direct numerical control* was chronologically the first to appear and involves the use of a large computer to control many separate NC machine tools. Advances in computer hardware resulted in smaller computers. One consequence of this trend toward miniaturization was the use of one computer for each machine tool. This was called *computer numerical control*.

The third extension of NC, *adaptive control,* was initiated around 1962. Just as the U.S. Air Force had sponsored much of the original research on numerical control, it also provided financial support for the initial work on adaptive control. This term has come to denote a machining system that measures one or more process variables (such as cutting force, temperature, horsepower, etc.) and manipulates feed and/or speed in order to compensate for undesirable changes in the process variables. Its objective is to optimize the machining operation, something that NC is unable to accomplish.

The fourth extension of NC which we shall discuss is the *industrial robot.* Basically, an industrial robot is an NC machine that performs materials handling functions rather than metal processing operations. Robots are capable of carrying out other tasks in addition to the handling of workparts. Examples include spotwelding and spray painting. The same type of numerical control technology used to operate machine tools is used in industrial robots to actuate a mechanical arm and hand. The robot can be programmed to perform a certain set of activities for one job; then it can be reprogrammed for a new job.

These four so-called extensions of numerical control will be discussed in detail in the following four sections of this chapter. At the end of the chapter, we shall attempt to summarize the topic of numerical control by examining some of the trends in NC.

## 9.2 DIRECT NUMERICAL CONTROL (DNC)

There are a number of problems inherent in conventional NC which have motivated machine tool builders to seek improvements in the basic NC system. Among the difficulties encountered in using conventional numerical control machines are the following:

1. *Part programming mistakes.* In preparing the punched tape, part programming mistakes are common. The mistakes can be either syntax or numerical errors, and it is not uncommon for three or more passes to be required before the NC tape is correct. Another related problem in part programming is to achieve the optimum sequence of processing steps. This is mainly a problem in manual part programming. Some of the computer-assisted part programming languages provide aids to achieving the best operation sequence.

2. *Punched tape.* Another problem is the tape itself. Paper tape is especially fragile, and its susceptibility to wear and tear causes it to be an

unreliable NC component for repeated use on the shop floor. More durable tape materials, such as Mylar and aluminum foil, are utilized to help overcome this difficulty. However, these materials are relatively expensive.

3. *Tape reader.* The tape reader that interprets the punched tape is generally acknowledged among NC users to be the least reliable hardware component of the machine. When a breakdown is encountered on an NC machine, the maintenance personnel usually begin their search for the problem with the tape reader.

4. *Controller.* The conventional NC controller unit is hardwired. This means that its control features cannot be easily altered to incorporate improvements into the unit.

5. *Management information.* The conventional NC system cannot provide timely information on operational performance to management. Such information might include piececounts, machine breakdowns, and tool changes.

It was with these problem areas in mind that the machine tool builders developed the concept of using the general-purpose computer to control NC machines. Their concept has come to be called direct numerical control, or DNC.

## DNC defined

Direct numerical control can be defined as a manufacturing system in which a number of machines are controlled by a computer through direct connection and in real time. The tape reader is omitted in DNC, thus relieving the system of its least reliable component. Instead of using the tape reader, the part program is transmitted to the machine tool directly from the computer memory. In principle, one computer can be used to control more than 100 separate machines. (One commercial DNC system boasts a control capability of up to 256 machine tools.) The DNC computer is designed to provide instructions to each machine tool on demand. When the machine needs control commands, they are communicated to it immediately.

Figure 9.1 illustrates the general DNC configuration. The system consists of four components.

1. Central computer.
2. Bulk memory, which stores the NC part programs.
3. Telecommunication lines.
4. Machine tools.

The computer calls the part program instructions from bulk storage and sends them to the individual machines as the need arises. It also receives data back from the machines. This two-way information flow occurs in real time, which

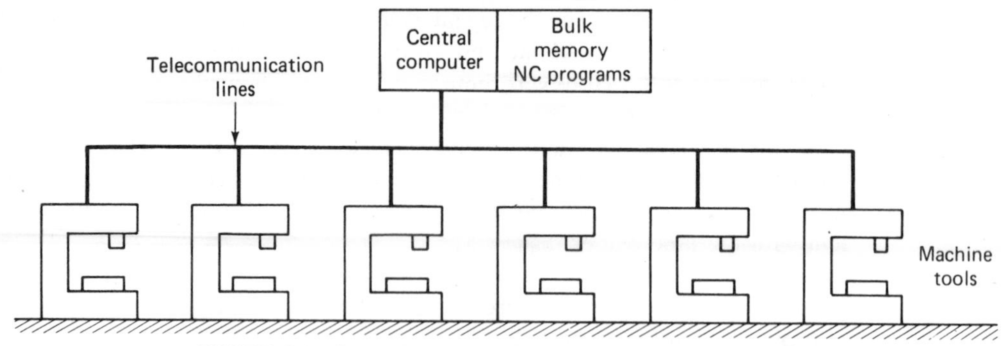


FIGURE 9.1 General configuration of direct numerical control (DNC) system.

means that each machine's requests for instructions must be satisfied almost instantaneously. Similarly, the computer must always be ready to receive information from the machines and to respond accordingly. The remarkable feature of the DNC system is that the computer is servicing a large number of separate machine tools, all in real time.

Depending on the number of machines and the computational requirements that are imposed on the computer, it is sometimes necessary to make use of satellite computers, as shown in Figure 9.2. These satellites are minicomputers, and they serve to take some of the burden off the central computer. Each satellite controls several machines. Groups of part program instructions are received from the central computer and stored in buffers. They are then dispensed to the individual machines as required. Feedback

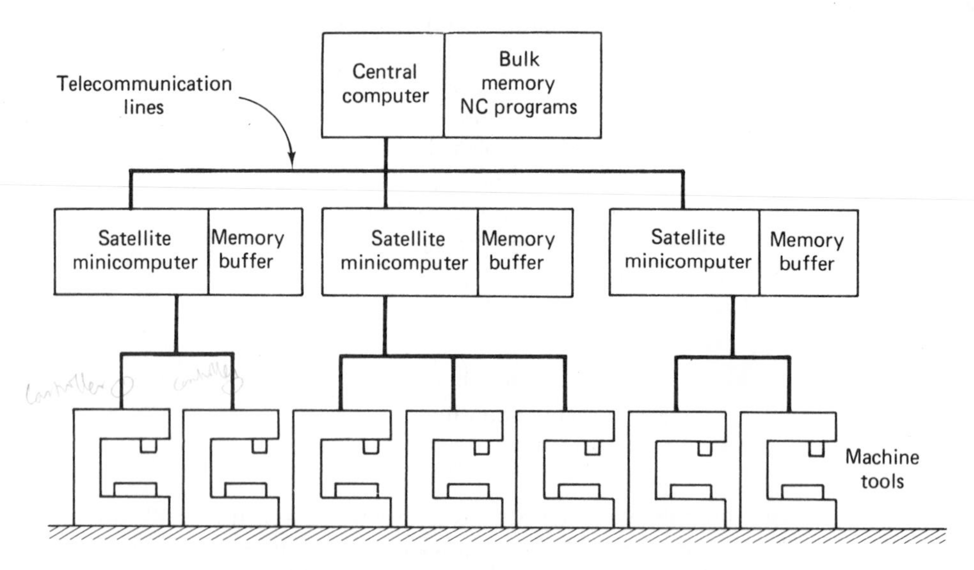


FIGURE 9.2 DNC with satellite minicomputers.

data from the machines are also stored in the satellite's buffer before being collected at the central computer.

## Two types of DNC

There are two alternative system configurations by which the communication link is established between the control computer and the machine tool. One is called a behind-the-tape reader system; the other configuration makes use of a specialized machine control unit.

BEHIND-THE-TAPE READER (BTR) SYSTEM. In this arrangement, pictured in Figure 9.3, the computer is linked directly to the regular NC controller unit. The replacement of the tape reader by the telecommunication lines to the DNC computer is what gives the BTR configuration its name. The connection with the computer is made between the tape reader and the controller unit—behind the tape reader.

Except for the source of the command instructions, the operation of the system is very similar to conventional NC. The controller unit uses two temporary storage buffers to receive blocks of instructions from the DNC computer and convert them into machine actions. While one buffer is receiving a block of data, the other is providing command instructions to the machine tool.

SPECIAL MACHINE CONTROL UNIT. The other strategy in DNC is to eliminate the regular NC controller altogether and replace it with a special machine control unit. The configuration is illustrated in Figure 9.4. This special MCU is a device that is specifically designed to facilitate communication between the machine tool and the computer. One area where this communication link is important is in circular interpolation of the cutter path. The special MCU configuration achieves a superior balance between accuracy of the interpolation and fast metal removal rates than is generally possible with the BTR system.

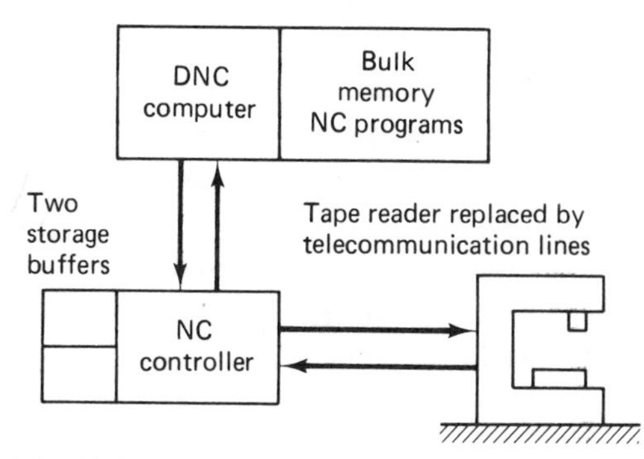


FIGURE 9.3 DNC with behind-the-tape reader (BTR) configuration.

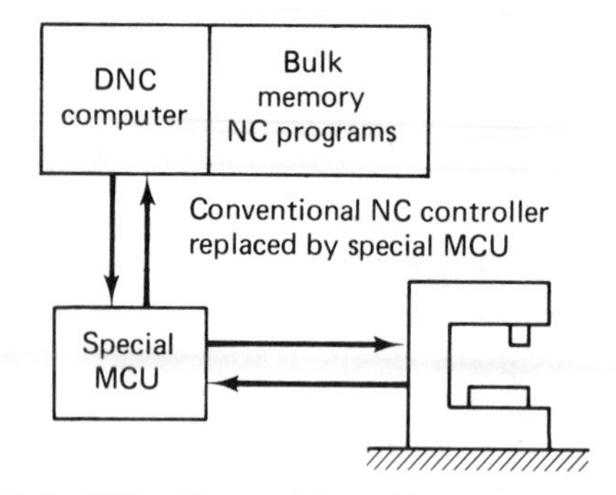


FIGURE 9.4 DNC with special machine control unit (MCU).

The special MCU is soft-wired, while the conventional NC controller is hardwired. The advantage of soft-wiring (the use of digital computer software rather than wired circuits) is its flexibility. Its control functions can be altered with relative ease to make improvements. It is much more difficult to make changes in the regular NC controller because rewiring is required.

At present, the advantage of the BTR configuration is that its cost is less, since only minor changes are needed in the conventional NC system to bring DNC into the shop. BTR systems do not require the replacement of the conventional control unit by a special MCU. However, this BTR advantage is a temporary one, since more and more NC machines are being sold with computer numerical control (CNC), to be discussed in the next section. The CNC controller serves much the same purpose as a special MCU when incorporated into a DNC system.

## Advantages of DNC

The advantages typically cited for DNC systems are as follows:

1. Time sharing—the control of more than one machine by the computer.
2. Greater computational capability for such functions as circular interpolation.
3. Remote computer location—the computer is located in a computer-type environment.
4. Elimination of tapes and tape reader at the machine for improved reliability.
5. Elimination of hardwired controller unit on some systems.
6. Programs stored as cutter location data can be post-processed for whatever suitable machine is assigned to process the job.

One obvious question concerning DNC is this: What happens if the computer breaks down? The answer is that production stops. In practice, this has not turned out to be too much of a problem simply because the central computer is so much more reliable than the conventional NC machine.

## 9.3 COMPUTER NUMERICAL CONTROL (CNC)

DNC is only one of two approaches in which the computer is used to control the NC machine. Chronologically, DNC came first. The initial DNC systems appeared commercially in the mid-to-late 1960s. Since then the physical size of the digital computer has been reduced at the same time that its computational capabilities have been increased. The result of these improvements has been the development of a new systems concept in NC: computer numerical control.

CNC is an NC system that utilizes a dedicated, stored-program computer to perform some or all of the basic numerical control functions. As of this writing, the typical CNC system uses a minicomputer as the controller unit. It is expected that in future generations of computer numerical control, microcomputers will become predominant.

Because a digital computer is used in both CNC and DNC, there is often confusion surrounding the two system types. Following are the principal differences between the DNC and CNC:

1. DNC computers distribute instructional data to, and collect data from, a large number of machines. CNC computers control only one machine, or a small number of machines.
2. DNC computers occupy a location that is typically remote from the machines under their control. CNC computers are located very near their machine tools.
3. DNC software is developed not only to control individual pieces of production equipment, but also to serve as part of a management information system in the manufacturing sector of the firm. CNC software is developed to augment the capabilities of a particular machine tool.

Except for the fact that a digital computer is used, CNC has more in common with conventional NC than it does with DNC. The external appearances of the NC and CNC machines are similar, and the part programs are entered in a similar manner. Punched tape readers are still the common device for entering the part program into the system. However, with conventional numerical control, the punched tape is cycled through the reader for every workpiece in the batch. With CNC, the program is entered once and then stored in the computer memory. Thus, the tape reader is used only for the original loading of the part program and data. Compared to regular NC, CNC offers additional flexibility and computational capability. New system options can be incorporated into the CNC controller simply by reprogramming the unit. Because of this reprogramming capacity, both in terms of part

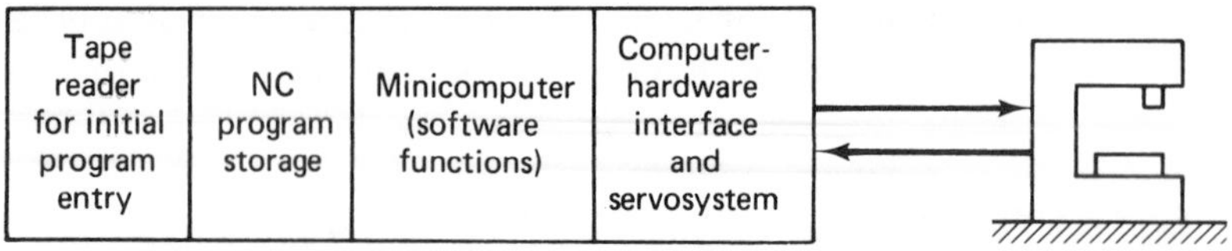

| Tape reader for initial program entry | NC program storage | Minicomputer (software functions) | Computer-hardware interface and servosystem |
|---|---|---|---|

FIGURE 9.5 General configuration of computer numerical control (CNC) system.

program and systems options, CNC is often referred to as softwired numerical control. This is a big distinction relative to the conventional hardwired NC controller. Figure 9.5 shows the general CNC system.

### Advantages of CNC

Among the advantages of computerized numerical control over regular NC are the following:

1. *The part program tape and tape reader are used only once to enter the program into memory.* This results in improved reliability, since the tape reader is commonly considered the least reliable component of a conventional NC systm.
2. *Tape editing at machine site.* The NC tape can be optimized during tape tryout at the site of the machine tool.
3. *Greater flexibility.* The most significant advantage over conventional NC is CNC's flexibility. New options can be added to the system easily and at relatively low cost.
4. *Metric conversion.* CNC can accommodate conversion of tapes prepared in units of inches into the new International System of units.
5. *Total manufacturing system.* CNC is more compatible with the use of a total manufacturing information system.

Considering this last point, what we will no doubt see in the future is more integration of CNC and DNC. The CNC computer will be used for machine tool control while management/manufacturing information about the performance of the process will be channeled to a central computer.

## 9.4 ADAPTIVE CONTROL MACHINING[1]

### Introduction

One of the principal reasons for using numerical control (including DNC and CNC) is that NC reduces the nonproductive time in manufacturing. This is accomplished through a reduction in the following elements, which constitute

[1]This section on adaptive control is based largely on Groover [7].

a significant portion of total production time:

1. Workpiece handling.
2. Setup of the job.
3. Lead times between receipt of an order and production.
4. Tool changes.
5. Operator delays.

Because these nonproductive elements are reduced relative to total production time, a larger proportion of the machine tool's time is spent in actually machining the workpart. Although NC has a significant effect on downtime, it can do relatively little to reduce the in-process time compared to a conventional machine tool. The most promising answer for reducing the in-process time lies in the use of adaptive control (sometimes abbreviated AC). Whereas numerical control guides the sequence of tool positions or the path of the tool during machining, adaptive control determines the proper speeds and/or feeds during machining as a function of variations in such factors as work-material hardness, width or depth of cut, air gaps in the part geometry, and so on. Adaptive control has the capability to respond to and compensate for these variations during the process. Numerical control does not have this capability. Accordingly, adaptive control should be utilized in applications where the following conditions are found:

1. The in-process time consumes a significant portion of the total production time.
2. There are significant sources of variability in the job for which adaptive control can compensate. In essence, adaptive control adapts speed and/or feed to these variable conditions.

Our discussion of adaptive control in this section will be limited to its application in the machining process. In Section 15.4 we shall treat adaptive control strategies at a more general level.

### Adaptive control defined

For a machining operation, the term *adaptive control* denotes a control system that measures certain output process variables and uses these to control speed and/or feed. Some of the process variables that have been used in adaptive control machining systems include spindle deflection or force, torque, cutting temperature, vibration amplitude, and horsepower. In other words, nearly all the metal-cutting variables that can be measured have been tried in experimental adaptive control systems. The motivation for developing an adaptive machining system lies in trying to operate the process more efficiently. The typical measures of performance in machining have been metal removal rate and cost per volume of metal removed.

The chronological development of machining adaptive control has been interesting. Starting in the early 1960s, the Bendix Research Laboratories began their attempts to develop an adaptive controller that could be used for metal machining and other processes. This work was sponsored by the U.S. Air Force. At about the same time, Cincinnati Milacron also initiated work on a similar system. What they both found was that it was extremely difficult to develop practical systems that could measure the true performance of the machining process. The reason was the general inability to measure the important process variables accurately in a machine shop environment. They also found that these initial systems were very expensive. Consequently, the adaptive control machines that were finally put into operation were somewhat less sophisticated (and less expensive) than the research adaptive systems developed earlier. The difference between the practical AC systems and the earlier research AC systems prompted the definition of two distinct forms of adaptive control for machining.

ADAPTIVE CONTROL OPTIMIZATION (ACO). These systems are represented by the early Bendix research on adaptive control machining. In this form of adaptive control, an index of performance is specified for the system. This performance index should be a measure of overall process performance, such as production rate or cost per volume of metal removed. The objective of the adaptive controller is to optimize the index of performance by manipulating speed and/or feed in the operation.

ADAPTIVE CONTROL CONSTRAINT (ACC). These are represented by the systems that were ultimately employed in production. In this form of adaptive control, constraint limits are imposed on the measured process variables. The objective of the adaptive controller is to manipulate speed and/or feed to maintain the measured variables at or below their constraint limit values.

Current-day adaptive control machining systems generally fall into the second category—adaptive control constraint systems. Basically, most ACO systems attempt to maximize the ratio of work material removal rate to tool wear rate. In other works, the index of performance is

$$\text{IP} = \text{a function of } \left(\frac{\text{MRR}}{\text{TWR}}\right)$$

where MRR = material removal rate

TWR = tool wear rate

The trouble with this performance index is that TWR cannot be measured on-line with today's measurement technology. Hence, the IP above cannot be monitored during the process. Eventually, sensors will be developed to a level at which the true process performance can be measured on-line. When this occurs, adaptive control optimization systems will become more prominent.

It was previously stated that AC should be applied in situations in which there are significant sources of process variability. Let us consider why a machining operation might be an attractive candidate for applying adaptive control.

### Sources of variability in machining

The following are the typical sources of variability in machining where adaptive control can be most advantageously applied. Not all of these sources of variability need be present to justify the use of AC. However, it follows that the greater the variability, the more suitable the process will be for using adaptive control.

1. *Variable geometry of cut in the form of changing depth or width of cut.* In these cases, feed rate is usually adjusted to compensate for the variability. This type of variability is often encountered in profile milling or contouring operations.
2. *Variable workpiece hardness and variable machinability.* When hard spots or other areas of difficulty are encountered in the workpiece, either speed or feed is reduced to avoid premature failure of the tool.
3. *Variable workpiece rigidity.* If the workpiece deflects as a result of insufficient rigidity in the setup, the feed rate must be reduced in order to maintain accuracy in the process.
4. *Tool wear.* It has been observed in research that as the tool begins to dull, the cutting forces increase. The adaptive controller will typically respond to tool dulling by reducing the feed rate.
5. *Air gaps during cutting.* The workpiece geometry may contain shaped sections where no machining needs to be performed. If the tool were to continue feeding through these so-called air gaps at the same rate, time would be lost. Accordingly, the typical procedure is to increase the feed rate, by a factor of two or three, when air gaps are encountered.

These sources of variability present themselves as time-varying and, for the most part, unpredictable changes in the machining process. We shall now examine how adaptive control can be used to compensate for these changes.

### A typical adaptive control system

A typical practical application of AC is in profile or contour milling jobs on a numerical control machine using feed as the controlled variable, and spindle deflection (to measure force) or horsepower, or both, as measured variables. It is common to attach an adaptive controller to an NC machine tool. Numerical control machines are a natural starting point for AC because of two reasons. First, NC machine tools often possess the required servomotors on the table axes to accept automatic control. Second, the usual kinds of machining jobs for which NC is used possess the sources of variability that make AC feasible. Several large companies have retrofitted their NC machines to incorporate AC capabilities. In fact, one company called Macotech Corporation in Seattle specializes in retrofitting for other companies.

The control strategy is of the ACC type rather than ACO type: constraint limits are established for the measured process variables. For example, if cutter deflection is the measured variable, the value of the maximum spindle deflection which the particular cutter and machine tool spindle can withstand is calculated. This value becomes the operating level of spindle deflection. Maximum production rates are obtained by running the machine at the highest feed rate consistent with this force level. Since force is dependent on such factors as depth and width of cut, the end result of the control action is to maximize metal removal rates within the limitations imposed by existing cutting conditions.

Figure 9.6 presents a schematic diagram of the adaptive control machining system. It operates on the principle of maintaining a constant cutter force

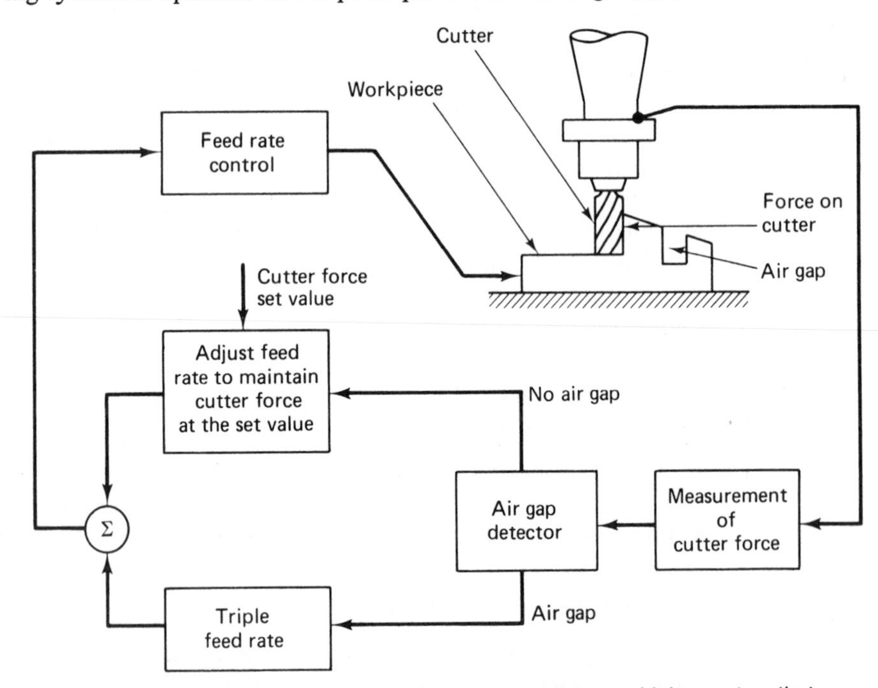


FIGURE 9.6 Configuration of typical adaptive control machining system that uses cutter force as the measured process variable.

during the machining operation. When the force increases due to increased workpiece hardness or depth or width of cut, the feed rate is reduced to compensate. When the force decreases, owing to decreases in the foregoing variables or air gaps in the part, feed rate is increased to maximize the rate of metal removal.

Figure 9.6 shows the presence of an air gap override feature which monitors the cutter force and determines if the cutter is moving through air or through metal. This is usually sensed by means of a low threshold value of cutter force. If the actual cutter force is below this threshold level, the controller assumes that the cutter is passing through an air gap. When an air gap is sensed, the feed rate is doubled or tripled to minimize the time wasted traveling across the air gap. When the cutter reengages metal on the other side of the gap, the feed reverts back to the cutter force mode of control.

More than one process variable may be measured in an adaptive control machining system. Originally, attempts were made to employ three measured signals in the Bendix system: temperature, torque, and vibration. Currently, the Macotech system has used both cutter load and horsepower generated at the machine motor. The purpose of the power sensor is to protect the motor from overload when metal removal rate is constrained by spindle horsepower rather than spindle force.

## Benefits of adaptive control in machining

A number of potential benefits accrue to the user of an adaptive control machine tool. The advantage gained will depend on the particular job under consideration. There are obviously many machining situations for which adaptive control cannot be justified.

1. *Increased production rates.* Productivity improvement was the motivating force behind the development of adaptive control machining. On-line adjustments to allow for variations in work geometry, material, and tool wear provide the machine tool with the capability to achieve the highest metal removal rates that are consistent with existing cutting conditions. This capability translates into more parts per hour. Given the right application, adaptive control will yield significant gains in production rate compared to conventional machining or numerical control. The potential for improvement in production rate is illustrated in Figure 9.7.

2. *Increased tool life.* In addition to higher production rates, adaptive control will generally provide a more efficient and uniform use of the cutter throughout its tool life. Because adjustments are made in the feed rate to prevent severe loading of the tool, fewer cutters will be broken.

3. *Greater part protection.* Instead of setting the cutter force constraint limit on the basis of maximum allowable cutter and spindle deflection, the force limit can be established on the basis of work size tolerance. In this

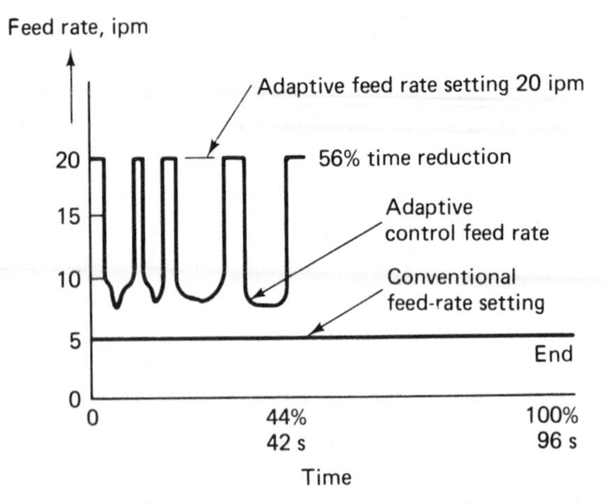


FIGURE 9.7 Two milling passes across same workpart, one with conventional feed-rate setting, the other with adaptive control. (Reprinted from Dallas [3].)

way, the part is protected against an out-of-tolerance condition and possible damage.

4. *Less operator intervention.* The advent of adaptive control machining has transferred the control over the process even further out of the hands of the operator and into the hands of management via the part programmer.

5. *Easier part programming.* A benefit of adaptive control which is not so obvious concerns the task of part programming. With ordinary numerical control, the programmer must plan the speed and feed for the worst conditions that the cutter will encounter. He or she may have to try out the program several times before being satisfied with the choice of conditions. In adaptive control part programming, the selection of feed is pretty much left to the controller unit rather than to the part programmer. The constraint limit on force, horsepower, or other variable must be determined according to the particular job and cutter used. However, this can often be calculated from known parameters for the programmer by the system software. In general, the part programmer's task requires a much less conservative approach than for numerical control. Less time is needed to generate the tape for a job, and fewer tryouts are necessary.

Several years ago a study was conducted for the U.S. Air Force to quantitatively evaluate the advantages of adaptive control over conventional NC machining [17]. The test was performed using a low-cost retrofitable adaptive control (LCRAC) system whose control mode was similar to that described earlier in this section as a typical AC system. The study was divided into three sequential stages, each of which compared the AC system against conventional NC. Different test parts and machines were used in the three

TABLE 9.1 Comparison of Metal Cutting Times—Conventional NC versus Low-Cost Retrofitable Adaptive Control (LCRAC) [17]

| | Metal cutting time (min) | | | |
|---|---|---|---|---|
| | *NC time* | *LCRAC time* | *Savings* | *Percent* |
| Aluminum | | | | |
| Phase I | 21 | 18 | 3 | 14 |
| Phase II | 83 | 49 | 34 | 41 |
| Phase III | 381 | 248 | 133 | 37 |
| Total aluminum | 485 | 315 | 170 | 35% |
| Steel | | | | |
| Phase I | 289 | 202 | 87 | 30 |
| Phase II | 102 | 75 | 27 | 26 |
| Phase III | 1151 | 731 | 400 | 35 |
| Total steel | 1542 | 1028 | 514 | 33% |
| Titanium | | | | |
| Phase I | 182 | 113 | 66 | 36 |
| Phase II | 72 | 37 | 34 | 48 |
| Phase III | 338 | 168 | 170 | 50 |
| Total titanium | 591 | 318 | 273 | 46% |
| Program total | 2618 | 1661 | 957 | 37% |

stages. Parts were made of aluminum, steel, and titanium. Table 9.1 is taken from the report and shows the overall savings from adaptive control to be 37%.

When adaptive control machining is applied to appropriate jobs, the economic savings can be substantial.

## 9.5 INDUSTRIAL ROBOTS

An industrial robot is a general-purpose, programmable machine possessing certain anthropomorphic characteristics. It is most typically used for parts handling tasks but can also be used in conjunction with a variety of manufacturing processes. The robot can be programmed to carry out a sequence of mechanical movements. It will perform that sequence over and over again until reprogrammed to perform some other motion cycle. General-purpose industrial robots are most likely to be economical and practical in applications with the following characteristics:

1. *Hazardous working conditions.* In job situations where there are potential dangers to a human operator or where the workplace is hot and uncomfortable, industrial robots are likely candidates for the job. Robots have become especially important in these types of job situations since the Occupational Safety and Health Act (OSHA) became effective in 1971.

2. *The job is repetitive.* Many manufacturing jobs fit this description. Even if the cycle is long and involves a sequence of many separate moves, an industrial robot may be feasible. One requirement is that the sequence of actions must not change from one cycle to the next.

3. *The workpart to be moved is heavy.* Some industrial robots are capable of lifting items as heavy as several hundred pounds.

A sampling of tasks performed by industrial robots would include the following more typical applications:

*Parts handling.* A large variety of pick-and-place jobs, moving workparts from one location and repositioning them at another location.

*Machine loading and unloading.* The types of production equipment involved include stamping presses, forge presses, die-casting machines, injection molding machines, and most types of metal-cutting machine tools. In many instances, the robot is set up to control and synchronize the operation of equipment.

*Spray painting.* The spray paint nozzle is attached to the robot's arm. The arm is programmed to move through a sequence of continuous-path motions to complete the painting operation.

FIGURE 9.8 Industrial robot used for loading and unloading workparts in milling operation. (Courtesy of Unimation, Inc.)

*Welding.* Both spot welding and continuous welding. Perhaps the most typical applications in this category are on body fabrication lines of the automobile manufacturers.

*Assembly.* In simple mechanical assembly, robots perform operations which are basically an extension of their pick-and-place motions.

Several applications of industrial robots are pictured in Figures 9.8 and 9.9.

Worldwide there are approximately 200 models of robots produced by nearly 100 manufacturers. The countries that are taking the lead in robot development and use seem to be Japan, the United States, the United Kingdom, and Italy [18]. The beginnings of the robot in the United States can be traced to the inventor George Devol, who began patenting his industrial robot designs in the early 1950s. In 1958, Devol licensed a firm to begin to fabricate several of the machines. Three prototypes were built in 1962, and by 1966, seventy robots had been built by manual methods. This firm was to become Unimation, Inc., one of the major producers today of industrial robots. Other important U.S. firms include AMF Versatran and Cincinnati Milacron. As of this writing there are an estimated 20,000 robots working in factories throughout the world.

FIGURE 9.9 Industrial robots used to perform spot welding operations on automobile assembly line. (Courtesy of Unimation, Inc.)

## Robot types and features

There are a number of ways to classify industrial robots. Among the ways we will consider are according to general physical configuration, point-to-point versus continuous-path, and the degrees of motion freedom possessed by the robot. Programming of the robot also represents a method of classification, but we will consider this topic in a separate subsection.

General Physical Configuration. There are two principal robot configurations: *polar* and *cylindrical.* The two types are illustrated in Figure 9.10. In the polar configuration the body of the robot pivots either horizontally or vertically or both. Attached to the body is an arm that is moved by the body motion. In turn, a gripping device or hand is attached to the arm. In the cylindrical configuration, the robot body is a vertical column that swivels around a vertical axis. The arm of the robot is in a horizontal orientation and can be made to move up or down and in or out with respect to the body.

Degrees of Freedom. There are six basic motions or degrees of freedom in the design of a robot. These six degrees of freedom are intended to emulate the versatility of motion possessed by the human operator. In current robot design there are up to six basic motions of which the robot is capable: three arm and body motions and three wrist movements. The mechanical hand movement is not usually considered one of the basic degrees of freedom.

*Arm and body motions:*

1. *Vertical traverse.* Up-and-down motion of the arm. In the polar configuration this is caused by movement of the body.
2. *Radial traverse.* Extension and retraction of the arm (in-and-out movement).
3. *Rotational traverse.* Rotation about the vertical axis (right or left swivel of the robot body).

*Wrist motions:*

4. *Wrist swivel.* Rotation of the wrist.
5. *Wrist bend.* Up-or-down movement of the wrist, which also involves a rotational movement.
6. *Wrist yaw.* Right-or-left swivel of the wrist.

Figure 9.10 pictures these basic motions for the polar and cylindrical robot configurations.

Although it does not count as one of the six basic motions, the robot's hand usually includes some form of mechanical action. It typically consists of a clamp or jaws or some other form of gripping device. This appendage of the

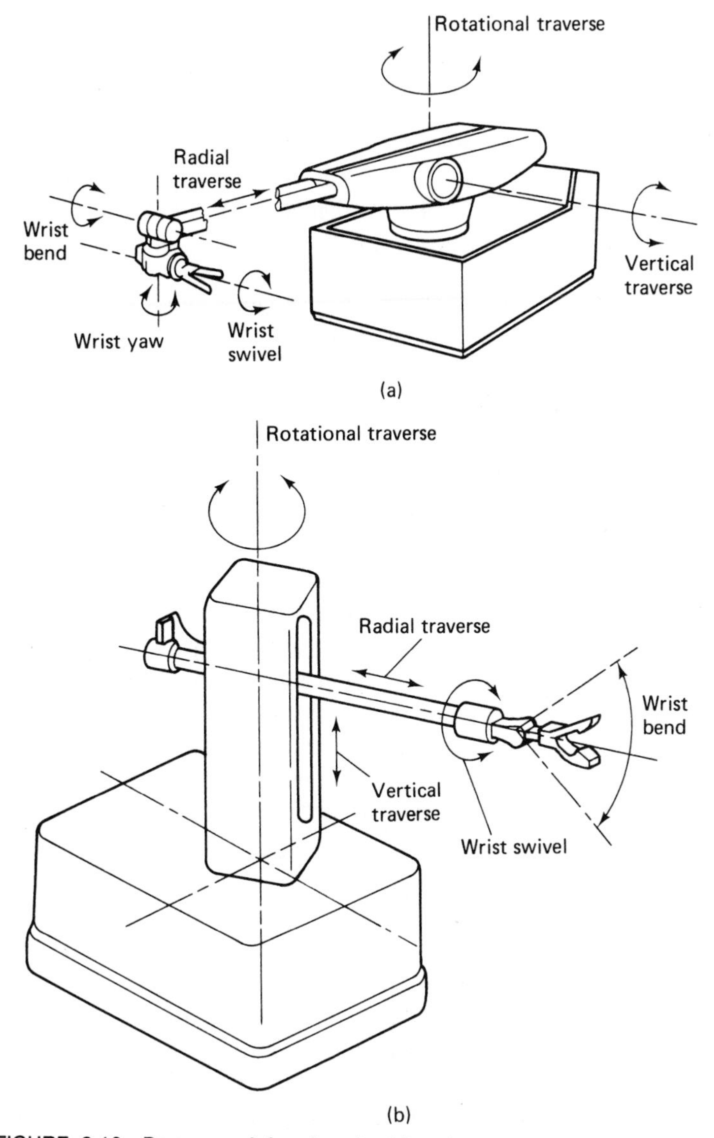


FIGURE 9.10 Degrees of freedom in (a) polar and (b) cylindrical robot configurations. (Reprinted from Whetham [17].)

robot must often be designed specifically for the task assigned to the robot. In a sense, it constitutes the special tooling that permits the general-purpose robot to do a specific job.

PTP VERSUS CONTOURING. As with NC machining systems, industrial robots can be classified as point-to-point or continuous-path.

PTP robot systems constitute an estimated 90% of the market. With PTP, the robot's movement is controlled from one point location in space to another. Each point is programmed into the robot's control memory and then played back during the work cycle. PTP motions would be quite satisfactory for most pick-and-place operations as well as machine loading and unloading jobs. It is also possible to synchronize the activities of the robot with other equipment. In such an arrangement, the robot's program is temporarily interrupted while it waits for the production equipment to complete its cycle.

Contouring robots possess the capacity to follow a closely spaced locus of points which describe a smooth compound path. The control of the path increases the memory requirements of the robot. Also, greater accuracy of arm position and movement are needed for many contouring applications. All of this means that continuous-path robots are more expensive than PTP types. However, in certain industrial operations, continuous-path control is necessary. Among these operations are paint spraying, continuous welding processes, laying down an adhesive bead along a complicated path, grasping an object that is moving, and other tasks.

## Programming of industrial robots

In the introduction to this chapter, an industrial robot was defined as an NC machine that performs material handling and other functions. Many robots use controls very similar to those utilized in NC machine tools. Other industrial robots make use of other programming and control schemes.

There are three principal ways to program a robot, depending on the type of memory system used in the robot:

1. *Manual method*. This programming method is used for simpler robots which have memories consisting of relay-type sequences or a pinboard matrix or some similar electrical–mechanical devices. The program is set up by fixing stops, setting switches, arranging wires, and so on. Most robots using this programming method do not have a full six degrees of freedom. Typically, the range is two to four axes of motion. Also, the number of programming steps is somewhat limited on these robots.

An extension of this programming method involves the use of punched cards or tapes containing the program. This increases the number of programming steps by an order of magnitude. Several models of robots in Japan using punched tape are capable of up to 1000 steps.

2. *Walkthrough*. More-complex robots use magnetic tape or discs as the memory unit. Recent models are even employing small computers as the memory unit. With these robots programming is carried out in either of two ways: walkthrough or leadthrough. In the walkthrough method, the programmer "teaches" the robot by manually moving the arm and hand through the sequence of motions or positions. At each move it signals the memory to record the position for future playback during the work cycle.

3. *Leadthrough.* This is the other programming scheme used for more sophisticated robots with magnetic memories. In the leadthrough method, the programmer power drives the robot through its sequence of moves by means of a control console. Each move and position is recorded in the robot's memory.

## 9.6 CURRENT TRENDS IN NC

It seems appropriate to summarize these three chapters on numerical control by discussing some of the current trends in the technology and what the future might hold.

The direct numerical control systems that were marketed in the late 1960s and early 1970s were extremely expensive. Their high cost, combined with an unfavorable economic climate at that time, caused businessmen to resist the temptation to plunge into the new DNC technology. Also, the DNC systems available then were somewhat rigid in terms of management reporting formats and hardware requirements. The more recent advent of CNC systems, together with lower-cost computers and improvements in software, have resulted in the development of hierarchical computer systems in manufacturing. In these hierarchical systems, CNC computers have direct control over the production machines and report to satellite minicomputers, which in turn report to other computers, and so on. There are advantages to this hierarchical approach over the DNC packages that were offered around 1970. The common theme in these advantages is flexibility. The information system can be tailored to the specific needs and desires of the firm. This contrasts with many of the early DNC systems, in which the reporting formats were fixed, in some cases providing more data than management wanted and in other cases omitting details that management needed. Another advantage of the hierarchy approach is the ability to gradually build the system instead of implementing the entire DNC configuration all at once. This piece-by-piece installation of the computer-integrated manufacturing system is a more versatile and economic approach. It permits changes and corrections to be made more easily as the system is being built. It also allows the company to spread the cost of the system over a longer time period and to obtain benefits from each subsystem as it is installed. The hierarchical computer arrangement embraces the DNC philosophy, which is to provide useful reports on production operations to management in real time. One might say that DNC has not really been replaced by this new approach; it has simply altered its physical form. Reference [13] provides a good application study of a modern DNC installation at General Electric Company's Evendale works. In Section 11.7, we will examine the general characteristics of the hierarchical configuration of computers in manufacturing.

Another development in DNC/CNC is the introduction of *flexible manufacturing systems,* or FMS [9]. An FMS is a group of NC machines or other mechanized workstations interconnected by a materials handling system, all controlled by computer. It is an attempt to combine the flexibility of NC with the efficiency of automated flow lines. Chapter 19 will concentrate on the topic of flexible manufacturing systems.

One advance on the horizon in part programming which may find widespread use in the future is *voice programming*—NC part programming by voice commands [20]. The part programmer reads the part drawing and states the coordinates and tool motions required to process the part. In effect, he explains the job to a computerized part programming system. In turn, the system recognizes the discrete wave pattern of different words spoken by the human voice and converts these into the punched NC tape. The system uses a computer with significant core memory in order to possess the required word-recognition capability. The advantage envisioned for voice programming is reduced tape preparation time. The part programmer should be able to talk his or her way through the NC commands more quickly than he or she can write them, hence increasing productivity in the part programming function.

In robotics, research and development efforts are proceeding to provide industrial robots with sensory capabilities, including vision ([14], [19]). The robot is equipped with a TV camera focused on the workplace. Pattern recognition techniques are then implemented using computer technology to identify part defects and improperly oriented components.

The reader may have noted that all the trends mentioned above involve the use of computer technology. The computer is becoming an important factor in manufacturing technology. Acknowledgment of the tremendous potential for the computer in the production environment brings us to the next major topic in this book on automation: computer-aided manufacturing.

## REFERENCES


[1] ARONSON, R. B., "Let the Robot Do It," *Machine Design,* November 27, 1975, pp. 54–59.

[2] CARLISLE, R., "All You Ever Wanted to Know about the Mac 40 Adaptive Control System," *N/C Commline,* November/December, 1978, pp. 24–25.

[3] DALLAS, D. B. (Editor), *Tool and Manufacturing Engineers Handbook,* 3rd ed., Society of Manufacturing Engineers, Dearborn, Mich. (McGraw-Hill Book Company, New York), 1976, Chapter 12.

[4] DALLAS, D. B., "The Robot Enters the System," *Manufacturing Engineering,* February, 1979, pp. 70–73.

[5] DAWSON, B. L., "A Computerized Robot Joins the Ranks of Advanced Manufacturing Technology," *Proceedings, Thirteenth Annual Meeting and Technical*

*Conference,* Numerical Control Society, March, 1976, Cincinnati, Ohio, pp. 339–356.

[6] GROOVER, M. P., "A New Look at Adaptive Control," *Automation,* Vol. 20, No. 4, April, 1973, pp. 60–63.

[7] GROOVER, M.P., "Adaptive Control and Adaptive Control Machining," *Educational Modules,* MAPEC, Copyright, Purdue Research Foundation, 1977. All rights reserved.

[8] HARRINGTON, J., *Computer Integrated Manufacturing,* Industrial Press, Inc., New York, 1973.

[9] KEARNEY & TRECKER CORP., *Understanding Manufacturing Systems* (a series of technical papers), Milwaukee, Wis.

[10] MATHIAS, R., "New developments in Adaptive Control of Milling Machines," *Proceedings, Thirteenth Annual Meeting and Technical Conference*, Numerical Control Society, March, 1976, Cincinnati, Ohio, pp. 157–166.

[11] MUNSON, G., "Flexible Automation," *N/C Commline,* September/October, 1978, pp. 28–31.

[12] PRESSMAN, R. S., and WILLIAMS, J. E., *Numerical Control and Computer-Aided Manufacturing,* John Wiley & Sons, Inc., New York, 1977.

[13] SCHAFFER, G., "Stepping Up to CAM," *American Machinist,* November, 1978, pp. 87–90.

[14] SMITH, T. C., "N/C's Stake in the Productivity Race," *N/C Commline,* September/October, 1978, pp. 20–27.

[15] TANNER, W. R. (Editor), *Industrial Robots,* Vol. I: *Fundamentals,* Vol. II: *Applications,* Society of Manufacturing Engineers, Dearborn, Mich., 1979.

[16] Unimation Inc., *Unimate Industrial Robot System Planbook*, Danbury, Conn.

[17] WHETHAM, W. J., "Low Cost Adaptive Control Unit Manufacturing Methods," *Technical Report AFML-TR-73-263*, Manufacturing Technology Division, Air Force Materials Laboratory, Air Force Systems Command, November, 1973.

[18] WINSHIP, J., "Robots in Metalworking," *American Machinist,* November, 1975, pp. 87–110.

[19] WINSHIP, J., "Update on Industrial Robots," *American Machinist,* January, 1979, pp. 121–124.

[20] "A Look at NC Today," *Manufacturing Engineering,* Vol. 73, No. 10, October, 1975, pp. 24–28.

[21] "Off-the-Shelf Automation," Staff Report, *Automation,* May, 1976, pp. 70–75.

## PROBLEMS

**9.1.** Discuss the probable suitability of adaptive control in each of the following machining situations:

(a) Drilling a series of 10-mm-diameter holes in aluminum. Each hole is to be 20 mm deep.

(b) Peripheral milling at a depth of cut of 3.0 mm across a steel bar of rectangular cross section. The steel is C1020. The dimensions of the bar are: length = 1.0 m, cross section 50 mm by 90 mm.

(c) End milling a cast-iron casting where there are several flats to be milled at various depths of cut. The casting is a gear box housing for a piece of earthmoving machinery.

**9.2.** A milling cut is to be taken across a workpiece that is 500 mm long. The average depth of cut along this length is 5 mm and the maximum depth is 8 mm. If the job is machined using NC, the feed rate must be set according to worst-case conditions for the maximum depth throughout the entire length. But if the job is done using adaptive control, the feed rate will be adjusted throughout the 500-mm length for differences in depth of cut. The feed rate will be changed in an inverse relationship to depth as provided by the following equation:

$$f = \frac{10}{d}$$

where $f$ = feed rate, mm/s
$d$ = depth of cut, mm

Hence, the feed rate using NC would be 1.25 mm/s throughout the entire length. The feed rate under adaptive control would average 2 mm/s but would vary, reaching as low a value as 1.25 mm/s when the depth was 8 mm. Determine how long the job would take to mill under each of the two alternatives: NC versus adaptive control.

**9.3.** A workpiece is to be machined in an end-milling operation. Thirty separate cuts are to be made on the piece, at various depths of cut. The accompanying table gives the number of cuts made at each depth, together with the cumulative length of cut at each depth.

| *Number of cuts* | *Depth of cut (mm)* | *Total length of cut (mm)* |
|---|---|---|
| 7 | 3 | 400 |
| 5 | 10 | 200 |
| 12 | 5 | 250 |
| 6 | 1 | 150 |
| 29 | Air gaps | 600 |

The air gaps occur both as a result of the workpiece geometry and when the tool is being positioned for the next cut.

If the workpiece is machined using NC, the feed rate would be set to allow for worst-case conditions (depth of cut = 10 mm). A feed rate of 1 mm/s would be used throughout the piece, including the air gaps.

If the workpiece is machined using adaptive control, the feed rate would be changed throughout the sequence of cuts according to depth. The control system monitors cutting force at the cutter and adjusts feed inversely with force as provided in the following equation:

$$f = \frac{1000}{F_c}$$

where $f$ = feed rate, mm/s
$F_c$ = cutter force, newtons (N)
1000 = cutter force set value for this cutter

The cutting force $F_c$ is assumed to vary directly with depth of cut as follows:

$$F_c = 100d$$

where $d$ is depth of cut in mm.

During air gaps the control system is programmed to feed the cutter at 15 mm/s.

(a) Determine how long the job would take to machine using NC.
(b) Determine how long the job would take to machine using adaptive control.
(c) What are the percent savings in time, adaptive control versus NC?
(d) Assume that the NC system could be programmed to account for the presence of air gaps so that the cutter could feed through air at 15 mm/s. How would this affect the total machining time under NC?
(*Note:* In your computations for this problem, ignore the effects of machine tool dynamics.)

**9.4.** In Problem 9.2 or 9.3, explain how an increase in the variability of depth of cut gives adaptive control the advantage over NC. If there were no variation in depth throughout the workpiece in these problems, would there be any advantage in using adaptive control?

**9.5.** An existing NC machine is being considered as a retrofit candidate for adaptive control. Retrofitting will cost $20,000. It is estimated that there will be a 37% average savings in machining time using adaptive control compared to the existing NC machining time. This estimate is based on the U.S. Air Force LCRAC Report [17]. The NC machine is currently operated three shifts—6000 h/yr. Time on the machine costs $25/h. The average job performed on this machine takes 100 h to complete and has a revenue value to the company of $3000/job. Of the 100 h/job, 60% represents nonproductive time, including setups, workpiece changing, and so on. The AC retrofitting will have no effect on this nonproductive time. The remaining average 40 h/job is machining time. The average programming cost per job is $300. It is estimated that there will be no change in programming cost per job if the retrofit is made.

(a) How much profit per job, if any, is the company currently making on the NC machine? Ignore any consideration of overhead costs not included in the data given above.
(b) Determine the increase in the number of jobs that can be run on the machine if the NC machine is retrofitted with adaptive control.
(c) Will the AC retrofit pay for itself in the first year of operation?

**9.6.** Three industrial robots are being considered to replace four manual welding operations on a major appliance housing fabrication line. Each robot will cost $40,000, including special tooling and programming for the welding operations. Annual maintenance and operating cost for each robot will be $5000. The current manual operators are paid at the rate of $17,000/year, including fringe benefits. Using a 5-year service life, $10,000 as the salvage value at the end of the service life, and a 10% rate of return, is the investment in robots worthwhile?

# part IV

# Computer-Aided Manufacturing

chapter 10

# Fundamentals of CAD/CAM

## 10.1 INTRODUCTION

The single most important trend in production automation is the increasing use of digital computers in the design and manufacture of products. CAD/CAM is the term used to describe this trend and it stands for computer-aided design and computer-aided manufacturing. CAD/CAM has been described as "the wave of the future in manufacturing" and "the new industrial revolution" and other similar expressions. The application of computers in design and manufacturing constitutes the most significant opportunity for substantial productivity gains in industry today. It is expected that this trend toward computerization will ultimately lead to the computer-integrated automatic factory before the end of this century.

A number of technological developments related to computer-aided manufacturing have been forecasted by leading world experts.[1] Among these forecasts are the following:

[1]Probably the best known of these manufacturing technology forecasts is the Delphi-type forecast by C.I.R.P. (International Institute for Production Engineering Research) of 1971. The University of Michigan and the Society of Manufacturing Engineers conducted similar studies in 1977.

1. Computer software for automation and optimization of all steps in the production of a workpart will be developed by 1980. At the time of this writing (January, 1979), most of this forecast has been realized. Software systems are available for automated NC part programming, control software is used in computer numerical control for machine tool table and spindle control, and computer software has been developed to optimize the machining process in adaptive control.

2. By 1985, entire manufacturing plants are expected to be automated and optimized under the control of a central computer.

3. By 1990, more than half of the machine tools built will be components in flexible manufacturing systems rather than "stand-alone" machines. These manufacturing systems will feature automatic workpart handling between machines, not unlike the flow line systems discussed in Chapters 4 and 5. The use of computers to control the machines and workpiece movement will also be a characteristic of these flexible manufacturing systems.

These technological developments represent complex computerized systems, many of the components for which we have already discussed in previous chapters (automated NC part programming, adaptive control machining, automatic workpart handling in flow line production systems, etc.). Computer-aided manufacturing can be viewed as a continuation in the development of industrial automation. This is the viewpoint adopted in this book. In the current chapter, we present a survey of CAD/CAM. The emphasis, of course, will be directed toward computer-aided manufacturing. Section 10.2 is an introductory discussion of computer-aided design. Section 10.3 provides a framework for dealing with the subject of CAM. In subsequent chapters of the book, the individual topics within this framework will be developed.

## 10.2 COMPUTER-AIDED DESIGN

Computer-aided design can be described as any design activity that involves the effective use of the computer to create or modify an engineering design. There are two basic reasons for using the computer in the design of a product:

1. *To increase the productivity of the designer.* This is accomplished by helping the designer to visualize the product and its component subassemblies and parts; and by reducing the time required in synthesizing, analyzing, and documenting the design.

2. *To create a data base for manufacturing.* In the process of creating the documentation for the product design (geometric specification and di-

mensions of the product and its components, material specifications for components, bill of materials, etc.), much of the required data base to manufacture the product is also created.

In this section, we will examine how the computer can be utilized to achieve these two purposes.

## The design process

The process of design is characterized by Shigley [20] as an iterative process, which consists of the following six phases:

1. Recognition of need.
2. Definition of problem.
3. Synthesis.
4. Analysis and optimization.
5. Evaluation.
6. Presentation.

Recognition of need involves the realization by someone that a problem exists for which some corrective action should be taken. This might be the identification of some defect in a current machine design by an engineer or the perception of a new product marketing opportunity by a salesperson. Definition of the problem involves a thorough specification of the item to be designed. This specification includes physical and functional characteristics, cost, quality, and operating performance.

Synthesis and analysis are closely related and highly iterative in the design process. A certain component or subsystem of the overall system is conceptualized by the designer, subjected to analysis, improved through this analysis procedure, and redesigned. The process is repeated until the design has been optimized within the constraints imposed on the designer. The components and subsystems are synthesized into the final overall system in a similar iterative manner.

Evaluation is concerned with measuring the design against the specifications established in the problem definition phase. This evaluation often requires the fabrication and testing of a prototype model to assess operating performance, quality, reliability, and other criteria. The final phase in the design process is the presentation of the design. This includes documentation of the design through drawings, material specifications, assembly lists, and so on. Essentially, the documentation requires that a design data base be created.

## Application of computers in the design process

The computer can beneficially be used in three phases of the design process:

1. Synthesis (engineering design).
2. Analysis and optimization (engineering analysis).
3. Presentation (preparation of the engineering drawings).

ENGINEERING DESIGN. Engineering design has traditionally been accomplished on drawing boards with the design being documented in the form of a detailed engineering drawing. Mechanical design includes the drawing of the complete product as well as its components and subassemblies, and the tools and fixtures required to manufacture the product. Electrical design is concerned with the preparation of circuit diagrams, specification of electronic components, and so on. Similar manual documentation is required in other engineering design fields (structural design, aircraft design, chemical engineering design, etc.). In each engineering discipline, the approach has traditionally been to manually synthesize a preliminary design and then to subject that design to some form of analysis. The analysis may involve sophisticated engineering calculations or it may involve a very subjective judgment of the aesthetic appeal possessed by the design. The analysis procedure identifies certain improvements that can be made in the design. As stated previously, the process is iterative. Each iteration yields an improvement in the design. The trouble with this iterative process is that it is time-consuming. Many engineering labor-hours are required to complete the design project.

Computer-assisted engineering design is performed by a CAD system rather than a single designer working over a drawing board. The CAD system is comprised of the following components [1]:

1. *The designer*.
2. *The hardware*—the computer and its peripheral equipment. (The operation of the computer and peripherals will be discussed in Chapter 11.)
3. *The software*—the general system software and the CAD software.
4. *The problem*—the design problem to be solved.

It is important to conceptualize a computer-aided design system in this fashion because the designer must work with the other system components much more closely than in the traditional approach in order to carry out the design process. In effect, the CAD system magnifies the powers of the designer. This has been referred to as the *synergistic effect*. The designer performs the portion of the design process which is most suitable to human intellectual skills (conceptualization, independent thinking); the computer performs the tasks best suited to its capabilities (speed of calculations, visual display, storage of large amounts of data), and the resulting system exceeds the sum of its components.

Modern CAD systems used in engineering design (synthesis phase) are based on *computer interactive graphics*. Computer graphics denotes a user-oriented system in which the computer is employed to create, transform, and display data in the form of pictures or symbols (see Figures 10.1 and 10.2). The user in the computer graphics design system is the designer, who communicates data and commands to the computer through a keyboard terminal. The computer communicates with the user via a cathode ray tube (CRT). The designer creates an image on the CRT screen by entering commands to call the desired software subroutines stored in the computer. The image is constructed out of basic geometric elements—points, lines, circles, and so on. It can be modified according to the commands of the designer—enlarged, reduced in size, moved to another location on the screen, rotated, and so on. Through these various manipulations, the required details of the design are formulated.

Some interactive design systems make use of a light pen together with the alphanumeric keyboard terminal to input data and commands to the system. The designer keys in the desired function and touches the CRT screen where the function is to be carried out. For example, a certain portion of the design image can be enlarged so that additional details can be "sketched" in. Two points displayed on the CRT screen can be connected with either solid or dashed lines. Many other functions are possible. The light pen feature has

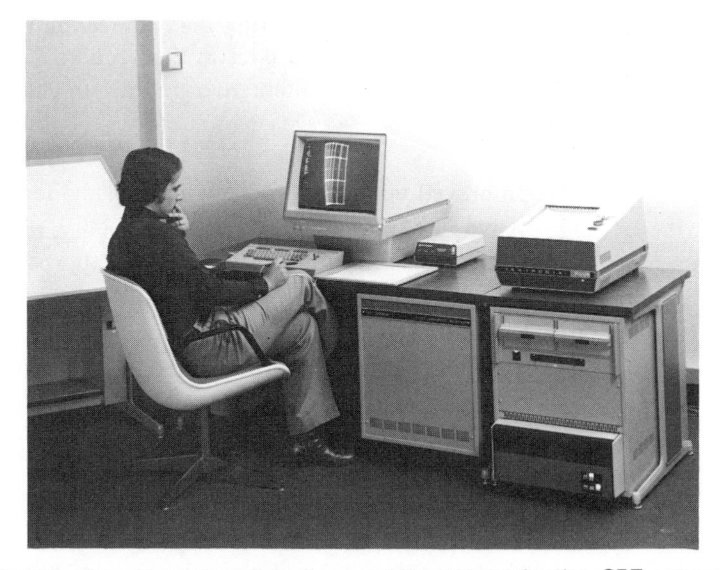

FIGURE 10.1 Computer interactive graphics system showing CRT screen, keyboard, and electronic pen. (Courtesy Tektronix, Inc.)

FIGURE 10.2 Close-up view of part design on CRT display. (Courtesy Gerber Scientific Instrument Company.)

been found to be a very fast and convenient mechanism by which the designer/draftsman can develop a new design. In effect, the interactive design system is an electronic tool that replaces the conventional drafting board technique.

The designer's creations are stored in digitized form in the computer's memory. These digital data can ultimately be converted into a "hard copy" of the design. The process of preparing the engineering drawing with the aid of the computer will be discussed later in this section. The benefits that can be achieved through the use of interactive graphics in engineering design are the following:

1. Improved visualization of the item being designed.
2. Ability to examine alternative designs in a relatively short period of time compared to the drawing board approach.
3. Ease in designing the mating parts that are to be assembled together in the product.
4. Capability to simulate the operation of the item being designed.
5. Ability to solve computational design problems conveniently and in real time.

It is believed that the use of interactive computer-aided design systems may increase productivity in engineering design by as much as tenfold. The reader interested in interactive computer graphics and CAD systems is referred to references [1], [3], [5], and [12].

ENGINEERING ANALYSIS. This is the phase in the design process which Shigley [20] refers to as analysis and optimization. In the formulation of nearly any product design, some engineering analysis is required. The analysis may involve stress–strain calculations, heat-transfer computations or the use of differential equations to describe the dynamic behavior of the system being designed. The computer can be used to aid in this analysis work. It is often necessary that specific programs be developed internally by the engineering analysis group to solve some particular design problem. An example of this might be a program based on the finite-element method to solve a heat-flow problem in some unusual component design. In other situations, commercially available general-purpose programs can be used to perform the engineering analysis. Examples of general programs would be simulation packages such as GASP (General Activity Simulation Program) and GPSS (General Purpose Systems Simulator) and statistical programs. More specific engineering analysis software has also been developed for commercial distribution. Examples of this software include ECAP (Electronic Circuit Analysis Program) developed by IBM Corporation, and programs for machine mechanism analysis such as ADAMS (Automatic Dynamic Analysis of Mechanical Systems).

PREPARATION OF ENGINEERING DRAWINGS. In future CAD/CAM systems, the engineering drawing may become obsolete. Today, it is an important means of documenting and communicating the designer's work. It is used by nearly every functional department in a manufacturing firm (not only design, but manufacturing and industrial engineering, machine operators and their foremen, tool design, production planning and control, sales, etc.). The techniques for producing and reproducing engineering drawings have improved over the years, and many of these modern techniques make use of the computer.

It was mentioned earlier that in interactive graphics, the design images on the CRT display could be stored in the form of digital data. Automated drafting machines are available which can convert these digital data into line drawings on paper.

One of the most common conversion devices is the electronic X-Y plotter, which is essentially a numerically controlled pen or pencil writing on a large piece of paper. The paper is attached to a large flat surface. On some drafting machines, the flat surface is a large horizontal table. On other models, the paper surface is in a nearly vertical orientation. Parallel tracks are located on two sides of the flat surface. A bridge is driven along these tracks to provide the $x$-coordinate motion. Attached to the bridge is another track, on which rides a writing head. Movement of the writing head relative to the bridge produces the $y$-coordinate motion. The writing head carries the pen or pencil, which can be raised or lowered to provide contact with the paper as desired. The control unit for the X-Y plotter accepts digitized data in the

FIGURE 10.3 Automated drafting machine. (Courtesy Gerber Scientific Instrument Company.)

form of punched tape or magnetic tape. Some drafting machines can be commanded directly by the computer in the CAD system. The size of the automated drafting tables ranges up to roughly 5 ft by 20 ft surface (approximately 1.5 by 6 m) with plotting accuracies approaching $\pm 0.001$ inch ($\pm 0.025$ mm). An automated drafting machine is shown in Figure 10.3.

The applications of these X-Y plotting systems are varied and go beyond the preparation of design engineering drawings. Some of these applications are listed below [8]:

Mechanical design drawings.

Printed circuit board layout and integrated circuit chip design. These drawings are used to prepare the artwork used in the manufacture of IC chips in the electronics industry [2].

Automobile body styling.

Aircraft wing and fuselage design.

Ship-hull design.

Geological survey contour maps.

Weather maps.

Road alignment and grade plots.

NC tool movement verification.

## Parts classification and coding systems

Parts classification and coding involves the grouping of similar part designs into classes, and relating the similarities by means of a coding scheme. Designers can use the classification and coding system to retrieve existing part designs rather than always redesigning new parts. There are several uses of such systems in manufacturing also, and we will postpone further discussion of this subject until Section 18.3.

### Creating the manufacturing data base

One of the main reasons for the interest in computer-aided design is that it offers the opportunity to develop the data base needed to manufacture the product. In the conventional manufacturing cycle practiced for so many years in industry, engineering drawings were prepared by design draftsmen and then used by manufacturing engineers to develop the process plan (i.e., the "route sheets"). The activities involved in designing the product were separated from the activities associated with process planning. Essentially, a two-step procedure was employed. This was both time-consuming and involved duplication of effort by design and manufacturing personnel. In the ideal CAD/CAM system, a direct link is established between product design and manufacturing. It is the goal of CAD/CAM not only to automate certain phases of design and certain phases of manufacturing, but also to automate the transition from design to manufacturing. Computer-based systems have been developed and are currently being developed which create not only much of the data and documentation for manufacturing but also in some cases the tools required to carry out the production operations.

One example of the fabrication of production tools is in the manufacture of integrated circuits (ICs) and printed circuit boards [2]. Complex integrated circuits are being produced on silicon chips no larger than $\frac{1}{4}$-in. square. Some of these ICs are sophisticated enough to be used as the central processing units in microcomputers (see Section 11.2 for a description of microcomputers). In the manufacture of integrated circuits, masks are required in the various stages of processing. Basically, the masks establish the circuit on the chip and must therefore be made with a high degree of accuracy. Nearly all producers of ICs use automated drafting machines of the type previously described to create the artwork for the masks. The integrated circuits are designed using interactive graphics, and the production tools (the masks) are fabricated using the digitized data created during design.

Another example of computer-based integration of design and manufacturing is in numerical control part programming. The mechanism that establishes the design/manufacturing interface is the computer interactive graphics system. The product designer uses the interactive graphics system to establish the geometry, dimensions, and tolerances for the various parts. These same design data can be displayed on the CRT for the process planner to use in preparing the NC part program. The part image can be displayed on the CRT screen while the part programmer calls for a circle representing the cutting tool to be displayed. Using a light pen or keyboard to provide commands to the system, the programmer directs the tool to move around the outline of the workpart. Multiple passes across the work, representing roughing and finishing cuts, can be programmed with relative ease. The computer stores all the tool movements so that the programmer can recall the motions from storage to check out the part program. In effect, the programmer has accomplished the same result as if the APT language had been used. The tool commands that are stored in memory can ultimately be used to prepare the NC punched tape. In future CAD/CAM systems, it will be possible, at least for certain workpart configurations, to automatically generate the NC part program directly from the design data. The construction of the tool path will be performed by the computer without the assistance of a part programmer.

Having established the logical connection between CAD and CAM, we shall direct our attention next to the central theme of this part of the book—computer-aided manufacturing.

## 10.3 COMPUTER-AIDED MANUFACTURING

Computer-aided manufacturing is defined by CAM-I[2] as follows:

> The effective utilization of computer technology in the management, control, and operations of the manufacturing facility through either direct or indirect computer interface with the physical and human resources of the company.

Applications of computer-aided manufacturing fall into two broad categories:

1. Direct applications, in which the computer is used either to monitor or to control the manufacturing operations. We shall refer to this category of CAM applications as "computer process monitoring and control."
2. Indirect applications, where the computer is used in support of the manufacturing activities in the plant, but there is no direct connection between the computer and the production process. Examples of these indirect applications include the use of the computer in production and inventory control, NC part programming, order scheduling, and development of work standards. We shall refer to this type of computer application as "CAM for manufacturing support."

### Computer process monitoring and control

Computer process monitoring and control involves a direct application of the computer in manufacturing. The computer is linked to the production operations. The reason computers lend themselves to this type of application is

[2]CAM-I stands for Computer-Aided Manufacturing—International, a nonprofit firm based in Arlington, Texas.

because any industrial or manufacturing process can be viewed as consisting of two features:

1. *The physical process.* This includes the mechanical hardware (the machine tool, processing equipment, fixtures, tooling, etc.). and the material being processed (the raw materials, chemicals, work-in-process, etc.).
2. *The flow of information.* In order to control the operation, information about it (machine status, process variables, economic performance, piece counts, etc.) must be processed and used to make control decisions.

The digital computer is highly adept at dealing with the second of these features by virtue of three fundamental capabilities:

1. Speed of computations and transactions.
2. Data storage capacity.
3. Decision-making capability (when properly programmed).

Manufacturing processes have evolved over the centuries. This evolution has resulted in production operations which are increasingly complex, and the problem of handling the flow of process information has become more demanding. The manual methods used to monitor and regulate industrial operations are being replaced by computerized techniques. For purposes of organization, we will divide the direct computer applications in manufacturing into the following categories:

1. Computer process monitoring.
2. Direct digital control.
3. Supervisory computer control.

These categories will be defined in this section and discussed in more detail in Chapters 12, 14, and 15, respectively.

COMPUTER PROCESS MONITORING. Computer process monitoring applies the computer to observe the process and associated equipment and to collect and record data from the operation. The computer is not used to directly control the process. Control over the operation remains in the hands of human operators. However, data collected and calculations performed by the computer can be used to guide the operator in the control of the process. Figure 10.4 illustrates the computer process monitoring function. There are three general types of data collected during production monitoring:

1. *Process data.* The input and output variables of the process are monitored to determine the overall status or performance of the operation. When the process variables are found to deviate from previously determined

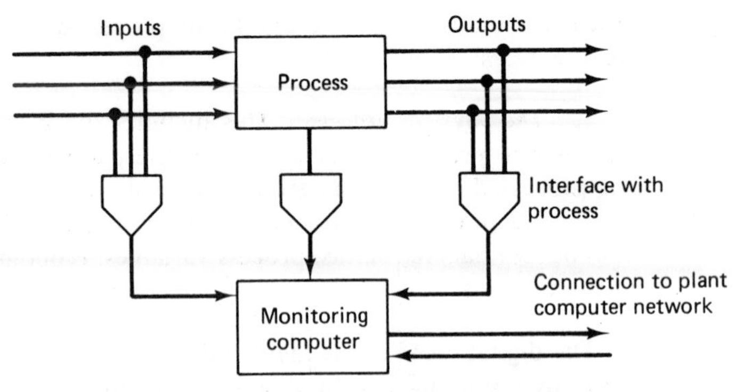


FIGURE 10.4 Computer process monitoring.

desired values, the computer signals the operator to take corrective action.

2. *Machine tool/equipment data.* Data can also be collected by the computer which relate to the status of the production machine tools and equipment. This type of data is valuable to help avoid machine breakdowns, to monitor equipment utilization, to determine optimum tool change schedules, to diagnose the cause of equipment breakdowns, and so on.

3. *Product data.* Because of government regulations, the company may be required under law to collect and preserve production data on its products. Computer monitoring of the process is the most convenient and accurate method for fulfilling these requirements. The firm may also want to collect product data for its own uses. Examples of these data include piece counts, production yield, and product quality attributes.

Collection of data from factory operations can be accomplished by any of several means. Shop data can be entered by the operators through manual entry terminals located throughout the plant. The data can also be collected automatically. One type of automated data collection system is called a "data logging system," which is designed to collect and store data for off-line analysis. Another category of automatic data collection is the "data acquisition system," in which there is a direct connection between the computer and the process interface devices. We will discuss these systems and other aspects of computerized production monitoring in Chapter 12.

Direct digital control and supervisory computer control involve applications in which the computer not only collects data from the manufacturing process, but also controls the manufacturing process.

DIRECT DIGITAL CONTROL. In tracing the evolution of industrial process control, manual regulation of the production operation came first. The operator had to determine how well the process was doing by means of his own senses. To help in this task, measuring devices were developed which gave the operator a more definite and accurate picture of the process. The operator was still required to make the necessary adjustments in the input variables to keep the operation running smoothly. The advent of analog control meant that these adjustments could be made automatically, without the human operator getting directly involved in every minor change that had to be made. With analog control, each individual output variable is monitored and changes are made in the corresponding input variable to maintain the output at a desired level. As industrial processes became more and more complex, the number of these control loops grew. In a large, modern industrial plant (e.g., oil refinery, chemical plant, paper mill, etc.), the number of control loops can easily exceed several hundred. When the digital computer was developed, it was only logical to consider the replacement of these analog loops by the computer.

Direct digital control, abbreviated DDC, replaces the conventional analog control units with the digital computer operating on a time-shared, sampled-data basis. The DDC computer performs calculations that simulate the operation of the analog control elements. One computer is used to service many control loops, so the computer is time-shared. Since the computer cannot continuously monitor each process variable, the variables are sampled periodically, with the sampling frequency being sufficient to approximate the performance of analog control.

The general configuration of a direct digital control application is shown in the schematic diagram of Figure 10.5. Chapter 14 presents a more complete discussion of DDC, including the kinds of calculations made, relative advantages over analog control, and the disadvantages of DDC.

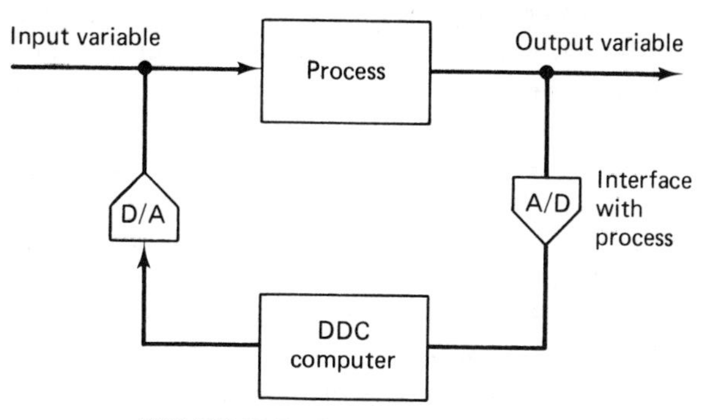


FIGURE 10.5 Direct digital control.

Supervisory Computer Control. In supervisory computer control, the computer is directly linked to the manufacturing process for the purpose of optimizing some overall performance objective of the process. The performance objective might be related to product cost, profit, yield, or some other economic goal. Supervisory control is a higher level of control than DDC as pictured in Figure 10.6. With direct digital control, the computer is applied to the individual control loops to regulate the process. With supervisory control,

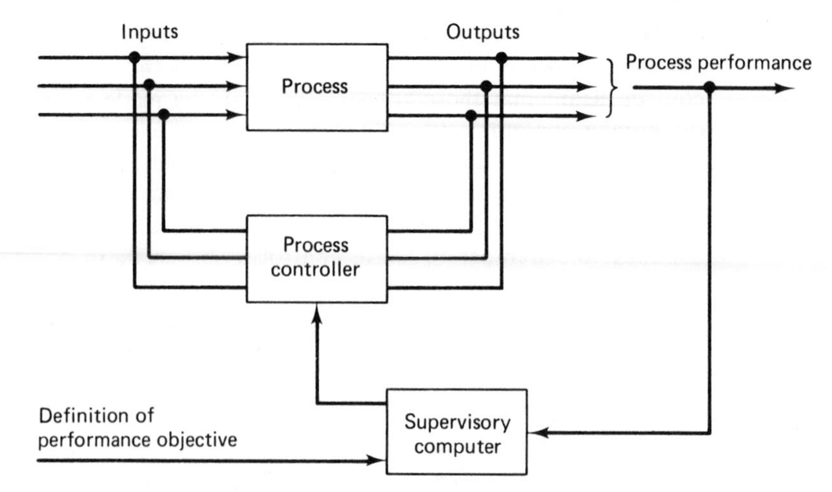


FIGURE 10.6 Supervisory computer control.

the computer is used to determine how the individual control loops should be manipulated to achieve the desired process performance objectives.

There are a variety of control strategies that can be utilized in supervisory computer control. The various categories are treated more thoroughly in Chapter 15, but a survey is presented below. We divide them into six basic strategies, as follows:

1. *Regulatory control.* The purpose of regulatory control is to maintain the performance output of the process at some desired level. To accomplish this objective, the individual control loop actions are calculated and implemented by the supervisory control computer.

2. *Feedforward control.* Feedforward control anticipates the effect of disturbances to the process and takes action to compensate for these disturbances before they can have a serious impact on process performance. In principle, feedforward control corrects for the disturbance before the output is affected.

3. *Preplanned control.* This term refers to a variety of applications in which the computer is used to command the process through a predetermined series of steps. In some of the applications, there is no explicit performance objective to be achieved. The computer is assigned the task of directing the processing sequence simply because it can perform this task more efficiently than alternative means. Examples of computerized preplanned control would include:

Direct numerical control (DNC) and computer numerical control (CNC), where the computer directs the processing steps of a machine tool. CNC and DNC were previously discussed in Chapter 9.

Program control, in which the computer is used to guide the process through production startups or shutdowns as well as changeovers from one product to the next.

Sequencing control, used to command the process through a sequence of on/off steps. This application of the computer can be found in the operation of automated flow lines (previously discussed in Chapters 4 and 5). The power feed at each station, transfer of workparts, and other functions requiring the activation of an on/off switch are performed by the supervisory computer.

4. *Steady-state optimal control.* Here the computer is used to calculate the optimum operating conditions at which to run the process. The term "steady state" is used to indicate that the operating conditions can be determined in advance of the production run, and no monitoring of the overall performance is required during the operation. This is not the case in the following strategy of adaptive control.

5. *Adaptive control.* This is a strategy for optimizing the process during the production run. The process is subjected to unpredictable time-varying disturbances. Adaptive control is used to monitor the performance of the process and to continually redetermine how optimum performance can be achieved in the face of these disturbances. Adaptive control is sometimes referred to as "self-optimizing" control. It is distinguished from steady-state optimal control by the fact that the process variables must be measured during the operation in order for the optimum to be determined.

6. *On-line search techniques.* These are used when optimum operating conditions cannot be determined by any of the foregoing strategies. All of the preceding strategies require a reasonably well formulated mathematical or procedural model of the process. When such a model cannot be determined, the optimum must be found through a systematic search of the possible operating conditions. In large, complex, industrial processes, the process control computer can be used to direct the search. In this type of application, a series of exploratory moves are made in the vicinity of the current operating point. The results of the exploration are used to indicate the changes in conditions that will improve process performance.

## CAM for manufacturing support

In addition to the direct applications of the computer for process monitoring and control, computers can also be used indirectly to serve a support role in the operations of the plant. When used in this way, the computer is not linked directly to the manufacturing process. Instead, the computer is used "off-line" to provide data and information for the effective management of the production activities. In this section we survey these applications of CAM for manufacturing support. A more complete treatment of many of the applications is presented in other chapters of the book.

The following is a listing to show the variety of indirect functions provided by the computer in plant operations:

COST ESTIMATING. The task of estimating the cost of a new product has been simplified in most industries by computerizing several of the key steps in the preparation of the cost estimate. The computer can be used to prepare cost estimates on new parts based on actual cost analyses of previously produced similar components. It can also be used to apply the appropriate direct and indirect labor rates for the sequence of planned operations on new designs. It then proceeds to sum the individual component costs from the engineering bill of materials to determine the overall product cost.

JOB COSTING. After the product is in production, time and cost data must be compiled to determine the actual manufacturing cost of the job. Industry practice today is to use the computer as an aid in compiling the individual operator time cards and calculating the product cost. Some plants have installed systems to collect data directly from the shop floor, including not only cost data but other types of production data as well. This type of data collection system falls within the scope of computer production monitoring which has been previously discussed.

COMPUTERIZED MACHINABILITY DATA SYSTEMS. One of the problems in operating a production machine shop is to determine the speeds and feeds for a given job. A number of computer programs have been developed to help determine the best cutting conditions. The objective in the computations is to determine the conditions (speed and feed) which either minimize the unit cost of the operation or maximize the production rate. The calculations are based on data obtained in either the shop or laboratory which relate tool life to cutting conditions. A more detailed discussion of computerized machinability data systems will be presented in Chapter 16.

COMPUTER-AIDED PROCESS PLANNING (CAPP). Process planning is concerned with the preparation of "route sheets," which list the sequence of operations and machines required to produce a workpart. Just as the part drawing is the documentation of the product design, the route sheet documents the process plan. The function of process planning is the traditional responsibility of manufacturing engineering or industrial engineering. It is a tedious and time-consuming task. For a complex workpart, it may take several days to prepare the route sheet. For several years, work has been underway to develop computerized systems for generating the sequence of processing steps. These systems rely on a parts classification and coding system that defines the attributes of the part (geometry, dimensions, material, etc.). From the definition of the part, the operations required to produce the part are determined by the computer. We will enlarge upon this discussion of computer-automated process planning in Chapter 18.

COMPUTER-ASSISTED NC PART PROGRAMMING. The subject of part programming for numerical control was discussed in Chapter 8. For complex workpart geometries, computer-assisted part programming represents a more efficient method for generating the NC punched tape than does manual part programming.

DEVELOPMENT OF WORK STANDARDS. The time study department has the responsibility for setting time standards on jobs performed by direct labor in the factory. Establishing standards by direct time study can be a tedious and time-consuming task. In the past few years, a number of computer-based systems have been commercially introduced for setting work standards. These computer packages are based on the use of standard data that have been developed for the work elements comprising a job. By summing the times for the separate work elements which are required to perform a new job, the computer program can be used to set the time standard for the job. A more comprehensive discussion of computerized work standards will be presented in Chapter 16.

COMPUTER-AIDED LINE BALANCING. Finding the best allocation of work elements among the stations of an assembly line is a large and difficult problem if the line is of significant size. Computer programs have been developed to solve the line balancing problem, the most familiar of which is probably CALB (Computer Assembly Line Balancing). We have previously treated line balancing in Chapter 6, and CALB was one of the programs discussed in Section 6.4.

PRODUCTION/INVENTORY CONTROL. The computer has found widespread use in the functions of production control and inventory control. These functions include the following:

Maintaining inventory records.
Automatic reordering of stock items when inventory has been depleted to the reorder point.
Labor and machine utilization reports.
Maintaining tool records and tool control.
Order scheduling.
Capacity planning.
Factory shop floor control.

Several of these areas are discussed in Chapter 17.

MATERIAL REQUIREMENTS PLANNING (MRP). A subject closely tied in with production and inventory control but often considered as a separate topic is material requirements planning (MRP). Orlicky [13] defines an MRP system in the following way:

> An MRP system consists of a set of logically related procedures, decision rules, and records designed to translate a "master production schedule" into time-phased net requirements, and the planned coverage of these requirements, for each component inventory item needed to implement the schedule.

Because of the large amount of data required in a material requirements planning system, the computer is a virtual necessity to implement MRP. We will be discussing material requirements planning in Chapter 17.

## Overview of CAM

In this chapter we have surveyed the field of CAD/CAM and have explored a variety of computer applications in manufacturing. Figure 10.7 is a block diagram that will help to put the subject of computer-aided manufacturing into perspective (at least the way the subject is organized in this book). Starting from the bottom of the diagram, the lowest block represents the manufacturing operations of the firm. As we have described, there are two categories of CAM applications: computer process monitoring and control, and CAM for manufacturing support. The three contiguous blocks represent the three types of monitoring and control applications: production monitor-

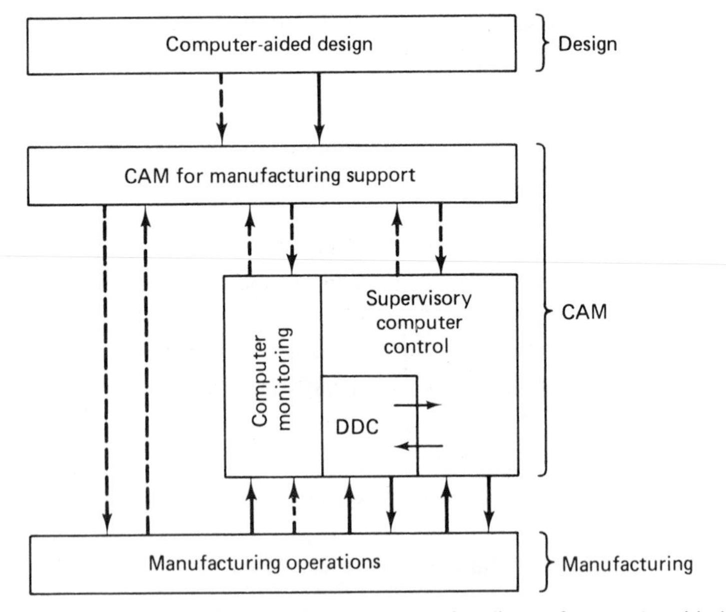


FIGURE 10.7 Block diagram showing various functions of computer-aided manufacturing and their relationships to design and manufacturing operations.

ing, direct digital control, and supervisory computer control. Production monitoring may have either an on-line (solid line) or an off-line (dashed line) connection with the manufacturing operations. In either case, the data flow is in only one direction, from the process to the computer. With DDC and supervisory control, the data flows in both directions. The computer that performs supervisory control may be linked directly to the analog devices that control the process. Or it may provide the input to the DDC system. These various possible signal flows are indicated in the block diagram. Solid lines are used to indicate the direct connection between the computer and the process.

The next block in Figure 10.7 represents CAM for manufacturing support. Production schedules, inventory control data, MRP reports, NC tapes, time standards, line balancing solutions, and other documents can all be generated by the computer to support production operations. Some form of information feedback is often required to periodically update these documents. This two-way communication link between the computer and the process is an off-line one, and is therefore represented by the dashed lines.

The highest block in the figure stands for the computer-aided design function. In most instances the information flow from design to manufacturing is off-line. However, as demonstrated earlier, there are some computer applications in which the link from CAD to CAM is direct (e.g., the generation of NC tool path data directly from the design data base). This direct link is shown as a solid line, whereas the off-line data flow is indicated by the dashed line.

The introduction to CAD/CAM presented in this chapter should establish a foundation for the reader to study the individual topics included in the field of computer-aided manufacturing. The computer is becoming pervasive in the operations of a modern manufacturing plant. It is no longer a luxury to use computers in manufacturing; it has become a necessity.

To understand computer-aided manufacturing, one must understand computers. The following chapter is intended to provide the reader with a basic exposition on computers: how they work and how they communicate with the production process. Chapters 12 through 15 are concerned with production monitoring and process control, with emphasis on how the computer is used to accomplish these functions. Topics related to CAM for manufacturing support are found in various sections throughout the book, but we concentrate on this theme in Chapters 16 and 17.

## REFERENCES

[1] ALLAN, J. J., *CAD Systems*, Proceedings of the IFIP Conference on Computer-Aided Design Systems held February, 1976, North-Holland Publishing Company, New York, 1977.

[2] CASSELL, D. A., *Introduction to Computer-Aided Manufacturing in Electronics*, John Wiley & Sons, Inc., New York, 1972.

[3] CHASEN, S. H., *Geometric Principles and Procedures for Computer Graphic Applications*, Prentice-Hall, Inc., Englewood Cliffs, N.J., 1978.

[4] "Computer Aided Design and Manufacturing" (a series of seven articles), *Manufacturing Engineering*, July, 1978, pp. 40–47.

[5] "Computer Aided Design and Manufacturing" (a series of six articles), *Manufacturing Engineering*, January, 1979, pp. 69–81.

[6] DALLAS, D. B. (Editor), *CAD/CAM and the Computer Revolution*, Society of Manufacturing Engineers, Dearborn, Mich., 1974.

[7] GETTLEMAN, K., "Computers Advancing World Manufacturing Technology," an interview with M. E. Merchant, *Modern Machine Shop*, June, 1978, pp. 123–131.

[8] HARRINGTON, J., *Computer Integrated Manufacturing*, Industrial Press, Inc., New York, 1973.

[9] HARRISON, T. J. (Editor), *Minicomputers in Industrial Control*, Instrument Society of America, Pittsburgh, Pa., 1978.

[10] KAUFMAN, R. E., "Mechanism Design by Computer," *Machine Design*, October 26, 1978, pp. 94–100.

[11] "Minicomputers that Run the Factory," *Business Week*, December 8, 1973, pp. 68–78.

[12] NEWMAN, W., and SPROUL, R., *Principles of Interactive Computer Graphics*, McGraw-Hill Book Company, New York, 1973.

[13] ORLICKY, J., *Material Requirements Planning*, McGraw-Hill Book Company, New York, 1975.

[14] POND, J. B., "On the Road to CAD/CAM," *Iron Age*, March 28, 1977, pp. 37–44; "The Road to CAD/CAM, Part II: Europe," *Iron Age*, April 25, 1977, pp. 39–44.

[15] *Proceedings, Design Automation Conference*, June, 1976, Institute of Electrical and Electronics Engineers, New York, 1976.

[16] REMBOLD, U., SETH, M. K., and WEINSTEIN, J. S., *Computers in Manufacturing*, Marcel Dekker, Inc., New York, 1977.

[17] SAVAS, E. S., *Computer Control of Industrial Processes*, McGraw-Hill Book Company, New York, 1965.

[18] SCHAFFER, G., "Computers in Manufacturing," *American Machinist*, April, 1978, pp. 115–130.

[19] SCHAFFER, G., "Stepping Up to CAM," *American Machinist*, November, 1978, pp. 87–90.

[20] SHIGLEY, J. E., *Mechanical Engineering Design*, 3rd ed., McGraw-Hill Book Company, New York, 1977.

[21] SMITH, C. L., *Digital Computer Process Control*, International Textbook Co., Scranton, Pa., 1972.

[22] TITUS, J. S., "CAD/CAM versus CAM/CAD: A Viable Alternative," *Tech Paper MS78-150*, Society of Manufacturing Engineers, Dearborn, Mich., 1978.

## 11.1 THE COMPUTER AND ITS PERIPHERAL EQUIPMENT

In this chapter we discuss the fundamentals of the digital computer with specific reference to its use with the manufacturing process. We entertain three questions in the presentation:

1. How do computers and computer systems work?
2. How does the computer collect data from the process?
3. How does the computer send command signals to the process?

Armed with the answers to these three questions, the reader should be prepared to consider the various strategies for applying computers in manufacturing.

## Computer architecture

The modern digital computer is an electronic machine that can perform mathematical or logical calculations and data processing functions in accordance with a predetermined program of instructions. The computer itself is often referred to as *hardware*, whereas the various programs of instructions are referred to by the term *software*. This section and the next will concentrate on the hardware. Section 11.3 is concerned with software.

There are three basic architectural components of a general-purpose digital computer:

1. Central processing unit (CPU).
2. Memory.
3. Input/output section (I/O).

The relationship of these three components is illustrated in the diagram of Figure 11.1. The *central processing unit* is often considered to consist of two subsections: a control unit and an arithmetic-logic unit (ALU). The *control unit* coordinates the operations of all the other components. It controls the input and output of information between the computer and the outside world through the I/O section, synchronizes the transfer of signals between the various sections of the computer, and commands the other sections in the performance of their functions. The *arithmetic-logic unit* carries out the arithmetic and logic manipulations of data. It adds, subtracts, multiplies, divides, and compares numbers according to programmed instructions. The *memory* of the computer is the internal storage unit. The data stored in this section are arranged in the form of words which can be conveniently transferred to the ALU or I/O section for processing. Finally, the *input/output section* provides the means for the computer to communicate

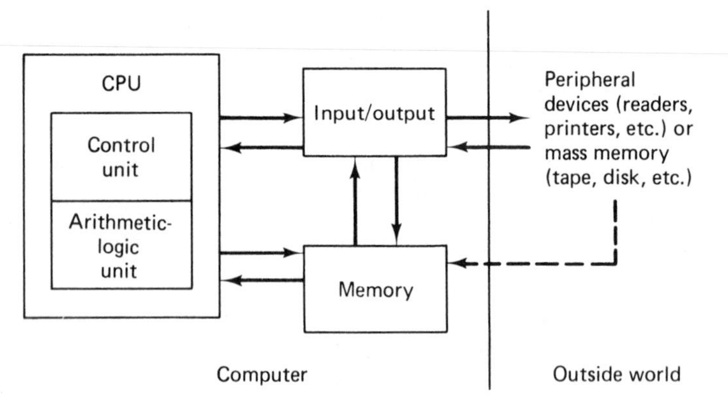


FIGURE 11.1 Basic structure of a digital computer.

with the external world. This communication is accomplished through peripheral equipment such as readers, printers, CRTs, and process interface devices. The computer may also be connected to mass or secondary memory (e.g., tapes, disks, etc.) through the I/O section of the computer.

We will examine each of these three main sections of the computer in more detail.

### Central processing unit (CPU)

The central processing unit controls, sequences, and synchronizes the activities of the other computer components and performs the various arithmetic and logic operations on the data. It performs these functions by means of *registers*. Computer registers are small memory devices that can receive, hold, and transfer data. Each register consists of binary cells to hold bits of data. The number of bits in the register establishes the word length the computer is capable of handling. The number of bits per word can be as few as 4 (microcomputers) or as many as 64 (large scientific computers).

The arrangement of these registers constitutes several functional areas of the CPU. A representative configuration is given in Figure 11.2. To

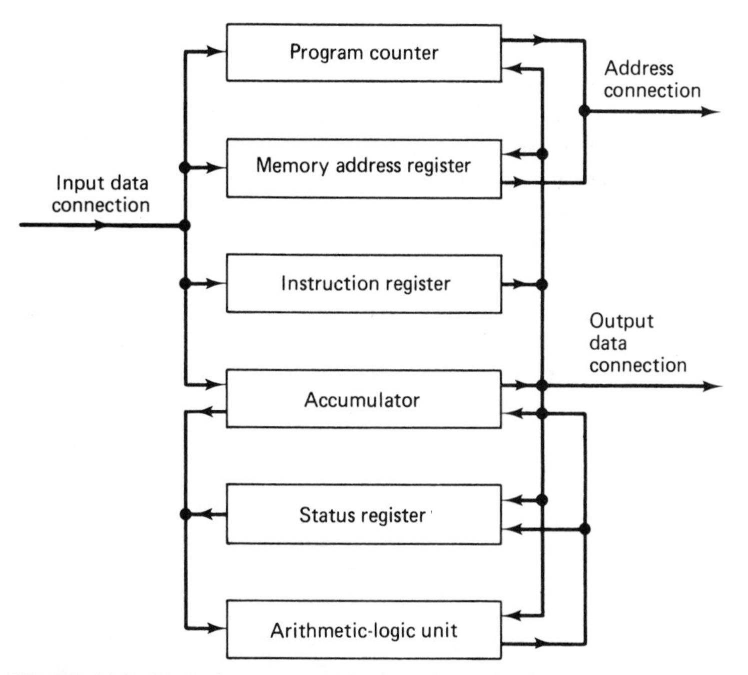


FIGURE 11.2 Typical arrangement of registers in the computer's CPU. (Adapted from Leventhal [10].)

accomplish a given sequence of programmed instructions, the functions of these register units would be as follows:

*Program counter.* The program counter holds the location or address of the next instruction. An instruction word contains two parts: an operator and an operand or a reference to an operand. The *operator* defines the type of arithmetic or logic operation to be carried out (additions, comparisons, etc.). The *operand* usually specifies the data on which the operation is to be performed. The CPU sequences the instructions to be performed by fetching words from memory according to the contents of the program counter. After each word is obtained, the program counter is incremented to go on to the next instruction word.

*Memory address register.* The location of data contained in the computer's memory unit must be identified for an instruction, and this is the function of the memory address register. This unit is used to hold the address of data held in memory. A computer may have more than a single memory address register.

*Instruction register.* The instruction register is used to hold the instruction for decoding. Decoding refers to the interpretation of the coded instruction word so that the desired operation is carried out by the CPU.

*Accumulator.* An accumulator is a temporary storage register used during an arithmetic or logic operation. For example, in adding two numbers, the accumulator would be used to store the first number while the second number was fetched. The second number would then be added to the first. The sum, still contained in the accumulator, would then be operated on or transferred to temporary storage, according to the next instruction in the program.

*Status register.* Status registers are used to indicate the internal condition of the CPU. A status register is a one-bit register (often called a *flag*). Flags are used to identify such conditions as logical decision outcomes, overflows (where the result of an arithmetic operation exceeds the word capacity), and interrupt conditions (used in process control).

*Arithmetic-logic unit (ALU).* The ALU provides the circuitry required to perform the various calculation operations and manipulations of data. The typical configuration of the arithmetic-logic unit is illustrated in Figure 11.3. The unit has two inputs for data, inputs for defining the function to be performed, data outputs, and status outputs used to set the status registers or flags (described above).

The arithmetic logic unit can be a simple adder, or its circuitry can be more complex for performing other calculations, such as multiplication and division. ALUs with simpler circuits are capable of being

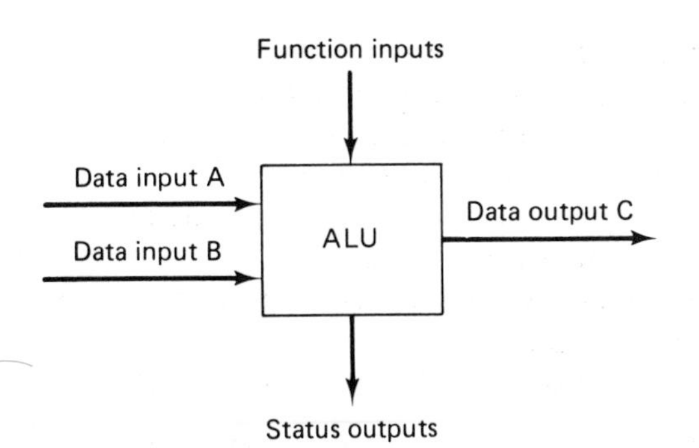


FIGURE 11.3 Typical structure of an arithmetic-logic unit.

programmed to perform these more complicated operations, but more computing time is required. The more complex arithmetic-logic units are faster, but these units are more costly.

Referring to Figure 11.3, the two inputs, A and B, enter the ALU and the logical or mathematical operation is performed as defined by the function input. Among the possible functions are addition, subtraction, increment by 1, decrease by 1, and multiplication. The ALU places the result of the operation on A and B in the output, C, for transfer to the accumulator.

### Main memory

The main memory is located in the computer and is used by the CPU for manipulation of data. The memory section consists of binary storage units which are organized into bytes (there are typically 8 bits per byte). A byte is a convenient size for the computer to handle. Computer words can typically be 4, 8, 12, 16, 32, or 64 bits long. Each word has an address in the memory. The CPU calls words from memory by referring to the word address. The time required to find the correct address and fetch the contents of that memory location is called the *access time*. Access time is an important factor in determining the speed of the computer. Access times range between $10^{-7}$ s (100 nanoseconds) to several microseconds. Random-access memories (RAMs) are generally used.

The main memory section stores all the instructions and data of a program. Thus, the CPU must transfer these instructions and data (in the form of words) to and from the memory throughout the execution of the program.

The main memory must be distinguished from mass memory, which is external to the computer (see Figure 11.1). The files of programs and data contained in mass memory are not directly available to the central processing unit. The contents of mass memory must be read into the computer (for storage in the main memory) through the input/output section. Hence, use of mass memory is a much more time-consuming process. This time can be reduced in some computer configurations by reading the contents of mass memory directly into the main memory without going through registers in the CPU. This link between external and internal memory is called *direct memory access* (indicated as an option by the dashed line in Figure 11.1).

There are a variety of types of memories used in computers, microprocessors, programmable controllers (see Section 11.5 for a description of programmable controllers), and other data storage devices. Table 11.1 presents a glossary of the terms used to define various memory types.

TABLE 11.1 Glossary of Terms Used to Define Various Types of Memories

*Read/Write [R/W] Memory*: Originally, this term was applied to magnetic-core memories in which the contents of the individual magnetic cores were erased during the normal read operation. Accordingly, the contents had to be read back into memory. Now the term R/W memory describes any memory in which the data contents can be readily altered.

*ROM [Read-Only Memory]*: This is a memory whose data contents are fixed permanently during fabrication of the memory device.

*PROM [Programmable Read-Only Memory]*: A memory device that can be programmed under special conditions (using either a high current or a laser beam to break certain connections and form the program). Once programmed, the memory contents cannot generally be altered.

*EPROM or EROM [Erasable PROM]*: A PROM whose contents can be erased by exposure to ultraviolet light.

*EAROM [Electrically Alterable ROM]*: This is a read-only memory, except that the data contents can be electrically changed. This, in effect, makes it a reprogrammable memory device except that rewriting the contents must usually be preceded by a separate erase procedure.

*RAM [Random-Access Memory]*: A read/write memory that can be read and reprogrammed during normal operation. The term applies to memories whose contents can be accessed directly and quickly without reading through memory locations in a sequential fashion.

## Input/output section

The function of the input/output section is to communicate with the peripheral devices used to operate the computer. There are two parts to this function. First, programs and data are read into the computer. The I/O section must interpret the incoming signals and hold them temporarily until they are placed in main memory or into the CPU. Second, the results of the calculations and data processing operations must be transmitted to the appropriate peripheral equipment.

In the traditional operation of the computer (e.g., scientific and engineering calculations, business data processing, manufacturing support functions), the computer must communicate with people via the peripheral devices. When the computer is used for process monitoring and control applications, the input/output section must communicate with the manufacturing process as well as people. In the following subsection we discuss the peripheral devices used for the more traditional applications of the computer. In Sections 11.4 and 11.5, the problem of interfacing the computer with the manufacturing process is presented.

### Peripheral equipment

The peripheral devices used to operate a large computer or minicomputer in conventional applications include:

> Electromechanical devices for transmitting information (programs and data) to and from the computer's input/output section. These devices include teleprinters, CRTs, card read/punch devices, and line printers.
>
> Mass memory units, including disks and magnetic tapes.

The following paragraphs describe these common devices.

TELEPRINTERS OR TELETYPES. The teleprinter is a relatively low cost but slow device for input/output operations with the computer. The printer speed of various models is 10 to 30 characters per second. The device can be connected to a remote computer by means of telephone lines. The appearance of a teleprinter is similar to that of a typewriter.

Teleprinters can be obtained with paper tape read and punch capability. Hence, the paper tape can be prepared in advance and fed into the computer more efficiently via a high-speed paper tape unit connected to the input/output section of the computer. These devices have a punched tape read capability of about 200 characters per second and a punch capacity of about 100 characters per second. By computer input/output standards, these speeds are still very slow.

The teleprinter is often used in process control as the link between the operator and the control computer. In this application, the requirement for computer-operator communication at high speeds is minimal. Hence, the relatively low cost of this peripheral device makes it attractive.

CATHODE RAY TUBE (CRT). The CRT display unit is capable of higher speeds of data transfer than is the teletype. We can distinguish between two types of cathode ray tubes. The first type is limited to character printing capability (no drawings) and is somewhat cost-competitive with the teleprinter. The second type also has the capability for graphic or pictorial display of data. Owing to the greater complexity of these vector drawing units, they are more expensive. In both types, the CRT units are used to display data that has been entered by the operator or received from the computer.

A CRT is convenient to use by the operator in developing programs because of its editing capabilities. It is a relatively easy task to either add or delete characters or lines or to make corrections to a program. The CRT can also be used by the operator to display files contained in computer memory. Both program editing and file interrogation represent applications in which the CRT is used in a conversational mode. Because of their convenience, cathode ray tubes are becoming very popular in computer operations, both in traditional applications and in CAD/CAM applications.

CARD READERS AND CARD PUNCHES. Another method of program and data entry to the computer is by cards. The program and data are prepared in advance on a keypunch and then the deck is submitted to the computer in a batch mode. The computer must therefore read the deck into its memory by means of a card read device. An electromechanical card read unit is capable of processing the cards at a rate of about 150 to 200 cards/min. Photoelectric card read devices are capable of rates faster than 1000 cards/min.

Some applications call for the computer to produce output in the form of cards. Card punches are available as peripheral devices for this purpose. These units are slower than card readers, with rates less than 100 cards per minute.

LINE PRINTERS. In a typical data processing shop, the volume of output is sufficient to justify a high-speed line printer. These units print entire lines at one cycle (80 to 132 characters/line) at rates that may exceed 1000 lines/min. These units are expensive and the cost is generally proportional to the printing rate. One of the most recent innovations in high-speed printers combines laser and xerography technologies to achieve print speeds of around 10,000 lines/min.

The need for high-speed printing equipment is virtually nonexistent in process control operations. However, line printers are required for many of the data processing applications to support manufacturing (e.g., preparation of production schedules, MRP reports, direct labor reports, etc.).

DISKS. A disk is one of several types of mass-storage devices used with digital computers. These secondary memory units serve to hold either programs or data for the computer.

Disks are direct access devices whereas the magnetic tapes described in the following subsection are sequential memory units. Because of the sequential nature in which memory contents are stored in magnetic tapes, access times are considerably slower than with disks.

Three types of disks are distinguished. In all cases, the information storage is provided by a flat circular disk, the surface of which is coated with magnetic material. The surface is arranged into tracks for data and program storage. The *moving-head disk* uses a single read/write head which is moved across the surface of the disk and positioned at the desired track. The time required for the moving head to be positioned makes this disk configuration relatively slow. Access times are therefore relatively long.

The *fixed-head disk* uses a read/write head for each track. This increases the speed with which data can be transferred, and access times are faster than for the moving-head disk. However, the cost of the fixed-head disk is greater.

The *floppy disk* is the most recent of the disk-type memory devices. The operation of the floppy disk is similar to the moving-head disk. It is fabricated of a flexible material (hence the name floppy disk) with magnetic coating on both sides. Because of its relatively low cost, it is becoming a convenient mass storage medium.

MAGNETIC TAPES. Magnetic tapes used for computer mass storage are similar to tape recorder tapes. The data are arranged in tracks or channels along the length of the tape. There are typically seven or nine parallel tracks across the $\frac{1}{2}$-in.-wide tape. The advantage of tapes is that they can hold significantly more data than disks. However, the disadvantage is that the time to access data on the tapes is considerably longer. Cassette tape units are a relatively recent innovation in mass storage hardware. Their appearance is much like that of audio tape cassettes. It is anticipated that digital cassette tapes may eventually replace punched paper tapes in process control applications, including numerical control machining.

## 11.2 COMPUTER SIZES[1]

Computers are generally considered to fall into three size categories, all of which can be based on roughly the same architecture described in Section 11.1. The three categories are:

1. The large general-purpose computer.
2. The minicomputer.
3. The microcomputer.

In the current section, we shall define and compare the three types and establish their relationship to computer-aided manufacturing.

### The three sizes defined

LARGE GENERAL-PURPOSE COMPUTERS. The large computer is distinguished by its cost, capacity, and function. The price of a new corporate-sized general-purpose computer runs into millions of dollars. The main

[1] This section is adapted from L. A. Leventhal, *Introduction to Microprocessors*, Prentice-Hall, Inc., Englewood Cliffs, N.J., 1978, pp. 1–17.

memory capacity is several orders of magnitude larger than the minicomputer, and the speed with which calculations can be made is roughly 10 times the speed of a minicomputer or microcomputer. These features result in a machine that is suited to the following two functions:

1. *Solving complex engineering and scientific problems.* Examples would include iterative calculation procedures often required in heat-transfer analysis, fluid dynamics analysis, or structural design analysis. These would be typical examples in computer-aided design (CAD) applications.
2. *Large-scale data processing.* Examples would include generation of production schedules and material requirements planning (MRP) reports, compiling production costs, summarizing direct labor costs, and so on. These are instances where the computer is used to support manufacturing activities.

MINICOMPUTERS. In the mid 1960s, smaller computers began to appear on the market. The name "minicomputer" was adopted to distinguish them from the larger computers. The smaller computers were the result of a trend in computer technology that still continues. The trend is toward miniaturization—packaging the same computer power into smaller volumes. This means that large computers could become even more powerful with each new generation, or that the same computing capability could be contained in a smaller machine. The minicomputer manufacturers chose the second alternative. The cost of a minicomputer can range from several thousand dollars up to around $50,000.

Minicomputers can be used for the same two general functions as large computers. However, the size of the problems and data processing must be scaled down to accommodate the reduced capacity of the minicomputer. Minicomputers can also be used for process monitoring and control. The characteristics of minicomputer applications are often similar to those of the microcomputer, and we will present these characteristics below.

MICROCOMPUTERS. A microcomputer uses a microprocessor as the basic central processing unit. The microprocessor consists of single integrated circuits contained on very small silicon bases, which are called LSI chips (LSI stands for "large-scale integration"). The LSI chips can be manufactured in large quantities very inexpensively. Hence, the cost of a simple microcomputer can be as low as $10 (the hand-held calculator). The microprocessor is capable of performing virtually all the functions of the conventional CPU (e.g., arithmetic-logic operations, fetch instructions or data from memory, etc.). Accordingly, the microprocessor can be connected to a memory unit and the appropriate input/output device(s) to form a microcomputer.

The relatively low cost of microcomputers and minicomputers has opened opportunities for new applications. These applications have the following characteristics:[2]

[2]Ibid., p. 4.

1. The computer is a system component. The overall system, which might be a piece of test equipment, a machine tool, or a banking terminal, uses the small computer much as it might a switch, power supply, or display. The computer may not even be visible from the outside.

2. The computer performs a specific task for a single system. It is not shared by different users as a large computer is. Instead, the small computer is part of a particular unit, such as a medical instrument, typesetter, or factory machine.

3. The computer has a fixed program that is rarely changed. Unlike a large computer, which may solve a variety of business and engineering problems, most small computers perform a single set of tasks, such as monitoring a security system, producing graphic displays, or bending sheets of metal. Programs are often stored in a permanent medium or read-only memory.

4. The computer often performs real-time tasks in which it must obtain the answers at a particular time to satisfy system needs. Such applications include machine tools that must turn the cutter at the right time to obtain the correct pattern or in missile guidance, where the computer must apply thrust at the proper time to achieve the desired trajectory.

5. The computer performs control tasks rather than arithmetic or data processing. Its primary function might be managing a warehouse, controlling a transit system, or monitoring the condition of a patient.

### Comparison of computer sizes

Leventhal [10] compares the features of four computer sizes ranging from a representative microcomputer to a large general-purpose data-processing system. The four examples are:

1. Intel MCS-80 microcomputer.
2. Computer Automation NAKED MINI, a small minicomputer.
3. Digital Equipment (DEC) PDP 11/45, a large minicomputer.
4. IBM 370/Model 168, a large computer.

The comparison of the four types is summarized in Table 11.2. The prices for the four computers range from $250 to $4.5 million. The cost of a microcomputer is low enough that it could be a component in products such as machine tools, laboratory instruments, cash registers, CRT terminals, automobiles, and others. The word length of the computer defines the number of bits (binary digits) that can be handled by the CPU at one time. The word length determines the amount of data that can be processed in a given time. Typical word lengths used in the three general categories of computers are given in Table 11.3.

TABLE 11.2 Comparison of Features of Four Computers

| | *IBM 370/168* | *DEC PDP 11/45* | *Computer automation* NAKED MINI | *Intel* MCS-80 |
|---|---|---|---|---|
| Cost | $4.5 million | $50,000 | $2500 | $250 |
| Word length (bits) | 32 | 16 | 16 | 8 |
| Memory capacity (8-bit bytes) | 8.4 million | 256K* | 64K | 64K |
| Processor add time | 0.13 μs | 0.9 μs | 3.2 μs | 2.0 μs |
| Maximum I/O data rate (bytes/second) | 16 million | 4 million | 1,400,000 | 500,000 |
| Number of general-purpose registers | 64 | 16 | 3 | 7 |
| Peripherals (from manufacturer) | All types | Wide variety | Disk, tape, card, line printer, CRT, cassette | Paper tape reader, floppy disk, PROM programmer |
| Software | All types | Wide variety | Operating system, assembler, FORTRAN, BASIC | Assembler, monitor, PL/M, editor |

*1K = 1024

*Source*: L. A. Leventhal, *Introduction to Microprocessors: Software, Hardware, Programming*, Prentice-Hall, Inc. Englewood Cliffs, N.J., 1978.

TABLE 11.3 Typical Word Lengths for Three Computer Sizes

| *Computer size* | *Word length (bits)* |
|---|---|
| Large computers | 32, 36, 60, 64 |
| Minicomputers | 12, 16, 32 |
| Microcomputers | 4, 8, 16 |

In terms of main memory capacity, the IBM 370/168 has roughly 30 times the capacity of the PDP 11/45 and 125 times the memory of the microcomputer. Memory capacity determines the program size and data storage capacity of the computer. Additional storage can be obtained from peripheral mass storage (e.g., disks, tape), but this increases the program execution time.

Typical execution times are given in the form of processor add times in Table 11.2 for the four computers.

The remaining criteria in the comparison (I/O rate, number of registers, peripherals, and software support) attest to the greater power and versatility of the large general-purpose computer. However, many monitoring and control applications in manufacturing do not require this data processing power. These applications are, in fact, ideally suited to the small minicomputers and microcomputers.

## 11.3 COMPUTER PROGRAMMING LANGUAGES

The instructions executed by the computer are in the form of binary words. In Section 8.2, we discussed the binary number system in the subsection on numerical control tape coding. It was demonstrated how the binary number system could be used to represent any number in the more familiar decimal system. We also saw how a group of binary digits could be used to represent alphabetic characters and other symbols. In NC programming, the arrangement of the binary digits on the NC tape provides the commands to operate the machine tool. Numerical control works because the machine tool control unit is capable of interpreting the particular arrangement of bits on the tape and translating these into machine motions. If the reader can accept this explanation of how NC works, much of the mystery surrounding the internal workings of the computer should be eliminated. Instead of transforming the group of bits into mechanical operations as in NC, the computer interprets the configuration of bits as an instruction to perform an electronic operation such as add, subtract, load into memory, and so on. The sequence of these binary-coded instructions establishes the program by which the computer performs its calculations and data manipulations. This computer program is, of course, analogous to the part program in NC by which the machine tool performs its metal-cutting operations.

The binary-coded instructions that computers can interpret are called *machine language*. Just as an NC part programmer would have difficulty reading the holes in a punched paper tape, computer programmers find it difficult to write programs in machine language. To complicate things even more, there are differences in machine languages among different computers. To facilitate the task of computer programming, higher-level languages are available. In all, there are three levels of computer programming languages that the reader should distinguish. These are:

1. Machine language.
2. Assembly language.
3. Procedure-oriented languages.

These three categories will be discussed in the following paragraphs.

| Address | Contents |
|---|---|
| 0 | 00100001 |
| 1 | 01000000 |
| 2 | 00000000 |
| 3 | 00010001 |
| 4 | 01100000 |
| 5 | 00000000 |
| 6 | 00000110 |
| 7 | 00001010 |
| 8 | 01111110 |
| 9 | 00010010 |
| 10 | 00100011 |
| 11 | 00010011 |
| 12 | 00000101 |
| 13 | 11000010 |
| 14 | 00001000 |
| 15 | 00000000 |
| 16 | 01110110 |

FIGURE 11.4 Machine language program. (Reprinted from Leventhal [10].)

### Machine language

Machine language (also sometimes called *object code*) consists of binary words that can be immediately interpreted and executed by the computer. An example of a machine language program is presented in Figure 11.4. The reader can readily appreciate how difficult and cumbersome it would be to program the computer using machine language. It is much easier to program in the next level of language: assembly language.

### Assembly language

The programmer finds it easier to use assembly language because the computer instructions are mnemonic—easier to remember. Each instruction is an abbreviation which usually resembles the operation that is to be performed by the computer. Some examples will illustrate this feature.

ADD = add.
SUB = subtract.
CLR = clear.
LDA = load accumulator.
LDX = load index register.
WAI = wait for interrupt.

An assembly language program is shown in Figure 11.5, using a different set of mnemonic abbreviations from the examples given above (different computers use different assembly languages).

```
        LXI   H, BLK1      ; Memory pointer 1 = Start of block 1
        LXI   D, BLK2      ; Memory pointer 2 = Start of block 2
        MVI   B, COUNT     ; Count = Length of blocks
TRANS:  MOV   A, M         ; Get element of block 1
        STAX  D            ; Move element to block 2
        INX   H
        INX   D
        DCR   B
        JNZ   LOOP
        HLT
```

FIGURE 11.5 Assembly language program. (Reprinted from Leventhal [10].)

For the computer to be able to understand the assembly language program, a means must be provided for converting assembly language into machine language. This conversion is accomplished by a special program called an *assembler*. The assembler takes the assembly language program as input and produces a machine language program as output. The assembly language program is called the *source code* and the machine language program is called the *object code*. Many process monitoring and control programs written for minicomputers and microcomputers are written in assembly language.

## Procedure-oriented languages

Assembly languages are considered to be low-level languages because the instructions are elementary and somewhat removed from the engineering and data processing problems to be solved. The programmer must be concerned with fairly basic data manipulations as illustrated in Figure 11.5. Another obstacle with assembly language programming is that the language is machine-dependent. The programmer must learn a new language with each new computer.

The procedure-oriented languages overcome many of the programming difficulties of the assembly languages. Procedure-oriented languages are considered to be "high-level languages" because the instruction sets are similar to algebraic and/or English statements. Also, the languages are usually machine-independent. As with assembly language, a special program called a *compiler* is required to translate the program written in the high-level language into the basic machine language of the particular computer.

Some of the important procedure-oriented computer languages are:

1. FORTRAN (FORmula TRANslator), which is used to solve engineering and scientific problems. Other procedure-oriented languages used in engineering are ALGOL, APL, BASIC, and PL/1.
2. COBOL (COmmon Business Oriented Language), which is suited to the data processing needs of business.

```
        DO 100   I = 1, 10
100     BLK1  (I)  = BLK2 (I)
```

FIGURE 11.6 FORTRAN program. (Reprinted from Leventhal [10].)

3. APT (Automatically Programmed Tool), which is used for numerical control part programming. Other NC part programming languages are ADAPT, AUTOSPOT, COMPACT II, EXAPT, SPLIT, and UNIAPT (refer to Sections 8.4 and 8.5).

Figure 11.6 shows a portion of a FORTRAN program. Readers familiar with this language will recognize that the two statements transfer the contents of one storage location called BLK2 (which contains 10 items of data) into a storage location called BLK1. These two FORTRAN statements replace the 10 lines of assembly language shown in Figure 11.5 and the 16 lines of machine language in Figure 11.4. The advantage of FORTRAN to the programmer is therefore obvious. The disadvantage of using the procedure-oriented languages is that the compiler, a necessary ingredient in using these languages, is expensive and takes up memory space in the computer. Therefore, the computer must be large enough to accommodate this additional need for memory. This is usually not much of a problem with large computers and large minicomputers. For smaller computers, often found in process control applications, it is a problem. To overcome this difficulty, FORTRAN control programs can be translated by a large computer to produce the object code (in machine language) for the mini- or microcomputer utilized in the application. Because FORTRAN is so universally familiar to technical people, it is frequently used for process monitoring and control despite its drawbacks in these applications.

## 11.4 THE COMPUTER–PROCESS INTERFACE

In Section 11.1, a variety of peripheral devices were discussed for interfacing the computer to its environment (e.g., teleprinters, CRTs, card read devices, etc.). These pieces of apparatus are required for the traditional data processing applications and engineering calculations. In computer process monitoring and control, a different problem is encountered—the problem of connecting the computer to the manufacturing process. Accordingly, there are two categories of computer interface in computer-aided manufacturing:

1. The data processing (DP) interface.
2. The manufacturing process interface.

Since we have already discussed the general characteristics of the computer and the traditional DP interface equipment, we shall devote the remainder of this chapter to problems and systems related to the computer–process interface.

### Characteristics of manufacturing process data

In order for the computer to be used in process monitoring and control, the computer must collect data from the manufacturing operation. If the computer is utilized to directly control the process, data (i.e., commands or instructions) must be communicated to the process. The data flowing back and forth between the computer and the process can be classified into three types:

1. Continuous analog signals.
2. Discrete binary data.
3. Pulse data or discrete data that are not restricted to binary. In other words, more than two values are possible.

CONTINUOUS ANALOG DATA. A continuous variable is one that assumes a continuum of values over time. The variable is uninterrupted as time proceeds (at least during the cycle of the manufacturing process). An analog variable is one that can take on any value within a certain limited range. The amplitude of the variable is not restricted to a discrete set of values. A continuous analog variable is one that possesses the attributes of both a continuous variable and an analog variable. These attributes are exhibited in Figure 11.7. Most industrial operations, in both discrete-parts

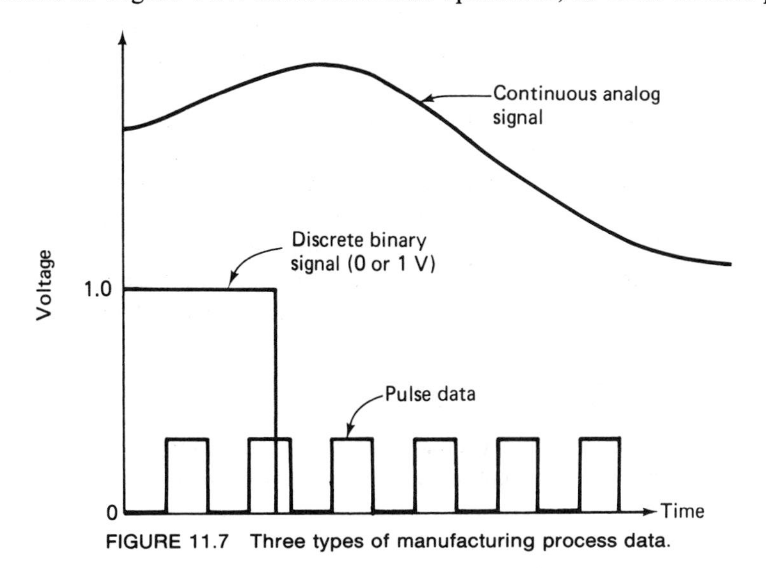


FIGURE 11.7 Three types of manufacturing process data.

manufacturing and the process industries, are characterized by continuous analog variables. Examples are force, temperature, flow rate, pressure, velocity, and so on. All of these variables are continuous with time during the process, and they can take on any of an infinite number of possible values within a certain range. As a practical matter, the number of values is limited only by the capacity of the measuring instrument to distinguish the different levels of the variables.

DISCRETE BINARY DATA. Discrete binary data are data that can assume either of two possible values, such as on or off, opened or closed, and so on. Examples of hardware whose status would be indicated as discrete binary data are switch or relay contacts, electrical motors, and flow valves. Such data might take on the following significance in a manufacturing operation:

To sense the presence or absence of a workpart at the proper workstation location.

To indicate whether the power feed drive of a transfer line drill head was working.

To signal that the flow valve of a chemical process was opened or closed.

The characteristics of a discrete binary signal are illustrated in Figure 11.7. In the figure, this type of signal is represented by one of two possible voltage levels (0 V or 1 V) whose status may change over time.

PULSE DATA OR DISCRETE DATA. Pulse data are a train of pulse signals as indicated in Figure 11.7. This type of data is used in digital transducers, such as digital tachometers and turbine flow meters. An electrical pulse train can be used to drive a stepping motor, which is found in a wide variety of computer control applications because of its compatibility with the digital computer. Discrete data are similar to discrete binary data except that the number of possible levels is not limited to two. A prime example of discrete data is piece counts. Pulse data are related to discrete data because the number of pulses in a pulse train can be counted within a certain time interval. Hence, pulse data can be converted into discrete data, and vice versa.

**EXAMPLE 11.1**

Let us examine the different classes of data that might be found in a metal-cutting NC machine tool application. Examples of continuous analog signals would be:

Cutting force or torque.

Cutting temperature.

Velocity of spindle rotation—assumed continuously variable.

Feed rate (table speed)—assumed continuously variable.

Among the discrete binary data in the operation would be:

Workpart in place in the fixture or not.

Cutting fluid on or off.

Critical dimensions machined within tolerance or not.

Machine tool under operator command or automatic cycle.

Machine tool operational or broken down (would be used to tabulate machine utilization).

Examples of pulse data or discrete data that might apply to the application are:

Pulse train to drive stepping motor for $x$-coordinate table position.

Pulse train to drive stepping motor for $y$-coordinate table position.

Pulse train indicating spindle rotational speed to be converted to discrete data for display on operator console.

Piece counts (production per shift).

## Process data input/output

To implement a system of computer process monitoring and control, the three classes of manufacturing data must be interfaced with the computer. For monitoring the process, a means must be provided for inputting the data to the computer. For process control, a method must be devised for output of command signals from the computer to the process. The general configuration of this computer–process interface is shown in Figure 11.8. There are six categories of interface representing inputs and outputs for the three types of process data. These categories are:

1. Analog-to-digital interface.
2. Contact input interface.
3. Pulse counters.
4. Digital-to-analog interface.
5. Contact output interface.
6. Pulse generators.

Certain types of discrete data would be entered through the data processing interface (depicted in Figure 11.8), perhaps by manual data entry terminals or CRTs located in the factory.

In the paragraphs that follow, we consider the six categories of interface between the computer and the manufacturing process.

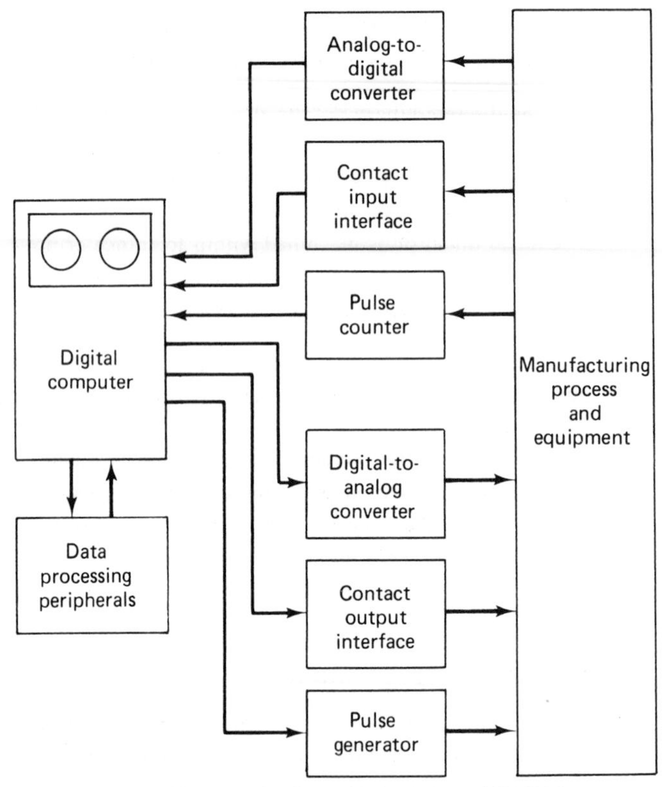


FIGURE 11.8 Computer–process interface.

ANALOG-TO-DIGITAL INTERFACE. The continuous analog signals must be converted into digital values in order to be used by the computer. The procedure for making this conversion typically involves the steps illustrated in Figure 11.9. These steps involve a variety of hardware devices, which can be enumerated as follows:

1. *Transducers.* Transducers are used to measure the continuous analog signals by converting the variable (such as temperature, flow rate, force, etc.) into a more convenient electrical signal (such as voltage or current). Transducers will be discussed in more detail in Section 11.5.

2. *Signal conditioning.* The electrical signal leaving the transducer may require conditioning. The reason for this requirement may be that the signal (usually voltage) contains random noise, in which case an *RC* filter would be used to smooth out the signal. Another reason for signal conditioning is that the transducer output may be of the wrong form. For example, a current signal would be converted into a voltage signal.

3. *Multiplexer.* The multiplexer is used to share the analog-to-digital converter (ADC) among many incoming signals. In effect, the multiplexer

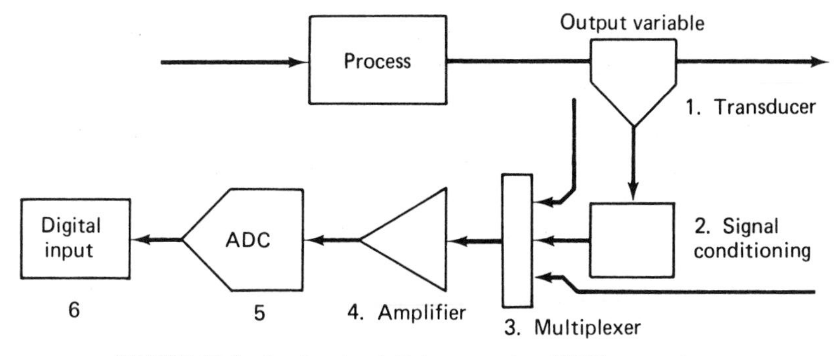


FIGURE 11.9 Analog-to-digital conversion (ADC) procedure.

samples each incoming signal at periodic intervals and sends the signal on to the ADC. Multiplexers will be discussed in Section 11.5.

4. *Amplifiers.* Amplifiers are used to scale the incoming signal either up or down to be compatible with the range of the analog-to-digital converter.

5. *Analog-to-digital converter* (*ADC*). The function of the ADC is to transform the incoming analog signal into its digital equivalent. The operation of analog-to-digital converters will be explained in Section 11.5.

6. *Digital input.* This is the computer's input/output section and its function is to accept the output of the ADC. A device called a limit comparator is sometimes inserted between the ADC and the input channel. The purpose of the limit comparator is to compare the incoming digitized signal with certain upper and lower limits. As long as the signal is within the limits, the data do not enter the computer. Hence, the advantage of using a limit comparator is that CPU time is required only if the measured variable has strayed outside the desired limits.

CONTACT INPUT INTERFACE. Discrete binary data are read into the computer through the contact input interface. This interface consists of relatively simple contacts that can be either opened or closed to indicate the status of the limit switches, motor pushbuttons, and valve positions associated with the process. The computer stores in memory the desired status of these contacts and periodically scans the actual status for comparison.

PULSE COUNTERS. Some measuring instruments, called digital transducers, generate a series of pulses as their output. Pulse counters are used to convert the pulse train into a digital quantity. This quantity is then entered into the computer through its input channel. We will discuss the operation of pulse counters in Section 11.5.

DIGITAL-TO-ANALOG INTERFACE. The three preceding interfaces have all been concerned with process data inputs to the computer. The digital-to-analog interface is one of three interfaces by which data and commands can be communicated back to the process. This interface converts digital data generated by the computer into a pseudo-analog continuous signal. We refer to this as a pseudo-analog signal because the digital output of the computer can be expressed with only a limited precision, which depends on the word length of the computer. We shall focus on this problem when we discuss digital-to-analog converters (DAC) in Section 11.5. A "data-hold circuit" is required to take the analog signal at the desired level until the next digitized value comes from the computer. A multiplexer is sometimes sandwiched between the DAC and the hold circuit to share the digital-to-analog converter.

CONTACT OUTPUT INTERFACE. In applications of computer monitoring only, the contact output interface would be used to turn on indicator lights and operator alarms. In some applications the computer might be programmed to shut down the process during emergencies through the contact output interface. In computer control applications, this subsystem is used to control solenoids, motors, alarms, and other similar devices. Alarms and indicator lamps can also be operated as in monitoring. The computer can be programmed to control the sequence of activities in the process through this contact output interface. We have previously referred to this mode of control as "sequencing control."

In the contact output interface, the computer sets the position of the contact in one of two states—on or off. The contact is maintained in that position until changed.

PULSE GENERATORS. These devices generate a pulse train as specified by the computer. The pulse train is typically used to operate stepping motors. In the pulse train, the pulses are of a certain amplitude and frequency to be compatible with the stepping motor.

## 11.5 INTERFACE HARDWARE

In the preceding section, the general configuration of the computer–process interface was discussed. In this section we examine some of the hardware devices that make up this interface.

### Transducers and sensors

A transducer is a device that converts one type of physical quantity into another type (commonly electrical voltage). The reason for making the conversion is that the converted signal can be more conveniently used or evaluated. Transducers are often called sensors when they are used to measure the value of a physical quantity. Transducers are of two general types:

1. Analog transducers.
2. Digital transducers.

ANALOG TRANSDUCER. This type of transducer produces a continuous analog signal such as electrical voltage or current. The signal can be interpreted as the value of the measured variable. To make the interpretation, a calibration procedure is required. The *calibration* of the measuring device is performed to establish the relationship between the measured variable and the converted output signal (e.g., voltage). Analog transducers are sometimes classified according to the level of the output signal. The two basic categories are low-level signals (e.g., the voltage signal would be measured in millivolts) and high-level signals (where the voltage signal would be greater than 1 V).

DIGITAL TRANSDUCERS. Digital transducers are measuring devices that produce a digital output signal. The digital signal may be in the form of a set of parallel status bits or a series of pulses that can be counted. In either case the digital signal represents the quantity to be measured. Digital transducers are finding an increased number of applications because of the ease with which they can be read when used as a stand-alone measuring instrument, and because of their compatibility with the digital computer.

DESIRABLE FEATURES OF SENSORS. Some of the desirable characteristics of transducers used for process monitoring and control applications are the following:

1. *High accuracy*. Accuracy is usually defined to mean that the measurement will contain no systematic positive or negative errors about the true value.

2. *High precision*. This means that the measurement will be made with little random variability or noise in the measured value. The distinction between accuracy and precision in measurement is illustrated in Figure 11.10.

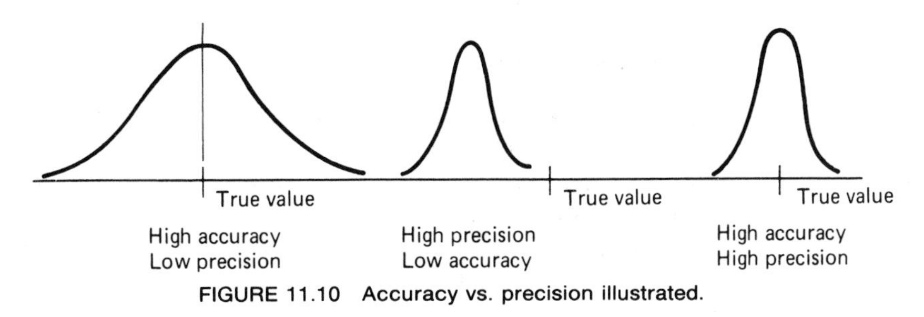


FIGURE 11.10 Accuracy vs. precision illustrated.

3. *Wide operating range.* The sensor should possess the attributes of accuracy and precision over a wide range of the physical variable being measured.

4. *Speed of response.* The sensor should be able to respond quickly to changes in the physical variable. Ideally, the time lag would be zero. The sensor must be capable of identifying hazardous operating conditions in the shortest possible time.

5. *Ease of calibration.* The measuring device should be easy to calibrate. It should not be subject to drift during use, which would necessitate frequent recalibrations of the sensor. *Drift* refers to the gradual loss of accuracy between the measured variable and the transduced signal (e.g., millivolts).

6. *High reliability.* The sensor should not be subject to frequent mechanical or electrical failures. It must be capable of operating in extreme environments characteristic of the process (e.g., high temperature, humidity, vibration, pressure, etc.).

7. *Low cost.* The cost to purchase (or fabricate) and install the sensor should be low relative to the worth of the information provided by the sensor.

There are few measuring devices that possess all of these desirable characteristics. A compromise must be made among these various features when selecting a transducer for a particular application.

COMMON TRANSDUCERS. Several of the common transducers and measuring devices are listed in Table 11.4. It is not the purpose of this book to present a complete description of these instruments. A much more detailed discussion of sensors and instrumentation principles is contained in references [3] and [18].

TABLE 11.4 Common Transducers and Measuring Devices

*Ammeter*—meter to indicate electrical current.

*Bourdon tube*—widely used industrial gage to measure pressure and vacuum.

*Chromatographic instruments*—laboratory-type instruments used to analyze chemical compounds and gases.

*Inductance-coil pulse generator*—transducer used to measure rotational speed. Output is pulse train.

*Linear-variable-differential transformer (LVDT)*—electromechanical transducer used to measure angular or linear displacement. Output is voltage.

*Manometer*—liquid column gage used widely in industry to measure pressure.

*Ohmmeter*—meter to indicate electrical resistance.

*Optical pyrometer*—device to measure temperature of an object at high temperatures by sensing the brightness of the object's surface.

TABLE 11.4 (continued)

*Orifice plate*—widely used flowmeter to indicate fluid flow rates.

*Photometric transducers*—a class of transducers used to sense light, including phototubes, photodiodes, phototransistors, and photoconductors.

*Piezoelectric accelerometer*—transducer used to measure vibration. Output is emf.

*Pitot tube*—laboratory device used to measure flow.

*Positive displacement flowmeter*—variety of transducers used to measure flow. Typical output is pulse train.

*Potentiometer*—instrument used to measure voltage.

*Pressure transducers*—a class of transducers used to measure pressure. Typical output is voltage. Operation of the transducer can be based on strain gages or other devices.

*Radiation pyrometer*—device to measure temperature by sensing the thermal radiation emitted from the object.

*Strain gage*—widely used transducer used to indicate torque, force, pressure, and other variables. Output is change in resistance due to strain, which can be converted into voltage.

*Thermistor*—also called a resistance thermometer; an instrument used to measure temperature. Operation is based on change in resistance as a function of temperature.

*Thermocouple*—widely used temperature transducer based on the Seebeck effect, in which a junction of two dissimilar metals emits emf related to temperature.

*Turbine flowmeter*—transducer to measure flow rate. Output is pulse train.

*Venturi tube*—device used to measure flow rates.

## Analog-to-digital converters

An analog-to-digital converter (ADC) is a device that converts an analog signal into digital form. It is sometimes called an encoder. The device that performs the reverse process, called a digital-to-analog converter (DAC), will be discussed in the following subsection.

Analog signals are usually continuous, as illustrated in Figure 11.11. Analog-to-digital conversion consists of three phases:

1. *Sampling*. The continuous signal is periodically sampled to convert it into a series of discrete-time analog signals. This sampling process is illustrated in Figure 11.11.

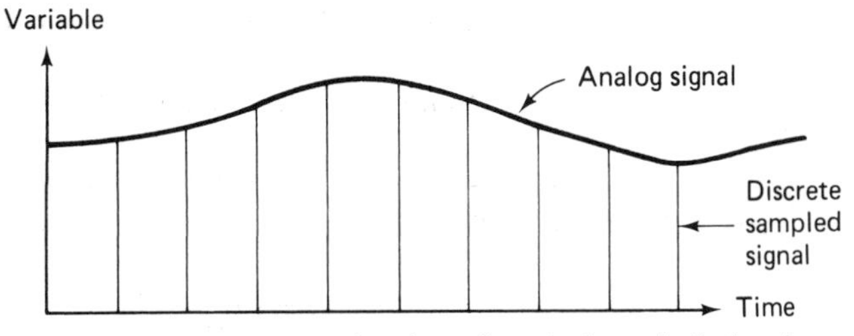


FIGURE 11.11 Analog signal vs. discrete (sampled) signal.

2. *Quantization*. Each discrete-time analog value must be assigned to one of a finite number of previously defined amplitude levels. These amplitude levels consist of discrete values of voltage ranging over the full scale of the ADC.

3. *Encoding*. The various amplitude levels obtained during quantization must be converted into digital code. This involves the representation of the amplitude level by a sequence of binary digits. The quantization and encoding of a discrete-time analog signal will be explained below when we discuss ADC resolution.

GENERAL CHARACTERISTICS OF AN ADC. In selecting and applying an analog-to-digital converter, the following basic considerations are important:

1. *Sampling rate*. This is the rate at which the continuous analog signal is sampled. Higher sampling rates mean that the continuous waveform of the analog signal can be more closely approximated. When the ADC is used with multiplexing to convert a significant number of process signals, a high sampling rate is desirable because it means that these signals can all be surveyed in a short period of time. In cases where the process variables change slowly, it is not necessary to sample the sensors with high frequency. The upper limit on the sampling rate is imposed by the conversion time of the ADC.

2. *Conversion time*. The conversion time of an ADC is the time interval between when the incoming analog signal is applied and when the digital output value has been established by the quantization process. In short, it is the time required by the ADC to perform its function. Conversion time depends on the type of procedure used to make the conversion (two of these procedures will be discussed below). It also is proportional to the number of bits used to define the value. The number of bits determines the resolution capability of the ADC.

3. *Resolution*. The resolution of an analog-to-digital converter refers to the precision with which the analog signal is evaluated. Since the signal must be represented in binary form, the precision is determined by the number of quantization levels, which in turn is determined by the bit capacity of the ADC and computer. The number of quantization levels is defined by

$$\text{number of quantization levels} = 2^N \tag{11.1}$$

where $N$ is the number of bits. The resolution of the ADC is usually defined as the reciprocal of the number of quantization levels:

$$\text{resolution} = 2^{-N} \tag{11.2}$$

The range of each quantization level, which is really the measure of precision, is determined by the full-scale range of the incoming analog signal. Hence, the spacing of each quantization level is given by

$$\text{quantization level spacing} = \frac{\text{full-scale range}}{2^N} \qquad (11.3)$$

The incoming signal must typically be amplified (either up or down) to a range of 0 to 10 V to be compatible with the analog-to-digital converter.

The error that results from the quantization process is called the *quantization error,* and it can be considered analogous to a roundoff error in numerical analysis. Using this roundoff analogy, the quantization error can be as large as one-half the quantization level spacing. This can be expressed

$$\text{quantization error} = \pm\tfrac{1}{2}\ \text{quantization level spacing} \qquad (11.4)$$

Quantization errors can be reduced by increasing the number of bits used in the ADC process.

## EXAMPLE 11.2

A continuous-voltage signal is to be converted by an ADC. The maximum possible range of the voltage signal is 0 to 10 V. Two ADCs are being considered for a process control application, one with a 4-bit capacity, the second with an 8-bit capacity. Determine the number of quantization levels, the resolution, the spacing of each quantization level, and the quantization error for the two alternatives.

For the 4-bit capacity, the number of quantization levels, according to Eq. (11.1), is 16. The resolution is $\frac{1}{16} = 0.0625 = 6.25\%$, by Eq. (11.2). The spacing of each quantization level is given by Eq. (11.3):

$$\text{quantization level spacing} = \tfrac{10}{16} = 0.625 \text{ V}$$

The quantization error would be one-half this value, or 0.3125 V.

For the 8-bit A/D converter, the number of quantization levels = 256. The resolution = 0.39%, the quantization level spacing = 0.0391 V, and the quantization error = 0.0195 V.

Clearly, the 8-bit capacity produces a more precise analog-to-digital conversion.

## EXAMPLE 11.3

For the 4-bit analog-to-digital converter of Example 11.2, show how a voltage signal of $\pm 10$ V range might be interpreted in binary form.

Of the four binary digits, we will use the first to indicate sign: 0 means a negative voltage and 1 means a positive voltage. The three remaining digits will be used to indicate successively increasing quantization levels. Since one binary digit is being used to designate the polarity of the signal, this leaves three digits for quantization of the full-scale 10 V. The quantization level spacing would therefore be $10 \times 2^{-3} = 1.25$ V. The series of bits 1000 might be used to represent the range 0 to

1.25 V and 1001 to represent the range 1.25 to 2.50 V, and so forth. The complete range of voltage signal values would be assigned binary numbers according to the following listing:

| *Voltage range (V)* | *Binary number* | *Voltage range (V)* | *Binary number* |
|---|---|---|---|
| 0 to +1.25 | 1000 | 0 to −1.25 | 0000 |
| +1.25 to +2.5 | 1001 | −1.25 to −2.5 | 0001 |
| +2.5 to +3.75 | 1010 | −2.5 to −3.75 | 0010 |
| +3.75 to +5.0 | 1011 | −3.75 to −5.0 | 0011 |
| +5.0 to 6.25 | 1100 | −5.0 to −6.25 | 0100 |
| +6.25 to +7.5 | 1101 | −6.25 to −7.5 | 0101 |
| +7.5 to +8.75 | 1110 | −7.5 to −8.75 | 0110 |
| +8.75 to +10.0 | 1111 | −8.75 to −10.0 | 0111 |

TYPES OF ADC. A variety of methods can be used to convert a continuous voltage signal into its digital counterpart. We will discuss only two of these methods.

1. Successive approximation method.
2. Integrating ADC method.

The reader interested in pursuing the subject of analog-to-digital conversion in more detail is directed to references [1] and [14].

The successive approximation method is the most common ADC technique. To convert an electrical voltage into binary code by the successive approximation method, trial voltages are successively compared to the input signal. The general scheme is illustrated in Figure 11.12. Suppose that the input signal to be encoded was 6.8 V. The first trial voltage might be 5.0 V. Comparing the input voltage with the trial voltage would yield a "1" if input exceeded trial voltage, and "0" if input was less than trial voltage. Each subsequent trial voltage would be one-half the preceding value. The digital encoded value would be developed by this successive comparison process as illustrated in Figure 11.12. The resolution of the encoding procedure depends on the number of bits used to define the value. On the other hand, better resolution leads to increased conversion times. A typical conversion time might be 9 microseconds for the 6-bit precision shown in Figure 11.12.

There are several types of integrating analog-to-digital converters. All of them operate by converting the input voltage into a time period that can be measured by a counter. We will explain the operation of the simplest of the integrating-type ADCs. This is called the *single-slope converter* and its operation is illustrated in Figure 11.13. The input voltage is compared with a voltage that increases linearly with time (it is a ramp function). A counter is used to measure the time required from the initiation of the comparison voltage until it reaches the value of the input signal. This time is proportional to the incoming voltage from the transducer. The resolution of this ADC method depends on the frequency of the pulse train during the time interval

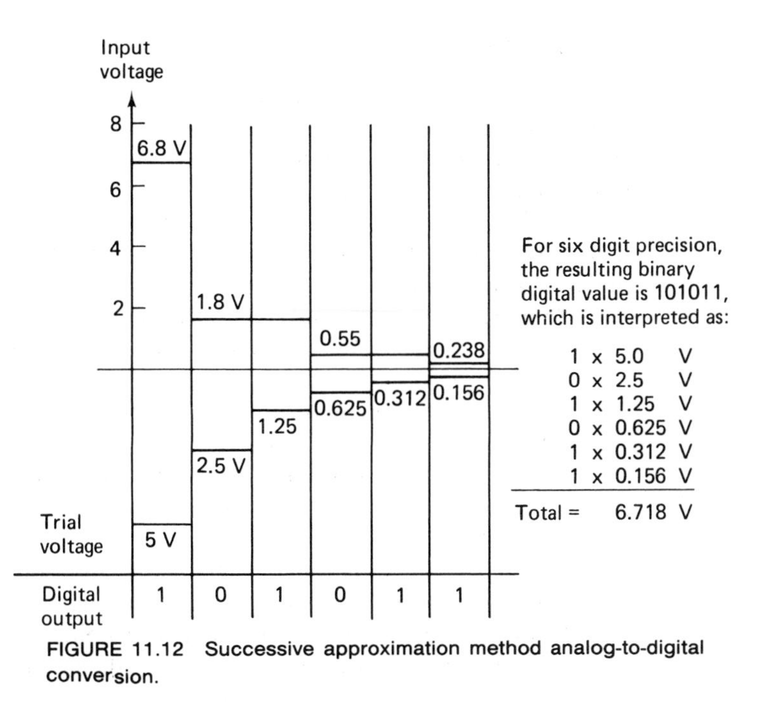


FIGURE 11.12 Successive approximation method analog-to-digital conversion.

of the comparison and the slope of the comparison ramp voltage. Higher pulse-train frequency and lower ramp slopes would provide greater precision in the conversion process. A typical conversion time for 6-bit precision would be about 14 $\mu$s. Hence, the single-slope integrating type ADC is somewhat slower than the successive approximation ADC method.

An improved version of the integrating ADC method is the *dual-slope converter.* Its advantage is greater accuracy and less susceptibility to random signal noise. The disadvantage is lower conversion speed.

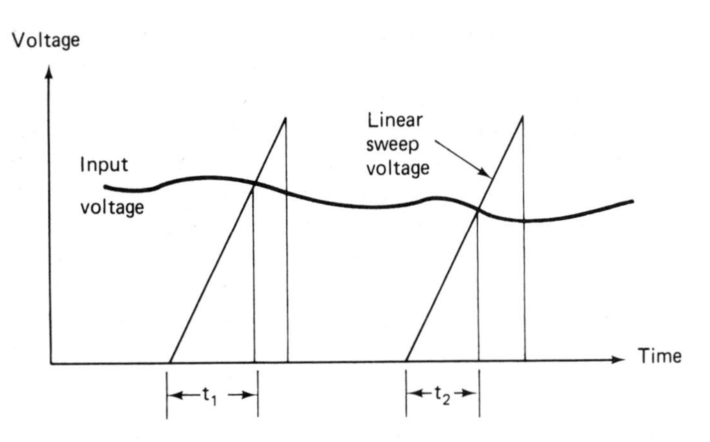


FIGURE 11.13 Single-slope-type integrating analog-to-digital conversion.

## Digital-to-analog converters

The function performed by a digital-to-analog converter (DAC) is the reverse of the ADC function. The process control computer receives data from the process via the analog-to-digital link, performs its control calculations, and sends commands back to the process. Since the computer operates on digital data and the process variables are continuous analog signals, the DAC is required to provide the communication link from the computer back to the process. Digital-to-analog converters may also be required to drive various types of recorders, plotters, and other electronic display units.

The DAC process can be viewed as consisting of two steps:

1. *Decoding.* This involves the conversion of digital data (from the computer) into sampled analog data.
2. *Data holding.* This step transforms the sampled data into a continuous analog signal (usable by the process or other analog device).

DECODING. The first step in the DAC procedure is to convert the binary digital data into its equivalent sampled analog signal format. This is accomplished by transferring the digital data to a binary register. This register controls a reference voltage source. The level of the output voltage depends on the status of the bits in the register. Each successive bit controls one-half the voltage of the preceding bit. Thus, the output voltage is determined as

$$V_o = V_{ref}\left[0.5B_1 + 0.25B_2 + 0.125B_3 + \ldots + (2^N)^{-1}B_N\right] \tag{11.5}$$

where $V_o$ = output of the decoding operation
$V_{ref}$ = reference voltage
$B_1, B_2, \ldots, B_N$ = status (0 or 1) of successive bits in the register.

DATA HOLDING. The holding device converts sampled data into a continuous analog signal. Data holding is an integral part of the digital-to-analog conversion process. The objective of the data hold is to approximate the envelope formed by the sampled data, as illustrated in Figure 11.14. Data-holding devices are sometimes classified according to the order of the extrapolation calculation used to determine the voltage output between sampling instants. The ideal envelope shown in Figure 11.14 would be a very high order extrapolator. The most common data extrapolator is the zero-order hold. In this case the output voltage between sampling instants is a sequence of step signals, as shown in Figure 11.14. We can express this voltage between sampling instants as a function of time very simply in the following way:

$$V(t) = V_o \tag{11.6}$$

where $V_o$ is the output voltage from Eq. (11.5).

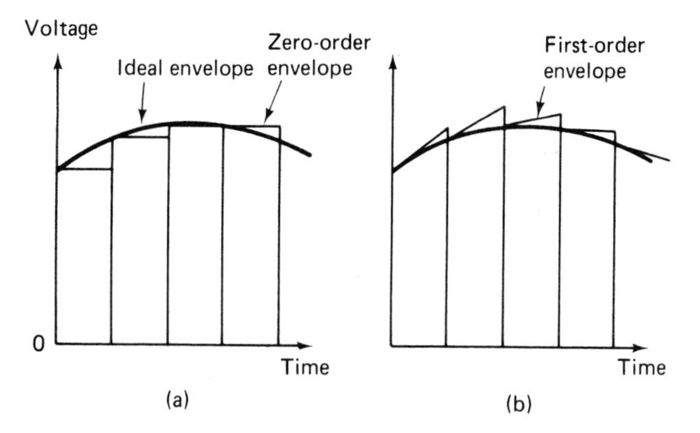


FIGURE 11.14 Data hold function: (a) zero-order hold; (b) first-order hold.

The first-order hold is less common than the zero-order hold but more closely approximates the true envelope of the sampled data values. With the first-order hold, the voltage $V(t)$ between sampling instants changes with a constant slope determined by the two preceding $V_o$ values. Expressing this mathematically,

$$V(t) = V_o + at \tag{11.7}$$

$$a = \frac{V_o - V_o(-\tau)}{\tau} \tag{11.8}$$

where $a$ = rate of change of $V(t)$
$\tau$ = time interval between sampling instants
$V_o(-\tau)$ = value of $V_o$ at the preceding sampling instant (removed in time by $\tau$)

Both the zero-order hold and the first-order hold are illustrated in Figure 11.4. Hold extrapolators of higher order are not common.

## EXAMPLE 11.4

A digital-to-analog converter uses a reference voltage of 100 volts and has six-binary-digit precision. In three successive sampling instants 0.5 second apart, the data contained in the binary register are:

| *Instant* | *Binary data* |
|---|---|
| 1 | 101000 |
| 2 | 101010 |
| 3 | 101101 |

Determine the decoder output voltage according to Eq. (11.5). Also, plot the voltage signal as a function of time between sampling instants 2 and 3 using a zero-order hold and a first-order hold.

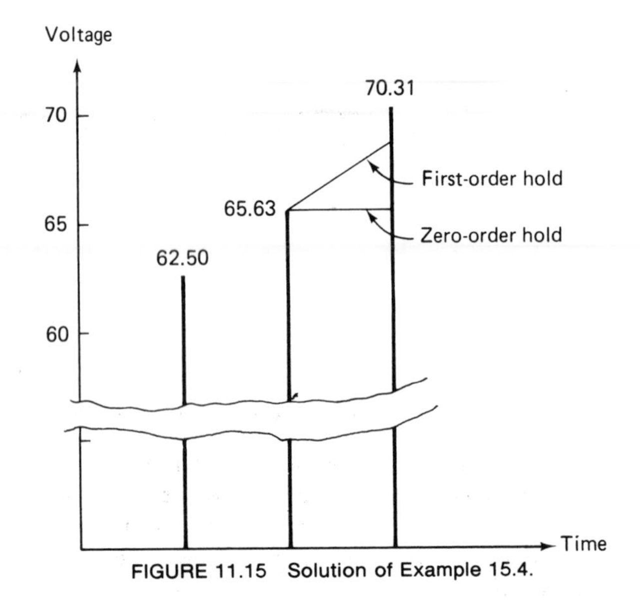


FIGURE 11.15 Solution of Example 15.4.

In sampling instant 1, the decoder output voltage would be

$$\begin{aligned} V_o &= 100[0.5(1)+0.25(0)+0.125(1)+0.0625(0) \\ &\quad +0.03125(0)+0.15625(0)] \\ &= 62.5 \text{ V} \end{aligned}$$

Correspondingly, $V_o = 65.63$ V for instant 2 and $V_o = 70.31$ V for instant 3.

The zero-order hold between instants 2 and 3 would produce a constant voltage of $V(t) = 65.63$ V. The first-order hold would produce a steadily increasing voltage. The slope can be determined from Eq. (11.8),

$$a = \frac{65.63 - 62.5}{0.5} = 6.25$$

and from Eq. (11.7),

$$V(t) = 65.63 + 6.25t$$

The solution to this example is displayed in Figure 11.15. Note that the first-order hold is superior in terms of anticipating the value of $V_o$ in the third sampling instant.

## Multiplexers

The multiplexer is a switching device connected in series with each input channel from the process. It is used to time-share the analog-to-digital converter (and associated amplifiers) among the incoming signals. The alternative to the use of the multiplexer would be to have a separate ADC for each transducer. This would be prohibitively expensive for a large installation with many inputs to the computer. Since the process variables need only be

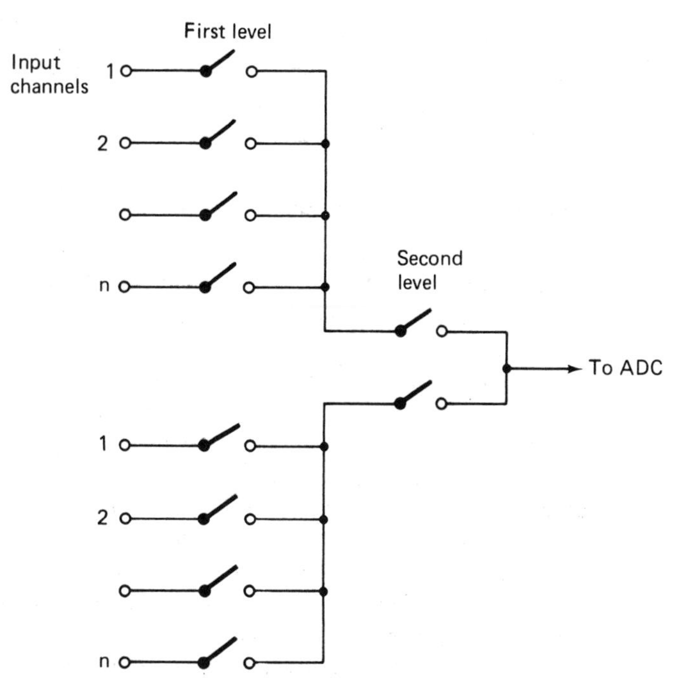


FIGURE 11.16 Traditional multiplexer configuration (two levels, single-ended multiplexers).

sampled periodically anyway, the multiplexer provides a very cost-efficient method of satisfying system design requirements.

A multiplexer consists of a set of switches arranged in the typical configuraion shown in Figure 11.16. The switches illustrated in the figure are single-pole, single-throw switches in which a common ground is utilized. This type of multiplexer is called a *single-ended multiplexer.* For two incoming leads from each transducer, the switches would be double-pole, single-throw types and the multiplexer would be called a *differential multiplexer.* The number of switches on a single multiplexer can range from 16 to more than 1000. To increase the number of input channels which can be connected to a single analog-to-digital converter, the multiplexers can be arranged into more than one level. This is shown in Figure 11.16. Each of the individual input channels can be addressed by the combination of switch positions at the two levels.

The order in which the incoming signals can be entered into the ADC can be random or sequential. With the random-order type, the individual input channels can be selected as needed by the control computer. With the sequential type, the input channels are sampled in a fixed sequence. The sequential type is less expensive than the random-order multiplexer.

The switches used in multiplexer construction are generally of three types: dry-reed contact relay, mercury-wetted contact relay, and digital logic devices. The dry-reed and mercury-wetted contacts can be used for low-level signals (millivolt range) and are limited to relatively slow sampling rates (perhaps 200 samples/s). The digital logic device is capable of much higher sampling rates (100,000 samples/s are possible).

### Pulse counters and pulse generators

As described in Section 11.4, some of the data flowing between the computer and the production operation are in the form of a pulse train, as illustrated in Figure 11.7. Certain types of transducers generate a series of pulses which must be read into the computer (see Table 11.4). The computer must also drive certain control devices (e.g., stepping motors) which respond to a train of pulses. To assist the computer in communicating with these hardware components, pulse counters and pulse generators are employed.

In the typical operation of a pulse counter, a register in the pulse counter is loaded with the number of pulses to be counted. At the start of the counting routine, a clock is initialized. As each pulse in the pulse train is received, the register subtracts one. When the register reaches zero, the clock time is recorded. The frequency of the pulse train is determined by dividing the number of pulses by the clock time.

A pulse generator is used to generate a pulse train as specified by the computer. There are two ways in which the pulses can be used. The first is where a certain number of pulses are to be generated. For example, the number of pulses might correspond to the desired $x$-coordinate position of an NC machine table. The second case is where a pulse train of a certain frequency is used to drive a stepping motor at a certain rotational speed.

The pulse generator performs its function by repeatedly opening and closing a contact, thus providing a sequence of discrete electrical pulses. The amplitude and frequency of the pulses are fixed to be compatible with the device to be controlled. One of the most common system components operated under pulse train command is the stepping motor.

### Stepping motors

The stepping motor is an electromechanical device driven by an electrical pulse train to produce a sequence of angular (rotational) movements corresponding to the number of pulses. Depending on the rate at which the pulses are transmitted to it, the stepping motor can be operated in either of two modes. At low pulse rates, the motor responds to each individual pulse by rotating through a certain angular position. In this mode, the motor can be commanded to rotate in either direction. At high pulse rates (the exact rate depends on the design and size of the motor), the stepping motor operates at virtually a smooth rotational speed. In this mode, the angular momentum of the motor is too great to be reversed without first decelerating the motor down to its single pulse mode. The dividing line between the single pulse mode and the smooth speed mode is called the *slewing speed.*

Commercially available stepping motors are generally classified into two types:

1. Permanent-magnet type.
2. Variable-reluctance type.

The *permanent-magnet* type uses a permanent-magnet rotor driven by an electromagnetic field. This type is basically a synchronous inductor motor and is usually built as a relatively slow speed device.

The *variable-reluctance* stepping motor operates on the interaction between a soft-iron rotor in an electromagnetic field. The variable-reluctance type is capable of relatively high speeds.

Stepping motors can be built with step angles ranging from as low as 1° (even smaller angles are possible) to a maximum of 120°. The allowable step angles on a stepping motor are determined by the relation

$$\alpha_s = \frac{2\pi}{n_s} \qquad n_s \geqslant 3 \tag{11.9}$$

where $n_s$ must be an integer value, and corresponds to the number of separate step angles in one complete revolution of the stepping motor. In a permanent-magnet stepping motor, $n_s$ is determined by the number of teeth on the rotor.

The angle of rotation of a stepping motor in response to a pulse train is given by

$$\text{angle of rotation} = P\alpha_s \tag{11.10}$$

where $P$ is the number of pulses. An alternative way of expressing this is

$$\text{angle of rotation} = f_p\, t\alpha_s \tag{11.11}$$

where $f_p$ = pulse rate (frequency of the pulse train), pulses/s
$t$ = duration of the pulse train

The rotational speed of the rotor in a stepping motor can be determined from the following equation:

$$S = \frac{60 f_p}{n_s} \tag{11.12}$$

where $S$ is the rotational speed in rpm.

**EXAMPLE 11.5**

The shaft of a stepping motor is attached to a worm gear which drives the table of a numerically controlled drill press. The worm gear has a spacing between gear teeth of 2.0 mm. The number of step angles in the stepping motor is 100.

(a) What is the size of the step angle on the motor?
(b) How many pulses are required to move the table exactly 150 mm from its current position?
(c) What is the linear travel speed of the table if the pulse train has a rate of 300 pulses/s?

The step angle size can be determined from Eq. (11.9).

$$\alpha_s = \frac{2\pi}{100} = 0.0628 \text{ rad} = 3.6°$$

To determine the answer to part (b), we must establish the relationship between linear table movement and rotation of the stepping motor. If the spacing between worm gear teeth is 2.0 mm, one complete revolution of the motor is required to move the table 2 mm. Hence, to move the table 150 mm, a total of 75 revolutions are required. Since 100 pulses are required for each revolution of the stepping motor, the number of pulses required to move the table 150 mm is

$$P = (75 \text{ rev})\ (100 \text{ pulses/rev}) = 7500 \text{ pulses}$$

Each pulse moves the table a distance of 2.0 mm/100 steps = 0.02 mm (0.00079 in.).
To obtain the linear travel speed of the table at a pulse rate of 300 pulses/s, we first obtain the rotational speed of the motor from Eq. (11.12):

$$S = \frac{60(300)}{100} = 180 \text{ rpm}$$

Since each rotation of the motor results in a table travel of 2.0 mm, the table moves at a speed of

$$(180 \text{ rev/min})(2 \text{ mm/rev}) = 360 \text{ mm/min}$$

(This corresponds to 14.17 in./minute.)

## Programmable controllers

For many years, the operation of production equipment was controlled by means of conventional relay control systems. Around 1969, the programmable controller (PC) was introduced as a substitute for relay logic systems. The PC is a sequential logic device which generates output signals according to logic operations performed on input signals. The sequence of instructions that determines the inputs, outputs, and logical operations constitutes the program. The original success of the programmable controller was due to its apparent similarity to relay control systems. Because wiring charts called *ladder diagrams* have been used for many years to set up relay panels, shop technical personnel are familiar with the use of these diagrams for wiring and maintenance of panels. Most PCs are programmed with the same type of ladder diagrams, so shop personnel have not had to learn a whole new programming system in order to use the PC. Advantages of programmable controllers over conventional relay controls include:

1. The PC can be reprogrammed. Conventional controls must be rewired and are often scrapped instead.
2. PCs take less space than relay control panels.
3. Maintenance of PCs is easier, and reliability is greater.
4. Programming the PC is often easier than wiring the relay control panel.
5. The PC can be connected to plant computer systems more easily than can relays.

A programmable controller consists of a power supply, a processor, and input/output. The power supply accepts external electrical input (e.g., 115 V or 230 V) and provides the required dc voltage to operate the processor and I/O devices. The processor consists of memory, equipment control program, and connectors for interfacing with the input/output devices. In turn, the I/O devices are attached to a mounting panel for wiring to the process equipment.

Basic control functions performed by a programmable controller include the following:

1. *Control relay function.* This involves the generation of an output signal from one or more inputs according to a particular logic rule.

2. *Timing functions.* An example would be the generation of an output signal a certain length of time after an input signal has been received. Another example would be to maintain an output signal for a certain length of time and then shut off.

3. *Counting functions.* The counter adds up the number of input contact closures and produces a programmed output when the sum reaches a certain count. The counter can then be reset.

4. *Arithmetic functions.* Addition, subtraction, and comparison capabilities are available on some PCs.

These are some of the more important functions that can be performed on programmable controllers. Different PC models have different capabilities.

Types of production machines controlled by programmable controllers include transfer lines; flow line conveyor systems; injection molding machines; grinding, welding, and stamping machines; and production test equipment.

In recent years, the use of large-scale integrated (LSI) technology in programmable controllers has reduced the differences between PCs and microcomputers/minicomputers. PCs can be interfaced with the computer in integrated factory CAM systems. The programmable controller is used to regulate certain aspects of the process. Minicomputers control other functions of the process and collect process data.

## 11.6 PROGRAMMING FOR COMPUTER CONTROL

In Sections 11.4 and 11.5, we considered the problem of interfacing the manufacturing process to the computer, and vice versa. In Section 11.3, the various types of programming languages were discussed. In the present section, we shall be concerned with the subject of computer programming for process monitoring and control. The reader is assumed to possess at least a basic understanding of computer programming, and FORTRAN programming in particular. For those readers without this background, there are many books on computer programming available.

### Requirements of control programming

There are several reasons why computer programming for process monitoring and control is different from programming for data processing and scientific/engineering calculations. The requirements of programming for computer control are the following:

1. *Timer-initiated events.* The computer must be capable of responding to events that are triggered by clock time. An example would be sampled-data values of process variables that must be collected at regular time intervals during process monitoring.
2. *Process-initiated interrupts.* The computer must be able to respond to incoming signals from the process. Depending on the relative importance of these signals, the computer might be required to interrupt its current program of calculations to perform higher-priority functions. A process-initiated interrupt is typically triggered by abnormal operating conditions, which indicates that some corrective control action is needed.
3. *Computer commands to process.* The process-initiated interrupt represents a communication from the process to the computer. This may be sufficient for process monitoring. But for process control, the computer system must have the software capability to direct the various process hardware devices that regulate the process in the desired manner.
4. *System- and program-initiated events.* These are events related to the computer system itself. When several computers are linked together in a computer network, data must be transferred back and forth between the

computers. This type of arrangement is common in CAM systems, where different computers occupy different levels in a pyramidal structure. Such an arrangement, called a hierarchical computer system, will be discussed in Section 11.7. Communication between computers falls into the category of a system-initiated event. Signals from peripheral devices such as a card reader or printer are also examples of system events. An example of a program-initiated event is when the program calls for data to be printed on some output device (e.g., teletype, CRT, line printer). The system and program events are really no different from the computer operations found in business and engineering applications. In process control, these events generally occupy a relatively low priority compared to process- or timer-initiated events.

5. *Operator-initiated events.* Also similar to conventional programming, the control system software must be capable of accepting input from operating personnel. This operator input may include any of the following types of items:

Request for printout of certain process variables or status data.

New programs or changes in existing programs.

Startup instructions or commands which may be part of the process operating routine.

Batch number or customer identification data to be associated with a certain production run.

The preceding control software requirements can be satisfied by means of a priority interrupt system.

### Interrupt system

Computers used for process control are equipped with an interrupt logic system. The purpose of the interrupt system is to permit program control to be switched between different programs or subroutines in response to different priority interrupts. When the interrupt is received, program control suspends execution of the current program and transfers to a predetermined storage location corresponding to the type of interrupt. Meanwhile, the location and status of the interrupted program is remembered so that its execution can be resumed when servicing of the interrupt has been completed.

The various interrupt conditions can be classified as external or internal. External interrupts are triggered by events that are external to the computer system. These include process-initiated interrupts and operator inputs. Internal interrupts are also called system interrupts because they are generated within the system itself. These include timer-initiated events, computer commands to the process, and other system- or program-related events.

Priority Interrupt Levels. The reason for using the interrupt system is to make the computer perform more important functions before it performs less important functions. For this reason, the various possible functions that the computer might be called upon to perform are classified according to priority level. A higher-level priority function can interrupt a lower-level function. The system designer must decide what level of priority should be attached to the various computer functions. A typical ranking of priority levels among control functions might be the following:

| *Priority level* | *Computer function* |
|---|---|
| 1 (lowest level) | Operator inputs |
| 2 | System interrupts |
| 3 | Timer interrupts |
| 4 | Commands to process |
| 5 (highest level) | Process interrupts |

However, this listing is intended only as a rough guide to priority levels. There may be different levels of interrupt priority within a certain category listed above. Some process interrupts may be more important than others. Certain system interrupts may take precedence over some of the process interrupts. Also, the computer interrupt system may have more or less than the five levels shown above.

A *single-level interrupt system* has only two modes of operation: normal mode and interrupt mode. The normal mode is interruptible but the interrupt mode is not. This means that overlapping interrupts are serviced on a first come/first served basis. This could lead to hazardous consequences if an important process interrupt was forced to wait in the queue while a series of unimportant operator and system interrupts were serviced.

A *multilevel interrupt system* has more than two operating modes as follows:

Normal mode.

Interrupt priority 1—interruptible by priority 2.

Interrupt priority 2—interruptible by priority 3.

And so on, for as many priority levels as are possessed by the system.

Figure 11.17 shows how the computer control system would respond to different priority interrupts under the single-level and multilevel interrupt designs. Programming for the multilevel system is somewhat more complicated than for the single-level configuration. However, the additional software expense is of low consequence compared to the risk of upsetting the process when high-priority interrupts are ignored.

Real Time Clock Interrupts. The timer-initiated interrupts are worthy of special mention. This is one of the most useful features of the

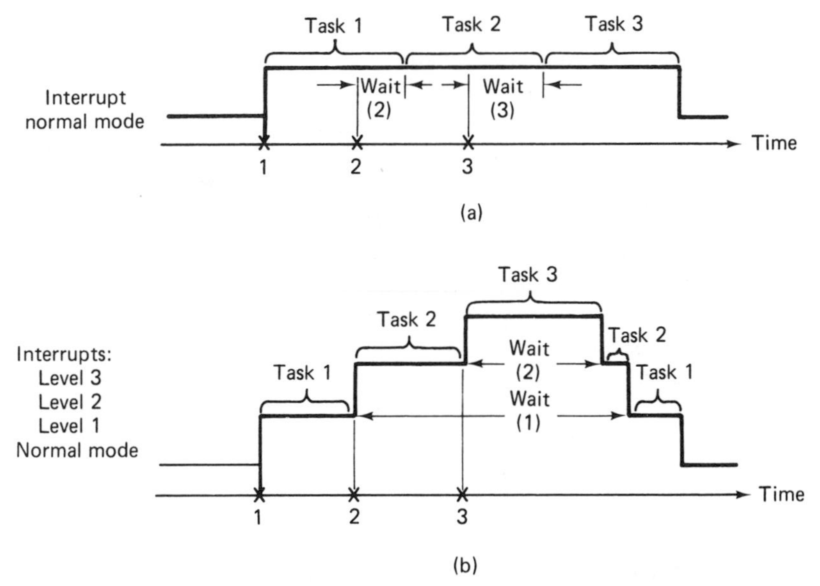


FIGURE 11.17 Response of the computer control system to different priority interrupts: (a) single-level interrupt system, (b) multilevel interrupt system.

control computer system. Timer-initiated events are driven by a real-time clock which basically marks time with the outside world. The real-time clock is a programmable control device that can generate an interrupt signal at regular time intervals. These time intervals can range from 100$\mu$s to several minutes, depending on the capability of the real time clock.

Some of the uses of the real time clock in computer control applications are the following:

1. To scan the variables of a manufacturing process at regular sampling intervals.
2. To initiate a recomputation of optimum conditions at which to operate a process. External conditions and raw materials change over time and the operating parameters of the process must be periodically fine-tuned.
3. To output data to the operator's console at various times during the production run.
4. To open and close contacts in sequencing control at appropriate times during the operating cycle.

## Control software

Control programming is typically done in either assembly language or in procedure-oriented languages such as FORTRAN. The advantage of assembly language is that the statements are often more suited to the elementary data manipulations and commands required in process monitoring and control. More efficient programs can be formulated in assembly language. Hence, these programs take less time to execute. This becomes important in highly repetitive computations that are often required in monitoring and control applications. The advantage of procedure-oriented languages such as real-time FORTRAN is that so many engineers and operating personnel are familiar with FORTRAN. Learning to program FORTRAN for real-time control does not require a drastic change from basic programming skills already possessed by these people. Simply stated, higher-level languages such as FORTRAN are easier and faster for the programmer to use than assembly languages. It is expected that as new software is developed specifically for process control applications, and as the trend continues toward more computer memory capacity per cost, procedure-oriented languages will become predominant for control programming.

It is beyond the scope of this book to present a detailed description of control programming. Our objective is to convey to the interested reader an appreciation of computer programming for process monitoring and control. An example will help to serve this objective.

### EXAMPLE 11.6

The program is written in FORTRAN and run on a DEC (Digital Equipment Corp.) PDP 11/34. The purpose of the program is to collect data on cutting temperature during a turning operation. A tool-chip thermocouple is used as the sensor. It uses the tool and the work material as the two dissimilar metals of thermocouple. The millivolt output of the thermocouple is fed through an analog-to-digital converter into the computer. The thermocouple output is sampled four times per second for 5 seconds. Then the 20 mV and temperature readings are printed out on a teletype, together with the averages of the 20 readings.

The program is shown in Figure 11.18 and a typical printout of results is displayed in Figure 11.19. The reader who is familiar with FORTRAN will recognize the structure and syntax of the statements. After the initializing statements (lines 1 through 8), the CALL ASICLN (LUN) in line 9 identifies the input unit assigned to the ADC. Literally, ASICLN(LUN) means "assign industrial controller to (logical unit number)." In line 10, the program calculates the scale factor to use for the analog-to-digital converter (refer back to the discussion on ADCs in Section 11.5 and Examples 11.2 and 11.3).

In lines 11 and 12, the program instructs the machine tool operator to "put the tool against the work, when done hit carriage return." The statement in line 15,

CALL AIRD (1, ICONT, IHOLD, ISTAT, LUN)

commands the computer to accept the analog input from the ADC. This provides a zero set point for subsequent measurements taken during the cut. In lines 20 and 21, this value of the "offset voltage" is programmed to be printed.

The program then instructs the operator to start taking measurements with statement: "When you want to start readings, hit carriage return." Lines 25 through 32 comprise a DO loop which calls for the analog input (CALL AIRD) to be read and stored in a storage location called IHOLD. This is done for 20 cycles (J = 20 in line

FIGURE 11.18 Process monitoring program (written in FORTRAN) of Example 11.6.

```
0001          PROGRAM L4
0002          DIMENSION ISTAT(2),IHOLD(1),ICONT(1)
0003          DIMENSION V(300)
0004          LOGICAL *1 DUMMY
0005          DATA ICONT/"140003/, LUN/1/
0006          DATA GAIN/200./
0007          J=20
0008          SUM = 0.
      C** ASSIGN A LOGICAL DEVICE NUMBER TO THE ICS UNIT
0009          CALL ASICLN(LUN)
      C** CALCULATE THE VOLTAGE CONVERSION FACTOR TO USE
0010          FACTOR = 10.24 / (32768.*GAIN)
0011          TYPE 1000
0012  1000    FORMAT (' PUT THE TOOL AGAINST THE WORK, WHEN DONE HIT <CR>',$)
0013          ACCEPT 1001,DUMMY
0014  1001    FORMAT (A1)
0015          CALL AIRD (1,ICONT,IHOLD,ISTAT,LUN)
0016  1       CONTINUE
0017          IF (ISTAT(1).EQ.0) GO TO 1
0019          OFFSET = FLOAT (IHOLD(1)) * FACTOR
0020          TYPE 1010,OFFSET
0021  1010    FORMAT (' THE OFFSET VOLTAGE IS ',F10.5)
0022          TYPE 1025
0023          ACCEPT 1001,DUMMY
0024  1025    FORMAT (' WHEN YOU WANT TO START READINGS HIT <CR>',$)
0025          DO 10 I=1,J
      C** INITIATE A VOLTAGE READING
0026          CALL AIRD (1,ICONT,IHOLD,ISTAT,LUN)
      C** WAIT UNTIL YOU HAVE  COMPLETED THE VOLTAGE READING
0027  49      IF (ISTAT(1).EQ.0) GO TO 49
      C** CALCULATE THE VOLTAGE
0029          V(I) = FLOAT (IHOLD(1)) * FACTOR-OFFSET
0030          CALL WAIT(15,1)
0031          SUM =SUM + V(I)
      C** IF J READINGS HAVE BEEN TAKEN, DROP THROUGH, ELSE, GO TO THE
      C**     BEGINNING OF THE LOOP
0032  10      CONTINUE
0033          DO 20 I=1,J
0034          TYPE 1021,I,V(I)* 1000,TEMP(V(I))
0035  1021    FORMAT (' READING ',I3,' IS ',F10.2, ' MV',2X,F6.1,2X,'DEG. F')
0036  20      CONTINUE
0037          AVG = SUM / 20.
0038          TYPE 1022,AVG*1000.,TEMP(AVG)
0039  1022    FORMAT(/,1X'THE AVG READING = ',F10.5 ,' MV',5X,F6.1,2X,'DEGREES')
0040          END


0001          FUNCTION TEMP(A)
0002          TEMP=-4 + 67.097*1000.*A + .9612 * ((A*1000.)**2.)
0003          RETURN
0004          END
```

7). The readings are taken at $\frac{1}{4}$-s intervals. This is defined by the statement in line 30,

CALL WAIT (15, 0)

This command makes the computer wait for 15 "ticks" before accepting the input. A "tick" in this computer system is $\frac{1}{60}$ s. Hence, 15 ticks equals $\frac{1}{4}$ s.

The remainder of the program is for printout of the data. Lines 33 to 36 print the 20 values of millivolt and corresponding cutting temperature values. The symbol

```
>RUN L4
PUT THE TOOL AGAINST THE WORK, WHEN DONE HIT <CR>
THE OFFSET VOLTAGE IS      0.00000
WHEN YOU WANT TO START READINGS HIT <CR>

READING    1 IS        12.88 MV   1019.2   DEG. F
READING    2 IS        12.93 MV   1023.8   DEG. F
READING    3 IS        12.95 MV   1026.1   DEG. F
READING    4 IS        12.97 MV   1028.4   DEG. F
READING    5 IS        12.95 MV   1026.1   DEG. F
READING    6 IS        12.95 MV   1026.1   DEG. F
READING    7 IS        13.00 MV   1030.7   DEG. F
READING    8 IS        12.95 MV   1026.1   DEG. F
READING    9 IS        13.05 MV   1035.3   DEG. F
READING   10 IS        13.07 MV   1037.6   DEG. F
READING   11 IS        13.02 MV   1033.0   DEG. F
READING   12 IS        13.02 MV   1033.0   DEG. F
READING   13 IS        13.05 MV   1035.3   DEG. F
READING   14 IS        13.07 MV   1037.6   DEG. F
READING   15 IS        13.10 MV   1039.9   DEG. F
READING   16 IS        12.88 MV   1019.2   DEG. F
READING   17 IS        13.17 MV   1046.8   DEG. F
READING   18 IS        13.13 MV   1042.2   DEG. F
READING   19 IS        13.13 MV   1042.2   DEG. F
READING   20 IS        13.15 MV   1044.5   DEG. F

THE AVG READING =      13.02125 MV       1032.7   DEGREES
TT2   --   STOP
>
```

FIGURE 11.19 Output of process monitoring program of Example 11.6.

V(I) is used for the 20 values of voltage. V(I) multiplied by 1000 converts the voltage measurements to millivoltage for printout. The term TEMP(V(I)) calls a function for converting the voltage output of the tool-chip thermocouple into the corresponding temperature. The conversion is based on an equation that was derived from the calibration curve for the particular tool work material combination. In this case, the calibration equation is

$$\text{TEMP} = -4 + 67.097\ \text{MV} + .9612\ (\text{MV})^2$$

where TEMP is the calculated cutting temperature and MV is thermocouple output. The function that performs this calculation is shown in Figure 11.18.

Lines 38 and 39 print the average values of millivolt and cutting temperature.

Figure 11.19 illustrates the output from the program during one cut of the turning operation. The teletype is located at the lathe, while the PDP 11/34 is remotely located in a separate room.

## 11.7 HIERARCHY OF COMPUTERS IN CAM

One of the trends that has developed in computer-aided manufacturing is that the computers in a corporation form a sort of pyramidal arrangement not unlike the pyramidal structure of management in a manufacturing firm. The term that has evolved to describe this computer pyramid is *hierarchical*

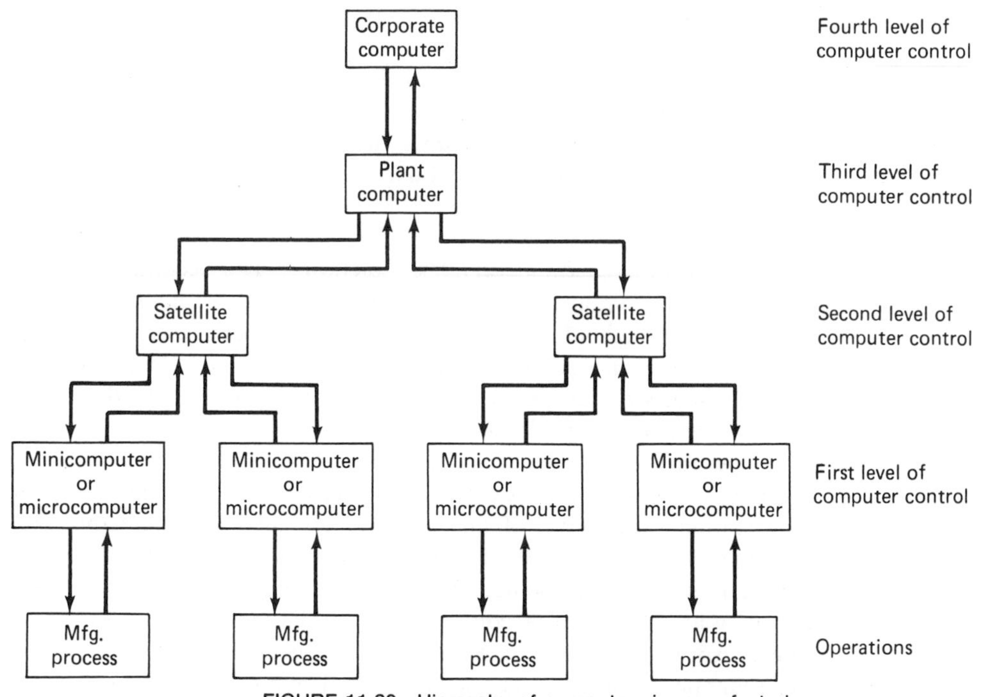


FIGURE 11.20 Hierarchy of computers in manufacturing.

*computer system.* The general configuration is illustrated in Figure 11.20. The various computers in the hierarchy are tied together by communication links. These individual links provide a computer–communications network for forwarding data and information up the chain of command from the manufacturing process all the way to the corporate computer. In the other direction, commands, schedules, and so on, are passed down to the individual production operations from above.

## Levels of computer control

Let us examine the various levels within the hierarchy of computers in a manufacturing plant.

FIRST LEVEL OF COMPUTER CONTROL. At the lowest level within the hierarchy are the computers that are connected directly with the process, either to monitor the process or to control it. The computers that make up this level are small: minicomputers and microcomputers. As the trend continues toward lower cost and greater computational power per hardware volume, there will be a growing use of microcomputers in production equipment.

The first-level computers are located in close proximity to the processes they monitor and control. In many cases, the computer is an integral component of the production machine tool (an example would be a CNC machine). The computer is dedicated to the process and performs a limited number of specific tasks. Basically, these tasks are to serve the process (monitoring and/or control functions) and to communicate with the second level in the computer hierarchy.

SECOND LEVEL OF COMPUTER CONTROL. The second level of computers are minicomputers. In the hierarchy, several of these minicomputers are located throughout the plant and report to a larger plant computer. For this reason, minis are sometimes referred to as "satellite" computers. To form the pyramidal structure, each satellite is linked to several smaller computers at the process level (first level of control).

The purpose of the second level of control is to serve in a supervisory capacity for its section of the plant. The supervisory function would include coordinating the activities of smaller computers under its command. Data are collected on the operation of the individual machine tools, inspection stations, and so on, and instructions are relayed back to the separate processes and stations. The types of control strategies discussed in Section 10.3 under "Supervisory Computer Control" are often executed at this level. For example, program control or sequencing control might logically be handled for many processes at this second level in the computer hierarchy. On-line search strategies or adaptive strategies might be appropriate for certain industrial processes at this level. On the other hand, some of the supervisory strategies are more likely to be found either at the first level or at the third level in the plant hierarchy. For instance, computer numerical control (CNC) is commonly exemplified by one minicomputer or microcomputer controlling one machine. This represents first-level control. By contrast, direct numerical control (DNC) is often accomplished at the plant level, or third level in the computer pyramid. The plant level computer uses the second-level computers as relay stations to provide blocks of instructions to the individual machine tools.

THIRD LEVEL OF COMPUTER CONTROL. The third level in the hierarchical structure is the central plant computer. Data from the various individual plant operations are collected and summarized to prepare timely reports for plant management. These reports may be on a daily, weekly, or monthly basis, depending on the nature of the information. Operating instructions are sent back to the production equipment through the second and first levels of control (e.g., DNC).

The computer at the plant level must be a larger, more complete data processing system. Its use must be shared between plant operations control and other business-related functions that are performed at the plant level. For example, payroll, cost accounting, production control, industrial engineering, and so on, would all have access to the plant computer.

FOURTH LEVEL OF COMPUTER CONTROL. The fourth level in the hierarchy is the corporate computer. Data are collected and compiled from the various plants in the corporation by the corporate-level computer. The communication with the individual plant computers can be achieved by means of telephone long lines. The purpose of the corporate computer in the hierarchy is to summarize plant operations and performance for the entire company. The corporate computer must be shared with the other departments at the corporate level: sales, marketing, research, design engineering, accounting, and so on.

The hierarchical structure has evolved to be the most effective and efficient arrangement for implementing computers in manufacturing. At one time, before minicomputers and microprocessors became so predominant, it seemed most feasible to use one large plant computer to handle all monitoring and control functions. Indeed, this mode of plant control has proven itself to be feasible in the process industries. In this configuration, direct communication is established between the single plant computer and the various processes in the plant. Today, in the manufacturing industries as well as the process industries, the evolution of computer technology has given the advantage to the computer hierarchy approach. The following is a listing of the important benefits of this approach to computer-aided manufacturing:

1. Computer control can be installed gradually rather than all at once. Each individual computer project can be justified on its own merits. The company is not required to make an all-or-nothing commitment to undertake a complete plant-wide computer system, which would be the case if a large single computer were selected. The advantage is that the firm is not faced with one large expense. The cost of the computer system can be spread over a number of years.
2. The hierarchical structure contains redundancy. In the event of a computer breakdown, other computers in the system are programmed to assume the critical tasks of the computer that is down.
3. Software development is more easily managed in the hierarchical configuration. Since the computers are separated in the pyramidal arrangement, programming for each project can be handled separately. Once the project is installed, changes in software are more easily accomplished, with less chance of disrupting the system.

The software issue is complex, with not all the advantages on the side of the hierarchical system. The minicomputers in the system have limited memory capacity, which results in programming difficulties. The small computers do not have the software support from the computer manufacturer that is usually available for larger computers. The problem of communication among the various minicomputers is often compounded by the differences in computer design. The software development costs of a new computerized system will often exceed the cost of the hardware. Hence, the software problem should be given a good deal of consideration in the planning of a CAM system and in the specification of hardware for the system.

## REFERENCES


[1] *Analog-Digital Conversion Handbook,* Analog Devices, Norwood, Mass., 1972.

[2] CADZOW, J. A., and MARTENS, H. R., *Discrete-Time and Computer Control Systems,* Prentice-Hall, Inc., Englewood Cliffs, N.J., 1970.

[3] DIEFENDERFER, A. J., *Principles of Electronic Instrumentation,* W. B. Saunders Co., Philadelphia, 1972.

[4] GARRET, P. H., *Analog Systems for Microprocessors and Minicomputers,* Reston Publishing Company, Inc., Reston, Va., 1978.

[5] HARRISON, T. J. (Editor), *Minicomputers in Industrial Control,* Instrument Society of America, Pittsburgh, Pa., 1978.

[6] HOLLO, F. R., "Programmable Controllers—Familiarity Breeds Success," *Automation,* February, 1974, pp. 68–72.

[7] HORST, R. L., "Hierarchical Application of Computers for Process Control," *Tech Paper MS76-740,* Society of Manufacturing Engineers, Dearborn, Mich., 1976.

[8] KLINE, R. M., *Digital Computer Design,* Prentice-Hall, Inc., Englewood Cliffs, N.J., 1977.

[9] LEE, T. H., ADAMS, G. E., AND GAINES, W. M., *Computer Process Control: Modeling and Optimization,* John Wiley & Sons, Inc., New York, 1968.

[10] LEVENTHAL, L. A., *Introduction to Microprocessors: Software, Hardware, Programming,* Prentice-Hall, Inc., Englewood Cliffs, N.J., 1978.

[11] LUTZ-NAGEY, R., "Control Systems," *Production Engineering,* June, 1978, pp. 42–50.

[12] MCGLYNN, D. R., *Microprocessors: Technology, Architecture, and Applications,* John Wiley & Sons, Inc., New York, 1976.

[13] OBERT, P., "The 12 Most Often Asked Questions about Programmable Controllers," *Instruments and Control Systems,* August, 1976, pp. 25–27.

[14] REMBOLD, U., SETH, M. K., and WEINSTEIN, J. S., *Computers in Manufacturing,* Marcel Dekker, Inc., New York, 1977.

[15] SAVAS, E. S., *Computer Control of Industrial Processes,* McGraw-Hill Book Company, New York, 1965.

[16] SCHAFFER, G., "Computers in Manufacturing," *American Machinist,* April, 1978, pp. 115–130.

[17] SMITH, C. L., *Digital Computer Process Control,* International Textbook Co., Scranton, Pa., 1972.

[18] SOISSON, H. E., *Instrumentation in Industry,* John Wiley & Sons, Inc., New York, 1975.

[19] TOGINO, K., "Hierarchical Approach to CAM System," *Tech Paper MS77-769,* Society of Manufacturing Engineers, Dearborn, Mich., 1977.

## PROBLEMS

**11.1.** A continuous voltage signal is to be converted into its digital counterpart by an analog-to-digital converter. The maximum voltage range is $\pm 30$ V. The ADC has a 12-bit capacity. Determine the number of quantization levels, the resolution, the spacing of each quantization level, and the quantization error for this ADC.

**11.2.** A voltage signal with a range of 0 to 115 V is to be converted by means of an ADC. Determine the minimum number of bits required to obtain a quantization error of (a) $\pm 5$ V maximum; (b) $\pm 1$ V maximum; (c) $\pm 0.1$ V maximum.

**11.3.** A digital-to-analog converter uses a reference voltage of 120 V dc and has eight-binary-digit precision. In one of the sampling instants, the data contained in the binary register are 01010101. If a zero-order hold is used in generating the output signal, determine the voltage output level.

**11.4.** A DAC uses a reference voltage of 80 V and has 6-bit precision. In four successive 1-s sampling periods, the binary data contained in the output register were 100000, 011111, 011101, and 011010. Determine the equation for the voltage as a function of time between sampling instants 3 and 4 using (a) a zero-order hold; (b) a first-order hold.

**11.5.** In Problem 11.4, suppose that a second-order hold was to be used to generate the output signal. The equation for the second-order hold is

$$V(t) = V_0 + at + bt^2$$

where $V_0$ is the starting voltage at the beginning of the time interval. For the binary data given in Problem 11.4, determine the values of $a$ and $b$ that would be used in the equation for the time interval between instants 3 and 4. Compare the first-order and second-order holds in anticipating the voltage at the fourth instant.

**11.6.** In a certain specially designed stepping motor, the number of step angles is 60.
(a) What is the size of the step angle on the motor?
(b) How many pulses are required to rotate the motor through five complete revolutions?
(c) If it is desired to rotate the motor at 75 rpm, what pulse rate would be required?

**11.7.** A stepping motor is to be used to drive the table of an NC drill press. The motor shaft will be attached to a worm gear with a gear tooth spacing of 3.0 mm. It is desired to be able to drive the table in linear increments of 0.015 mm.

(a) To achieve this precision of movement, how many step angles are required on the stepping motor?

(b) Using your answer to part (a), what pulse rate is required to drive the table at a feed rate (table speed) of 100 mm/min?

# Computer Process Monitoring

## 12.1 INTRODUCTION

Computer process monitoring consists of an installation in which the computer is connected to the manufacturing process and associated equipment, either directly or indirectly, to observe the operation. The monitoring function has no direct effect on the mode of operation except that the data provided by monitoring may result in improved supervision of the process. The industrial process is not regulated by commands from the computer. Any use that is made of the computer to improve process performance is indirect, with human operators acting on information from the computer to make changes in the plant operations. The flow of data between the process and the computer is in one direction only—from process to computer.

The data collected during computer monitoring of manufacturing operations fall into three general categories:

1. Process data.
2. Equipment data.
3. Product status data.

In addition to computer monitoring in production, computers are also used in laboratory testing situations to record data and results from experiments.

### Process data

The status of any industrial process can be characterized by the technological variables associated with that process. Examples of these variables would be temperatures, pressures, forces, concentrations of chemicals, and so on. By monitoring the values of these technological variables, the general performance of the process can be assessed. In many industrial operations, the variables should be maintained at certain desired levels throughout the process cycle. When the actual values deviate from these desired levels, the computer is programmed to signal the operator that corrective action should be taken.

### Equipment data

The second function of computer process monitoring is to collect data related to the status of the production equipment.

Computer monitoring of equipment status can ensure against undesirable or even dangerous combinations of equipment settings. For example, excessive vibration of rotating equipment would provide a warning of impending machine breakdown. Without computer monitoring, this type of hazard might go undetected by the manual operator until failure occurred.

### Product data

The third category of data collected during computer process monitoring is product status data. The computer can be employed to gather on-line data on product quality, yield, production rate, and so on. Other off-line data can be compiled at the computer by manual entry devices. This may include piece counts, quality characteristics determined from manual inspection operations, operator job times, and other information that would be difficult to obtain automatically.

### Laboratory testing

In addition to computer monitoring of production operations, effective use can be made of the computer to monitor experiments performed in either the laboratory or the production plant. Many experimental situations involve one or more of the following data collection situations:

1. The recording of variable measurements and test results over long periods of time. This would be the case in experiments in which each set of

test conditions requires a considerable time span to complete. Examples would be found in chemical processes with slow reaction times.

2. The necessity to take large amounts of data over a relatively short period of time. Physical processes that occur rapidly would fit into this category. There is the need to generate precise, accurate measurements of the variables as a function of time. The variables change too rapidly to be recorded by human observers. Some form of automatic data collection is required. An example of this type of situation in manufacturing would be in a press operation or a forging process, where the actual cutting or forming operation often takes less than a second. The forces, stresses and strains in the material, temperatures, and even the velocity of the punch or ram change very rapidly over a short time span.

3. The need to observe several variables (perhaps many variables) simultaneously and to relate them over time. In attempting to develop a mathematical model for a complex physical process, it is often necessary to take measurements on more than the number of variables that will be included in the model. The reason for this is that the particular variables that will ultimately be used in the model cannot always be identified initially. By performing a statistical analysis on the data (regression analysis is commonly used), those variables that have the strongest effect on the dependent variable can be identified and incorporated into the model.

In each of these situations, computerized data collection is appropriate. The computer monitoring system can be programmed to sample the process periodically over long time spans or to take large amounts of data quickly over short time spans. The data can then be formatted for statistical analysis or summarized in a desirable report style.

## 12.2 TYPES OF PRODUCTION MONITORING SYSTEMS

In this section, we discuss some of the system configurations used for production monitoring. Most of the hardware components used to build these systems have already been described in Sections 11.4 and 11.5. These components include manual entry terminals, transducers and sensors, analog-to-digital converters, multiplexers, real-time clocks, and other devices for collecting and processing data from the factory. Modern production monitoring systems also include, of course, the computer to regulate the various data collection hardware elements and to transform the data into useful information.

This section is organized into the following four topic areas:

1. Manual entry systems.
2. Data logging systems.
3. Data acquisition systems.
4. Multilevel scanning.

These topics represent various systems for collecting data on production operations. The systems are not necessarily mutually exclusive. A given computer monitoring system may include several of the features in the topic areas listed above. For example, data acquisition systems may include manual entry devices and automatic sensing instruments for collecting data.

### Manual entry systems

Some types of shop floor data are most conveniently communicated to the computer by means of manual data entry stations. Workpiece counts, order progress status, machine downtime, and product quality results are examples of data that can be entered by means of human operators located in the shop. In discrete-parts manufacturing, manual entry stations constitute the most common factory data collection systems in use today. The terminals range in complexity from simple pushbutton devices to sophisticated typewriter keyboard units with built-in microprocessor and data storage capability. Factory workers must be trained to key in the proper data, including operator identification number, the type of data being reported, and the data itself in correct format. Some terminals are equipped with telephones for voice communication with shop foremen or maintenance personnel to report emergency situations.

FIGURE 12.1 IBM 5230 data entry terminal.

Shop floor data entered from manual entry stations can be communicated directly to the computer. This is sometimes referred to as an "on-line" system. The alternative is to use an "off-line" configuration, in which the data are collected and stored in a data storage device for subsequent processing by the computer. An example of the latter alternative is the IBM 5230 Data Collection System [5], which consists of data entry stations, time entry stations, and a controller. The data entry stations capture information from the factory, such as production piece counts, inventory counts, material and job status data, quality inspection data, job times (for costing purposes), and tool usage data. Factory management defines the type of information to be collected by the system. The time entry stations are located at entrance and exit doors and serve the function of time clocks to punch in and out. The controller unit collects and stores the data from the various entry stations. The shop records can be stored on diskettes for processing by the plant computer to generate the desired reports. Figure 12.1 shows a factory operator using one of the data entry stations.

**EXAMPLE 12.1**

Burron Medical Products, Inc., is a manufacturer of medical supplies to hospitals, medical distributors, and other companies. It pioneered the disposable syringe. Annual sales approach $10 million and the company employs approximately 400 people, 300 of which are direct labor. The number of individual items produced runs into the millions. Burron Medical Products uses an IBM 5230 Data Collection System to monitor its inventory and production activities. The U.S. Food and Drug Administration requires traceability of the products made by Burron. Accordingly, three data entry terminals are located throughout the factory and warehouse to record the necessary data to comply with FDA requirements. Figure 12.1 illustrates one of these terminals.

Operator-entered data include item number, lot identification number, lot quantities, and customer identification for lots shipped. About 300 transactions per day are entered through the terminals and stored in the controller unit, which is located in the central computer room. A diskette is used as the data storage medium. Once each day, the data are transferred to the company's computer to generate reports. These reports include a daily shipping record, finished goods status, and a stores transactions register, all of which satisfy Burron's own inventory control requirements as well as the FDA requirement for product traceability.

## Data logging systems

A *data logger* (DL) is a device that automatically collects and stores data for off-line analysis. Strictly speaking, the data could be analyzed by a person without the aid of computer. Our interest here is in data logging systems that operate in conjunction with computers. Data loggers can be classified into three types [3]:

1. Analog input/analog output.
2. Analog input/analog and digital output.
3. Analog and digital input/analog and digital output.

Type 1 can be a simple one-channel strip chart recording potentiometer for tracking temperature values using a thermocouple as the sensing device. Types 2 and 3 are more sophisticated instruments, which have multiple input channels and make use of multiplexers and ADCs to collect process or experimental test data from several sources. The DL can be interfaced with tape punches, magnetic tape units, teletypes and printers, plotters, and so on. They can also be interfaced with the computer for periodic transfer of data.

A *programmable data logger* (PDL) is a device that incorporates a microprocessor as part of the system. The microprocessor serves as a controller to the data logger and can be programmed by means of a keyboard. The programmable data logger can easily accommodate changes in rate or sequence of scanning the inputs. The PDL can also be programmed to perform such functions as data scaling, limit checking (making certain the input variables conform to prespecified upper and lower bounds), sounding alarms, and formatting the data to be in a compatible and desirable format with its interface devices.

### Data acquisition systems

The term *data acquisition system* (DAS) normally implies a system that collects data for direct communication to a central computer. It is therefore an on-line system, whereas the data logger is an off-line system. However, the distinction between the DL and the DAS has become somewhat blurred as data loggers have become directly connected to computers.

Data acquisition systems gather data from the various production operations for processing by the central computer. The basic data can be analog or digital data which are collected automatically (transducers, ADCs, multiplexers, etc.), or it can be submitted by means of manual entry terminals by human operators. It is a factory-wide system, as compared with data loggers, which are often used locally within the plant. The number of input channels in the DAS is therefore typically greater than in the DL system. For the data logger the number of input channels might range between 1 and 100, while the data acquistion system might have as many as 1000 channels or more. The rate of data entry into the DL system might be 10 readings per second for multiple-channel applications. By contrast, the DAS would have to be capable of a data sampling rate of up to 1000 per second. Because of these differences, the data acquisition system would be considerably more expensive than the data logger.

Data acquisition systems can be arranged into hierarchical configurations (refer to Section 11.7), in which several minicomputers or microcomputers are located throughout the plant. These intermediate-level computers serve as relay stations to collect and summarize shop data before sending them to the plant computer. They also serve as controllers for the data acquisition subsystems. In this capacity they perform many of the same functions as the microprocessors of programmable data loggers (control over the individual data measuring devices, alarm sounding, limit checking). Other functions include real-time data analysis and decision making.

### Multilevel scanning

In the data acquisition system, it is possible for the total number of monitored variables to become quite large. Although it is technically feasible for all these variables to be monitored through multiplexing, some of the signals would not be needed under normal operating conditions. In such a situation it is convenient to utilize a multilevel scan configuration, as illustrated schematically in Figure 12.2. With multilevel scanning, there would be two (or more) process scanning levels, a high-level scan and a low-level scan. When the process is running normally, only the key variables and status data would be monitored. This is the high-level scan. When abnormal operation is indicated by the incoming data, the computer switches to the low-level scan, which involves a more complete data logging and analysis to ascertain the source of the malfunction. The low-level scan would sample all the process data or perform an intensive sampling for a certain portion of the process that might be operating out of tolerance.

The computer monitoring system can also be programmed to permit the operator to request a low-level scan of the process.

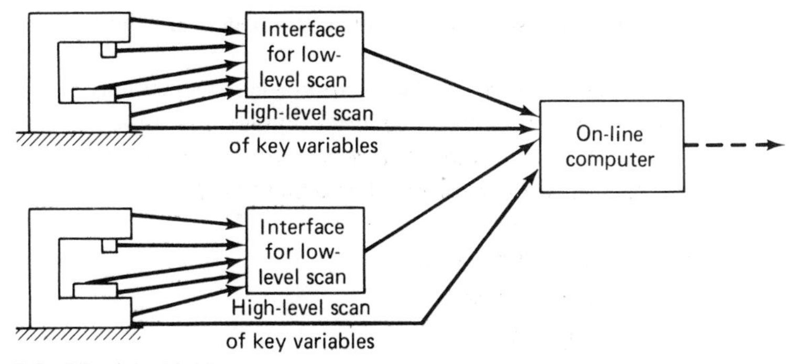


FIGURE 12.2 Multilevel scan arrangement in computer process monitoring.

## 12.3 INFORMATION PROVIDED BY MONITORING

In order to achieve a return on investment for the computer monitoring system, the information provided must be more valuable than the cost of the collection system. Generally, the value of the monitored information is measured in terms of the improvement in production performance. The objective of better process performance constitutes the principal justification for installation of a computer process monitoring system in most cases. In addition, there are also instances where the firm is required to keep data relating to details of its manufacturing operations. Burron Medical Products of Example 12.1 was required to satisfy FDA product traceability regulations. In these instances, a computer monitoring system may represent the most feasible method of compiling the data.

We shall organize the types of information provided by computer process monitoring into three basic classifications, depending on how the information is used:

1. Operations management.
2. Maintenance.
3. Manufacturing records.

The three classes overlap to some extent. Information on machine breakdowns which is compiled primarily for maintenance personnel is of interest to production management. It may also be necessary to maintain records of machine breakdowns for product warranty purposes, since the equipment failures can affect quality of output.

### Monitoring for operations management

Computer production monitoring comprises one of the main sources of data in the company's management information system. The kinds of information in which management would be interested are the following:

PRODUCTION. The computer can record current production levels, and can be programmed to generate reports to compare current with past production rates or to predict future production. Management would be interested in such production performance measures as:

Production rates.
Piece counts.
Scrap rates.
Worker overtime.
Product costs.
Inventory levels and their relationships to current production.
Production schedule overruns.

Management reports can be provided either in summary form at periodic intervals or on a real-time basis as demanded.

MACHINE EFFICIENCY. Another area of importance to operations management is equipment utilization and efficiency. The following questions would be of interest to the plant supervision personnel:

At what level of efficiency did the production equipment perform during the most recent reporting period?

What were the reasons for machine breakdowns? How much production time was lost for each reason?

How much electrical power (and cost) was consumed by each production machine? Can a plant-wide strategy be developed to limit the peak power load by alternating the use of certain pieces of equipment?

PRODUCT QUALITY. A firm's reputation is often based on the quality of the products it manufactures. The computer can be used to monitor and record the various parameters of the product that define its quality, and to generate reports to management that summarize these quality data.

## EXAMPLE 12.2

An example of a computerized quality monitoring and inspection system is the internal combustion engine test system illustrated in Figures 12.3 and 12.4. The system was engineered and built by Scans Associates, Inc., of Livonia, Michigan, and consists of 14 computer-controlled test stations served by an automated conveyor system. Engines to be tested are placed on a specially designed pallet at the conveyor loading station. The conveyor carries the pallet with engine to the first unoccupied test station. The engine and pallet are drawn into the station, where they are automatically positioned, clamped, and sealed.

In the test stand, the engine is cold-tested (air-motored rather than fuel-powered). The computer-controlled inspection procedure consists of the following tests:

1. Water jacket leak test.
2. Oil cavity leak test.
3. Air motoring test, during which the engine is operated at two different speeds by feeding compressed air through the spark plug holes in proper firing rotation. Oil pressure is automatically checked for an acceptable low limit. An inspector watches the sight glass for signs of leakage from combustion chamber to water cavity during this sequence.
4. Engine timing is automatically set.

If an engine should fail to pass any of the tests, it is automatically conveyed to a manual inspection and repair station. Assuming that the condition can be corrected, the engine is then returned to the automatic inspection system for retesting. Engines that pass the inspection procedure are carried by the conveyor to an unloading station.

Figure 12.3 shows an overall view of the engine test system. Figure 12.4 shows an engine in one of the test stands, positioned and clamped for the inspection sequence. The computer-controlled test system is capable of a throughput rate of more

FIGURE 12.3 Overall view of quality monitoring system described in Example 12.2 for automatic inspection of V-8 engines. (Courtesy of Scans Associates.)

FIGURE 12.4 Engine test stand of quality monitoring system described in example 12.2. (Courtesy of Scans Associates.)

than 200 engines per hour with only five trained inspectors. It can also accept eight different engine models on two different block styles. Floor space required for the system is about one-fourth of the space needed for equivalent conventional testing facilities.

PROCESS TECHNOLOGY IMPROVEMENT. Because of its capacity to collect large quantities of process data, the computer can be used to study and learn about a particular manufacturing operation. The success of many industrial processes is based on operator skill and experience. Without detracting from the credit that is due the operators, it is always good to establish a strong technology base for any production process. Dedicated computer monitoring of a given process can provide data whose analysis will lead to the derivation of basic relationships and equations that govern the process. Although laboratory studies of a process under idealized conditions are certainly valuable, it is often necessary to perform the investigation in the factory environment to obtain realistic data. Computer monitoring makes this data collection task feasible. Advancement in process technology and higher productivity are the desired outcomes of this type of investigation.

### Monitoring for production maintenance

Another area related closely to operations management is the function of maintaining and repairing the production equipment. Some of the tasks that can be included in computer monitoring to aid maintenance personnel are the following:

CYCLE-TIME ANALYSIS. Certain manufacturing operations have a fixed cycle time or a process cycle time that should vary only between fixed limits. Examples would be automated transfer lines and assembly machines as well as certain types of chemical operations. Computer monitoring of the cycle time can indicate variances beyond the acceptable limits which may be a signal of trouble in the equipment. For example, erratic cycle times on a transfer machine may be due to a limit switch or other device not functioning properly.

EQUIPMENT STATUS MONITORING. The computer can be used to observe the status of the equipment components: motors, solenoids, limit switches, relays, and so on. The computer software compares the actual status with the desired status for the current phase of the manufacturing operation. The maintenance personnel are alerted to any deviations so that repairs can be effected or replacement of a defective component can be made.

TOOL MANAGEMENT. This application involves machining operations and other manufacturing processes that utilize tooling. In metal cutting, the tool is used up during the operation. The life of the cutting tool has an expected value in terms of the number of workparts (or machine cycles) that can be machined. Computer monitoring can be used to track the number of machine cycles and to signal the operator when a tool change is needed.

Additional options can be programmed into the computer monitoring system. Optimum tool change schedules can be developed for transfer lines and other machines that utilize multiple tools in the operation. Analyses can be made to determine which tools fail unexpectedly most frequently, thus leading to improved tool designs for those cutting operations.

BREAKDOWN DIAGNOSIS. In many industrial processes, the production equipment is extremely sophisticated and complex. When malfunctions or breakdowns occur, it is often difficult for the maintenance people to immediately identify the source of the trouble. The computer can be used to help diagnose the problem. If a short circuit has occurred, a low-level scan of the equipment followed by computer interpretation of the data may isolate the short circuit. If a sequence of processing steps has not followed in the proper order, the monitoring program may be able to logically determine which machine component is malfunctioning. This type of breakdown analysis may significantly reduce equipment downtime, thus increasing machine efficiency.

### EXAMPLE 12.3

Production monitoring for a large number of machine tools is reported by Berry of Bendix Corporation Industrial Controls Division [1]. The computer monitoring system was installed for a total of 160 automatic screw machines in one plant. Both manual data entry stations as well as a computerized data acquisition system were used for each of the machines. Control of the individual machines was by conventional relay panels. Objectives of the computer monitoring system were:

Automatic collection of piece count data through monitoring of the complete machine cycles. The piece count data included identification of part numbers.

Compiling of production statistics on run time, idle time, downtime, and tool change time.

Alarms for machine failure.

Production scheduling and reporting on a real-time basis.

The hardware components of the system included:

Interface hardware for each machine control panel.

Manual data entry terminals at each machine.

Data acquisition network from the machines to the computer.

Multiplexers and addressing equipment for orderly presentation of data to the computer.

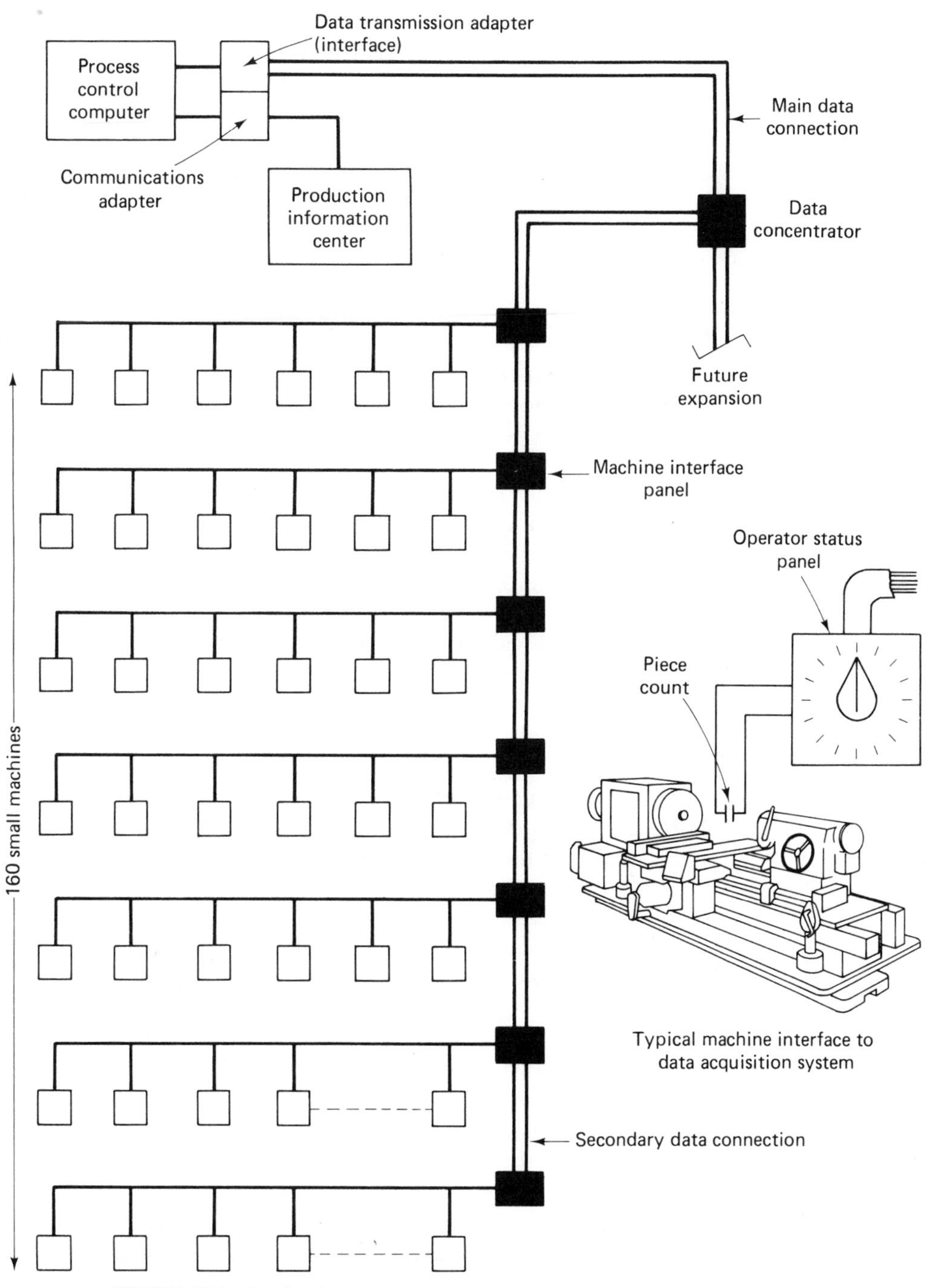


FIGURE 12.5 Production monitoring system of Example 12.3. (Reprinted by permission from Berry [1].)

Computer interface to the data acquisition network.

Computer peripherals for communicating to the system users.

This computer monitoring installation was primarily limited to production scheduling and operations management. It contained few options to help maintenance personnel. However, the monitoring system was justifiable, owing to the large volume of workparts involved. A diagram of the production monitoring system is presented in Figure 12.5.

## Monitoring for manufacturing records

Because of government regulations or product warranty considerations, the firm may be required to collect and preserve manufacturing records for long periods of time. Production monitoring by the computer probably constitutes the most reliable and accurate method to accomplish this chore. The manufacturing data may later be used as legal evidence in product liability suits. Production records can also be useful in product recall situations. It may be necessary to know the serial numbers of the products on which a certain engineering design change became effective, so that only the affected products are called back for correction of a design defect. This recall problem seems to have received widespread publicity in recent years in the automobile industry.

## Other considerations in production monitoring

The main objective of computer process monitoring is to provide information that can be used to improve productivity and reduce manufacturing costs. To achieve this objective, the information (reports, statistical analyses, downtime alarms, diagnostic analyses, etc.) must be presented to the intended user in a manner that can be easily comprehended. Berry has cataloged some of the common pitfalls that sometimes occur in generating information from computer production monitoring systems [1]:

Generating too much information. This sometimes makes it difficult for operating personnel to distinguish the important information from the unimportant.

Including too many abbreviations in reports to operating personnel. This tends to perpetuate the notion among manufacturing people that computer people speak a different language.

Printing numbers and data only, without proper identification of the problems or potential problems that may be indicated by the numbers.

Not designing the system to permit users to request information they desire from the monitoring system.

Not allowing first-line production supervisors to have access to certain information generated by the system for upper management.

Telling the user that report formats and information content of the system cannot be altered. One of the advantages of the use of digital computers is flexibility of system design. If the information generated by the computer does not meet with user approval and acceptance, the system is liable not to succeed. It is better to take the extra time to program the system to report information desired by the user than to run the risk that the system will not be utilized.

In subsequent chapters we shall consider how the computer can be used not only to collect data from the production process but also to control the process. We will first consider the general problem of process control—what is commonly called linear feedback control theory.

## REFERENCES

[1] BERRY, S. A., "Techniques in the Application of Computers to Industrial Monitoring," *CAD/CAM and the Computer Revolution,* Society of Manufacturing Engineers, 1974, pp. 221–239.

[2] BURKE, G., JR., Burron Medical Products, Inc., personal communications, 1979.

[3] BROWN, J., "Choosing between a Data Logger (DL) and Data Acquisition System (DAS)," *Instruments and Control Systems*, September, 1976, pp. 33–38.

[4] FORD, R. F., "Production Monitoring with Computers," *Automation*, November, 1972, pp. 38–42.

[5] IBM Publication G580-0072-0: "A Data Collection System from IBM to Help Increase Operational Productivity," The IBM 5230 Data Collection System.

[6] KESSLER, E., "Production Recording for Profit Control," *Production Engineering,* July, 1978, pp. 44–47.

[7] REMBOLD, U., SETH, M. K., and WEINSTEIN, J. S., *Computers in Manufacturing*, Marcel Dekker, Inc., New York, 1977, Chapter 8.

[8] SAVAS, E. S., *Computer Control of Industrial Processes*, McGraw-Hill Book Company, New York, 1965, Chapter 2.

[9] Scans Associates, Inc., "Engine Testing Equipment," *Bulletin 6502A*, Livonia, Mich., 1975.

[10] SMITH, C. L., *Digital Computer Process Control*, Intext Educational Publishers, Scranton, Pa., 1972, Chapter 1.

[11] *Report on Machine Tool Computer Control and Monitoring*, Buhr Computer Systems Group, Bendix Corp.

# chapter 13

# Process Control Fundamentals—Modeling and Analysis

## 13.1 INTRODUCTION

In the previous chapter, we discussed how the digital computer can be used to observe the manufacturing process and to provide useful information by which the process performance can be improved. However, simply observing the production operation does not make full use of the computer's capabilities. The next logical step beyond process monitoring is process control. The computer can be utilized not only to collect data from the process, but to manipulate the process under its direct command, without the human operator acting as intermediary.

As indicated in Chapter 10, two methods of using the computer are available to control industrial processes:

Direct digital control (DDC).
Supervisory computer control.

Before discussing these methods, we shall attempt to place the subject of digital computer process control in perspective by first examining the more traditional topic of linear control theory. This will be the focus of the current chapter.

Industrial processes are incompatible with the digital computers that are supposed to control them. Digital computers operate on data that are digital and discrete, whereas most industrial processes are characterized by variables that are analog and continuous. As discussed in Sections 11.4 and 11.5, this is the reason why analog-to-digital and digital-to-analog conversion is necessary in computer control. Before digital computers, analog devices were used for process control in industrial plants. The development of analog controllers represented a major advance in industrial automation. Many of the same concepts and strategies used in conventional analog control are used today in computer control (direct digital control in particular). To understand how computers are used to control production operations, it is appropriate that we comprehend the underlying principles of process control.

The analysis of analog controllers and the processes with which they operate are based on the use of linear differential equations. Such equations constitute the foundation for linear control theory.

Linear control theory is concerned with the analysis of systems that can be modeled by linear ordinary differential equations. Examples of the use of differential equations can be drawn from nearly every technical field. These include mechanical systems (spring–mass–damper), electrical circuits (resistance–inductance–capacitance circuits), chemical reactions, thermal systems (the Fourier equation of heat transfer), hydraulic mechanisms (fluid power systems), and many others. We shall consider some of these examples in subsequent sections of this chapter.

Linear differential equations are preferred over nonlinear ones because they are generally much easier to solve. Unfortunately, real-life physical relationships are not always linear. Nevertheless, enough of the real world obeys linearity that linear control theory is a powerful tool in the study of process control.

## 13.2 PROCESS MODEL FORMULATION

### Types of models

One of the first steps in the study and analysis of a physical process (including most industrial processes) is to develop a mathematical model to represent the process.

There are two basic methods of formulating a mathematical model of a physical process. The first method is to analytically derive the relationship between the process variables of interest. *Analytical models* are obtained based on fundamental assumptions and principles about the phenomenon being studied. An example would be the familiar Newton's equation for constant mass (specifically *Newton's second law*):

force = mass × acceleration

Since acceleration is the second derivative of displacement (let $y$ represent displacement), we could express Newton's second law as

$$F = M\frac{d^2y}{dt^2} \tag{13.1}$$

where $F$ is force, $M$ is mass, and $d^2y/dt^2$ = acceleration. The reader will recognize this as a differential equation. A differential equation is any algebraic equality that contains one or more derivatives.

The second method for formulating a mathematical model of a physical process is to base the model on experimental data collected from observing the process. This is called an empirically derived model or an *empirical equation*. No particular concern is given to the underlying principles about the process. An equation is determined which best fits the observed data. Statistical techniques, such as regression analysis, are often used to find the best equation. An example of an empirical equation that should be familiar to manufacturing people is the *Taylor tool life equation*:

$$VT^n = C \tag{13.2}$$

where $V$ is the cutting speed, $T$ is the tool life, and $n$ and $C$ are constants in the equation. Taylor developed this equation based on experimental data of cutting speed and corresponding tool life. For a particular tool, work material, and cutting conditions, he found that certain values of $n$ and $C$ in Eq. (13.2) gave the best "fit" to the data.

Mathematical models of physical processes can also be developed by combining the two approaches above. A basic form of the equation is hypothesized from theoretical considerations. Then data are collected from the actual process to determine the most accurate parameter values for the model. This hybrid approach has much merit and is often used in industrial process model formulation.

There are other ways to classify mathematical models. We can distinguish physical models and economic models. *Physical models* define the relationships among the physical variables in the process. *Economic models* are used to characterize the relationships among the economic factors (cost, profit, rate of production) in the process. They can often be employed to define the economic performance of the process under consideration. In the present chapter, our attention will be directed primarily at physical process models. In Chapter 15, we shall examine process performance objectives. At that time consideration will be given to economic models of manufacturing operations.

Let us consider how a mathematical model might be formulated using differential equations. We will begin with a relatively simple, yet illustrative, example.

## Hydraulic system

Figure 13.1 represents a physical system of the type chemical engineers might have to consider in the design of an industrial flow process. As shown in the drawing, let $x_1$ represent the flow rate into the tank and $x_2$ represent the rate of flow out. Appropriate units might be liters per minute. The volume in the tank would be determined by the cross-sectional area of the tank, $A$ (which we assume to be constant throughout the height), and the level of fluid in the tank. We will use the symbol $h$ to symbolize this liquid level ($h$ stands for head).

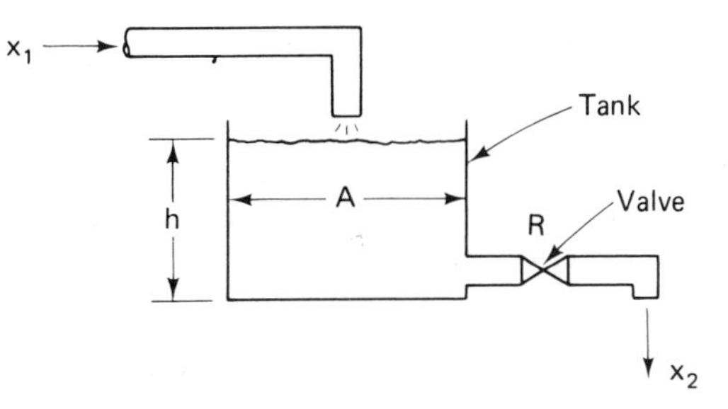


FIGURE 13.1 Hydraulic system.

The "continuity" law specifies for a system such as this one that the flow rate into the tank minus the flow rate out of the tank must equal the rate of liquid volume increase (or decrease) in the tank. This can be expressed in the form of a differential equation:

$$x_1 - x_2 = A\frac{dh}{dt} \tag{13.3}$$

Figure 13.1 shows that the outlet stream is restricted by a valve of resistance $R$. If the flow through the valve is laminar (rather than turbulent flow), the hydraulic resistance is equal to the head (pressure) divided by the flow rate through the valve. This relationship can be expressed as

$$R = \frac{h}{x_2} \quad \text{or} \quad x_2 = \frac{h}{R} \tag{13.4}$$

By substituting Eq. (13.4) into Eq. (13.3), we can express the differential equation as follows:

$$x_1 - \frac{h}{R} = A\frac{dh}{dt}$$

or

$$AR\frac{dh}{dt}+h=Rx_1 \tag{13.5}$$

Equation (13.5) is a first-order linear differential equation. The variable $h$ is the dependent variable in the system and $x_1$ is the input. Time $t$ is the independent variable. The input can change over time in a variety of ways. For purposes of mathematical analysis, the form of the input can be a step function or some type of oscillation, such as a sine wave. These two possibilities are shown in Figure 13.2. Other idealized mathematical functions can also be used to test the behavior of the system.

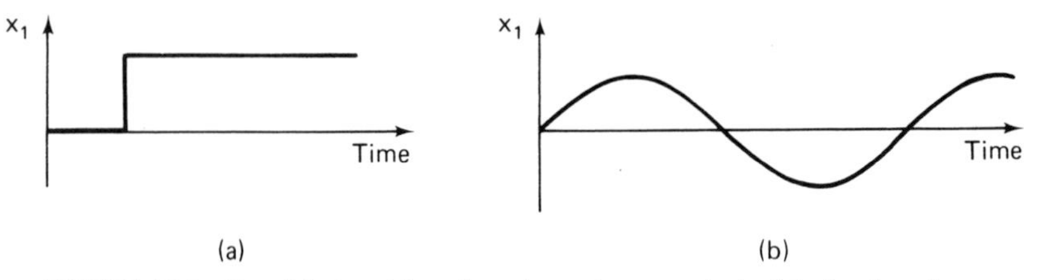


FIGURE 13.2 Possible input functions in systems analysis: (a) step function; (b) sine-wave function.

Let us consider the response of the hydraulic system to the step input shown in Figure 13.2(a). The solution of Eq. (13.5) would begin with the formulation of the characteristic equation, which is

$$ARs+1=0$$

where $s$ stands for the differential operator. The root of the characteristic equation is

$$s=\frac{-1}{AR}$$

We shall define the term

$$\tau=AR$$

Hence, the complimentary function is

$$h_c=C_1e^{-t/\tau}$$

Since the input to the system is a step function, it is a constant value, $X_1$, beyond time zero. The steady-state solution to Eq. (13.5) would therefore be a constant whose first derivative is zero.

$$AR(0)+h_p=RX_1$$
$$h_p=RX_1$$

The complete solution would consist of the complementary function plus the particular integral:

$$h = h_c + h_p = C_1 e^{-t/\tau} + RX_1 \tag{13.6}$$

Since the initial conditions are that $h=0$ when $t=0$,

$$0 = C_1(1) + RX_1$$

then

$$C_1 = -RX_1$$

and, therefore,

$$h = RX_1(1 - e^{-t/\tau}) \tag{13.7}$$

where $RX_1$ is specified to be a constant value in this case. The response represented by Eq. (13.7) is shown in Figure 13.3.

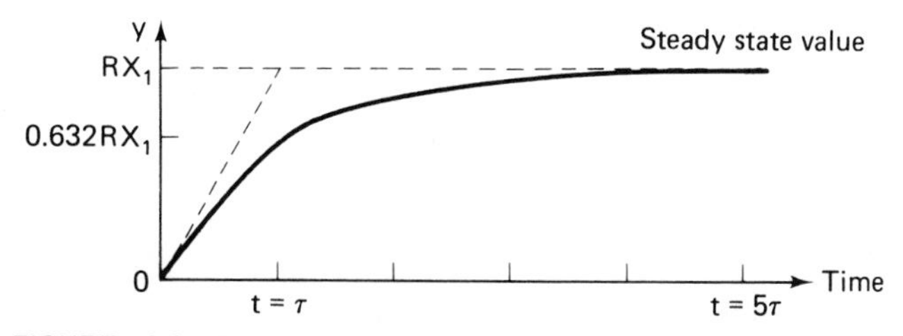


FIGURE 13.3 Response of a first-order system to a unit step input.

In Eq. (13.6), the term $RX_1$ is called the steady-state response of the system. It represents the value that the dependent variable will assume after sufficient time for any transient or startup effects to disappear. The steady-state value is shown in Figure 13.3 as the horizontal line at the level $RX_1$.

The term $C_1 e^{-t/\tau}$ in Eq. (13.6) is called the transient solution or transient response of the system. Because of the nature of the function $e^{-t/\tau}$, the transient response will die out as time increases. The term $\tau$ is referred to as the time constant in the solution of the first-order linear differential equation. Its value depends on the parameters in the equation, in this case $\tau = AR$. Its effect is to determine the length of time taken before the transient effects disappear. To see this effect, consider the term in parentheses in Eq. (13.7):

$$1 - e^{-t/\tau}$$

When $t=\tau$, the output has reached

$$1 - e^{-1} = 0.632$$

of its steady-state value. At $t=2\tau$, the output has reached

$$1-e^{-2}=0.865$$

of its final value. When $t=5\tau$, the output has achieved

$$1-e^{-5}=0.993$$

of the final value. It can be seen that the transient solution plays a minor role after $t=5\tau$. For all practical purposes, the output has completed its response to the step input.

First-order linear differential equations similar to Eq. (13.5) are quite common in modeling physical systems. The parameters of the equations would have different physical interpretations, but the basic form of the equation would be the same. This mathematical model is a basic building block of many complex physical and industrial processes.

Let us consider another building block, the second-order linear differential equation, by resorting to a familiar mechanical example.

## Spring–mass–damper system

Shown in Figure 13.4 is the typical spring–mass–damper arrangement. There are three components, all interrelated to produce a single mechanical system. A mass $M$ is connected to two solid surfaces, by a spring of constant $K_s$ on the one side and a dashpot whose damping coefficient is $K_d$ on the other side. The behavior of the system is characterized by the displacement of the mass, which is symbolized by the variable $y$.

When the mass is displaced, there are forces introduced that affect the motion of the mass. Thus, the variable $y$ is dynamic; that is, it is a function of time and the function can be expressed in the form of a differential equation. The forces acting on the mass come from three sources. First, as the mass moves, it undergoes acceleration and deceleration. The resulting force is defined by Newton's second law of Eq. (13.1) as

$$M\frac{d^2y}{dt^2}$$

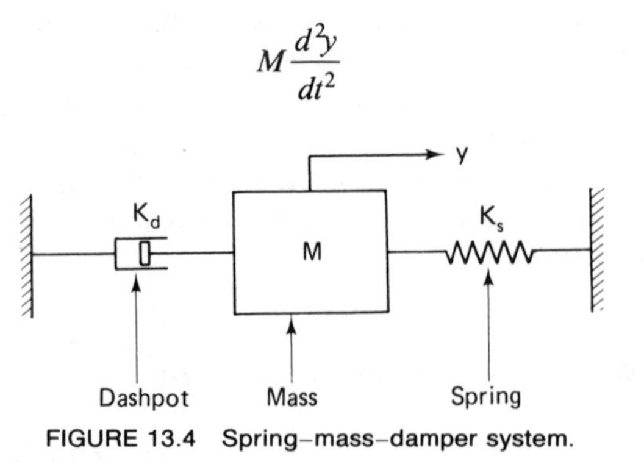


FIGURE 13.4 Spring–mass–damper system.

The second force results from the displacement of the spring. As the spring is either stretched or compressed, it exerts a force on the mass that is proportional and opposite to the displacement. (There are, of course, limits to this proportionality, but we will assume that the displacement is within those limits.) The force from the spring can be written as

$$K_s y$$

The third force component results from the action of the dashpot. The dashpot is a device that exhibits a tendency to resist motion. It is a frictional device (viscous friction) which gives a force proportional to velocity. Velocity is the first derivative of displacement $y$, so the damping force can be expressed as follows:

$$K_d \frac{dy}{dt}$$

Assuming that we have considered all the forces on the mass, we can write the equation of motion for the spring–mass–damper system by summing the forces:

$$\Sigma\,\text{forces} = M\frac{d^2y}{dt^2} + K_d\frac{dy}{dt} + K_s y = 0 \tag{13.8}$$

This second-order differential equation states that the system is not influenced by any stimuli that are external to itself. Our interest in the application of differential equations to model systems is usually motivated by the presence of some outside stimulus on the system. For example, if an initial displacement $x$ were given to the spring, the differential equation would be written

$$M\frac{d^2y}{dt^2} + K_d\frac{dy}{dt} + K_s y = K_s x \tag{13.9}$$

The system described by Eq. (13.9) is represented in Figure 13.5, in which the displaced end of the spring replaces the fixed wall.

In the equation, the term $K_s x$ is the driving force for the mechanical system. We refer to $x$ as the input to the system, and $y$ is the output that

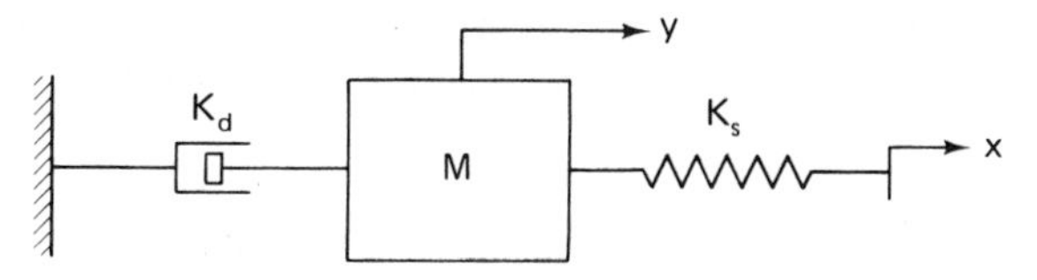


FIGURE 13.5 Spring–mass–damper system with positive displacement given to end of spring.

responds to the input. As with the previous first-order example, the form of the input can be a step function or some other more complex motion. For purposes of illustration, the step input is useful, and we will assume that the term $K_s x$ in Eq. (13.9) is of this type.

To solve, the characteristic equation is

$$Ms^2 + K_d s + K_s = 0 \tag{13.10}$$

where $s$ stands for the differential operator. The two roots to this characteristic equation are

$$s_{1,2} = \frac{-K_d}{2M} \pm \frac{\sqrt{K_d^2 - 4MK_s}}{2M} \tag{13.11}$$

The motion of the mass, as defined by the variable $y$, will depend on the relative values of the parameters of the system: $M$, $K_d$, and $K_s$. It turns out that four general outcomes are possible:

1. No damping in the system.
2. Underdamped system.
3. Critically damped system.
4. Overdamped system.

Let us consider the conditions that lead to each of these outcomes and what the typical response looks like for each case.

NO DAMPING. In this case the damping coefficient, $K_d = 0$ (the dashpot is removed in Figure 13.5). In this case the characteristic equation becomes

$$Ms^2 + K_s = 0$$

and the roots of the equation are

$$s_{1,2} = \pm j\sqrt{\frac{K_s}{M}}$$

where $j$ in this equation means that the roots are imaginary. This means that the behavior of the system is oscillatory and is described by a term called the *natural frequency* of the system. This is symbolized by $\omega_n$, where

$$\omega_n = \sqrt{\frac{K_s}{M}}$$

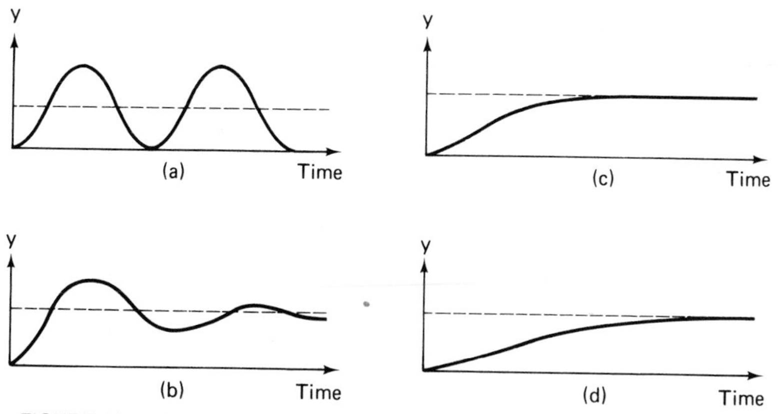


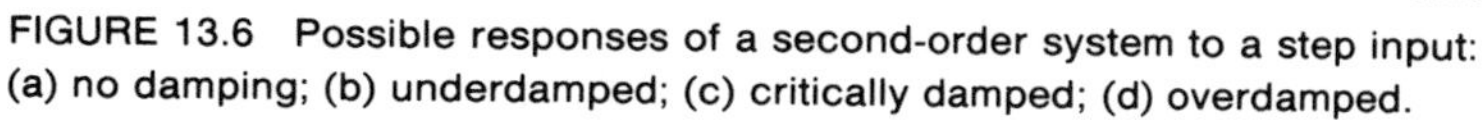


FIGURE 13.6 Possible responses of a second-order system to a step input: (a) no damping; (b) underdamped; (c) critically damped; (d) overdamped.

The solution of Eq. (13.9) assuming $x$ is a step input is

$$y = C_1 \sin \omega_n t + C_2 \cos \omega_n t + X \tag{13.12}$$

where $C_1$ and $C_2$ are constants that depend on initial conditions. The general appearance of this response is shown in Figure 13.6(a).

UNDERDAMPED SYSTEM. If a relatively small amount of damping is present, it means that the square-root value in Eq. (13.11) would be imaginary. Then the roots could be rewritten:

$$s_{1,2} = -a \pm j\omega_d$$

where

$$a = \frac{K_d}{2M}$$

$$\omega_d = \frac{\sqrt{4MK_s - K_d^2}}{2M}$$

The term $\omega_d$ is called the *damped natural frequency* of the underdamped system. The response of $y$ under these conditions would be

$$y = e^{-at}(C_1 \sin \omega_d t + C_2 \cos \omega_d t) + X \tag{13.13}$$

whose typical shape would appear as in Figure 13.6(b).

CRITICALLY DAMPED SYSTEM. This is the special case that results when

$$K_d^2 = 4MK_s$$

It means that the characteristic equation is

$$s_{1,2} = \frac{-K_d}{2M} \pm 0$$

and the solution of the differential equation in this case is

$$y = C_1 e^{-at} + C_2 t e^{-at} + X \tag{13.14}$$

The shape of this response is shown in Figure 13.6(c). It can be seen that a critically damped system responds to a step input in a relatively short time compared to the other responses shown in Figure 13.6.

OVERDAMPED SYSTEM. Conditions for the overdamped system occur when

$$K_d^2 > 4MK_s$$

The roots of the characteristic equation are

$$s_{1,2} = -a \pm b$$

where

$$a = K_d/2M, \text{ as before}$$

$$b = \frac{\sqrt{K_d^2 - 4MK_s}}{2M}$$

The system response in an overdamped case is

$$y = C_1 e^{(-a+b)t} + C_2 e^{(-a-b)t} + X \tag{13.15}$$

and the typical response shape is shown in Figure 13.6(d). There is no oscillation in this response, but the time required for the system to reach a steady-state value is longer than for the critically damped case.

Although the spring–mass–damper system analyzed above seems somewhat academic, its general characteristics are frequently encountered in real life. For example, the suspension system on an automobile is basically a spring–mass–damper mechanical system, although somewhat more complex than that analyzed above. There are other systems and components of systems that exhibit the characteristic of damping and mechanical elasticity. And, of course, all mechanical systems possess mass. The damping and elasticity are not always provided by objects that resemble a dashpot or a spring. For example, a cutting tool will deflect under the force of the machining operation, sometimes causing vibration, which is called "chatter." A rotating shaft that is driven on one end and must transmit motion to the other end will exhibit the characteristics of the idealized spring–mass–damper system.

### Concluding comments on modeling

In the two examples of first-order and second-order systems, the constants in the differential equations could be determined by taking measurements on the components of the system. In Eq. (13.5), the value of $A$ could be determined simply by measuring the area of the tank in Figure 13.1. The parameter $R$ could also be estimated by observing the pressure-flow characteristics of the valve. In Figure 13.5 the spring, mass, and damper were treated as separate components so that the values of $K_s$, $M$, and $K_d$ could be evaluated for Eq. (13.9). For many physical systems, the parameter values can be determined for the mathematical models simply by measuring the system components.

However, in many other physical situations, the constants in the model cannot be so easily evaluated. In these cases, the parameters must be determined by taking measurements of the behavior of the system responding to some known input. In effect, we deduce the model parameters by empirical means. Most often the general form of the model can be developed from theoretical considerations. Then, the constants of the model are evaluated based on observations of the system's performance. There are other occasions when the form of the differential equation cannot be formulated with confidence. When this occurs, the model must be derived on a completely empirical basis. The response data are compared against various hypothesized models, and the model that gives the best agreement is used.

Whatever the situation, mathematical models can be developed to represent the behavior of physical systems. Industrial processes are composed of several interconnected physical systems. By linking together the individual physical systems, the industrial process is constructed. By linking together the equations of the individual systems, the mathematical model of the industrial process is formed. In the next section we will consider a convenient method for dealing with this problem of building large process models from their component parts. We will also consider how these large models can then be reduced to a basic input/output relationship for the process.

## 13.3 TRANSFER FUNCTIONS AND BLOCK DIAGRAMS

In this section we develop the concepts of the transfer function and the block diagram. Both are important in the traditional theory of linear control systems.

## Transfer functions

The *transfer function* for a linear system or component of a system is defined as the ratio of the output to the input.[1] We will examine how the transfer function would be determined for a system component. Then we will consider how the transfer function would be obtained for an entire system or process.

TRANSFER FUNCTION FOR A SYSTEM COMPONENT. The transfer function for a component or a relatively simple system would be obtained as follows:

1. Determine the equation that expresses the behavior of the system. The equation must be linear if it is to be used as an element in a block diagram. It may be a linear differential equation or a linear equation, such as a proportionality relationship.
2. If the equation is a differential equation, write the equation in the differential operator format using the $s$-variable.
3. Rewrite the equation as the ratio of the output to the input.

Several examples will serve to illustrate how this procedure is accomplished. First, consider the hydraulic system of Section 13.2. Equation (13.5) expressed the differential equation for the system:

$$AR\frac{dh}{dt} + h = Rx_1$$

This is rewritten in the $s$-operator notation:

$$ARsh(s) + h(s) = Rx_1(s)$$

or

$$(ARs+1)h(s) = Rx_1(s)$$

Finally, the ratio of the output over the input is expressed:

$$\frac{h(s)}{x_1(s)} = \frac{R}{ARs+1} \tag{13.16}$$

which is the transfer function for this system. The original variables $h$ and $x_1$ were functions of time $t$. However, when the differential operator notation is used, these variables are transformed from functions of $t$ to functions of $s$. Readers who are familiar with differential equations will recognize $s$ as the Laplace variable.

[1]To be precise, it is defined as the ratio of the Laplace transform of the output to the Laplace transform of the input, given that all initial conditions are zero. The Laplace transform makes use of the differential operator $s$, which we shall explore more fully in the next section.

As a second example, consider the spring–mass–damper system developed in the previous section. The differential equation was given by Eq. (13.9):

$$M\frac{d^2y}{dt^2}+K_d\frac{dy}{dt}+K_sy=K_sx$$

Using the $s$-operator format, it is rewritten

$$Ms^2y(s)+K_dsy(s)+K_sy(s)=K_sx(s)$$

or

$$(Ms^2+K_ds+K_s)y(s)=K_sx(s)$$

The transfer function is obtained from the ratio of output to input:

$$\frac{y(s)}{x(s)}=\frac{K_s}{Ms^2+K_ds+K_s} \tag{13.17}$$

As a third illustration, consider a proportional relationship in which the input is simply multiplied by some gain factor to obtain the output. Such a relationship would be typical of various transducers in which a mechanical displacement or speed is converted into an electrical voltage signal, as shown in Figure 13.7. The basic equation which governs this relationship is

$$e=Kx$$

where $e$ stands for the voltage and $x$ is displacement. The basic input/output relationship is

$$\frac{e(s)}{x(s)}=K \tag{13.18}$$

where the variables are expressed as functions of $s$, to be consistent with the transfer functions of the other components in the system.

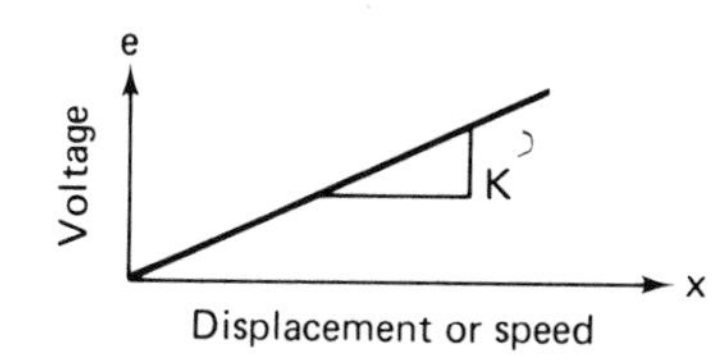


FIGURE 13.7 Proportional relationship.

TRANSFER FUNCTION FOR THE ENTIRE SYSTEM. The transfer functions for the various components of a system can be grouped together to form the transfer function of the entire system. Its interpretation is the same as for an individual component. It expresses the relationship between the input of the system and its output.

The method we will use to determine the system transfer function is block diagram algebra.

## Block diagrams

A block diagram is a pictorial method of portraying the interrelationships among components of a physical system or process. It is composed of blocks and arrows between the blocks. The arrows represent the flow of signals or variables. The blocks represent the operations that are performed on the signals. Each block contains a transfer function that defines exactly how the input signal is mathematically transformed into the output signal. Also, signals can be summed together, and one signal can be used as an input to more than a single block.

The basic elements of a block diagram are as follows:

1. *Block*. This represents a given system component and contains the transfer function for the component.

2. *Arrows*. The arrows indicate the direction in which the signals or variables flow. Figure 13.8 shows a block containing a transfer function. The arrows into and out of the block show the input and output of the component.

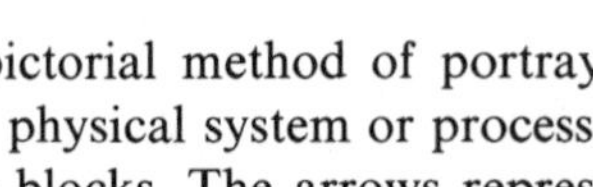

$x_1(s)$ → $\frac{R}{ARs + 1}$ → $h(s)$

FIGURE 13.8 Block diagram symbols: block and arrows (input and output arrows) for Eq. (13.16).

3. *Summing point*. The summing point is illustrated in Figure 13.9. It can be used to add or subtract signals as shown in Figure 13.9(a) and (b). More than two signals can flow into a summing point, but only one signal can flow out.

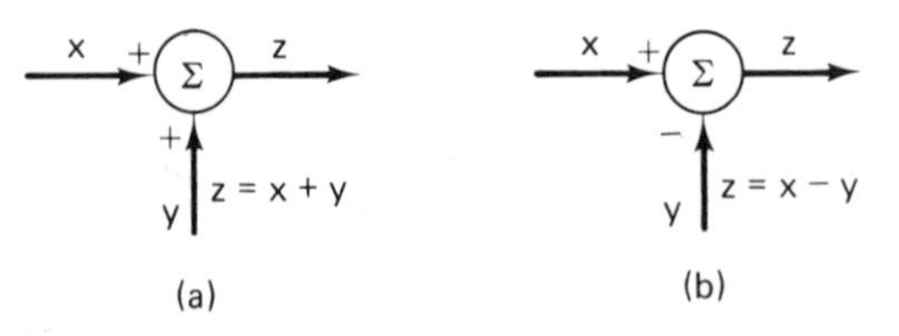


FIGURE 13.9 Block diagram symbols: summing points for addition (a) and subtraction (b).

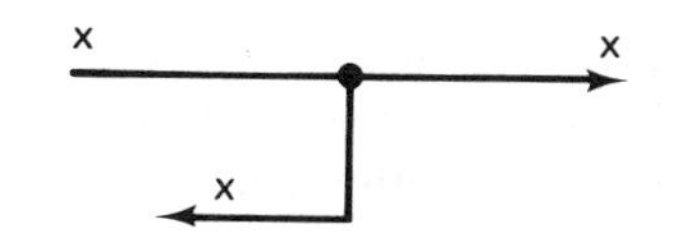


FIGURE 13.10 Block diagram symbols: takeoff point.

4. *Takeoff point.* The takeoff point permits a variable to be used in more than one place. It is shown in Figure 13.10. The variable is not changed in value when it proceeds through a takeoff point to two or more destinations (blocks and/or summing points).

An example of a block diagram is presented in Figure 13.11. The diagram shows the basic form of a feedback control system with input $x$ and output $y$. A takeoff point is used to feed the $y$ signal through the block indicated by $H$. $H$ represents the mathematical operation performed on the $y$ signal. $H$ is the transfer function for the block. The output of block $H$ is subtracted from the input signal $x$, and the difference is fed into block $G$. $G$ represents the transfer function that transforms the difference signal into the output $y$. All the elements of a block diagram—blocks, arrows, summing point, and takeoff point—are present in Figure 13.11.

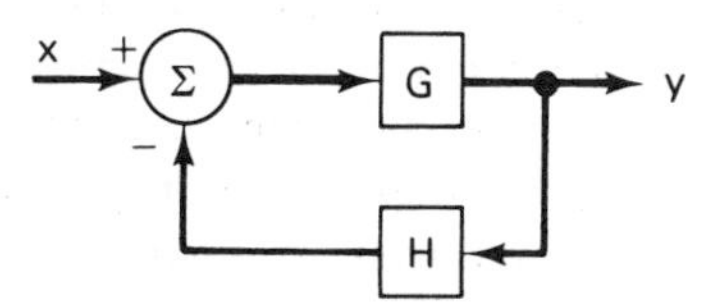


FIGURE 13.11 Block diagram of the basic form of a feedback control system.

Figure 13.11 presents an opportunity to define some of the terminology of feedback control systems. In a feedback control system, the controlled variable is compared with the input and any difference between them is used to drive the controlled variable toward its desired value. In Figure 13.11, $y$ is the controlled variable. The feedback system is also referred to as a closed-loop system, since the block diagram takes on the appearance of a closed loop. By contrast, an open-loop system is a system without feedback. In Figure 13.11, $G$ is the forward transfer function and $H$ is the feedback transfer function. The transfer function for the entire feedback control system can be determined from block diagram algebra.

## Block diagram algebra

It is possible to take a large complicated block diagram and reduce it to a single block. The single block contains the transfer function for the entire system represented by the original diagram. The process of reducing the diagram is called *block diagram algebra* and is based on traditional algebra.

Let us use the block diagram of Figure 13.11 to illustrate. The signals entering the summing point are the input signal $x$ and the signal $Hy$. Exiting the summing point is the signal

$$x - Hy$$

This is multiplied by the transfer function $G$ to obtain the output of the system:

$$\begin{aligned} y &= G(x - Hy) \\ &= Gx - GHy \\ y + GHy &= Gx \\ y(1 + GH) &= Gx \end{aligned}$$

Writing this as a ratio of $y$ to $x$:

$$\frac{y}{x} = \frac{G}{1 + GH}$$

Hence, we have the transfer function of the feedback control system pictured in Figure 13.11. To be technically correct, we should remember that this ratio is defined using the $s$-operator notation. Hence, $y/x$ should be $y(s)/x(s)$. The transfer functions $G$ and $H$ may represent complicated expressions in $s$.

By performing algebraic operations such as the above, various arrangements of blocks, summing points, and takeoff points interconnected by arrows can be reduced to a single block. It is not necessary to proceed through an algebraic manipulation similar to the one above for every block diagram. Instead, certain basic arrangements have been cataloged, and we present some of the more important block diagram equivalences below.

1. *Blocks in series.* These are also called cascaded blocks and are pictured in Figure 13.12(a). Components connected in series are equivalent to a single block whose transfer function is the product of the transfer functions of the individual components.

2. *Blocks in parallel.* Components connected in parallel as shown in Figure 13.12(b) are equivalent to a single block whose transfer function is the sum of the individual components' transfer functions.

3. *Elimination of a basic feedback loop.* This is the algebraic reduction worked for the basic feedback loop of Figure 13.11. The general form is presented in Figure 13.12(c).

4. *Moving a summing point.* In the reduction of a complex block diagram, it is generally desirable to move summing points to the left. This is illustrated in Figure 13.12(d).

5. *Moving a takeoff point.* It is sometimes desirable to move takeoff points to the right in block diagram reduction. This is shown in Figure 13.12(e).

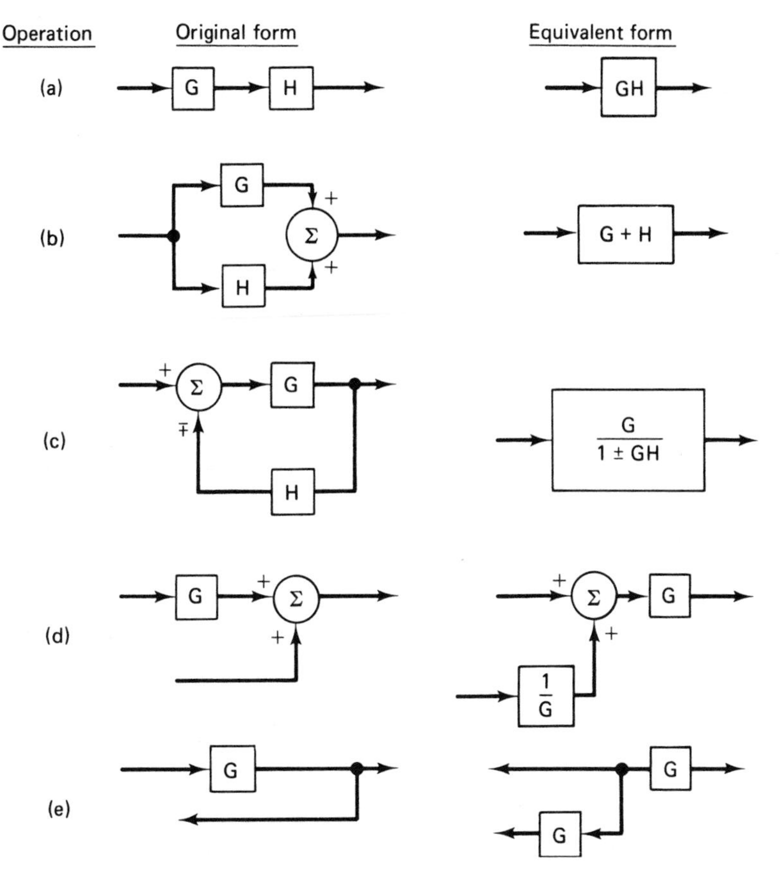


FIGURE 13.12 Basic operations in block diagram algebra: (a) blocks in series; (b) blocks in parallel; (c) elimination of a feedback loop; (d) moving a summing point; (e) moving a takeoff point.

STEPS IN BLOCK DIAGRAM REDUCTION. The reduction of a large complex block diagram to a single block using the equivalences shown in Figure 13.12 should proceed in the following general sequence:

1. Combine all series block arrangements into single blocks according to Figure 13.12(a).
2. Combine all parallel block arrangements into single blocks according to Figure 13.12(b).
3. Convert basic feedback loops into equivalent single blocks according to Figure 13.12(c).
4. Shift summing points to the left and takeoff points to the right according to Figure 13.12(d) and (e).

An example should help to illustrate the use of block diagram algebra in the simplification of a somewhat complex diagram.

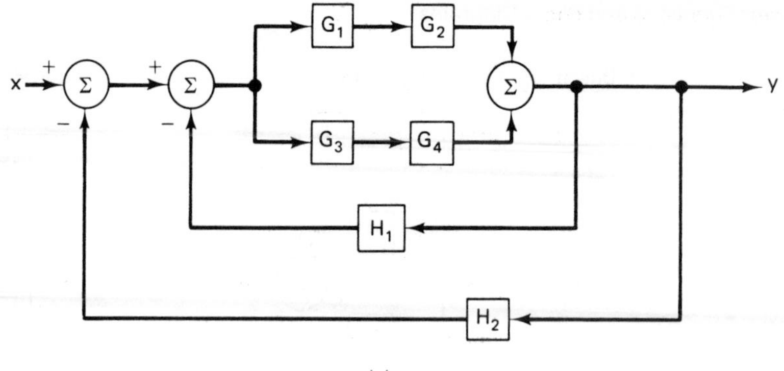


(a)

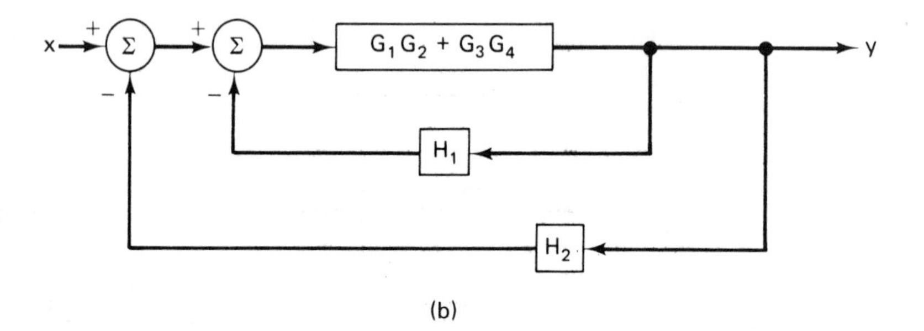


(b)

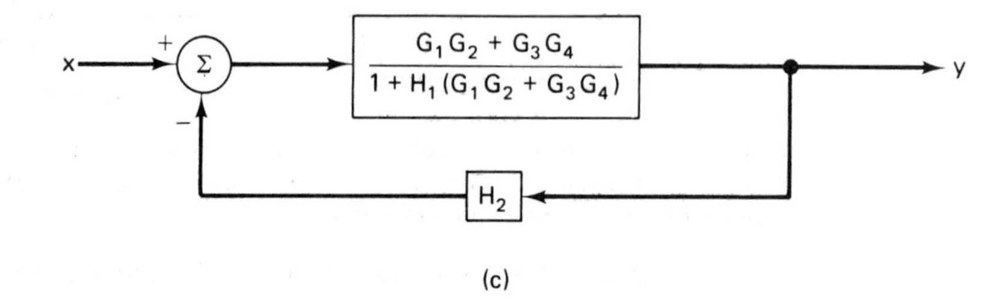


(c)

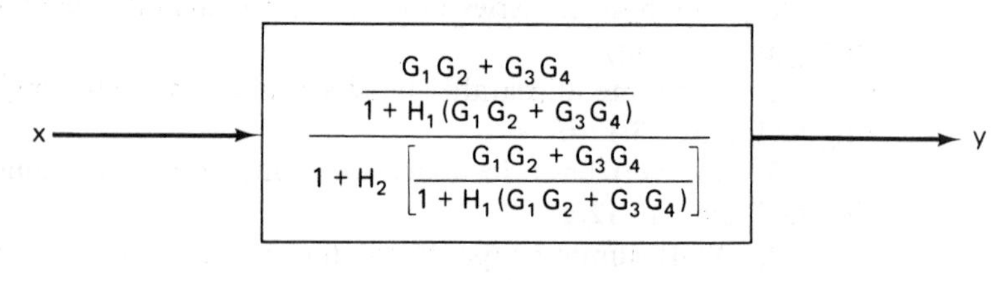


(d)

FIGURE 13.13 Example 13.1 block diagram: (a) initial block diagram; (b) block diagram after series and parallel reduction; (c) block diagram after first feedback loop elimination; (d) transfer function for the system.

## EXAMPLE 13.1

Given the block diagram of Figure 13.13(a), we want to reduce this to a single equivalent block in order to determine the system transfer function.

The solution is illustrated in parts (b), (c), and (d) of Figure 13.13. The first step is to combine series blocks, then parallel blocks. This is shown in part (b). Next, the inside feedback loop is converted into an equivalent block, Figure 13.13(c). Finally, the outside feedback loop is eliminated to provide the transfer function for the complete system.

## EXAMPLE 13.2

Figure 13.14 presents a sketch of two cascaded tanks. In each tank some chemical reaction takes place by the addition of small traces of ingredients to the basic fluid flowing through the system. Our purpose is to model the physical flow portion of the system rather than the chemical reactions that take place. The flow into the first tank is to be treated as the input and the head (or level) in the second tank is to be considered here as the system output.

The problem is to develop the block diagram for the system and then find the transfer function by block diagram algebra.

The reader will no doubt notice the similarities between this flow system and the hydraulic example we used to illustrate a first-order linear differential equation. The equations that govern this industrial flow process are the following:

$$x_1 - x_2 = A_1 \frac{dh_1}{dt}$$

$$x_2 = \frac{h_1}{R_1}$$

$$x_2 - x_3 = A_2 \frac{dh_2}{dt}$$

$$x_3 = \frac{h_2}{R_2}$$

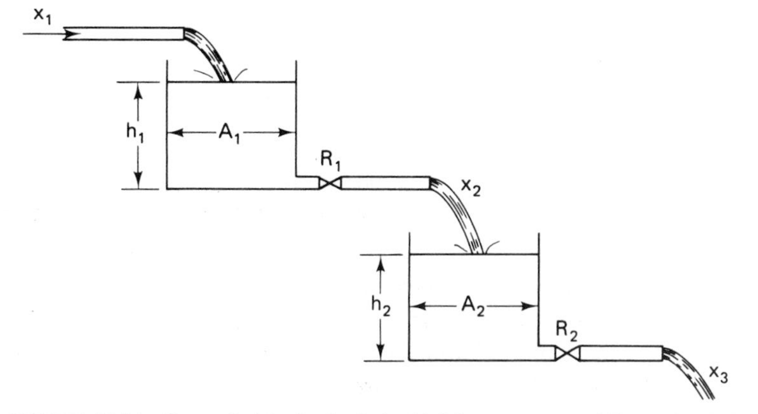


FIGURE 13.14 Cascaded tanks for industrial flow process of Example 13.2.

where $x$ = various flow rates
$h$ = head in each of the tanks
$R$ = hydraulic resistance

The first and third equations are differential equations and can be expressed in terms of the differential operator, $s$.

$$x_1 - x_2 = A_1 s h_1$$

$$x_2 - x_3 = A_2 s h_2$$

From the four equations, the block diagram can be constructed as shown in Figure 13.15(a). Although this system contains feedback loops, the relationship between the input $x_1$ and the defined output $h_2$ is basically an open-loop relationship. There is no feedback of the $h_2$ signal for comparison with the input $x_1$.

Figure 13.15(b) shows the transfer function for the industrial flow process. This expression could be reduced to a simpler form by algebraic manipulation.

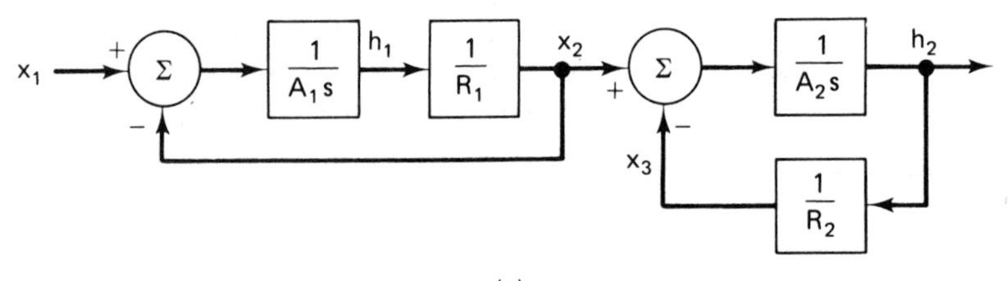


(a)

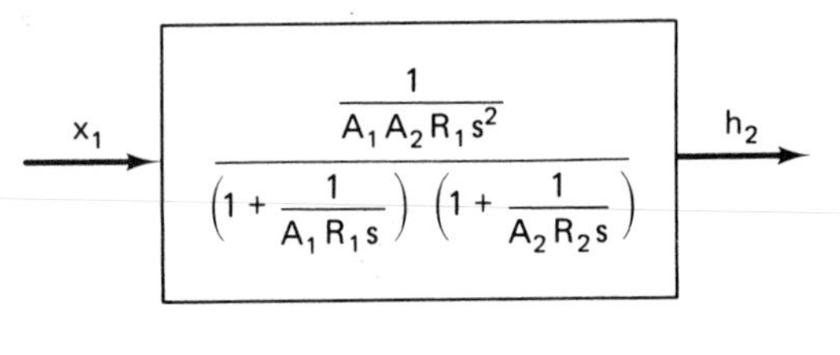


(b)

FIGURE 13.15 Solution to Example 13.2: (a) block diagram of industrial flow process; (b) transfer function determined after block diagram algebra.

MULTIPLE INPUTS TO THE SYSTEM. Many physical systems have more than one input. The approach for reducing the block diagram of a system with multiple inputs is to treat each input separately and independently. Following is the procedure:

1. Set all inputs equal to zero except one.
2. Determine the transfer function by block diagram algebra using that input alone.
3. Repeat steps 1 and 2 for other inputs.
4. Add algebraically the products of transfer functions and their respective inputs to obtain the effect of all inputs acting simultaneously.

We will illustrate this procedure on a simple example.

### EXAMPLE 13.3

Consider the control system of Figure 13.16(a), which has two inputs, $x_1$ and $x_2$. We want to determine the system response $y$ that results from these two inputs. First, we consider the case where $x_2$ is zero and $x_1$ is nonzero, shown in Figure 13.16(b). The transfer function for this arrangement is presented beneath the block diagram.

The next step is to set $x_1 = 0$ and assume that $x_2$ is nonzero, as shown in Figure 13.16(c). Again, the transfer function for this system is given below the block diagram.

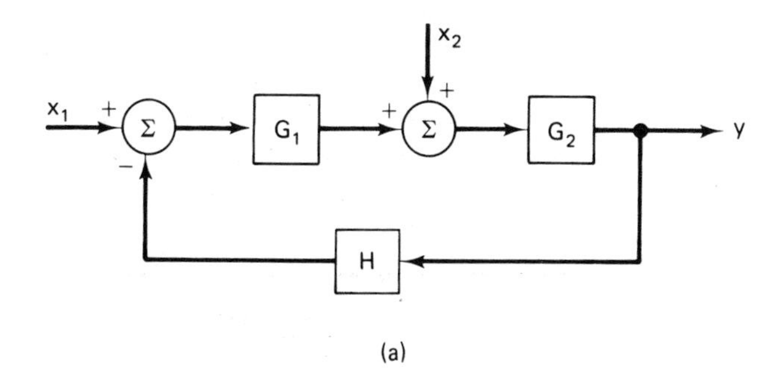


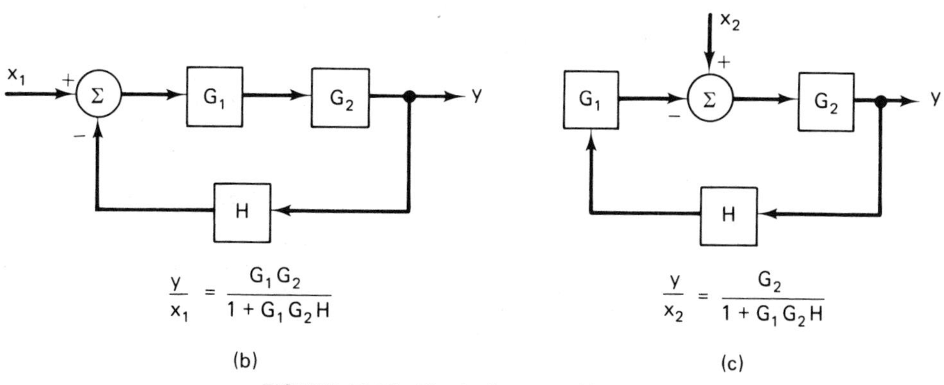


FIGURE 13.16 Block diagram with multiple inputs.

To determine the total system response, simply add the separate response of each input. Each separate response is the input multiplied by its corresponding transfer function.

$$y = \frac{G_1 G_2}{1 + G_1 G_2 H} x_1 + \frac{G_2}{1 + G_1 G_2 H} x_2$$

## EXAMPLE 13.4

Our fourth example of block diagram modeling is a case of multiple inputs and also a more realistic illustration of process modeling than that in Example 13.3. It is based on a research paper of Jaeschke, Zimmerly, and Wu [5]. The paper deals with the use of cutting temperature to control speed in a turning operation. The system measures cutting temperature by means of a tool-chip thermocouple (the two dissimilar metals forming the thermocouple junction are the tool and the chip coming off the workpiece). The voltage signal is amplified and fed back to be compared with a reference voltage. If cutting temperature is too high, speed is reduced. If cutting temperature is below the desired level, speed is increased. A schematic diagram of the control system is shown in Figure 13.17. The authors distinguish between two cases in the operation of the system. The first case is when speed is increasing. The block diagram for this situation is illustrated in Figure 13.18(a). The second case is when speed is decreasing, and the block diagram is shown in Figure 13.18(b). The difference in the two cases is that different components are used in each case. In Figure 13.18(a), the control system, consisting of integrator, amplifier, and Adjusto-Spede unit, must provide a positive torque to increase the cutting speed. In Figure 13.18(b), the cutting torque tends to decrease the speed and the control system components are not brought into action. Hence, two block diagrams must be used to completely describe the behavior of this control system.

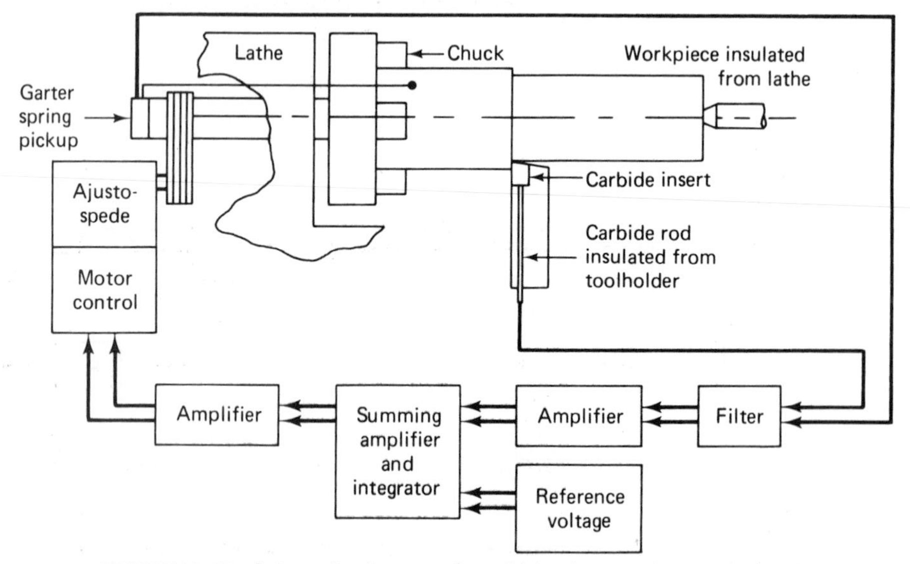


FIGURE 13.17 Schematic diagram of machining temperature control system of Example 13.4.

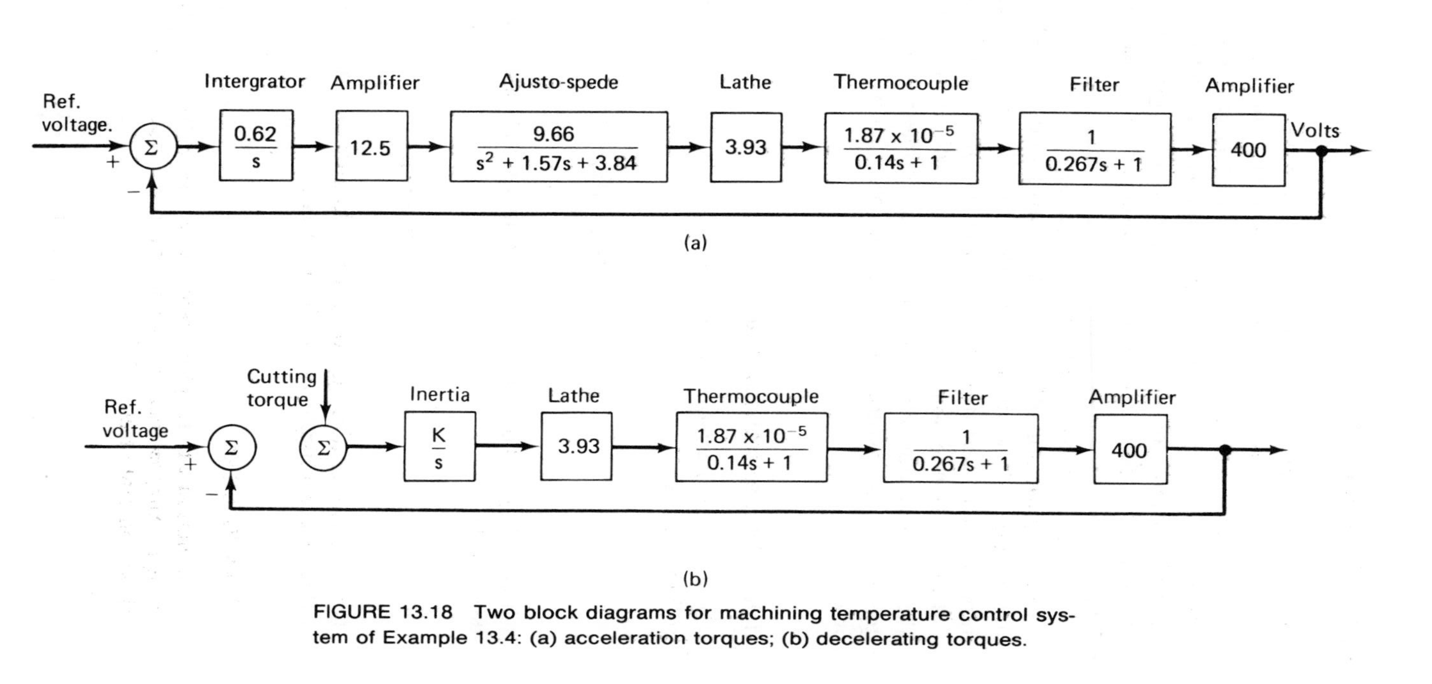


FIGURE 13.18 Two block diagrams for machining temperature control system of Example 13.4: (a) acceleration torques; (b) decelerating torques.

## 13.4 LAPLACE TRANSFORMS

In previous sections of this chapter, we have used the differential operator notation to express a differential equation. The symbol $s$ has been used to represent the mathematical operation of taking the derivative of any dependent variable with respect to time $t$. In this and the following sections, we will examine more closely the variable $s$ and its association with the Laplace transform to solve differential equations and to analyze linear control problems.

The Laplace transform is used to transform a differential equation which is a function of time $t$ into a corresponding algebraic equation which is a function of the variable $s$. When this is accomplished, we say that the function has been transformed from the time domain or $t$-domain into the $s$-domain. The reason for taking the trouble to transform the problem from the $t$- to the $s$-domain is to facilitate the problem's solution. It is usually easier to solve a complicated control system problem by using the Laplace transformation.

### The Laplace transform defined

The *Laplace transform* of $f(t)$, where $f(t)$ is a function of time, is defined in the following way:

$$F(s) = \int_0^{\infty} f(t)e^{-st}\,dt \tag{13.19}$$

where $F(s)$ is the Laplace transform of $f(t)$ and $s$ is a complex variable defined by

$$s = \sigma + j\omega \tag{13.20}$$

The transformation is, of course, limited to functions $f(t)$ for which the integral according to Eq. (13.19) exists.

**EXAMPLE 13.5**

Let us illustrate the Laplace transformation given by Eq. (13.19) by means of a simple example. Suppose that we wish to take the Laplace transform of the step function whose constant value is $C$.

$$f(t) = \begin{cases} C & t \geqslant 0 \\ 0 & t < 0 \end{cases}$$

Then, by Eq. (13.19),

$$F(s)=\int_0^\infty Ce^{-st}\,dt=\left.\frac{-Ce^{-st}}{s}\right|_0^\infty$$

$$=\frac{-C(0)}{s}-\frac{-C(1)}{s}=\frac{C}{s}$$

The Laplace transform may be determined for a variety of different functions of $t$, so its application is of great value in control systems analysis.

TABLE OF LAPLACE TRANSFORMS. To avoid the necessity of performing the integration defined by Eq. (13.19) for every function of $t$ in a particular problem, tables of common Laplace transforms have been prepared. These appear in books of mathematical tables and other references. We present some of the more familiar Laplace transforms in Table 13.1.

The last two entries in the table are the transforms for the first and second derivatives. In writing a differential equation, it is the usual practice to write the dependent variable alone (e.g., $y$) rather than stating explicitly that it is a function of $t$[e.g., $y(t)$]. However, when the variable is transformed into the $s$-domain, we must indicate this by using the form $y(s)$ to show that $y$ is a function of $s$. We have conformed to this convention in the last two entries of Table 13.1. The first derivative $dy/dt$ is shown in the left-hand column, where

TABLE 13.1 Table of Common Laplace Transforms

| *Time function f(t)* | *Laplace transform F(s)* |
|---|---|
| Unit step 1 | $\frac{1}{s}$ |
| Unit ramp $t$ | $\frac{1}{s^2}$ |
| Polynomial $t^n$ | $\frac{n!}{s^{n+1}}$ |
| Exponential $e^{-at}$ | $\frac{1}{s+a}$ |
| Sine wave $\sin\omega t$ | $\frac{\omega}{s^2+\omega^2}$ |
| Cosine wave $\cos\omega t$ | $\frac{s}{s^2+\omega^2}$ |
| $te^{-at}$ | $\frac{1}{(s+a)^2}$ |
| $e^{-at}\sin\omega t$ | $\frac{\omega}{(s+a)^2+\omega^2}$ |
| $e^{-at}\cos\omega t$ | $\frac{s+a}{(s+a)^2+\omega^2}$ |
| First derivative $\frac{dy}{dt}$ | $sy(s)-y(t=0)$ |
| Second derivative $\frac{d^2y}{dt^2}$ | $s^2y(s)-sy(t=0)-\frac{dy(t=0)}{dt}$ |

$dy/dt$ is a function of time $t$. In the right-hand column, the corresponding Laplace transform is written:

$$sy(s)-y(t=0)$$

The first term contains the differential operator $s$ and the variable $y(s)$, which is now a function of $s$. The second term is the initial value of $y$ at time $t=0$.

Similarly, the Laplace transform of the second derivative of $y$ is written

$$s^2y(s)-sy(t=0)-\frac{dy(t=0)}{dt}$$

The $s^2$ in the first term indicates the second derivative. The second term is $s$ multiplied by the value of $y$ at $t=0$. The third term is the value of $dy/dt$ at time $t=0$.

### EXAMPLE 13.6

Given the initial conditions that $y=5$ and $dy/dt=2$ at $t=0$, write the Laplace transform of the differential equation:

$$\frac{d^2y}{dt^2}+3\frac{dy}{dt}+y=0$$

The Laplace transform of the first term would be

$$s^2y(s)-5s-2$$

The transform of the second term would be

$$3[sy(s)-5]$$

and the transform of the third term would simply be $y(s)$. Hence, the complete Laplace transform of the differential equation above would be developed as follows:

$$s^2y(s)-5s-2+3sy(s)-15+y(s)=0$$

$$s^2y(s)+3sy(s)+y(s)-5s-17=0$$

$$(s^2+3s+1)y(s)=5s+17$$

$$y(s)=\frac{5s+17}{s^2+3s+1}$$

## The inverse transformation

In the typical application of Laplace transforms, the differential equation is converted from the time domain to the $s$-domain. It is then manipulated algebraically into an appropriate solution form. However, to be of value the solution must then be transformed back into the time domain. This procedure is called the *inverse transformation*. It is accomplished by making use again of Table 13.1. In other words, the transformations indicated in Table 13.1 can proceed in either direction.

### EXAMPLE 13.7

We shall demonstrate the use of the Laplace transform and the inverse transform in the solution of the following differential equation:

$$\frac{dy}{dt} + 5y = 0$$

where the initial conditions are that $y=3$ and $dy/dt=0$ at time $t=0$.

The Laplace transform would be written:

$$sy(s) - 3 + 5y(s) = 0$$
$$(s+5)y(s) = 3$$
$$y(s) = \frac{3}{s+5}$$

Now applying the inverse transform (the fourth entry in Table 13.1), we get the solution for $y$ as a function of time:

$$y = 3e^{-5t}$$

The solution for $y$ has a value of 3.0 when $t=0$ and decays exponentially toward zero as time increases.

PARTIAL FRACTION EXPANSION. The Laplace transform of the solution to the differential equation is often expressed as the ratio of two polynomials:

$$y(s) = \frac{a_0 + a_1 s + a_2 s^2 + \dots + a_m s^m}{b_0 + b_1 s + b_2 s^2 + \dots + b_n s^n} \tag{13.21}$$

The particular polynomial form may not appear as one of the entries in a table of standard Laplace transforms. Hence, it would be difficult to directly obtain the inverse transform of $y(s)$. It is therefore desirable to reduce the form of Eq. (13.21) to a more convenient, yet equivalent form. One method of accomplishing this reduction is provided by partial fraction expansions.

The procedure begins by first factoring the denominator of Eq. (13.21), to form

$$y(s) = \frac{a_0 + a_1 s + a_2 s^2 + \dots + a_m s^m}{(s+r_1)(s+r_2)\dots(s+r_n)} \tag{13.22}$$

where $-r$, $-r_2, \dots, -r_n$ are the roots of the denominator obtained by setting the denominator equal to zero. It is assumed that it is possible to factor the denominator of Eq. (13.21) into its roots even if some of the roots are imaginary or complex numbers.

The second step in the partial fraction expansion is to write Eq. (13.22) in the following form:

$$y(s)=\frac{C_1}{s+r_1}+\frac{C_2}{s+r_2}+\ldots+\frac{C_n}{s+r_n} \tag{13.23}$$

We will assume for the moment that each root is distinct. That is, none of the roots are equal.

The third step is to find the values of $C_1, C_2, \ldots, C_n$ in Eq. (13.23). This is accomplished by means of the following set of equations:

$$C_1=\frac{a_0+a_1s+a_2s^2+\ldots+a_ms^m}{(s+r_1)(s+r_2)\ldots(s+r_n)}(s+r_1)\Bigg|_{s=-r_1} \tag{13.24}$$

$$C_2=\frac{a_0+a_1s+a_2s^2+\ldots+a_ms^m}{(s+r_1)(s+r_2)\ldots(s+r_n)}(s+r_2)\Bigg|_{s=-r_2} \tag{13.25}$$

$$\vdots$$

$$C_n=\frac{a_0+a_1s+a_2s^2+\ldots+a_ms^m}{(s+r_1)(s+r_2)\ldots(s+r_n)}(s+r_n)\Bigg|_{s=-r_n} \tag{13.26}$$

After each of the constants has been determined in Eq. (13.23), the terms in the equation should be in the form of one of the common Laplace transforms. Converting the solution from the $s$-domain to the $t$-domain is a straightforward matter involving the inverse transforms in Table 13.1.

## EXAMPLE 13.8

To illustrate the partial fraction expansion method, let us use the same basic differential equation from Example 13.7 except that a forcing function equal to a unit step will be on the right-hand side of the equation.

$$\frac{dy}{dt}+5y=1$$

where $y=3$ and $dy/dt=0$ at $t=0$.

The Laplace transform would be written

$$sy(s)-3+5y(s)=\frac{1}{s}$$

$$(s+5)y(s)=\frac{1}{s}+3$$

$$=\frac{1+3s}{s}$$

$$y(s)=\frac{1+3s}{s(s+5)}$$

The roots of $s(s+5)=0$ are $s=0$ and $s=-5$. We wish to find the equivalent expression:

$$y(s)=\frac{C_1}{s}+\frac{C_2}{s+5}$$

The values of $C_1$ and $C_2$ are obtained by using Eqs. (13.24) and (13.25).

$$C_1=\frac{1+3s}{s(s+5)}(s)\bigg|_{s=0}=\frac{1+3(0)}{0+5}=\frac{1}{5}$$

$$C_2=\frac{1+3s}{s(s+5)}(s+5)\bigg|_{s=-5}=\frac{1+3(-5)}{-5}=\frac{14}{5}$$

Hence,

$$y(s)=\frac{1/5}{s}+\frac{14/5}{s+5}$$

Using Table 13.1 for the inverse transform, we get

$$y(t)=\frac{1}{5}+\frac{14}{5}e^{-5t}$$

The method above is appropriate for the case when there are no repeated roots in the denominator of Eq. (13.22). Suppose, upon factoring the denominator, that two of the roots are equal. Let the equal roots be $r_1$ and $r_2$. That is, $r_1=r_2$, so Eq. (13.22) could be written as follows:

$$y(s)=\frac{a_0+a_1s+a_2s^2+\ldots+a_ms^m}{(s+r_1)^2(s+r_3)\ldots(s+r_n)} \tag{13.27}$$

In this event, the partial fraction expansion corresponding to Eq. (13.23) would be

$$y(s)=\frac{C_1}{(s+r_1)^2}+\frac{C_2}{s+r_1}+\frac{C_3}{s+r_3}+\ldots+\frac{C_n}{s+r_n} \tag{13.28}$$

Evaluation of constants $C_3$ through $C_n$ would involve the same calculation procedure as before. Equations of the form of Eqs. (13.24) through (13.26) would be used. However, to evaluate $C_1$ and $C_2$ in Eq. (13.28) for the repeated roots, the following formulas must be used:

$$C_1=\frac{a_0+a_1s+a_2s^2+\ldots+a_ms^m}{(s+r_1)^2(s+r_3)\ldots(s+r_n)}(s+r_1)^2\bigg|_{s=-r_1} \tag{13.29}$$

$$C_2=\frac{d}{ds}\left[\frac{a_0+a_1s+a_2s^2+\ldots+a_ms^m}{(s+r_1)^2(s+r_3)\ldots(s+r_n)}(s+r_1)^2\right]_{s=-r_1} \tag{13.30}$$

## EXAMPLE 13.9

The case of repeated roots will be illustrated with the following equation in the $s$-domain:

$$y(s)=\frac{s+2}{(s+3)^2(s+1)}$$

The equation is written in the form of partial fractions:

$$y(s)=\frac{C_1}{(s+3)^2}+\frac{C_2}{(s+3)}+\frac{C_3}{s+1}$$

From Eq. (13.29),

$$C_1=\frac{s+2}{(s+3)^2(s+1)}(s+3)^2\bigg|_{s=-3}=\frac{-3+2}{-3+1}=\frac{1}{2}$$

From Eq. (13.30),

$$C_2=\frac{d}{ds}\left[\frac{s+2}{(s+3)^2(s+1)}(s+3)^2\right]_{s=-3}$$

$$=\frac{d}{ds}\left[\frac{s+2}{s+1}\right]_{s=-3}$$

$$=\frac{(s+1)-(s+2)}{(s+1)^2}\bigg|_{s=-3}=\frac{(-3+1)-(-3+2)}{(-3+1)^2}=-\frac{1}{4}$$

and from Eq. (13.26) we can calculate $C_3$.

$$C_3=\frac{s+2}{(s+3)^2(s+1)}(s+1)\bigg|_{s=-1}=\frac{-1+2}{(-3+1)^2}=\frac{1}{4}$$

Hence, we obtain

$$y(s)=\frac{0.5}{(s+3)^2}-\frac{0.25}{s+3}+\frac{0.25}{s+1}$$

whose inverse transform gives $y$ as a function of time:

$$y=0.5te^{-3t}-0.25e^{-3t}+0.25e^{-t}$$

Equations (13.29) and (13.30) can be extended to cover the general case of any number of repeated roots. However, we will not consider the general case here, preferring to leave this for books whose exclusive concern is with linear control theory. Two repeated roots constitutes the highest number of equal roots we will encounter in this book.

## 13.5 CONTROL ACTIONS

The block diagram of a conventional feedback control system was illustrated in Figure 13.11. The $G$-block is the forward transfer function. In a typical process control application the forward transfer function would consist of two main components, as illustrated in Figure 13.19. The first is the controller unit, whose transfer function is symbolized by $C(s)$. The second component is the transfer function of the process itself, symbolized by $P(s)$. It is desired to control the output of $P(s)$ by using the appropriate control action $C(s)$. By comparing Figures 13.11 and 13.19, the reader can see that the following relationship is true:

$$G(s) = C(s)P(s) \tag{13.31}$$

The conventional feedback control system compares the system output $y$ with the input $x$. Since the $y$ signal may be a different type than the input signal (e.g., $y$ is displacement while $x$ is voltage), it may be necessary to use a transducer, represented by $H(s)$, to convert the output signal into a form that is compatible with the input. The summing block in Figure 13.19 represents the comparison between $x$ and $y$. Any difference between the two is used as the input to the controller. Thus the actuating signal for $C(s)$ in the block diagram is the difference $e(s) = x(s) - H(s)y(s)$. The particular way in which $C(s)$ manipulates this signal is referred to as the control action. The output of the controller is $x_p(s)$, which becomes the input to the process $P(s)$. The controller unit must be designed so that its control action is to correct any deviations from the desired level of the output. The selection of the best control action $C(s)$ depends on the nature of $P(s)$ and $H(s)$.

There are four basic control actions, which are commonly used either alone or in some combination. These are:

1. Proportional control action.
2. Integral control action.
3. Derivative control action.
4. On-off control action.

We will discuss these four basic types in this section and then examine how to select the appropriate control action in the following sections.

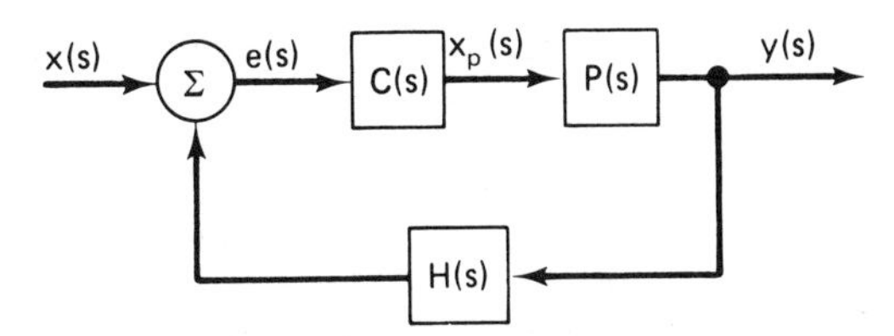


FIGURE 13.19 Block diagram for typical process control application.

## Proportional control

In proportional control, the controller output $x_p$ is proportional to the input $e$. This is expressed in equation form as

$$x_p = Ke$$

where $K$ is the constant of proportionality. For purposes of control system analysis, this can be expressed as a transfer function:

$$\frac{x_p(s)}{e(s)} = K \tag{13.32}$$

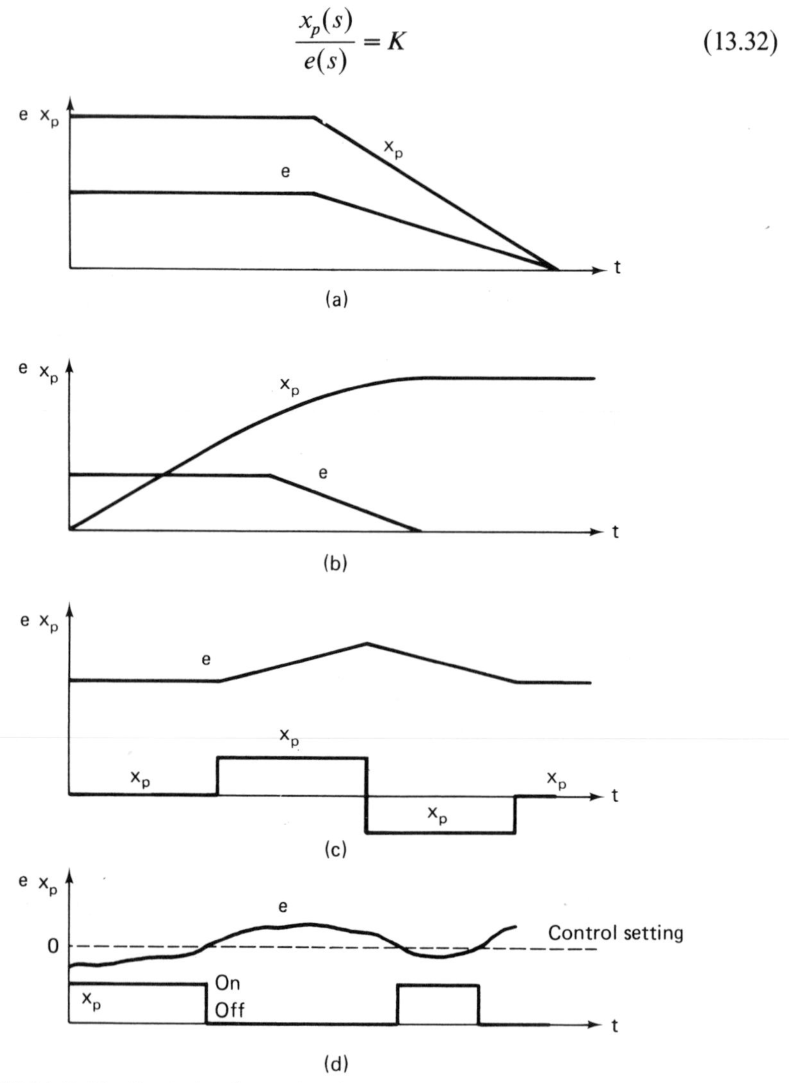


FIGURE 13.20 Control actions of various controllers ($e$ = input to controller, $x_p$ = output of controller): (a) proportional control; (b) integral control; (c) derivative control; (d) on–off control.

The value $K$ is often called the *gain* of the proportional controller.

The relationship between the error signal $e$ and the output signal $x_p$ is shown in Figure 13.20(a). Both signals are shown as a function of time. The nature of this control action is that as long as the error persists, the controller will continue to produce a corrective signal.

## Integral control

An alternative form of control action is integral control. In this case, the output of the controller is proportional to the time integral of the actuating signal $e$. This can be written mathematically as

$$x_p = K \int e dt$$

where $K$ is the gain of the integral controller. Rewriting this in transfer function notation, the integral control action becomes

$$\frac{x_p(s)}{e(s)} = \frac{K}{s} \tag{13.33}$$

The relationship between $e$ and $x_p$ is illustrated in Figure 13.20(b). If the error signal is positive, the correction signal $x_p$ will occur at an increasing rate. The rate of increase will be proportional to the level of the error signal. If the error signal is zero, the signal $x_p$ will be constant. In order for the correction signal $x_p$ to decrease, a negative error $e$ must be present. This results in the tendency for integral control action to overshoot the desired value of the output, which in turn produces an oscillatory response on the part of the process output.

## Derivative control

In derivative control, the value of $x_p$ is proportional to the rate of change of $e$. This can be expressed as follows:

$$x_p = K \frac{de}{dt}$$

where $K$ is the gain of the derivative controller. The corresponding transfer function for the derivative controller would be

$$\frac{x_p(s)}{e(s)} = Ks \tag{13.34}$$

The relationship between $x_p$ and $e$, both plotted over time, is shown in Figure 13.20(c). When the error signal $e$ is steady, the derivative control produces an output signal $x_p = 0$. When $e$ is increasing or decreasing, the controller provides for a signal that is proportional to the rate of increase or decrease. Hence, when $e$ is increasing at a steady rate, $x_p$ is a constant positive value. When $e$ is decreasing at a constant rate, $x_p$ is a negative value.

The disadvantage of derivative control, as illustrated in Figure 13.20(c), is that an error can exist without any corrective action being taken. In the first time interval of Figure 13.20(c), $e$ is a constant positive value. However, it is not until that error signal begins to change that we see the derivative controller coming into action. For this reason, it is not common to use derivative control by itself. Instead, it is generally used in parallel with one of the two previous control actions.

The advantage of derivative control is that it tends to anticipate the occurrence of a deviation from the desired output level. As soon as the $e$ signal deviates from a steady-state level, the derivative control action begins to make a correction.

### On–off control or two-position control

In many control systems, it is satisfactory for the controller to operate at either of two levels rather than over a continuous range as in the three preceding control actions. Typically, the two levels are on or off. More generally, this type of control action is referred to as *two-position control.* The reason is that the two levels may be other than on and off. For instance, the controller may operate at either of two constant speeds, or forward/reverse, and so on.

Perhaps the most familiar application of on–off control is in the operation of home heating systems. The controller unit is the thermostat. The occupant of the home sets the thermostat at the desired temperature level. If the actual room temperature is below the thermostat setting, the furnace (or alternative heating unit) is turned on by the thermostat. After a while, the room begins to heat up. When the room temperature finally reaches (or slightly exceeds) the thermostat setting, the furnace is turned off. As the room cools down, the thermostat finally turns the furnace back on. The home heating system cycles back and forth between the on and off positions.

The on–off control action is illustrated in Figure 13.20(d). When the error exceeds the control setting, the controller sets the value of $x_p$ to the "off" value. When $e$ is below the control setting, $x_p$ is set to the "on" position. Recall that $e$ = the difference between the input $x$ and the feedback signal from $y$. Accordingly, the control setting will usually be set at a value of zero.

The disadvantage with two-position control is that the controller response cannot be matched to the magnitude of the error signal. The control action is either too much or none at all. Depending on the sensitivity of the system, the controller may potentially cycle back and forth between the two positions at a frequency that is too high. To compensate for this problem, the controller is often provided with two limits. The lower limit may turn the controller on while the upper limit turns it off. In this mode of operation, the controlled variable is regulated within a range of values rather than at one given level. The advantage of the two-position controller is its low cost and simplicity. In many control situations, this type of control action is quite adequate.

On–off control is more cumbersome to analyze by traditional linear control theory because its control action represents a discontinuous function. For this reason, the system must be divided into two problems, one in which the controller is on, the other with the controller off. This was illustrated in Example 13.4.

## Combinations of control actions

Combinations of the proportional, integral, and derivative actions are frequently used to achieve the desired control over the process. The resulting transfer functions include the following:

1. Proportional plus integral:

$$\frac{x_p(s)}{e(s)} = K_1 + \frac{K_2}{s} \tag{13.35}$$

2. Proportional plus derivative:

$$\frac{x_p(s)}{e(s)} = K_1 + K_3 s \tag{13.36}$$

3. Proportional, integral, and derivative:

$$\frac{x_p(s)}{e(s)} = K_1 + \frac{K_2}{s} + K_3 s \tag{13.37}$$

### EXAMPLE 13.10

Let us examine the effect of using the three control actions—proportional, integral, and derivative—on the hypothetical process control situation shown in Figure 13.21. The process is represented by the transfer function

$$\frac{5}{s+2}$$

The feedback transfer function is a proportional gain of 0.1, and the contoller transfer function is represented by $C(s)$. Using block diagram algebra, the system transfer

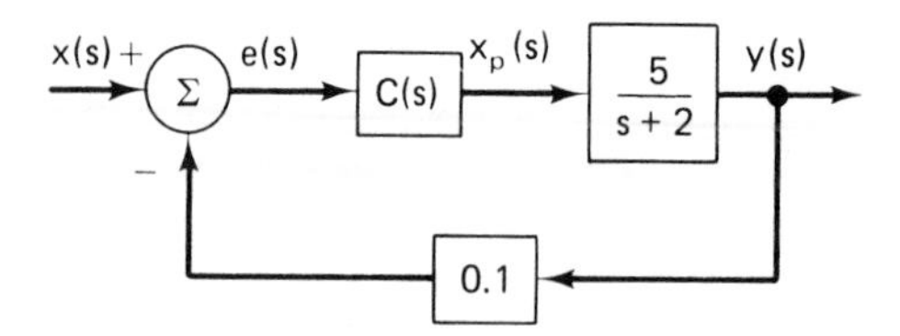


FIGURE 13.21 Block diagram for system of Example 13.10.

function is determined to be

$$\frac{y(s)}{x(s)}=\frac{5C(s)}{s+2+0.5C(s)}$$

The system transfer function can be obtained for each of the three control actions by substituting the particular controller transfer function $C(s)$ into the equation above. We will use a gain of $K=2$ for all three control actions.

1. Proportional control: $C(s)=2.0$

$$\frac{y(s)}{x(s)}=\frac{5(2)}{s+2+1}=\frac{10}{s+3}$$

2. Integral control: $C(s)=2/s$

$$\frac{y(s)}{x(s)}=\frac{10/s}{s+2+1/s}=\frac{10}{s^2+2s+1}$$

3. Derivative control: $C(s)=2s$

$$\frac{y(s)}{x(s)}=\frac{10s}{s+2+s}=\frac{5s}{s+1}$$

It can be seen that the system transfer functions are quite different for the three different control actions.

We will next examine the effect of a step input of $x=4.0$ on the system output $y$. This can be done by multiplying the Laplace transform of the input by the transfer function and then converting the resulting value of $y(s)$ back into the time domain. The transform of the input is $x(s)=4/s$.

1. For the proportional controller:

$$y(s)=\frac{4}{s}\frac{10}{s+3}=\frac{40}{s(s+3)}$$

$$=\frac{13.33}{s}-\frac{13.33}{s+3}$$

$$y=13.33(1-e^{-3t})$$

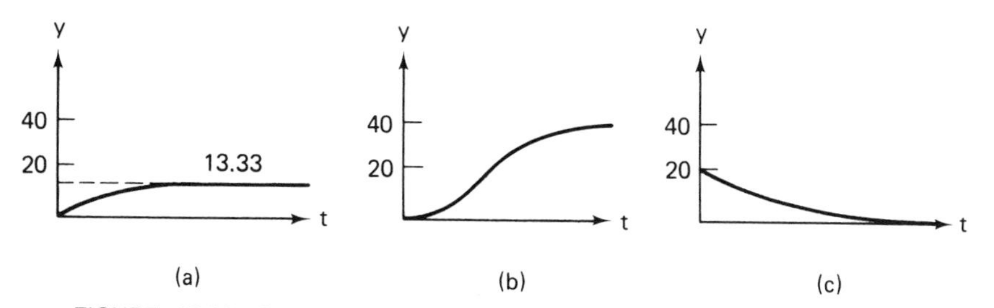


FIGURE 13.22 System responses to three controller types of Example 13.10: (a) proportional control; (b) integral control; (c) derivative control.

2. For the integral controller:

$$y(s) = \frac{4}{s}\frac{10}{(s+1)^2} = \frac{40}{s(s+1)^2}$$

$$= \frac{40}{s} - \frac{40}{(s+1)^2} - \frac{40}{s+1}$$

$$y = 40(1 - te^{-t} - e^{-t})$$

3. For the derivative controller:

$$y(s) = \frac{4}{s}\frac{5s}{s+1} = \frac{20}{s+1}$$

$$y = 20e^{-t}$$

The three responses are clearly different. For the proportional control, the value of $y$ quickly reaches a steady-state value of 13.33. In the case of integral control, a steady-state value of $y=40.0$ is achieved but at a much slower rate than for the proportional control. It should be mentioned that in both of these cases the gain $K$ could be altered to achieve a given steady-state output level for $y$. For the derivative controller, the steady-state value for $y$ is zero. The three solutions are plotted in Figure 13.22.

## 13.6 LINEAR SYSTEMS ANALYSIS

In the previous sections the presentation has been focused on the modeling of linear process control systems (linear differential equations and block diagrams) and on one of the important mathematical tools (Laplace transforms) to facilitate solution of these systems. We have also discussed several control actions (proportional, integral, derivative, and on–off control) which are commonly used in process control. In this section, our emphasis shifts from mathematical descriptions of linear control systems to mathematical analysis and design considerations. The current section covers several preliminary yet important topics in systems analysis. The following section presents the root-locus method for analyzing linear control systems, and the last section is concerned with the design aspects of systems engineering.

The analysis of any control system must include the following performance characteristics of the system:

1. *Stability*. Is the control system stable? If not, how can it be made stable? If it is stable, there are varying degrees of stability and it is important to know the extent of system stability.

2. *Steady-state performance*. What is the performance of the control system, as characterized by its output, after the transient response has disappeared? Does the output achieve some desirable steady-state value?

3. *Transient performance*. What is the system's transient response when subjected to a change in input or a disturbance? Does the system respond quickly to a change in input? Does the system correct itself quickly from the effect of some outside disturbance? Does the output oscillate and, if so, do the oscillations continue indefinitely?

These are the three principal areas of concern in control systems analysis. In this chapter we are limiting our discussion to linear systems. The analysis of nonlinear systems is far more complicated and we properly leave this to books devoted to the subject [4].

## System stability

*System stability* means that the system response will not tend toward a value of infinity when subjected to a noninfinite input. In a practical control system, an infinite response value would never be reached. Instead, the output would reach some limiting value or the system would self-destruct, whichever happens first. It is important to design the system so that it is stable. Accordingly, it is important in systems analysis to be able to recognize the characteristics of a system which cause instability.

A *stable system* is one in which the transient response decays as time increases. An *unstable system* is one in which the transient response increases as time increases. Two examples of system responses will illustrate the case of instability. In the response

$$y = C - C_1 e^{at} \tag{13.38}$$

the second term will increase (in a negative direction) indefinitely as time $t$ goes to infinity. In the following response:

$$y = C - e^{at}(C_1 \sin \omega t + C_2 \cos \omega t) \tag{13.39}$$

the second term will oscillate as a result of the sine and cosine terms, but the amplitude of oscillation will increase indefinitely because of the positive exponential term. In each of these equations, the first term represents the steady-state response and the second term represents the transient response.

For each of these hypothetical differential equation solutions, the transient response can be traced back to the characteristic equation. For the solution represented by Eq. (13.38), the characteristic equation in the Laplace domain would be

$$s-a=0 \tag{13.40}$$

so the root of the equation is $s=+a$. If the forcing function in the differential equation is a step input, the resulting solution would be of the form given by Eq. (13.38).

For the response indicated in Eq. (13.39), the characteristic equation would be

$$s^2-2as+(\omega^2+a^2)=0$$

or

$$(s-a+j\omega)(s-a-j\omega)=0 \tag{13.41}$$

for which the roots are

$$s=a-j\omega \quad \text{and} \quad s=a+j\omega$$

Again, a step input to the system would provide a solution, as given in Eq. (13.39).

For Eq. (13.40) the positive root ($s=a$) leads to the unstable solution in Eq. (13.38). In the case of Eq. (13.41), the real part of the complex roots ($s=a-j\omega$ and $s=a+j\omega$) are positive, and this leads to the instability of the solution given in Eq. (13.39). This leads us to the following conclusion:

> For a linear system to be stable, the roots of the characteristic equation must all be negative real numbers or complex numbers with negative real parts.

This statement is true for the first- and second-order systems represented by Eqs. (13.40) and (13.41), and it is also true for higher-order systems as well. For a linear system with 50 roots, the foregoing stability criterion remains valid. Even if only one of the 50 roots is positive and real, the system will be unstable.

Some systems are said to be *marginally stable.* This is the case in which the characteristic equation has complex roots with real parts equal to zero. An example of such a characteristic equation would be

$$s^2+\omega^2=0 \tag{13.42}$$

This yields roots $s = \pm j\omega$, and the solution of this system for a step input would be

$$y = C + C_1 \sin \omega t + C_2 \cos \omega t \tag{13.43}$$

This solution would continuously oscillate about a mean value of $y = C$. The oscillations would never die out. Accordingly, such a system would be of limited practical value in most situations, because the output never achieves a steady-state constant value.

Marginal stability (another name for this is *limited stability*) represents the dividing line between a stable system and an unstable system.

It should be mentioned that it is not always easy to verify system stability by using the negative real root criterion. The source of the difficulty is that the characteristic equation must be factored in order to determine whether the real roots are negative or the complex roots have negative real parts. Factoring the characteristic equation becomes difficult for high-order differential equations. There are other criteria that can be used to assess system stability, but these are beyond the scope of this survey on linear control theory.

## Steady-state performance

Assume that it has been determined that the process control system of interest is stable. The next concern in the analysis is to determine the steady-state performance of the system. There is an important theorem in control systems analysis which permits the determination of steady-state response. It is called the *final value theorem* and it uses the Laplace transform of the system output. This allows us to find the steady-state value of a function $f(t)$ without requiring that the inverse transform be determined. There is another theorem, called the *initial value theorem*, which is a companion to the final value theorem. The initial value theorem is not nearly as useful as the final value theorem. However, we present both theorems in the following subsections.

FINAL VALUE THEOREM. The final value of a function of time, $f(t)$, is the steady-state value and is given by

$$\lim_{t \to \infty} f(t) = \lim_{s \to 0} sF(s) \tag{13.44}$$

where $F(s)$ is the Laplace transform of $f(t)$. This assumes that

$$\lim_{t \to \infty} f(t)$$

exists.

**Initial Value Theorem.** The initial value of a function of time, $f(t)$, is given by

$$\lim_{t\to 0} f(t) = \lim_{s\to\infty} sF(s) \qquad (13.45)$$

Mathematically, the time variable $t$ is assumed to approach zero from a positive position.

### EXAMPLE 13.11

Let us illustrate these two theorems by considering the three systems from Example 13.10. Recall that there were three different control units used on the process and that the input to each system was a step function of value 4.0.

For the system that used the proportional controller, the solution in the $s$-domain was

$$y(s) = \frac{40}{s(s+3)}$$

According to the final value theorem, the steady-state solution would be

$$y(t=\infty) = \lim_{s\to 0} \frac{40s}{s(s+3)} = \frac{40}{0+3} = 13.33$$

The initial value theorem yields

$$y(t=0+) = \lim_{s\to\infty} \frac{40s}{s(s+3)} = 0$$

For the system using the integral control action

$$y(s) = \frac{40}{s(s+1)^2}$$

Using the final value theorem,

$$y(t=\infty) = \lim_{s\to 0} \frac{40s}{s(s+1)^2} = 40$$

From the initial value theorem,

$$y(t=0+) = \lim_{s\to\infty} \frac{40s}{s(s+1)^2} = 0$$

Finally, in the case of the derivative controller,

$$y(s) = \frac{20}{s+1}$$

From the final value theorem,

$$y(t=\infty)=\lim_{s\to 0}\frac{20s}{s+1}=0$$

Using the initial value theorem yields

$$y(t=0+)=\lim_{s\to\infty}\frac{20s}{s+1}=20$$

Each of these results agrees with the previous solutions obtained for $y$ in the time domain.

## Transient performance

A third characteristic of interest in linear systems analysis is the transient response of the system. There are several properties of the transient response that are of concern in systems analysis.

1. *Response time.* Time required for the system to reach steady state. One property that is sometimes used to characterize this time is called the *settling time*, defined as the time required for the output to achieve within 2% (sometimes 5% is used) of its final value when the system is subjected to a step input. It is generally desirable for the settling time to be as short as possible.
2. *Overshoot.* This response occurs when the output temporarily exceeds the desired steady-state value. The characteristic overshoot is shown in Figure 13.23(a). Technically, overshoot is defined as the maximum deviation of the response above the steady-state value.
3. *Oscillatory behavior.* Although the system may be stable according to the previous stability criterion, its response may tend to oscillate for an extended period of time. This is considered undesirable. Oscillatory transient response is illustrated in Figure 13.23(b).

To investigate these and other characteristics of the transient behavior, a number of analysis techniques are available. First, the analyst might obtain

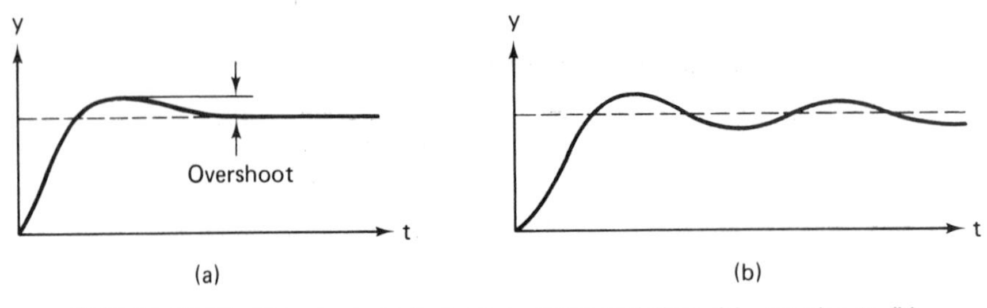


FIGURE 13.23 Transient performance characteristics: (a) overshoot; (b) oscillatory behavior.

the solution to the differential equations for the system. This is usually difficult for systems of higher than second order. To aid in solving such systems, the analog computer can be employed. In addition, digital computer programs that simulate the operations of the analog computer have been developed and are commercially available. These "analog simulators" possess many added features not available on most conventional analog computers. Examples of these features are on-off functions, multipliers, dividers, and other special functions. Hence, it is possible to obtain solutions to system models which are nonlinear. Analytical solutions to nonlinear models are often quite difficult to obtain.

It is cumbersome to develop the best design for a control system using the method of direct solution of the differential equation. Designing the system requires trying alternative controller actions (Section 13.5) and different system gains. Each alternative configuration leads to a new differential equation, and each equation must be solved.

Another technique for linear systems analysis which avoids much of this labor is the *root-locus method.* The root-locus method is a graphical technique that analyzes the roots of a system as a function of system gain. It does this without actually solving the differential equation. In addition, the method provides much insight into the stability characteristics and transient behavior of the system. We shall devote the final two sections of the chapter to a presentation of the root-locus method and how it can be used as an aid in process control design.

## 13.7 THE ROOT-LOCUS METHOD

In Section 13.3, the basic form of the feedback control system was described (Figure 13.11). The transfer function of this control system was defined as

$$\frac{y(s)}{x(s)}=\frac{G(s)}{1+G(s)H(s)} \tag{13.46}$$

Even for complicated feedback systems, it is generally possible to reduce the block diagram to the basic form shown in Figure 13.11 so that the transfer function of the type given by Eq. (13.46) can be written.

The essence of the root-locus method is contained in the denominator of the right-hand side of Eq. (13.46). It can be shown that the roots of the characteristic equation for the system must satisfy the equation

$$1+G(s)H(s)=0 \tag{13.47}$$

This expression can be rewritten in the form

$$G(s)H(s)=-1 \tag{13.48}$$

where $G(s)H(s)$ is called the *open-loop transfer function.*

$G(s)H(s)$ is, of course, a function of the variable $s$. In the definition of the Laplace transform in Section 13.4, the variable $s$ was defined as a complex number,

$$s=\sigma+j\omega$$

where $\sigma$ is the real part of the complex number and $\omega$ the imaginary part. Any given value of $s$ can be plotted in a coordinate system defined by the complex plane. In the complex plane, the abscissa is the $\sigma$-axis (real axis) and the ordinate is the $j\omega$-axis (imaginary axis).

Referring back to the open-loop transfer function $G(s)H(s)$, the values of $s$ that cause $G(s)H(s)=0$ are called the *zeros* of $G(s)H(s)$. Also, the values of $s$ that cause $G(s)H(s)=\infty$ are called the *poles* of $G(s)H(s)$. The poles and zeros of a function $G(s)H(s)$ can be plotted in the complex plane on a *pole–zero map*. The location of a pole of $G(s)H(s)$ is indicated on the pole–zero map by a cross ($\times$), and the zero locations are denoted by a circle (O).

In Section 13.5, we showed how the forward transfer function $G(s)$ was composed of a controller unit $C(s)$ and a process $P(s)$ for a typical process control application. Common control actions include proportional control [$C(s)=K$], integral control [$C(s)=K/s$], and derivative control [$C(s)=Ks$], plus combinations of these types. All of these control actions contain a gain factor $K$. The best value of $K$ must be determined in order to achieve the desired performance from the control system.

Since $G(s)H(s)$ is a function of $C(s)$ and since $C(s)$ contains the gain factor $K$, this means that the roots which satisfy Eq. (13.47) will vary depending on $K$. The locations of the roots in the complex plane will change as the value of $K$ is changed. The locus of these roots as a function of the gain $K$ is called the *root locus*, and the associated analysis to determine the loci is called the root-locus method.

### How to plot the root-locus

In order to determine the loci of roots for a given function $G(s)H(s)$, there are several rules to follow. We shall present the most fundamental of these rules, which permit the reader to plot the root locus for a limited variety of open-loop transfer functions.

1. *Magnitude requirement.* The basic expression of the magnitude requirement for the root locus is embodied in Eq. (13.48):

$$G(s)H(s)=-1$$

All points on the root locus must satisfy this requirement.

2. *Angular requirement.* An alternative way of expressing Eq. (13.48) in the complex plane is the following:

$$G(s)H(s) = 1 \underline{/n180^\circ} \qquad n = \pm 1, \pm 3, \pm 5, \ldots \qquad (13.49)$$

This is the angular requirement that must be satisfied by all points on the root locus. As we shall see later, this does not necessarily mean that all points on the root locus must be on the real axis ($\sigma$-axis).

An alternative way of expressing this angular requirement is:

$$\Sigma\theta_z - \Sigma\theta_p = n180^\circ \qquad n = \pm 1, \pm 3, \pm 5, \ldots \qquad (13.50)$$

where $\theta_z$ is the angle between any point on the root locus and the zero, and $\theta_p$ the angle made between any point on the root locus and the pole.

3. *Number of loci.* A root-locus plot will have a number of loci, or branches, equal to the number of poles of $G(s)H(s)$.

4. *Symmetry of the root-locus plot.* The root-locus plot will be symmetrical about the real axis ($\sigma$-axis).

5. *Starting points and end points.* The loci of roots begin at the poles of $G(s)H(s)$ and end at either the zeros of $G(s)H(s)$ or at infinity. The corresponding values of gain factor $K$ are $K=0$ at the poles of $G(s)H(s)$ and $K=\infty$ at the zeros of $G(s)H(s)$ or at infinity.

6. *Loci on the real axis.* Branches of the root locus lie on the real axis only to the left of an odd number of poles and/or zeros. For example, if poles existed at $-2$ and $-8$ and a zero was located at $-5$, the root locus would exist on the real axis from $-2$ to $-5$, and from $-8$ to $\infty$.

7. *Asymptotes.* The branches of the root-locus tend toward a set of straight-line asymptotes which begin at a point on the real axis. This point is called the *center of asymptotes* and is symbolized by $\sigma_c$. In determining the root-locus plot, it is helpful to compute $\sigma_c$ by means of the following formula:

$$\sigma_c = \frac{\Sigma P - \Sigma Z}{n_p - n_z} \qquad (13.51)$$

where $P$ and $Z$ represent the numerical values of the poles and zeros, respectively, and $n_p$ and $n_z$ are the numbers of poles and zeros.

The number of asymptotes emanating from the center of asymptotes, $\sigma_c$, is given by the difference between $n_p$ and $n_z$:

$$\text{number of asymptotes} = n_p - n_z \qquad (13.52)$$

Finally, the asymptotes are oriented at different angles with respect to the real axis. The angles will be denoted by the symbol $\alpha$, where

$$\alpha = \frac{n180^\circ}{n_p - n_z} \qquad n = \pm 1, \pm 3, \pm 5, \ldots \qquad (13.53)$$

Note that the multiple possible values of $n$ ($\pm 1, \pm 3, \pm 5, \ldots$) may allow $\alpha$ to take on more than a single value. Each of the asymptotes has its own angle $\alpha$.

As an illustration, consider

$$G(s)H(s) = \frac{K(s+1)}{s(s+2)(s+5)}$$

There are three poles and one zero, so there are two asymptotes according to Eq. (13.52). By Eq. (13.51), the center of asymptotes is located at

$$\sigma_c = \frac{(0-2-5)-(-1)}{3-1} = -3.0$$

The angles made by the two asymptotes are given by Eq. (13.53):

$$\alpha = \frac{180}{2}, \qquad \frac{3(180)}{2}$$

That is,

$$\alpha = 90^\circ \quad \text{and} \quad 270^\circ$$

8. *Breakaway points.* A *breakaway point* $\sigma_b$ is a point on the real axis where two branches of the root locus leave the real axis. A *break-in point* is a point on the real axis where two branches arrive at the real axis. The problem of locating breakaway points and break-in points in root-locus is the same.

The location of a breakaway point (or a break-in point) is found by solving the following equation:

$$\sum^{n_p} \frac{1}{\sigma_b - P} = \sum^{n_z} \frac{1}{\sigma_b - Z} \tag{13.54}$$

The summation process is carried out for the $n_p$ values of $P$ and the $n_z$ values of $Z$. To illustrate, consider the transfer function

$$G(s)H(s) = \frac{K}{s(s+5)}$$

To find the breakaway point(s) we must solve for $\sigma_b$ in Eq. (13.54), which, for this transfer function, is

$$\frac{1}{\sigma_b} + \frac{1}{\sigma_b + 5} = 0$$

The right-hand side is zero because there are no zeros in the transfer function.

$$\sigma_b + 5 + \sigma_b = 0$$
$$2\sigma_b = -5$$
$$\sigma_b = -2.5$$

The branches of the root locus leave the real axis at $\sigma_b = -2.5$.

9. *Departure and arrival angles*. The root locus leaves a complex pole with a certain angle, called the *departure angle*. The angle is measured against a line through the pole parallel to the real axis. Similarly, the root locus arrives at a complex zero with a certain angle, called the *arrival angle*. To accurately plot the root locus, it is necessary to determine the departure and/or arrival angles.

The departure angle for a pole or the arrival angle for a zero can be obtained by applying the angular requirement defined by Eq. (13.50) to a point very close to the pole under consideration. To illustrate, suppose that we wanted to determine the departure angles for the poles of

$$G(s)H(s) = \frac{K}{(s+1+j2)(s+1-j2)}$$

The poles are illustrated in Figure 13.24. Consider the pole at $-1+j2$. Pick a sample point near the pole and determine the angles made between that point and all zeros and poles. The angle between the sample point and the pole at $-1+j2$ is the unknown angle of departure. Since there are no zeros, the application of Eq. (13.50) reduces to

$$\Sigma\theta_z - \Sigma\theta_p = 0 - (90° + \theta) = \pm 180°$$

$$90 + \theta = 180°$$

$$\theta = 90°$$

The root-locus branch departs from the upper pole with an angle of 90°. Similarly, the reader can show that the other root-locus branch departs from the lower pole with an angle of −90°. This result could also be obtained by symmetry (rule 4).

Several example problems will serve to demonstrate how the foregoing rules are used to construct the root locus for a given transfer function.

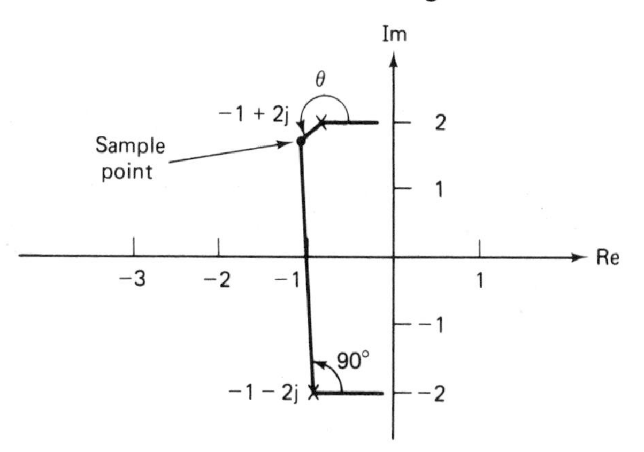


FIGURE 13.24 Application of rule 9 for determining departure and arrival angles in root-locus analysis.

## EXAMPLE 13.12

Construct the root-locus plot for the open-loop transfer function

$$G(s)H(s)=\frac{K}{(s+1)(s+3)(s+6)}$$

From rules 3 and 5, we know that there will be three branches to the root locus, that they will begin at $-1$, $-3$, and $-6$, and that they will all end at infinity. From rule 6, the root locus exists on the real axis between $-1$ and $-3$, and between $-6$ and $-\infty$. The asymptotes can be determined from rule 7. There are $n_p - n_z = 3-0=3$ asymptotes. Their center of asymptotes is located on the real axis at

$$\sigma_c=\frac{(-1-3-6)-(0)}{3-0}=-3.33$$

The angles made by the asymptotes are given by Eq. (13.53).

$$\alpha=\frac{180}{3},\quad \frac{3(180)}{3},\quad \frac{5(180)}{3}$$

$$\alpha=60°,\quad 180°,\quad \text{and}\quad 300°$$

The breakaway point can be found from Eq. (13.54) of rule 8.

$$\frac{1}{\sigma_b+1}+\frac{1}{\sigma_b+3}+\frac{1}{\sigma_b+6}=0$$

The solution yields a value of $\sigma_b=-1.88$. Since there are no complex poles or zeros, rule 9 does not apply.

The root locus for this example is plotted in Figure 13.25. Closed-loop poles start at $(-1,-3,-6)$ for $K=0$ and follow the three branches as $K$ approaches infinity.

## EXAMPLE 13.13

Construct the root locus for the open-loop transfer function

$$G(s)H(s)=\frac{K(s+1)}{(s+2-j)(s+2+j)}$$

There are two branches according to rule 3. One branch begins at the pole at $-2+j$ and the other at $-2-j$. The branches arrive at the real axis somewhere between $-1$ and $-\infty$, from rule 5. One branch goes to $-1$, the other to $-\infty$. To determine the break-in point, rule 8 must be used:

$$\frac{1}{\sigma_b+2-j}+\frac{1}{\sigma_b+2+j}=\frac{1}{\sigma_b+1}$$

The solution is $\sigma_b=-2.414$.

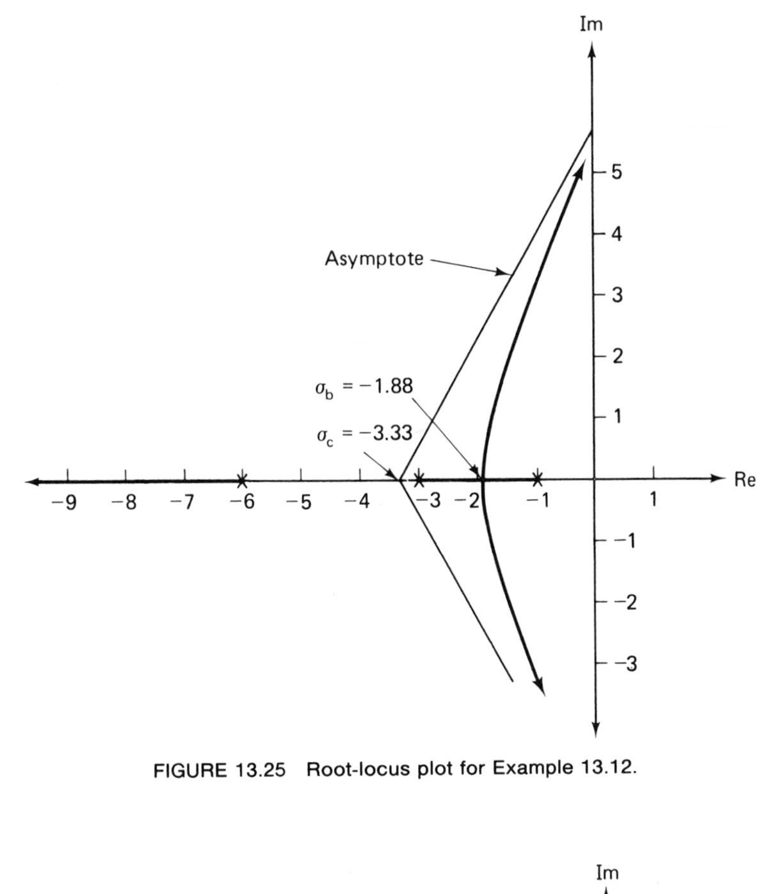


FIGURE 13.25 Root-locus plot for Example 13.12.

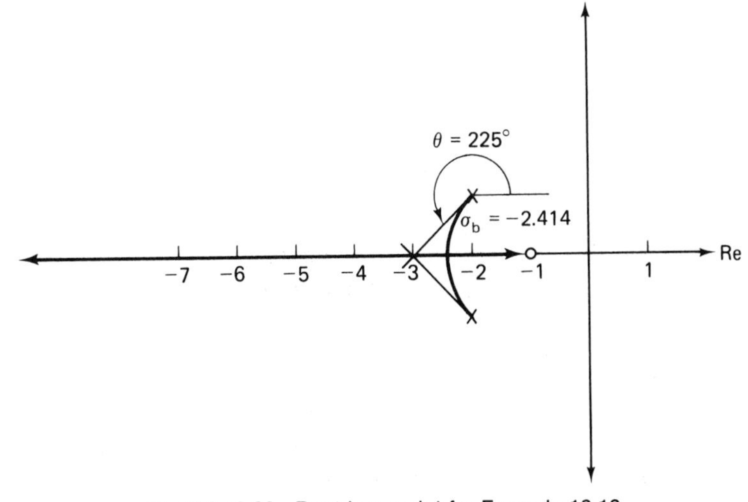


FIGURE 13.26 Root-locus plot for Example 13.13.

Rule 9 provides the method for determining the departure angles from the two complex poles. Picking a sample point near the pole at $-2+j$, Eq. (13.50) becomes the following:

$$\Sigma\theta_z - \Sigma\theta_p = 135 - (90 + \theta) = 180°$$

which leads to an angle of departure

$$\theta = -135° = 225°$$

The root-locus plot for this problem would appear approximately as shown in Figure 13.26 on page 395.

## 13.8 SYSTEM DESIGN

In the design of a process control system, the root-locus method can be used to decide on the system gain for achieving certain desired response characteristics. What makes this possible is that the root-locus plot shows the variation of the poles of the closed-loop system as a function of the gain $K$. In this section we shall demonstrate the relationships between the root locus and some of the important system specifications. Then we will explore the ways in which the root-locus plot can be utilized to design the system to meet the desired specifications.

### Root-locus relationships

There are a number of design objectives that are typically established for a feedback control system. The general criteria of system performance discussed previously in Section 13.6 are:

1. *Stability of the system.* The system must be stable for a variety of inputs.
2. *Steady-state performance.* The response of the system must achieve the desired value within a certain allowable error.
3. *Transient performance.* This criterion includes such specifications of the system as speed of response, the overshoot of the response, and the oscillatory behavior, which is determined by the system's damping characteristics.

These general criteria are related to many of the dynamic response characteristics we have considered in previous sections of this chapter. For example, to achieve a desired speed of response, the time constants of the system would have to be set below some specified value. Or, to obtain a desired oscillatory response, the damping characteristics of the system would have to be specified. To ensure against instability, the roots of the system characteristic equation could not be allowed to become positive (positive real parts in the case of complex roots). All these characteristics can be interpreted from the root-locus plot.

First, consider system stability. For the linear feedback system to be stable, all roots of the characteristic equation must have negative values. On the root-locus plot, negative roots, or negative real parts of complex roots, must appear to the left of the imaginary axis. Hence, the regions of stability and instability for a given system are as illustrated in Figure 13.27. For the system to be stable, the gain factor $K$ would have to be specified so that all the roots of the characteristic equation are on the left half of the complex plane.

The speed of response of the system is determined by the largest time constant appearing in the characteristic equation. The interpretation of the time constant $\tau$ was given in Section 13.2. On the root-locus plot, lines parallel to the imaginary axis represent lines that have constant $1/\tau$ value. Hence, they can be considered as lines of fixed time constant value. For example, suppose that the following specification were established for a given control system: the maximum allowable time constant must be less than or equal to $\frac{2}{3}$ s. In terms of the root locus, this means that all the roots of the system must lie to the left of the line located at $\sigma = -\frac{3}{2} = -1.50$, as shown in Figure 13.27.

Two additional terms were defined in Section 13.2: the undamped natural frequency $\omega_n$, and the damped natural frequency $\omega_d$. Both of these characteristics relate to the oscillatory behavior of the system response. Lines that are parallel to the real axis, as shown in Figure 13.27, are lines of

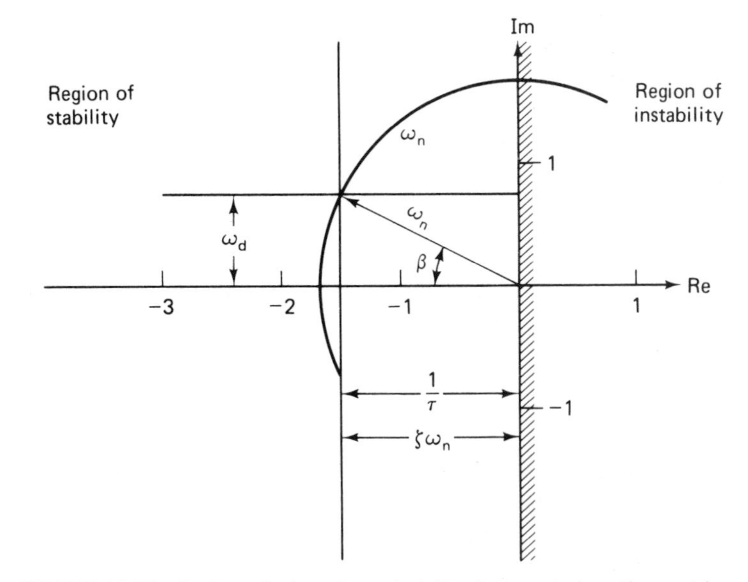


FIGURE 13.27 System design characteristics interpreted on the root-locus plot.

constant $\omega_d$. Constant $\omega_n$ values are indicated in the root-locus plot by circles about the origin. This is also illustrated in Figure 13.27. Although it is not desirable for a system to operate at its natural frequency since this implies no damping, it is nevertheless possible to graphically define the term in the complex plane.

One of the design specifications sometimes used for the damping characteristics of a system is the damping ratio, symbolized by the Greek letter $\zeta$. It is defined as the ratio of the actual damping of the system to the damping that would occur if the system were critically damped. Referring back to Eq. (13.11), the damping of a second-order system is characterized by the damping coefficient $K_d$. In a critically damped second-order system,

$$K_d^2 = 4MK_s$$

or

$$K_d = 2\sqrt{MK_s}$$

Hence, the damping ratio can be defined mathematically as

$$\zeta = \frac{K_d}{2\sqrt{MK_s}} \tag{13.55}$$

When the damping ratio is equal to unity ($\zeta = 1.0$), the system is critically damped. When $\zeta < 1.0$, the system will exhibit an oscillatory response; and when $\zeta > 1.0$, the response will be overdamped and there will be no oscillations.

The damping ratio can be used to relate the damped natural frequency $\omega_d$ of the system to its undamped natural frequency $\omega_n$.

$$\omega_d = \omega_n\sqrt{1-\zeta^2} \tag{13.56}$$

Also, the natural frequency can be related to the time constant by means of the damping ratio.

$$\frac{1}{\tau} = \zeta\omega_n \tag{13.57}$$

Hence, vertical lines on the complex plane representing constant $1/\tau$ values also represent constant $\zeta\omega_n$ values.

In the use of the root locus, constant $\zeta$ values are displayed as straight lines radiating from the origin. This is shown in Figure 13.27. The angle that relates the inclination of the damping ratio is $\beta$, defined as

$$\cos\beta = \zeta \tag{13.58}$$

It is clear that Eqs. (13.56) through (13.58) do not provide for values of $\zeta$ greater than unity.

## Root-locus design approaches

The root-locus method provides for two principal approaches to the system design problem. The two approaches are:

1. *Gain factor compensation.* System design specifications can sometimes be satisfied by setting the gain factor to an appropriate value.
2. *Addition of compensating elements.* When gain factor compensation does not achieve the desired results, the appropriate use of compensating elements in the closed-loop system will often improve system performance.

Generally, these approaches can be applied by making the proper selection of the controller (the most common control actions were discussed in Section 13.5) and/or by setting the gain factor to an appropriate value.

Both of the methods will be discussed and illustrated by means of example problems in the following paragraphs.

GAIN FACTOR COMPENSATION. This is the simpler approach. The objective is to achieve the desired system performance by selecting the best value of the gain factor $K$. The root-locus plot shows how the system's closed-loop poles vary as a function of $K$. Hence, the closed-loop poles can be located on the root-locus plot so as to achieve the required specifications by choosing the appropriate $K$ value.

### EXAMPLE 13.14

Let us consider the feedback system whose block diagram is presented in Figure 13.28. The closed-loop transfer function for this system is

$$\frac{y(s)}{x(s)} = \frac{G(s)}{1+G(s)H(s)} = \frac{\dfrac{2K}{(s+1)(s+3)(s+6)}}{1+\dfrac{K}{(s+1)(s+3)(s+6)}}$$

$$= \frac{2K}{(s+1)(s+3)(s+6)+K}$$

$$= \frac{2K}{s^3+10s^2+27s+18+K}$$

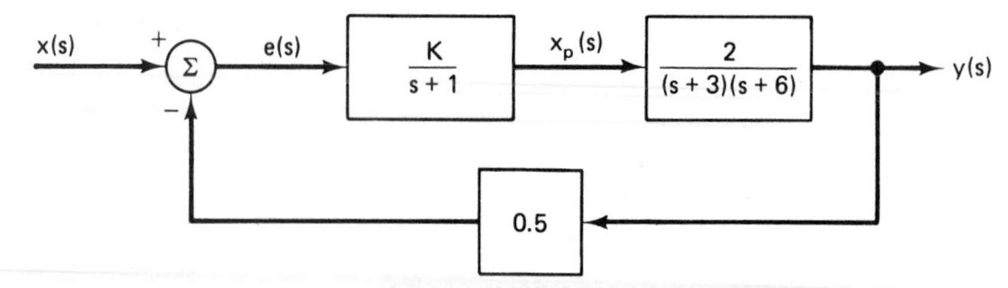


FIGURE 13.28 Block diagram for Example 13.14.

The open-loop transfer function is given by

$$G(s)H(s)=\frac{K}{(s+1)(s+3)(s+6)}$$

The reader may recall that this was the open-loop transfer function for Example 13.12 for which the root-locus plot was given in Figure 13.25.

The specifications for this system are that the maximum value of time constant is $\tau=0.80$ s (we want a relatively fast response from the system). It is also specified that there is to be no oscillation in the response.

***Solution:***

The time constant specification of $\tau=0.80$ means that $1/\tau=1.25$. To satisfy the design specification, the roots of the characteristic equation must lie on the root-locus plot to the left of $\sigma=-1.25$. Since a nonoscillatory response is required, the roots must all lie on the real axis. From the root-locus plot of Figure 13.25, the breakaway point lies at a value of $\sigma_b=-1.88$. Accordingly, the first root that satisfies the specifications lies on the real axis somewhere in the range $\sigma=-1.25$ to $\sigma_b=-1.88$. The second root lies between $\sigma_b=-1.88$ and $\sigma=-3.0$. And the third root will be on the real axis to the left of the pole at $\sigma=-6.0$.

Let us strive for the most favorable response in terms of speed of response. Our selection of the first root would be a value of $s=-1.88$ so as to minimize the time constant. The time constant $\tau=1/1.88=0.53$ s. To determine the value of $K$ that corresponds to $s_1=-1.88$, we make use of the magnitude requirement, Eq. (13.48):

$$G(s)H(s)=-1$$

$$\frac{K}{(s_1+1)(s_1+3)(s_1+6)}=-1$$

$$\begin{aligned}K&=-(s_1+1)(s_1+3)(s_1+6)\\&=-(-1.88+1)(-1.88+3)(-1.88+6)\\&=-(-0.88)(1.12)(4.12)\\&=4.06\end{aligned}$$

The second root also lies at the breakaway point $s_2=-1.88$. The third root has a value to the left of the pole at $\sigma=-6.0$, and must also satisfy the foregoing magnitude requirement when $K=4.06$:

$$\frac{K}{(s_3+1)(s_3+3)(s_3+6)}=-1$$

$$(s_3+1)(s_3+3)(s_3+6)=-4.06$$

Solving for $s_3$, we obtain a value for the third root of $s_3=-6.24$.

Thus, the root-locus analysis has provided a value of $K=4.06$, which satisfies the design requirements and gives a maximum time constant of $\tau_1=0.53$ s, well below the desired 0.80 s specification. The second time constant is also $\tau_2=0.53$ s, since both roots were located at the breakaway point. And the third time constant $\tau_3=1/6.24=0.16$ s.

It would be of instructional value to the reader to relate this solution back to the closed-loop system. Let us determine the response of the system to a step input $x=5$. The closed-loop transfer function is

$$\frac{y(s)}{x(s)}=\frac{8.12}{s^3+10s^2+27s+22.06}$$

We have already determined the roots of the denominator:

$$\frac{y(s)}{x(s)}=\frac{8.12}{(s+1.88)^2(s+6.24)}$$

Multiplying the transfer function by the Laplace transform of the input gives the Laplace transform of the output, $y(s)$:

$$y(s)=\frac{5(8.12)}{s(s+1.88)^2(s+6.24)}$$

Using the method of partial fraction expansion, we obtain

$$Y(s)=\frac{C_1}{s}+\frac{C_2}{(s+1.88)^2}+\frac{C_3}{s+1.88}+\frac{C_4}{s+6.24}$$

$$=\frac{1.84}{s}-\frac{4.95}{(s+1.88)^2}-\frac{1.50}{s+1.88}-\frac{0.34}{s+6.24}$$

Table 13.1 provides the inverse transforms to obtain $y(t)$:

$$y(t)=1.84-4.95te^{-1.88t}-1.50e^{-1.88t}-0.34e^{-6.24t}$$

## EXAMPLE 13.15

Suppose in Example 13.14 that the minimum possible setting of the gain factor was $K=10$. What will be the effect of this limitation on system performance? Will the system be stable if $K=10$?

***Solution:***

According to root-locus rule number 5, the value of $K$ increases as we move farther away from the poles of the root-locus plot. When the value of $K$ was 4.06, two of the roots of the characteristic equation were located at the breakaway point of the root locus. With a value of $K = 10.0$, these two roots will be located on the root locus away from the real axis.

To determine the value of these roots, start with the magnitude requirement, which is an expression of the system characteristic equation,

$$\frac{10}{(s+1)(s+3)(s+6)} = -1$$

or

$$s^3 + 10^2 + 27s + 28 = 0$$

Solving for the roots,

$$s_1 = -1.742 + 1.123j$$

$$s_2 = -1.742 - 1.123j$$

$$s_3 = -6.515$$

The approximate positions can be located in the root-locus plot of Figure 13.25.

Although we will not go through the calculations to determine the response to the same step input used in Example 13.14, we can tell from the roots of the characteristic equation that the response will be oscillatory and slower than in the previous example. The nature of this response is determined by roots $s_1$ and $s_2$, which are complex. The imaginary part, $\pm 1.123j$, gives the root its oscillatory behavior, while the real part, $-1.742$, determines the speed of response. The time constant associated with these roots is $\tau = 1/1.742 = 0.57$ s. The time constant associated with root $s_3$ is $1/6.515 = 0.15$ s.

We can determine the damped natural frequency from the root-locus plot of Figure 13.25. The vertical distance of roots $s_1$ and $s_2$ above and below the real axis gives the value of $\omega_d = 1.123$ rad/s. The undamped natural frequency associated with roots $s_1$ and $s_2$ is given by the distance between the origin and either of these two points.

$$\omega_n = \sqrt{(1.742)^2 + (1.123)^2} = 2.07 \ rad/s$$

From Eq. (13.56), we can determine the damping ratio for the system:

$$1.123 = 2.07\sqrt{1 - \zeta^2}$$

$$\zeta = 0.84$$

We can utilize Eq. (13.57) as a check on this value:

$$1.742 = \zeta(2.07)$$

$$\zeta = 0.84$$

Because of the negative real parts of the complex roots $s_1$ and $s_2$ and because $s_3$ is negative and real, the system is stable.

**EXAMPLE 13.16**

For the same basic system of the two preceding examples, let us determine the limiting gain factor $K$ for which stability will be maintained. To do this, we can examine the points at which the two root-locus branches cross the imaginary axis. These points will correspond to the two characteristic equation roots of the form

$$s_1, s_2 = 0 \pm j\omega$$

Therefore, we can write two equations with the two unknowns $K$ and $\omega$ as follows:

$$(-j\omega)^3 + 10(-j\omega)^2 + 27(-j\omega) + 18 + K = 0$$

and

$$(j\omega)^3 + 10(j\omega)^2 + 27(j\omega) + 18 + K = 0$$

Solving for the two unknowns, we obtain

$$\omega = 5.196$$
$$K = 252$$

Hence, the first two roots will be

$$s_1 = +5.196j$$
$$s_2 = -5.196j$$

The third root will be real and negative and will have a value that satisfies

$$s_3^3 + 10s_3^2 + 27s_3 + (18 + 252) = 0$$

That value is

$$s_3 = -10.0$$

With a gain factor of $K = 252$, the response of the system to a step input will continually oscillate. This results from roots $s_1$ and $s_2$, which contain no negative real portions to make the response decay with time.

ADDITION OF COMPENSATING ELEMENTS. It may not be possible to meet system specifications simply through adjustment of the gain factor $K$. When this is the case, the desired performance can sometimes be obtained by the addition of a compensating element to the closed-loop system. The purpose of this new element is to change the pole–zero configuration of the root locus into a more desirable form. Basically, the compensating element possesses a transfer function that changes the system transfer function in a beneficial way.

Although there are a number of different objectives in applying compensating elements, we will consider just one type in this survey: cancellation compensation. In this arrangement, the purpose of the compensating element is to cancel one or more of the poles (or zeros) of the existing system. In essence, the new element replaces an undesirable pole (or zero) with a desirable one. This has the consequence of improving overall system performance. Let us examine a hypothetical example to illustrate this general approach.

## EXAMPLE 13.17

Figure 13.29 shows the block diagram for a system that must have a time constant of 0.25 min or less (time units in this example are minutes, not seconds).

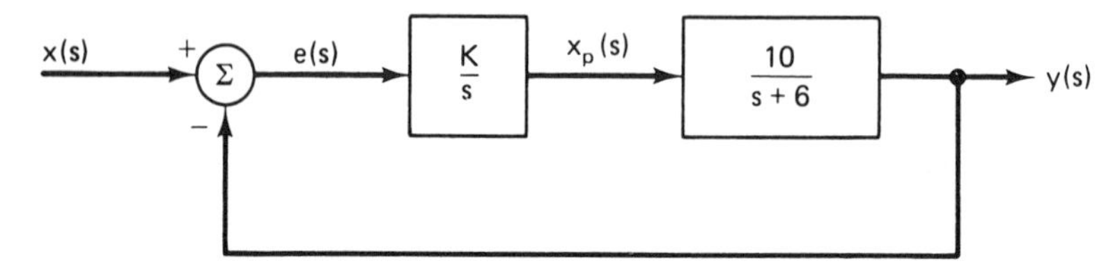


FIGURE 13.29 Block diagram for Example 13.17.

***Solution:***

The root-locus plot for the system as given is presented in Figure 13.30. It can be seen that the minimum possible time constant for this system is $\tau = 0.333$ min, which corresponds to the breakaway point $\sigma_b = -3.0$.

To meet system specifications, a compensating element must be added to the system to change the root-locus plot into a more desirable form. Consider the addition of the following compensating element, as illustrated in Figure 13.31:

$$\frac{s}{s+2}$$

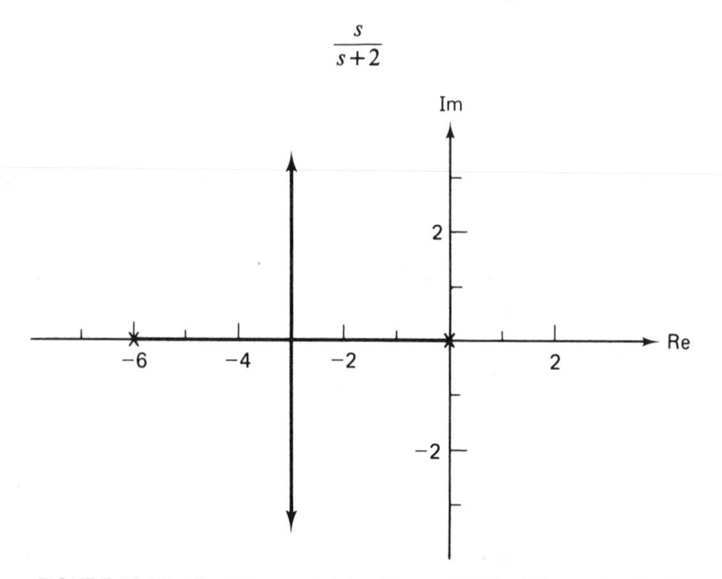


FIGURE 13.30 Root-locus plot for Figure 13.29 of Example 13.17.

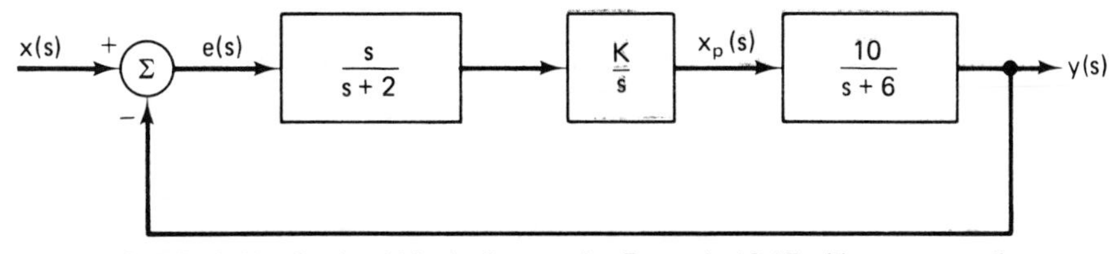


FIGURE 13.31 Revised block diagram for Example 13.17 with compensating element added.

This new element would change the open-loop transfer function of the system from

$$G(s)H(s)=\frac{10K}{s(s+6)}$$

into

$$G(s)H(s)=\frac{10K}{s(s+6)}\frac{s}{s+2}=\frac{10K}{(s+6)(s+2)}$$

The corresponding root-locus plot would be shifted to the left, as illustrated in Figure 13.32. In this case, a time constant of $\tau=0.25$ min is possible if the roots of the characteristic equation are both located at $s=-4.0$. The corresponding value of the gain factor can be determined from the magnitude requirement to be $K=0.4$.

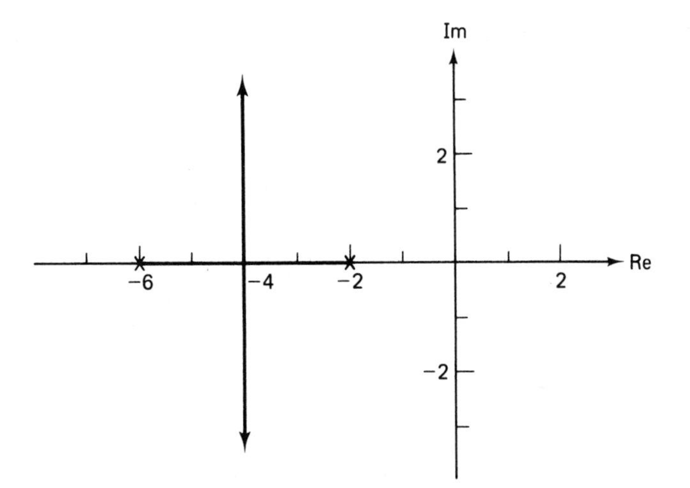


FIGURE 13.32 New root-locus plot for revised block diagram shown in Figure 13.31 of Example 13.17.

## REFERENCES

[1] D'Azzo, J. J., and Hoopis, C. H., *Linear Control System Analysis and Design,* McGraw-Hill Book Company, New York, 1975.

[2] Dorf, R. C., *Modern Control Systems,* Addison-Wesley Publishing Company, Inc., Reading, Mass., 1974.

[3] HARRISON, H. L., and BOLLINGER, J. G., *Introduction to Automatic Controls,* International Textbook Co., Scranton, Pa., 1963.

[4] GIBSON, J. E., *Nonlinear Automatic Control,* McGraw-Hill Book Company, New York, 1963.

[5] JAESCHKE, J. R., ZIMMERLY, R. D., and WU, S. M., "Automatic Cutting Tool Temperature Control," *International Journal of Machine Tool Design Research,* Vol 7, pp. 465–475, 1967.

[6] KUO, B. C., *Automatic Control Systems,* 3rd ed., Prentice-Hall, Inc., Englewood Cliffs, N. J., 1975.

[7] RAVEN, F. H., *Automatic Control Engineering,* 3rd ed., McGraw-Hill Book Company, New York, 1978.

## PROBLEMS

**13.1.** A mechanical measuring device used in a certain manufacturing process is believed to possess the response characteristics of a first-order system. That is, the mathematical model for the measuring device is a first-order linear differential equation. Data have been collected over a 20-s period at 5-s intervals on the input to the device, $x$, the output, $y$, and the rate of change of the output, $dy/dt$:

| *Time (s)* | $x$ | $y$ | $dy/dt$ |
|---|---|---|---|
| 0 | 5.3 | 2.0 | 0.26 |
| 5 | 5.7 | 2.2 | 0.26 |
| 10 | 5.9 | 2.4 | 0.22 |
| 15 | 5.4 | 2.6 | 0.04 |
| 20 | 4.9 | 2.5 | −0.02 |

From these data, determine the values of the parameters in the model. Specifically, for the differential equation:

$$A\,dy/dt + By = x$$

determine the values of $A$ and $B$ that are consistent with these data.

**13.2.** Solve the differential equation determined in Problem 13.1 for a step input $x = 1.0$ which starts at time $t = 0$. Assume that $y = 0$ at $t = 0$. What is the time constant for the system?

**13.3.** The speed control unit for operating the spindle on a certain machine tool is being studied to determine its response characteristics. As part of the study, measurements were made of the actual spindle speed in revolutions per second. The input to the control unit is voltage. Given in the table are the collected data when the speed control unit was subjected to a step input of 20 V at time $t = 0$.

| *Time (s)* | *Input voltage (v)* | *Output-spindle speed (rev / s)* |
|---|---|---|
| 0 | 20 | 0 |
| 0.5 | 20 | 3.32 |
| 1.0 | 20 | 5.23 |
| 1.5 | 20 | 6.05 |
| 2.0 | 20 | 6.25 |
| 2.5 | 20 | 6.21 |
| 3.0 | 20 | 6.12 |
| 4.0 | 20 | 6.01 |
| 5.0 | 20 | 5.99 |
| 10.0 | 20 | 6.00 |

(a) Plot the data on a piece of graph paper.
(b) Characterize the response as probably falling into which one of the following categories: (1) first order; (2) second order, no damping; (3) second order, underdamped; (4) second order, critically damped; (5) second order, overdamped.

**13.4.** For the data of Problem 13.3, determine the appropriate form of the differential equation that could be used as a mathematical model to describe the operation of the speed control unit. What additional data would you need to determine the parameter values in the differential equation?

**13.5.** Reformulate your answer to Problem 13.2 as a transfer function using the $s$-differential operator notation.

**13.6.** Reformulate your general answer to Problem 13.4 as a transfer function using the $s$-differential operator notation.

**13.7.** A mechanical device used in a forge press operation has the following differential equation of motion:

$$16.3\frac{d^2y}{dt^2}+87.5\frac{dy}{dt}+221y=F$$

where F is the forcing function and $y$ represents the response of the device.
(a) Determine the roots of the characteristic equation. Will the response be oscillatory?
(b) Determine the damped natural frequency $\omega_d$.

**13.8.** Rewrite the following differential equations as transfer functions, where $x$ represents the input and $y$ represents the output of the transfer function:

(a) $5\frac{dy}{dt}=2x$

(b) $3\frac{dy}{dt}+7y=x$

(c) $\frac{d^2y}{dt^2}+3\frac{dy}{dt}+2.5y=1.2x$

(d) $\frac{d^2y}{dt^2}+2\frac{dy}{dt}+8y=\frac{dx}{dt}+3x$

**13.9.** For the following set of equations, change each equation into its corresponding $s$-operator equation; then construct the block diagram that relates the three equations.

$$\frac{dz}{dt} + 2z = w$$

$$\frac{dy}{dt} + 7y = 3z$$

$$w = x - 0.3y$$

**13.10.** Construct the block diagram that combines the following set of equations expressed in the $s$-notation:

$$w = x - y$$

$$v = w - z$$

$$z(s+6) = v(s+2)$$

$$y(s^2 + 6s + 8) = z$$

Let $x$ be the input to the system and $y$ be the output.

**13.11.** Reduce the block diagrams in Figure P13.11 to a simple transfer function by means of block diagram algebra.

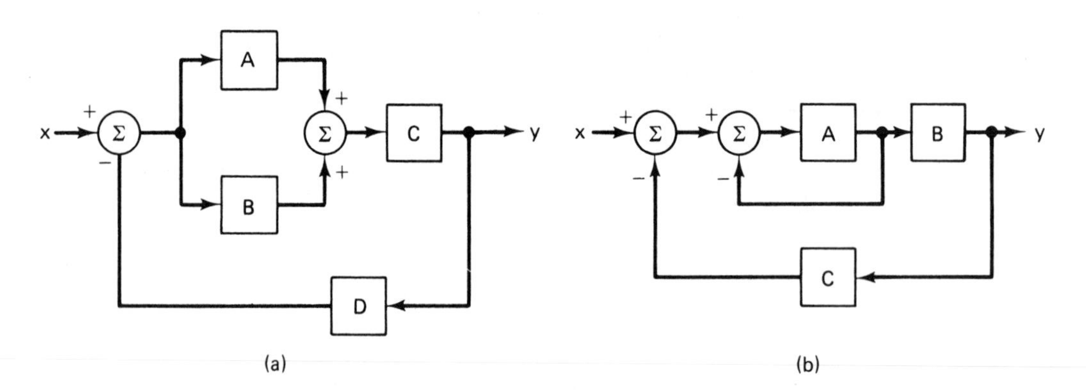


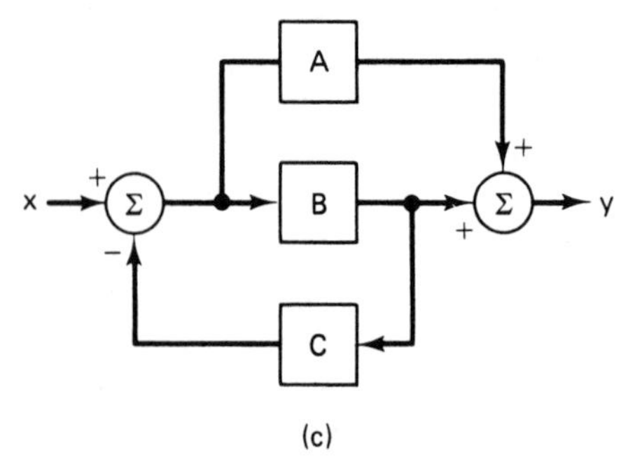


FIGURE P13.11

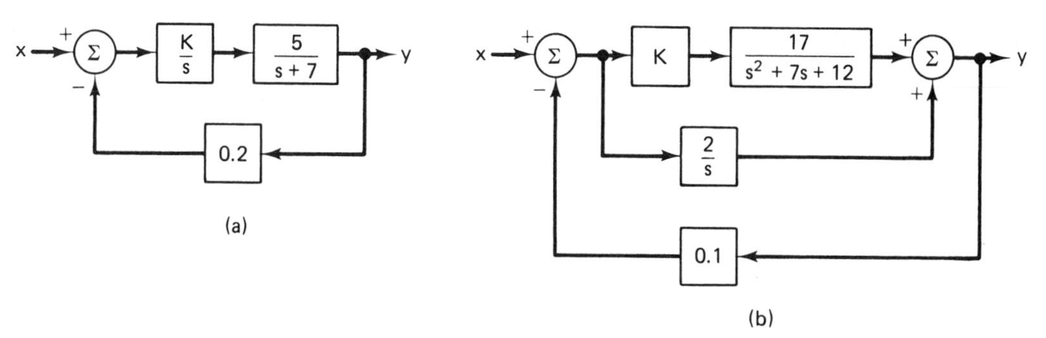


FIGURE P13.12

**13.12.** Reduce each of the block diagrams in Figure P13.12 into a single transfer funtion by means of block diagram algebra.

**13.13.** Reduce the block diagram determined in Problem 13.9 into a single transfer function with $x$ = input and $y$ = output.

**13.14.** Reduce the block diagram determined in Problem 13.10 into a single transfer function with $x$ = input and $y$ = output. Express this transfer function as a differential equation.

**13.15.** The following set of equations expresses the relationships that exist for a certain production process:

$$u = x - w$$
$$\frac{Ku}{s} + z = v$$
$$15v = y(s+6)$$
$$w(s+8) = 0.3y$$

The two inputs to the process are $x$ and $z$. The variable $y$ is the process output. Variables $u$, $v$, and $w$ are internal to the process.

(a) Construct the overall block diagram for the production process.

(b) Determine the transfer function for input $x$ and output $y$.

(c) Determine the transfer function for input $z$ and output $y$.

**13.16.** In Problem 13.15, if the input $x$ is a step input of value $x = 3.0$ and the input $z$ is a sine-wave function $z = 2\sin 5t$, write the Laplace transform of the output $y$. Assume that the initial conditions are zero.

**13.17.** Given the initial conditions that $y$ and its derivatives are all zero at $t = 0$, write the Laplace transform of the following differential equations:

(a) $\dfrac{d^2y}{dt^2} + 5\dfrac{dy}{dt} + 2y = 8$

(b) $2\dfrac{dy}{dt} + 7y = 3\sin 3t$

(c) $\dfrac{d^2y}{dt^2} + 8y = 5\cos t$

(d) $\dfrac{d^2y}{dt^2} + 6\dfrac{dy}{dt} + 8y = 2t$

(e) $7\dfrac{dy}{dt} + 3y = 5 + 2t$

**13.18.** Expand the following functions of $s$ into partial fractions by the methods of Section 13.4:

(a) $y(s)=\dfrac{3}{s^2+7s+10}$

(b) $y(s)=\dfrac{2s+1}{s(s^2+4s+3)}$

(c) $y(s)=\dfrac{5s}{s^2+8s}$

**13.19.** Expand the following functions of $s$ into partial fractions by the method of Section 13.4:

(a) $y(s)=\dfrac{s+1}{(s^2+4s+4)}$

(b) $y(s)=\dfrac{4}{s^2(s+7)}$

**13.20.** Find the inverse transform by the methods of Section 13.4 for the following:

(a) $y(s)=\dfrac{1}{s+1}+\dfrac{2}{s+3}-\dfrac{1}{s+2}$

(b) $y(s)=\dfrac{2}{s}+\dfrac{0.1}{s^2}+\dfrac{2}{s+5}$

(c) $y(s)=\dfrac{5}{s}+\dfrac{2}{s^2+4}$

(d) $y(s)=\dfrac{6}{s+5}+\dfrac{3}{(s+2)^2}$

(e) $y(s)=\dfrac{3}{s}+\dfrac{2s}{s^2+3}-\dfrac{1}{s^2+6s+9}$

**13.21.** Find the inverse transforms for parts (a) through (c) of Problem 13.18.

**13.22.** Find the inverse transforms for parts (a) and (b) of Problem 13.19.

**13.23.** Given the initial conditions $y=3$ and $dy/dt=0$ at $t=0$, write the Laplace transform of the following differential equation:

$$\frac{dy}{dt}+3y=4$$

Solve the differential equation by using the inverse transform method.

**13.24.** Given the initial conditions $y=0$, $dy/dt=0$, and $d^2y/dt^2=0$, write the Laplace transform of the following differential equation:

$$2\frac{d^2y}{dt^2}+10\frac{dy}{dt}+12.5y=16$$

Solve the differential equation by using the inverse transform method.

**13.25.** Use the initial value theorem and the final value theorem to determine the values of $y$ at time $t=0$ and $t=\infty$, respectively, for the following:

(a) $y(s)=\dfrac{5}{s(s+5)}$

(b) $y(s)=\dfrac{13}{s^2+18}$

(c) $y(s)=\dfrac{s+2}{s(s^2+9s+14)}$

**13.26.** Consider the process control situation illustrated in Figure 13.21 except that the process has the following transfer function:

$$\frac{4}{s+5}$$

The feedback transfer function is a proportional gain of 0.2. It is desired to achieve a steady-state response of $y=50$ when the input $x$ is a step function of value $x=20$ by selecting the appropriate gain factor $K$ using a proportional control $C(s)=K$. Determine the appropriate value of $K$ that will achieve the desired response value. Plot the response as a function of time on a piece of graph paper.

**13.27.** Construct the root-locus plots for each of the following open-loop transfer functions:

(a) $G(s)H(s)=\dfrac{K}{(s+2)(s+8)}$

(b) $G(s)H(s)=\dfrac{K}{s(s+2)(s+8)}$

(c) $G(s)H(s)=\dfrac{K(s+3)}{s^2(s+7)}$

(d) $G(s)H(s)=\dfrac{K}{s(s+1)(s+3)(s+4)}$

**13.28.** Construct the root-locus plots for each of the following open-loop transfer functions:

(a) $G(s)H(s)=\dfrac{K}{(s+1)(s^2+4s+5)}$

(b) $G(s)H(s)=\dfrac{5K(s+1)}{s^3+5s^2+6s}$

**13.29.** For each of the root-locus plots in Problem 13.27, determine the value of $K$ at which the system becomes unstable if the root-locus plot indicates that instability is possible.

**13.30.** Assume that the block diagram in Figure P13.30 represents the model for some industrial process (time units are seconds):

(a) Determine the open-loop transfer function for this system.
(b) Determine the closed-loop transfer function for the system.
(c) Plot the root locus for this system.

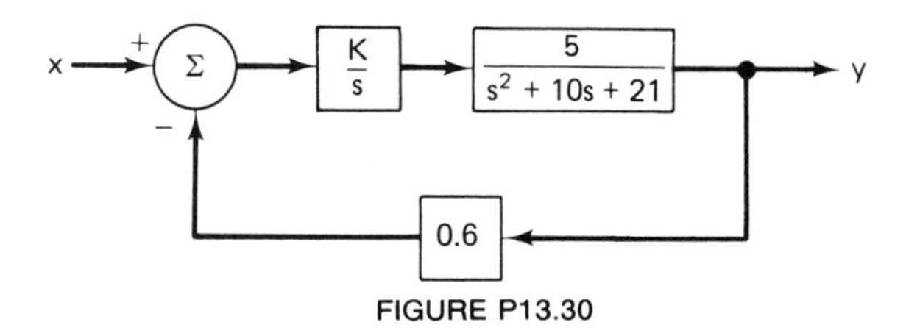


FIGURE P13.30

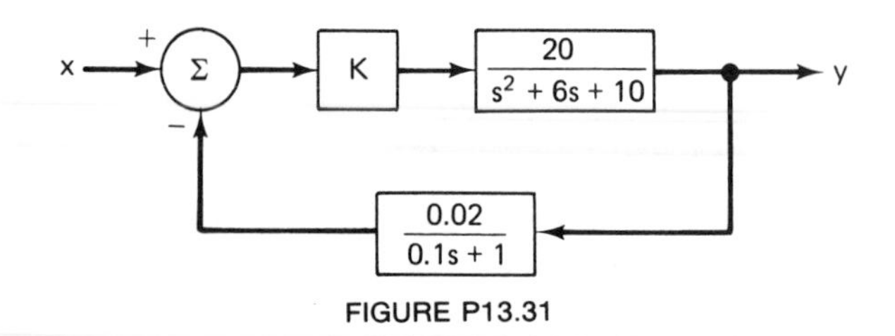


FIGURE P13.31

**13.31.** Assume that the block diagram in Figure P13.31 represents the model for some industrial process:

(a) Determine the open-loop transfer function for this system.
(b) Determine the closed-loop transfer function for this system.
(c) Plot the root locus.

**13.32.** Determine the value of $K$ at which the system in Problem 13.30 will become unstable. Also, determine the largest possible value of $K$ for which the system will have no complex roots.

**13.33.** Consider the industrial process of Problem 13.30. The specifications on this system are that the maximum value of the time constant is to be $\tau = 1.5$ s and that there is to be no oscillation in the response. Is this performance possible to achieve in this system and, if so, what is the corresponding value of the gain factor $K$?

**13.34.** Suppose in Problem 13.33 that the minimum possible setting of the gain factor is $K = 4$. What will be the effect of this limitation on system performance? Will the system response be oscillatory if $K = 4$?

**13.35.** In Example 13.14, the steady-state response for a step input of $x = 5.0$ is $y = 1.84$. Suppose that the desired value of the output is $y = 3.0$.

(a) Would you recommend that this be achieved by increasing the value of $K$ in the solution to this sample problem? If not, why not?
(b) How could the desired output of $y = 3.0$ be achieved?

**13.36.** Consider the industrial process of Problem 13.31. The specification for the system is that the maximum value of the time constant is to be $\tau = 0.40$ s. Determine the value of the gain factor $K$ that achieves this response. Determine the damping ratio for the system. Compute the damped natural frequency for this system.

**13.37.** Assume that the block diagram in Figure P13.37 represents the model for a certain industrial process.

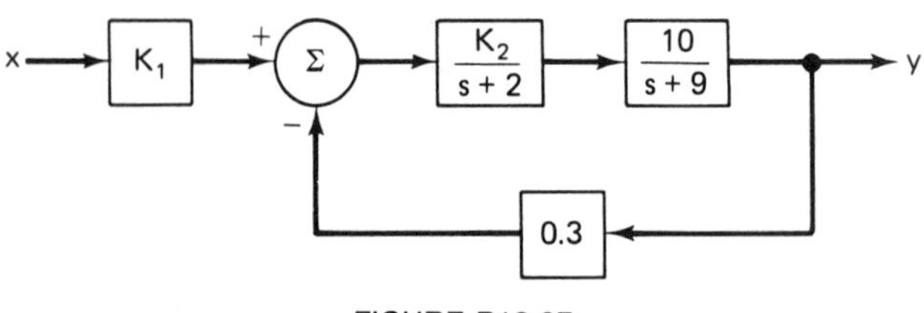


FIGURE P13.37

(a) Determine the value of the gain factor $K_2$ that achieves a minimum time constant of $\tau = 0.25$ s. Use the root-locus method.

(b) It is desired for the steady-state response to be $y = 5.0$ when the input is a step function $x = 2.0$. Determine the value of the gain factor $K_1$ which achieves this steady-state response.

(c) Write the equation that expresses $y$ as a function of time when $x$ is a step input of value $x = 2.0$.

# chapter 14

# Direct Digital Control

## 14.1 INTRODUCTION

As indicated in the introduction to the previous chapter, industrial processes and manufacturing operations are characterized by variables that are continuous and analog. Another feature of most processes is that there are many output variables to be controlled. A complex industrial process may have more than 1000 variables to be monitored and regulated. An oil refinery would be an example of such a complex system. Each pair of input/output variables represents a control loop. Before digital computer control, analog controllers were used to regulate the individual loops of an industrial process.

Direct digital control (DDC) involves the replacement of the conventional analog control devices with the digital computer. The regulation of the process is accomplished by the digital computer on a time-shared, sampled-data basis, rather than by many individual analog elements, each working in a continuous dedicated fashion. With DDC, the computer calculates the desired values of the input variables, and then these calculated values are applied *directly* to the process. This direct link between the computer and the process is the reason for the name "direct digital control."

DDC was originally perceived as a more efficient means of carrying out the same types of control actions as the analog elements that it replaced. However, the analog devices were somewhat limited in terms of the mathematical operations that could be performed (proportional control, integral control, derivative control, and combinations of these). The digital computer is considerably more versatile with regard to the variety of control calculations that it can be programmed to perform. Hence, direct digital control offers not only the opportunity for greater efficiency in doing the same job than analog control, it also opens up the possibility for increased flexibility in the type of control action, as well as the option to reprogram the control action should that become desirable.

In this chapter we shall explain how direct digital control works in comparison to analog control, the components of a DDC installation, and its advantages and disadvantages. The use of the computer simply to perform the DDC function (as it has been defined above) is somewhat anachronous today. The computer is capable of performing higher-level process control of the types that will be discussed in Chapter 15. For now, we limit ourselves to computer control at the process interface level.

## 14.2 ANALOG CONTROL

The three basic types of control action—proportional, integral, and derivative—typically performed by analog controllers were described in Section 13.5. Their mathematical operations were defined by Eqs. (13.32) through (13.34), respectively. These three control actions are often combined into a single equation to provide the necessary degree of regulation over the particular process variable. Equation (13.37) expressed the transfer function for this proportional–integral–derivative control:

$$\frac{x_p(s)}{e(s)} = K_1 + \frac{K_2}{s} + K_3 s \tag{14.1}$$

This equation can be restated in the time domain as follows:

$$x_p(t) = K_1 e + K_2 \int_0^t e\,dt + K_3 \frac{de}{dt} \tag{14.2}$$

This general expression for the three-mode control can be used to represent any one of the single control actions simply by setting the appropriate gain factors ($K_1$, $K_2$, or $K_3$) to zero. For instance, integral control can be obtained from Eq. (14.2) by setting $K_1$ and $K_3$ equal to zero. Similarly, any combination of two control actions can be obtained from Eq. (14.2) by making the third gain factor zero.

The mechanism by which the process variables are altered according to Eq. (14.2) depends on the particular system. If the variable signals are electrical voltages, these electrical signals can be multiplied, integrated, or differentiated by various circuits composed of operational amplifiers. Where the variables are not electrical signals, the mechanical, pneumatic, hydraulic, thermal, and so on, variables can be transduced into electrical form to facilitate the control calculations.

A typical control loop operated under analog control might be represented as in Figure 14.1. The process has multiple inputs and outputs, and the control loop shown might be one of many needed for complete process control. To measure the output variable, a transducer is used to convert the signal from its current physical form into an electrical signal. The transducer typically operates in conjunction with some type of measuring instrumentation that might be used to display the signal to the operator and to record the output variable over time. The measured variable is compared against a set point that has been entered by the operator. The dial in Figure 14.1 represents one method by which the operator might enter the set-point value. The difference between the set point and the measured process variable is the error signal used by the analog controller. This control unit calculates the desired adjustment to be made in the input variable which will reduce the error to zero. The subject matter of Chapter 14—process control fundamentals—constitutes the theoretical basis for determining the circuitry required in the analog controller. The controller circuitry would be designed to perform calculations on the error signal indicated by Eq. (14.1). More complicated manipulations of the error signal might also be possible with analog control, but typically the control strategy is based on a combination of proportional, integral, and derivative control. Many control loops require only on–off

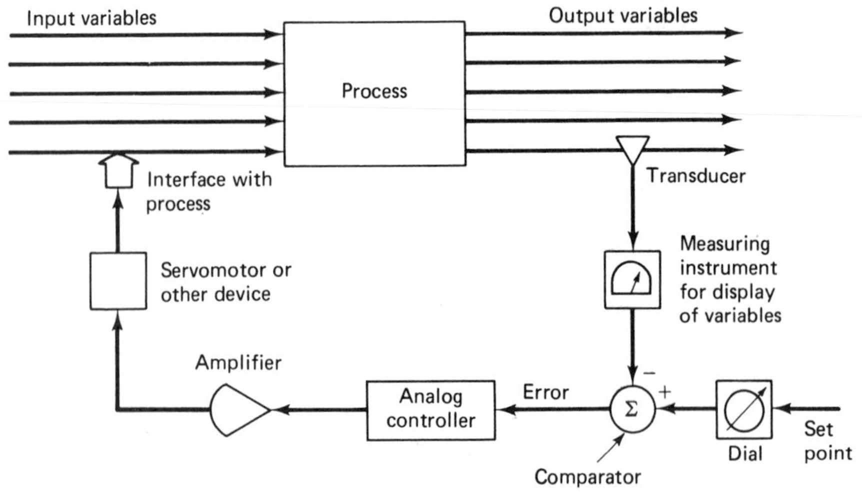


FIGURE 14.1 Typical control loop.

control, in which case a switching circuit would be used in place of the analog device.

The output of the analog control device is an electrical signal which must be interfaced with the process. Amplification of the signal may be required so that it can be utilized by a servomotor or other device. For example, the servomotor might alter the valve position in a fluid-flow process, either to increase or decrease the rate of flow to the process.

There are several features of analog control discussed in the preceding paragraphs which are relevant in this chapter on direct digital control:

1. The typical process has multiple control loops—in some cases, a large number of process variables must be controlled.
2. For each control loop, a set of hardware components, similar to those indicated in Figure 14.1, must be provided.
3. The analog control elements used in each control loop are capable of performing only a limited number of mathematical operations on the error signal [the types of operations are limited typically to those defined in Eq. (14.2)].
4. The analog controller is hardwired, meaning that the control actions performed by the analog controller cannot be easily changed. The gain factors $K_1$, $K_2$, and $K_3$ in Eq. (14.2) can be "tuned" to the process, but the basic operation of the controller cannot be altered except by rewiring.

## 14.3 DIRECT DIGITAL CONTROL

Considering the preceding features of analog process control, the idea behind direct digital control is to replace many of the individual hardware elements of each analog control loop with a digital computer operating in a time-sharing mode. Each control loop would be serviced by the same digital computer. Not all of the hardware components of the analog system can be replaced by the computer. Indeed, some of the control loops may be more appropriate for analog control. However, some of the elements illustrated in Figure 14.1 would be suitable candidates for replacement in DDC. Also, the introduction of direct digital control requires several additional system components not needed in an analog control system.

### Components of a DDC system

The components of the analog feedback loop that would be possible items for replacement by DDC components are:

1. *Analog controller.* Instead of performing the computations with the analog controller, the digital computer can be programmed to simulate the

analog control actions. The digital computer would be capable of more complex control functions than the typical proportional–integral–derivative control actions. Also, the control strategy could be changed if this were to become necessary, simply by reprogramming the digital computer.

2. *Recording and display components of the measuring instrument.* The measurement of the process variables can be displayed on a teletype or cathode ray tube (CRT) at the computer console with DDC.

3. *Set-point dial.* The individual process variable set points can be entered by the operator or by the technical supervisors at the computer console.

4. *Comparator.* Comparison of the desired set-point value against the measured process variable now becomes a calculation performed by the DDC computer.

The elements shown in Figure 14.1 which cannot be replaced are:

1. *Transducers and sensors.* The measurement of the process variables must still be taken by the appropriate sensors.

2. *Servomotors and process interface devices.* The hardware elements that make the adjustments in the input variables under computer command are still required. However, the nature of this equipment will be different. For example, instead of employing a servomotor, which is ideally suited for use with the analog controller, a stepping motor might be substituted. Other changes in process interface hardware might be appropriate for use with DDC, but the need for interface devices of some form is unavoidable.

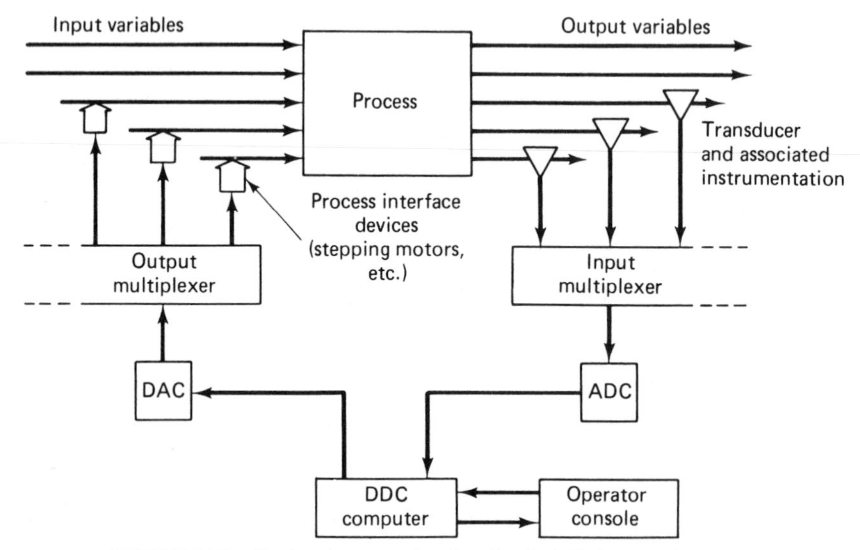


FIGURE 14.2 Basic components of a direct digital control system.

The additional hardware components required in a direct digital control system are as follows:

1. Analog-to-digital converters.
2. Digital-to-analog converters.
3. Multiplexers.

These devices were described in Chapter 11.

A block diagram showing the various components of a DDC system is presented in Figure 14.2. Several feedback loops are illustrated to demonstrate the concept of multiplexing.

## Control actions in DDC

In analog control, the mathematical operations are performed on continuous analog signals. With direct digital control, these operations must be approximated by finite-difference calculations. The finite-difference approximation of Eq. (14.2) is

$$x_{pn} = K_1 e_n + K_2 \Delta t \sum_{i=0}^{n} e_i + \frac{K_3}{\Delta t}(e_n - e_{n-1}) \tag{14.3}$$

where $x_{pn}$ = value of the process input as determined by the controller for time period $n$
$e_n$ = value of the error signal at time period $n$
$\Delta t$ = sampling interval = $t_n - t_{n-1}$
$K_1, K_2, K_3$ = constants

As the sampling interval $\Delta t$ becomes smaller, Eq. (14.3) more closely approximates the continuous function in Eq. (14.2). To put this timing into perspective, suppose that the control computations of Eq. (14.3) required 100 micro-seconds (0.0001s) to complete for each loop. Further, suppose that the DDC computer were required to service 100 control loops. How often would the computer be able to sample each loop? In other words, how long is the sampling interval $\Delta t$? The answer is that the computer would be capable of sampling each control loop every (100)(0.0001 s) = 0.01 s in theory. Some time would be lost by the multiplexer, ADC, and DAC, so the actual sampling interval would be slightly greater than this. Except for processes with extremely short time constants, this sampling interval would be more than adequate.

Let us examine how closely the finite-difference calculation used in DDC would approximate the operation of a continuous analog controller. Instead of considering the entire proportional–integral–derivative control, we will limit our example to proportional control only.

## EXAMPLE 14.1

In Example 13.10, we applied a proportional controller to the process pictured in Figure 13.21. Using a proportional gain factor of $K=2.0$, the continuous transfer function was

$$\frac{y(s)}{x(s)} = \frac{10}{s+3}$$

When a step function input of value $x=4.0$ was applied to this system, the resulting response was

$$y(t)=13.33(1-e^{-3t}) \tag{14.4}$$

Now consider the same process except that a direct digital controller is used for control rather than an analog controller. The abbreviated version of Eq. (14.3) which includes only proportional control would be

$$x_{pn}=K_1 e_n \tag{14.5}$$

The resulting DDC system is pictured in Figure 14.3. Note the similarities as well as the differences between this arrangement and the system of Figure 13.21. The feedback signal (equal to $0.1y$) is fed into an ADC and the value is compared with the set-point value of $x=4.0$. The calculated difference is the error value $e_n$. This error value is multiplied by 2.0, corresponding to the proportional gain factor. The resulting value is then converted back to analog form. A data hold circuit maintains this value throughout the time interval until the next sampling instant. The output of the data hold circuit is the input to the process, $x_{pn}$. Data hold circuits were described in Section 11.5. The response $y(t)$ is determined by the process transfer function $[5/(s+2)]$ acting on the step input $x_{pn}$. Since $x_{pn}$ changes value only at the end of each time interval instead of continuously as with analog control, the response $y$ will be different.

Let us use a value of $\Delta t=0.1$ s and assume that the starting value of $y$ is zero at $t=0$. What would be the difference in response between analog control and DDC?

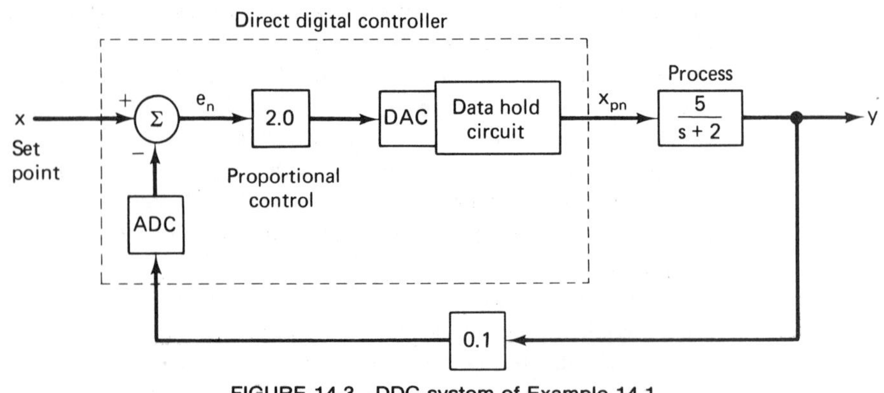


FIGURE 14.3 DDC system of Example 14.1

Under analog control, the value of $y$ at $t=0.1$ can be determined from Eq. (14.4):

$$y=13.33(1-e^{-0.3})=3.455$$

Throughout the time interval from $t=0$ to $t=0.1$ s, the value of $x_p$ continuously changes as the error term $e$ is gradually reduced from its initial value of

$$e=4-0.1y=4.0$$

Under direct digital control the response would be different because $x_p$ maintains the same value throughout the time interval $t=0$ to $t=0.1$. We can see how this would work by examining the output of each major component in the system of Figure 14.3. At the start, the input $x=4.0$ and $y=0$. Hence, the resulting error value would be $e_n=4.0$. After multiplying by 2.0 and converting to a continuous step value through the data hold circuit, $x_{pn}=8.0$. To the process, it is a (continuous) step function over the 0.1-s time interval. Hence, the value of $y(s)$ in the $s$-domain is

$$y(s)=\frac{8}{s}\frac{5}{s+2}=\frac{20}{s}-\frac{20}{s+2}$$

Finding the inverse transform of this function,

$$y(t)=20(1-e^{-2t})$$

At time $t=0.1$ s,

$$y=20(1-e^{-0.2})=3.625$$

which is different from $y=3.455$ obtained after 0.1 s using analog control. At the start of the second time interval ($t=0.1$ to $t=0.2$), the error signal in the analog control system has a value

$$e=4.0-0.1(3.455)=3.6546$$

Under DDC, the error has a value

$$e_n=4.0-0.1(3.625)=3.6375$$

Consider the response of the two alternative systems during the second interval. Under analog control the response, according to Eq. (14.4), would be

$$y=13.33(1-e^{-0.6})=6.014$$

With DDC, the proportional control algorithm calculates the value of $x_{pn}$ for the second interval to be

$$x_{pn}=2.0(3.6375)=7.275$$

This becomes the step input to the process which continues throughout the time interval $t=0.1$ to $t=0.2$ s. We leave it as an exercise for the reader to compute the response $y(t)$ during this interval. It should be noted that the initial conditions at the start of this second interval are $y=3.625$ and $dy/dt=32.75$.

A final comment is in order before we leave this example. With a time interval of $\Delta t=0.1$ s, the difference in response between the two systems is not significant. As the sampling interval is increased, the difference in response will become larger. Smaller $\Delta t$ values will bring the DDC response closer into agreement with the analog system response.

## 14.4 ADVANTAGES AND DISADVANTAGES OF DDC

There are several advantages which direct digital control is alleged to possess over analog control. Some of these advantages have been mentioned in previous sections of this chapter.

1. *More sophisticated control.* It is easier to program a more complex control strategy with DDC than with analog elements. Use of the digital computer also facilitates safety checks, trend analyses, and other tests deemed desirable.
2. *Control flexibility.* The computational algorithms in the digital control computer can more easily be reprogrammed to "fine-tune" the overall process operation.
3. *Equipment costs.* With timesharing, the installation costs of a DDC system should be less than analog control for a process with multiple feedback loops. Conceptually, there should be a break-even point in terms of the number of control loops needed to justify the use of DDC. If the number of required control loops exceeds this break-even point for a given process, direct digital control would have the advantage. If the number of feedback loops is small, analog control would be more economical.
4. *Reliability.* The components used in a direct digital control system are generally regarded to have better reliability than their analog counterparts.

Unfortunately, not all of these advantages accrue to every installation of direct digital control. Among the disadvantages and problems which were encountered in many of the early attempts in DDC are the following:

1. *Requirements for backup system.* Although the components used in DDC may be more reliable, the effect of a total computer failure on the operation of the process could be catastrophic. As a result, DDC systems are often designed with an analog control backup system, thus eliminating any hardware savings that might have resulted.
2. *Computer programming costs.* In many of the early DDC installations, the manpower efforts required to program and "tune" the system

exceeded the estimated requirements. As a result, the actual installation costs of the system were larger than expected. It seems to be very common that software development costs for computer systems are underestimated.

**EXAMPLE 14.2**

An example of direct digital control is described by Parsons et al.[1] The plant is an ethylene plant of the Phillips Petroleum Company located at Sweeny, Texas. The plant produces 500 million pounds of product annually. A total of 180 control loops are required to operate the plant. In the installation of DDC, it was decided to control 120 of these loops by computer, leaving 60 control loops for analog control. Since the operations personnel did not yet have complete confidence in the computer control system, the 60 loops were left on analog control as a precaution against complete breakdown of the DDC system. Also, approximately 25 of the 120 DDC loops had analog backup for the same reason. This arrangement was considered to be a minimum control configuration to keep the plant operating in the event of a computer failure.

The technical performance of the DDC system was rated a success. The plant experienced only 13 h of downtime in 32 months of DDC operation. However, from an economic standpoint, only a marginal return was achieved from the DDC installation compared to an equivalent analog system.

The practice of simply using the digital computer to imitate the characteristics of analog controllers seems to represent a transitional phase in computer process control. Direct digital control alone is difficult to justify in terms of reduced costs. However, the use of the computer in process control applications can be supported by improvements in the overall performance of the manufacturing operation. It is to this problem of overall process improvement and optimization that we address ourselves in Chapter 15.

## REFERENCES

[1] Cadzow, J. A., and Martens, H. R., *Discrete-Time and Computer Control Systems*, Prentice-Hall, Inc., Englewood Cliffs, N.J., 1970.

[2] Lee, T. H., Adams, G. E., and Gaines, W. M., *Computer Process Control: Modeling and Optimization*, John Wiley & Sons, Inc., New York, 1968, Chapter 3.

[3] Pike, H. E., "A Survey of Direct Digital Control," General Electric Company Report No. 67-C-477, 1967.

[4] Raven, F. H., *Automatic Control Engineering*, McGraw-Hill Book Company, New York, 1978, Chapter 10.

[1]J. R. Parsons, M. W. Ogleby, and D. L. Smith, "Performance of a Direct Digital Control System," presented at the 25th Annual ISA Conference, Philadelphia, October 26–29, 1970, summarized with permission of Instrument Society of America, © 1970.

[5] SAVAS, E. S., *Computer Control of Industrial Processes*, McGraw-Hill Book Company, New York, 1965, Chapter 11: "Direct Digital Control" by J. F. Hornor.

[6] SMITH, C. L., *Digital Computer Process Control*, International Textbook Co., Scranton, 1972, Chapter 1.

[7] WARE, W. E., "Direct Digital Control," *Instruments and Control Systems*, Vol. 38, June, 1965, pp. 79–83.

[8] WILLIAMS, T. J., "What to Expect from Direct Digital Control," *Chemical Engineering*, March 2, 1964, pp. 97–104.

## PROBLEMS

**14.1.** Solve Example 14.1 but use a sampling time interval of $\Delta t = 0.2$ s instead of 0.1 s. Compare $y(t)$ at $t = 0.2$ s under DDC and under analog control.

**14.2.** From Example 14.1, solve for the response $y(t)$ during the time interval between $t = 0.1$ s and $t = 0.2$ s. Use the sampling interval $\Delta t = 0.1$ s.

**14.3.** Use the finite-difference approximation of an analog integral controller to solve Example 13.10 in a manner similar to that of Example 14.1. The discrete approximation of an integral controller is

$$x_{pn} = K_2 \Delta t \sum_{i=0}^{n} e_i$$

Use values of $K_2 = 2.0$ and $\Delta t = 0.1$ s. Compare your results for the DDC response with that of the analog integral controller of Example 13.10. Use just one time interval.

**14.4.** The computer for a proposed direct digital control system has an initial cost of \$35,000. It is expected that software costs for the system will be about \$20,000 in the first year and \$3000 each year thereafter. It is to be used on a new industrial process that will have approximately 200 control loops. It is estimated that the savings per control loop resulting from the use of DDC will be \$600. Annual maintenance costs of the DDC system will be about \$5000, while annual maintenance costs of analog control system will be an average of \$40 per control loop—\$8000 for a complete analog system. The rate of return the company uses for investments of this type is 10%. Determine the number of control loops that must be installed with direct digital control if the DDC system is to pay for itself. (In other words, determine the break-even point.) Allow a 3-year period for the analysis.

# Supervisory Computer Control

## 15.1 INTRODUCTION

The two previous chapters were concerned with the problem of controlling the individual feedback loops of the process. By regulating each separate process variable, control of the overall process can be achieved. Control is obtained over each feedback loop by means of a set point for the loop. The question which then arises is: How are the set-point values determined?

### Supervisory computer control defined

The answer to this question leads us to define the term "supervisory computer control." Supervisory computer control denotes a computer process control application in which the computer determines the appropriate set-point values for each control loop in order to optimize some performance objective of the entire process. The performance objective of the process might be maximum production rate, minimum cost per unit of product, yield, or some other objective that pertains to the process. Based on a mathematical model of the process which is programmed into the computer, the computer calculates the set-point values that optimize the objective function. Adjustments in the set-points are then implemented in the control loops of the process in either of two ways:

1. *Analog control.* If the individual feedback loops are controlled by analog devices, the control computer is connected to these devices. The set-point adjustments are made through the appropriate interface hardware between the computer and the analog elements.
2. *Direct digital control.* If the feedback loops operate under direct digital control, the supervisory control program provides the set-point values to the DDC program. Both the supervisory control program and the direct digital control program can be contained in the same computer, or they can be in separate computers in a hierarchical configuration.

In addition to set-point adjustments in the control loops, the supervisory computer may also be required to exercise control over certain discrete variables in the process. Examples of this function would include starting or stopping motors, opening valves, setting switches, solenoids, and so on. This type of control is one we have previously called "on/off" control, in which the variables can be in either of two possible states. When regulation of the industrial operation consists exclusively of performing a sequence of these on/off steps in a predetermined order, this type of control is sometimes called *sequencing control*. We shall consider sequencing control to be a subset of supervisory control. Most industrial operations contain a mixture of analog and discrete variables. Accordingly, the control computer is called on to perform a combination of sequencing control and set-point control.

### Where to use supervisory computer control

Supervisory computer control has frequently been associated with the process industries: chemicals, petroleum, steelmaking, and so on ([12], [13]). In these applications, the processes to which supervisory control has been most successfully applied have been large and complex. In these cases, the volume of production is large enough and the efficiency of the process under conventional control is poor enough to justify the cost of installing the computer and developing the required software.

In recent years, the use of the digital computer in process monitoring and control applications has expanded to include many production areas outside the process industries. One of the biggest growth areas has been in discrete-parts manufacturing: metal machining, pressworking, electronic components manufacturing, assembly, and so on. Many of these applications are limited to monitoring functions, but in other cases the potential has been demonstrated for using the computer for control and optimization ([3], [11],

[15]). The characteristics of the production processes are different in discrete-parts manufacturing than in the process industries. In discrete-item production, the output is measured in number of parts rather than in gallons or tons. The operations are typically less complex with fewer variables. The processing time is usually of short duration compared to the process industries.

Along with the differences there are similarities between the processes in discrete-parts manufacturing and in continuous-process industries. In both cases, there are economic objectives to be achieved. Many manufacturing operations are complex enough that manual optimization methods are inadequate. For example, the metal-cutting operation is characterized by a fairly large number of process variables which determine the overall performance of the operation. These include cutting speed, feed, depth of cut, tooling, and raw materials. Although the value added per operation in discrete-item production is less than in continuous processing, the total volume of production in many manufacturing plants is significant enough and the productivity low enough to justify an investment in computer control.

Two factors contribute to the appeal of the computer in discrete-parts manufacturing. First, the physical size and cost of the digital computer have decreased dramatically in recent years. Today, it is not as difficult to justify the investment in computer hardware for process control as it was 10 to 15 years ago. Second, experience in developing software for process control has grown, thus reducing the programming costs for new process control projects. In companies with large numbers of machine tools, the same process control software can be adapted to more than one machine, so that the cost of software development per machine becomes affordable.

The basic guideline in deciding where to use the computer for supervisory and optimal control is this: wherever the cost of computer hardware and software can be justified through an improvement in process performance.

## Structural model of a manufacturing process

In Chapter 13, our attention was directed at the development and use of mathematical models of physical processes. In the current chapter we are more concerned with the economic performance of an industrial process rather than with its physical behavior.

To provide a framework for studying the various strategies that can be used in supervisory computer control, let us examine the technological structure that typifies nearly all manufacturing processes.

Most production operations are characterized by a multiplicity of dynamically interacting process variables. These variables can be cataloged into two basic types, input and output variables. However, there are different kinds of input variables and the same is true for output variables.

First, consider how input variables might be classified:

1. *Controllable input variables.* These are sometimes called *manipulative variables*, because they can be changed or controlled during the process. In a machining operation, it is technologically possible to make adjustments in speed and feed during the operation. Not all machine tools possess this capability. In a chemical process, the controllable input variables may include flow rates, temperature settings, and so on.

2. *Uncontrollable input variables.* Variables that change during the operation but which cannot be manipulated are defined as uncontrollable input variables. In chemical processing, the starting raw chemicals may be an uncontrollable input variable for which compensation must be made during the process. In machining, examples would be tool sharpness, work-material hardness, and workpiece geometry. The reader may argue that some of these examples can be controlled. However, from a control systems viewpoint, it is more appropriate to consider them as uncontrollable during the process.

3. *Fixed variables.* A third category of input to the process is the fixed variable. These are conditions of the setup, such as tool geometry and workholding device, which can be changed between operations but not during the operation. Fixed inputs for a continuous chemical process would be tank size, number of trays in a distillation column, and other factors that are established by the equipment configuration.

It should be evident to the reader that the classification of variables will be different for different processes. For example, feed rate may be a controllable variable for one machine tool and a fixed variable for another machine tool. It depends on the capability of the machine.

The other major type of variable in a manufacturing process is the output variable. It is convenient to divide output variables into two types:

1. *Measurable output variables.* The defining characteristic of this first type is that it can be measured on-line during the process. Examples in a machining operation would be tool forces, vibration, power, and temperature. Other output variables that cannot be measured on-line, at least not with the current state of sensor technology, are surface finish and tool wear.

2. *Performance evaluation variables.* These are the measures of overall process performance and are usually linked to either the economics of the process or the quality of the product manufactured. Examples of performance evaluation variables in machining might be: cost per unit produced, production rate, some given measure of product quality such as surface finish or part size, and so on. Examples in the continuous-process industries include yield, cost per gallon, and cost per ton.

The structural relationships between these different input and output variables are illustrated in Figure 15.1. The measurable output variables are

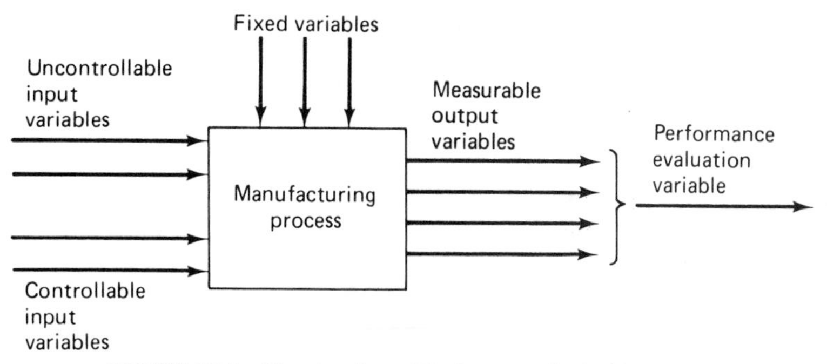


FIGURE 15.1 Structural model of a manufacturing process.

determined by the input variables. The performance of the process, as indicated by the performance evaluation variable, is determined by the measurable output variables. To assess process performance, the performance evaluation variable must be calculated from measurements taken on the output variables.

In feedback control, whether implemented by analog devices or DDC, the objective is to regulate the measurable output variables. In supervisory computer control, the objective is to control and optimize the performance evaluation variable for the process. This general hierarchy of control is illustrated schematically in Figure 15.2.

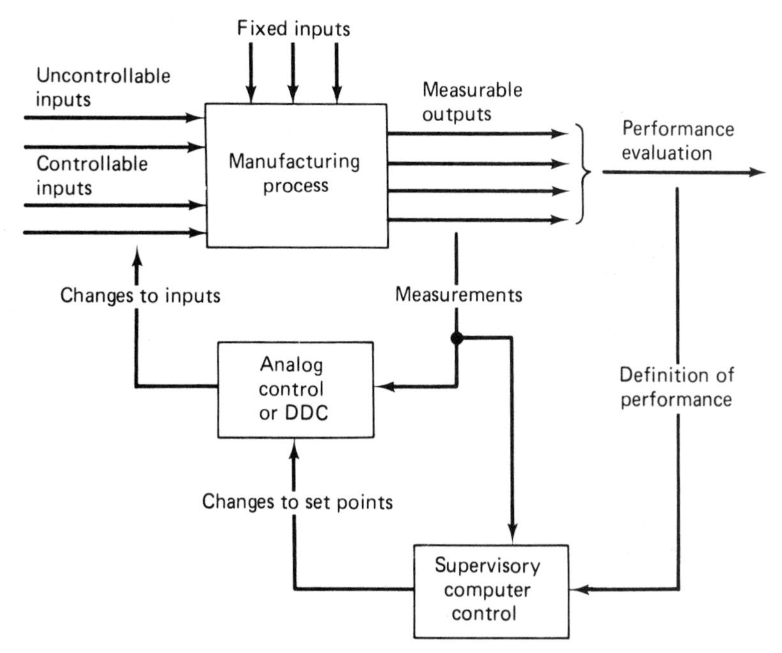


FIGURE 15.2 Hierarchy of supervisory control and feedback control.

Within the structural model of a manufacturing process, there are usually constraints or limits on the values the variables can assume. For example, there are practical limits on the rate of flow through a pipe of a given diameter, there is a maximum possible spindle speed for a given machine tool, and there are limits on the forces that can be endured by the moving elements of a piece of processing equipment. These constraints impose limits on the achievement of optimal performance in a manufacturing process.

## 15.2 CONTROL STRATEGIES IN SUPERVISORY COMPUTER CONTROL

There are a variety of control strategies that can be employed in supervisory computer control. The choice of strategy depends on the process and the performance objectives to be achieved through process control. In this section, we shall discuss the following control strategies:

1. Regulatory control.
2. Feedforward control.
3. Preplanned control.
4. Steady-state optimal control.
5. Adaptive control.
6. On-line search techniques.

In subsequent sections of this chapter we will examine more closely strategies 4, 5, and 6.

### Regulatory control

It is sometimes sufficient in the control of a complex industrial process to maintain the performance evaluation variable at a certain level, or within a given tolerance band of that level. This would be appropriate in situations where performance was measured in terms of product quality, and it was desired to maintain the product quality at a particular level. In a chemical process, this quality level might be the concentration of the final chemical product. The purpose of supervisory control would be to maintain that quality at the desired constant value during the process. To accomplish this purpose, set points would be determined for individual feedback loops in the process and other control actions would be taken to compensate for disturbances to the process. Regulatory control is analogous to feedback control except that feedback control applies to the individual control loops in the process. We are using the term "regulatory control" to describe a similar control objective for the overall process performance.

### Feedforward control

The trouble with regulatory control (the same problem is present with feedback control) is that compensating action is taken only after a disturbance has affected the process output. As indicated in Chapter 13, an error must be present in order for any control action to be initiated. An error means that the output of the process is different from the desired value.

In feedforward control the disturbances are measured before they have upset the process, and anticipatory corrective action is taken. In the ideal case, the corrective action compensates completely for the disturbance, thus preventing any deviation from the desired output value. If this ideal can be reached, feedforward control represents an improvement over feedback control.

The essential features of a feedforward control system are illustrated in Figure 15.3. The feedforward control concept can be applied to the individual measurable output variables in the process or to the performance evaluation variable for the entire process. The disturbance is measured and serves as the input to the feedforward control elements. These elements compute the necessary corrective action to anticipate the effect of the disturbance on the process. To make this computation, the feedforward controller contains a mathematical or logical model of the process which includes the effect of the disturbance. Feedforward control by itself does not include any mechanism for checking that the output is maintained at the desired level. For this reason, feedforward control is usually combined with feedback control as pictured in Figure 15.3. The feedforward loop is especially helpful when the

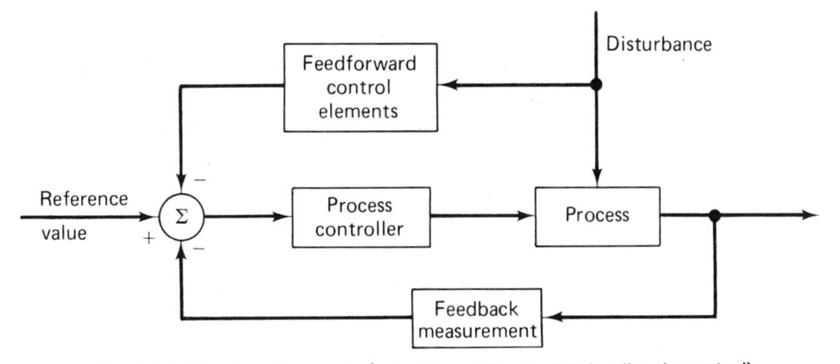


FIGURE 15.3 Feedforward control (combined with feedback control).

process is characterized by long "response times" or "dead times" between inputs and outputs. Feedback control alone would be unable to make timely corrections to the process.

### Preplanned control

The term *preplanned control* refers to the use of the computer for directing the process or equipment to carry out a predetermined series of operation steps. The control sequence must be developed in advance to cover the variety of processing conditions that might be encountered. This control strategy usually requires the use of feedback control loops to make certain that each step in the operation sequence is completed before proceeding to the next step. However, feedback information may not be necessary in every control command provided by the computer. In some cases, preplanned control does not qualify as a type of supervisory control because there may not be an explicit economic objective to be optimized. However, the objective of improved performance is implicit in the preplanned control strategy.

The name "preplanned control" is not universally applied throughout all areas of industry. Other terms are used to describe control strategies which are either identical or similar to preplanned control. What follows is a listing of some of the terms most frequently used.

COMPUTER NUMERICAL CONTROL. CNC was described in Chapter 9. Essentially, it involves the use of the computer to direct a machine tool through a program of processing steps. As such, it is a form of preplanned control. Direct numerical control (DNC), although not the same as CNC, involves a similar control sequence.

PROGRAM CONTROL. This term is used in the process industries. It involves the application of the computer to start up or shut down a large complex process, or to guide the process through a changeover from one product grade to another. It also refers to the computer's use in batch processing to direct the process through the cycle of processing steps.

With program control the object is to direct the process from one operating condition to a new operating condition and to accomplish this in minimum time. There are often constraints on this minimum time objective, so the strategy of program control is to determine the best trajectory of set-point values that is compatible with the constraints. In batch processing, there may be a sequence of operating conditions or states through which the process must be commanded.

The paper industry provides an example of program control. In the manufacture of various grades of paper, a slightly different operating cycle is required for each grade. The process control computer is programmed to govern the process through each phase of the operating cycle for any grade of paper produced.

SEQUENCING CONTROL. This class of preplanned control consists of guiding the process through a sequence of on/off type steps. The variables under command of the computer can take on either of two states, typically "on" or "off." In sequencing control, the process must be monitored to make sure that each step has been carried out before proceeding to the next step.

An example of the application of sequencing control is in automated production flow lines. The sequence of workstation power feed motions, parts transfer, quality inspections which may be incorporated in the line, and so on, are all included under computer control. In addition, the computer may be programmed to perform diagnostic subroutines in the event of a line failure, to help identify the cause of the downtime occurrence. Tool change schedules may also be included as one of the computer functions. The operators are directed by the computer when to change cutters. The alternative to the use of digital computers for sequencing controls is to use electromechanical relays, which require considerably larger floor space. One of the disadvantages of the electromechanical relay control is that extensive rewiring is needed to accommodate model changeovers. With the digital computer, the sequencing logic can be reprogrammed with relative ease.

### Steady-state optimal control

The term "optimal control" refers to a large class of control problems. We shall limit its meaning in this discussion to open-loop systems. That is, there is no feedback of information concerning the output. Instead, two features of the system must be known in advance. First, the measure of system performance must be defined. This is sometimes termed the *objective function* or index of performance or similar name. In essence, it can be considered analogous to our previously defined "performance evaluation variable." Second, the mathematical relationships between the output and the input must be defined. Hence, the solution of the control problem involves finding the value (or values) of the input variable (or variables) that optimizes the process performance. We shall discuss the steady-state optimal control problem further in Section 15.3.

### Adaptive control

In a sense, adaptive control represents a combination of feedback control and optimal control. As with optimal control, an index of performance must be defined for the system which indicates overall system performance. As with feedback control, measurements of the process are taken to determine adjustments in the controllable input variables.

In control theory, an adaptive control system is one which operates in an environment that changes over time in an unpredictable fashion. The system must compensate for this unpredictable environment by monitoring its own performance and regulating some portion of its control mechanism to improve the performance. Adaptive control is sometimes referred to as a *self-optimizing system*, which reflects the blend of both feedback and optimizing features. Adaptive control will be considered in Section 15.4.

### Search techniques

Adaptive control requires a fairly well-defined process model in order to work. The general relationships must be known among the performance evaluation variables, the measurable outputs, and the controllable inputs to the process.

If the process model cannot be formulated, owing to lack of knowledge about the process, a systematic search procedure can be used to try to optimize the performance. With the typical search routine, the performance variable is evaluated at the current operating conditions. Then, exploratory moves are made in the operating conditions to determine what changes will improve the process performance. These changes are carried out and new exploratory moves are made. The procedure is repeated until no further improvements can be made. We will examine several of the search techniques in Section 15.5.

## 15.3 STEADY-STATE OPTIMAL CONTROL

In the previous section, optimal control was classified as an open-loop control system. What this means is that there is no feedback measurement from the performance evaluation variable for the process. The optimal control problem must therefore be solved based on the following well-defined attributes of the system:

1. *Performance evaluation variable.* This measure of system performance is also called the objective function, index of peformance, or figure of merit. Basically, it represents the overall indicator of process performance that we desire to optimize by solving the optimal control problem. Among the performance objectives typically used in optimal control are cost minimization, profit maximization, production-rate maximization, quality optimization, least-squares-error minimization, and process-yield maximization. These objectives are general and must be specified to suit the particular application.
2. *Mathematical model of the process.* The relationships between the input variables and the objective function (measure of process performance) must be mathematically defined. The model is assumed to be valid throughout the operation of the process. That is, there are no disturbances that might affect the final result of the optimization procedure. This is why we refer to the problem as steady-state optimal control. The mathematical model of the process may include constraints on some or all of the variables.

These constraints limit the allowable region within which the objective function can be optimized.

With these two attributes of the process defined, the solution of the optimal control problem consists of determining the values of the input variables that optimize the objective function. To accomplish this task, a great variety of optimization techniques are available to solve the steady-state optimal control problem. The following list of optimization techniques demonstrates the wide variety of methods that are available. Suitable references to each technique are cited:

Differential calculus, [19].
Lagrange multiplier technique, [19].
Linear programming , [16] and [17].
Geometric programming, [19].
Dynamic programming, [1].
Calculus of variations, [6].
The optimum principle, [19].

These approaches have all been applied to problems in the class we are calling steady-state optimal control problems. Some of the approaches can be applied manually. Others are more appropriate for computer solution when employed to solve problems of practical size.

It is not feasible (nor is it within the scope of this book) to present all of these techniques in this section. The objective of our discussion on steady-state optimal control is to survey some of the methods listed to demonstrate the general problem area. The interested reader could devote a career to the study of optimization techniques. We have only a few pages to devote.

## Use of differential calculus

The steady-state optimal control problem and its solution can be demonstrated by means of the differential calculus. The problem is to solve some function for its maximum or minimum value. Suppose that $z$ is the function of $x$ we wish to maximize.

$$z = f(x) \tag{15.1}$$

It is known that $f(x)$ possesses a maximum and that the function is unimodal —that is, it has but a single peak. Then, to find the value of $x$ that maximizes $z$, take the partial derivative of $z$ with respect to $x$ and set the derivative equal to zero.

$$\frac{dz}{dx} = 0 \tag{15.2}$$

Solve Eq. (15.2) for $x$. The application to process control can be illustrated by an example. A minimization problem will be used rather than one in maximization, as above.

### EXAMPLE 15.1

The process of interest is shown in Figure 15.4 as a very elementary block diagram. The transfer function for the process is

$$\frac{y(s)}{x(s)} = \frac{50}{s+10} = \frac{5}{0.1s+1}$$

which shows that the dynamics of the process are represented by a first-order lag.

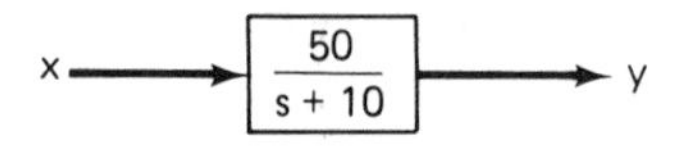


FIGURE 15.4 Process of Example 15.1.

The objective function for this process is to minimize $z$, where $z$ is a least-squares-error function. That is, the objective is to

$$\text{minimize } z = (Y-y)^2$$

where $z$ = objective function
$y$ = output variable of the process
$Y$ = desired steady-state value of $y$, which we will assume is 20; $Y=20$

Note that the two attributes of the steady-state optimal control problem are present: the measure of system performance, $z$, and the mathematical model of the process expressed as the transfer function, $y(s)/x(s)$.

Our concern is with the steady-state solution, so we will use an input $x$ to the process which has a constant value. To achieve this constant value, we will assume a step function input. Hence,

$$x(s) = \frac{x}{s}$$

where $x$ becomes the unknown constant value. Hence,

$$y(s) = \frac{5x}{s(0.1s+1)}$$

According to the final value theorem [Eq. (13.44)], the steady-state value of $y$ is

$$y(t=\infty) = \lim_{s\to 0} \frac{5xs}{s(0.1s+1)} = 5x$$

The objective is to

$$\text{minimize} \quad z = (20-y)^2$$
$$z = (20-5x)^2$$
$$z = 400 - 200x + 25x^2$$

This is the specific function of $x$ that was expressed in general form as Eq. (15.1). To find the value of $x$ that minimizes $z$, Eq. (15.2), is applied.

$$\frac{dz}{dx}=0-200+2(25)x=0$$
$$50x=200$$
$$x=4$$

Since $x$ is the independent input variable, we can now set $x=4$ and thereby minimize the performance criterion. Note that we have done so in an open-loop fashion (no feedback information about $y$ was used in setting $x=4$).

Stripping the problem to its bare essentials, we have a steady-state process model

$$y=5x$$

and an objective to make $y=20$. Given these two pieces of information, it seems quite obvious that $x$ must equal 4. Perhaps the reader perceived this during the statement of the problem. If so, Example 15.1 may have seemed like a rather involved procedure to determine an answer that was obvious from the start. Most optimal control problems are not as simple as this one. What we have attempted to demonstrate here is the general structure of the optimal control problem, not its many potential complexities.

## EXAMPLE 15.2

For a second illustration, still using the differential calculus, we will borrow an example problem from the book on optimization by Wilde and Beightler.[1] The example is a hypothetical manufacturing plant that produces a certain chemical by means of the five-stage process shown in Figure 15.5. Raw-material gases are mixed with recycled unreacted gases and brought up to operating pressure in a compressor. The mixture passes through a reactor, where the chemical product is produced, and is then isolated in the separator stage. The unreacted gases are then returned to the mixing stage by a recirculating compressor. The two variables that control the performance of the process are:

$x_1$ = operating pressure (measured in atmospheres) achieved in the second stage
$x_2$ = recycle ratio, which is the ratio of recirculated unreacted material to raw material entering the process

In the operation of this plant the objective is to minimize the total annual operating cost $z$. This cost includes direct operating expenses, plant and equipment capital costs, and so on. The influence of $x_1$ and $x_2$ on this annual cost can be summarized in the

[1]D. J. Wilde, and C. S. Beightler, *Foundations of Optimization*, Prentice-Hall, Inc., Englewood Cliffs, N.J., 1967, pp. 11–21.

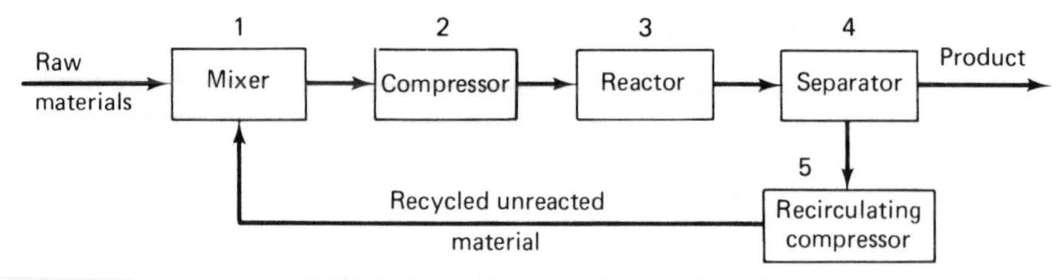


FIGURE 15.5 Process of Example 15.2.

following relationship:

$$z = 1000x_1 + \frac{4(10^9)}{x_1 x_2} + 2.5(10^5)x_2$$

Of course, the annual cost would also be determined by the amount of raw materials, but it is assumed that the plant will produce a certain quantity of final product, so the quantity of raw materials is predetermined.

The differential calculus approach for the two variable process would be to take the partial derivatives of the objective function $z$ with respect to $x_1$ and $x_2$ and set them equal to zero. This would provide two equations in two unknowns, $x_1$ and $x_2$, which could be solved to find the optimum set of operating conditions.

$$\frac{\partial z}{\partial x_1} = 1000 - 4(10^9)/x_1^2 x_2 = 0$$

$$\frac{\partial z}{\partial x_2} = -4(10^9)/x_1 x_2^2 + 2.5(10^5) = 0$$

Solution of the two equations gives

$$x_1 = 1000 \quad \text{and} \quad x_2 = 4$$

If this were a real problem in process control, the difficult part of its solution would not be in determining the optimal values of the input variables. The difficulty would lie in developing the underlying mathematical model for the process. This difficulty is a common one in supervisory process control situations. Great effort and expense are often required to acquire the data necessary to derive an accurate model of a given process.

## Linear programming

We will conclude the discussion of steady-state optimal control with a technique that is normally associated with the field of operations research rather than control systems. However, the linear programming approach is ideal for large-scale systems where the problem is to allocate limited resources in the pursuit of some linear objective function. The technique can be used either for maximization problems or minimization problems.

All linear programming problems possess the following characteristics:

1. *Linear objective function.* The objective in the problem is to maximize or minimize some linear objective function

$$z = a_1x_1 + a_2x_2 + \cdots + a_nx_n \tag{15.3}$$

2. *Linear constraints.* The optimization of the objective function is constrained by limited resources or other requirements of the problem. These limitations are expressed in the form of linear constraints, which are of three basic types:

a. Less-than-or-equal-to constraints:

$$b_1x_1 + b_2x_2 + \cdots + b_nx_n \leqslant b_0 \tag{15.4}$$

b. Greater-than-or-equal-to constraints:

$$b_1x_1 + b_2x_2 + \cdots + b_nx_n \geqslant b_o \tag{15.5}$$

c. Equal-to constraints:

$$b_1x_1 + b_2x_2 + \cdots + b_nx_n = b_0 \tag{15.6}$$

In any given linear programming problem, there may be a mixture of the three types. However, less-than-or-equal-to constraints are usually associated with maximization problems and greater-than-or-equal-to constraints are usually associated with minimization problems.

3. *Nonnegativity requirement.* The solution variables must take on either positive or zero values. Negative values are not allowed.

Within this framework there is a wide variety of different application areas, including problems that fall within our definition of steady-state optimal control. There are also various solution methods that can be applied to the linear programming problem, some of which are appropriate for the digital computer. We shall illustrate the linear programming problem by means of a simple example that contains only two variables. Realistic problems would have many variables. The reason for selecting a two-variable problem is so that the solution can be obtained by a graphical method. Readers interested in pursuing the more powerful approaches, such as the simplex method, should consult a text on operations research ([14], [16], and [17]).

### EXAMPLE 15.3

One of the most familiar problems in linear programming is the "product-mix" problem. Let us construct a fictional manufacturing control problem which involves the use of several limited resources to try to maximize profit. The process under consideration is greatly oversimplified. It consists of two stages, as shown in Figure 15.6. Stage A is an automated casting machine into which flows the raw material, a certain grade of metal. The output of stage A consists of the two base parts for the two products made in this process. The two products will be identified simply by numbers 1 and 2. Stage B is an assembly line where manual operators perform several machining and assembly operations to each of the products.

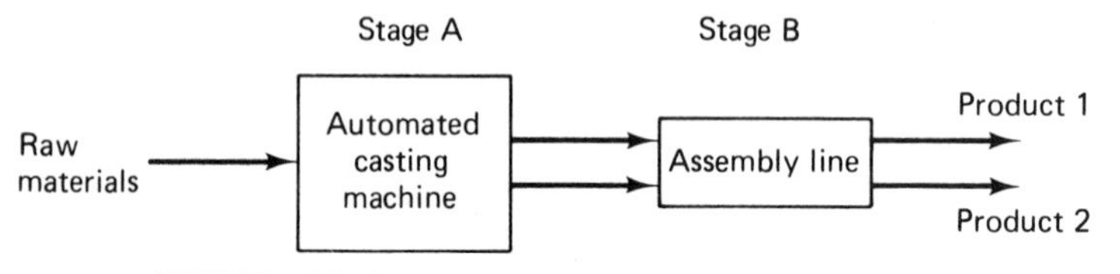


FIGURE 15.6 Manufacturing process of Example 15.3.

A profit can be made by the company from either of the two products. As close as the firm's accounting department can figure, the unit profit on product 1 is \$10.00 and the unit profit on product 2 is \$20.00. Hence, the total profit enjoyed by the company from operating this two-stage manufacturing process is

$$z = 10x_1 + 20x_2$$

where $x_1$ = number of units of product 1 produced
$x_2$ = number of units of product 2 produced

The objective, of course, is to maximize the total profit, represented by $z$.

The production manager is smart enough to realize that profit will not necessarily be maximized simply by manufacturing only product 2. More labor resources are required for each unit of product 2 produced. Following are the constraints on the problem. First, 1 unit of raw material is needed for each unit of product 1, and 1 unit of raw material is needed for each unit of product 2. The total amount of raw material that can be processed through stage A each day is 9 units. This constraint can be expressed mathematically as

$$1x_1 + 1x_2 \leq 9$$

Next, there is a labor constraint on the assembly line. One hour of labor is required for each unit of product 1, and 3 labor-hours are required for each unit of product 2. Total labor-hours available per day is 15 (two persons at $7\frac{1}{2}$ h each). This can be written

$$1x_1 + 3x_2 \leq 15$$

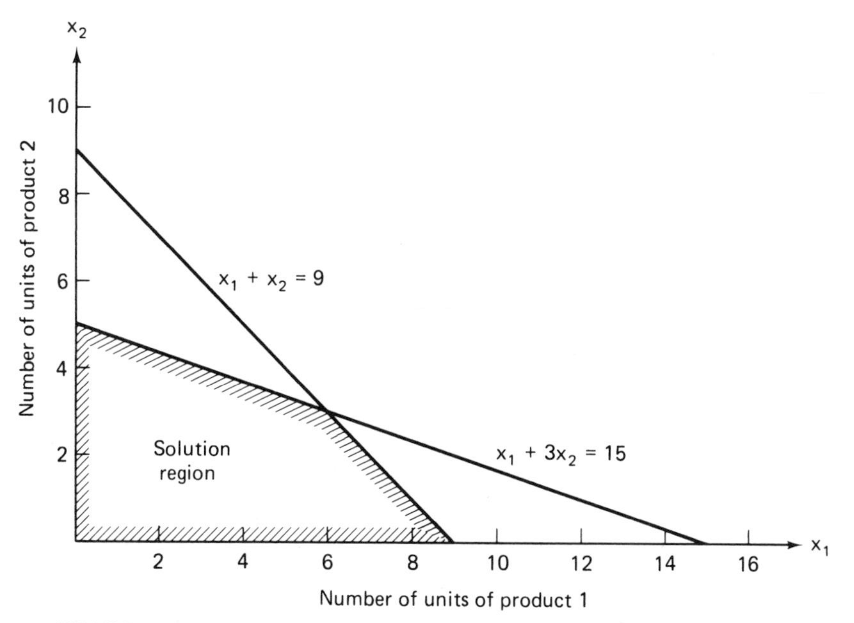


FIGURE 15.7 Two constraints plotted to form the allowable solution region for Example 15.3.

In the graphical method of solving the linear programming problem, these two constraints are plotted as the two lines

$$x_1 + x_2 = 9$$

$$x_1 + 3x_2 = 15$$

The two lines are shown in the graph of Figure 15.7. Since both of the constraints are of the less-than-or-equal-to type, expressed by Eq. (15.4), the area to the left and below these two lines constitutes the allowable region within which the solution must lie. The values of $x_1$ and $x_2$ must be positive, so the solution falls within the crosshatched area shown in Figure 15.7.

The objective of the problem is to find the combination of $x_1$ and $x_2$ values that maximizes

$$z = 10x_1 + 20x_2$$

To determine this maximum point, a series of constant-profit lines can be drawn on the same graph as the constraints. This is illustrated in Figure 15.8. Constant-profit lines have the same value of $z$ for all combinations of $x_1$ and $x_2$ on the line. When the constant-profit lines are superimposed on the constraint region, it can be seen that the optimum point is at $x_1 = 6$ and $x_2 = 3$ when $z = \$120$ per day. No other combination of $x_1$ and $x_2$ will yield a higher profit. To test this statement, we might try the largest possible $x_2$ value ($x_2 = 5$) within the convex polygon that defines the allowable region.

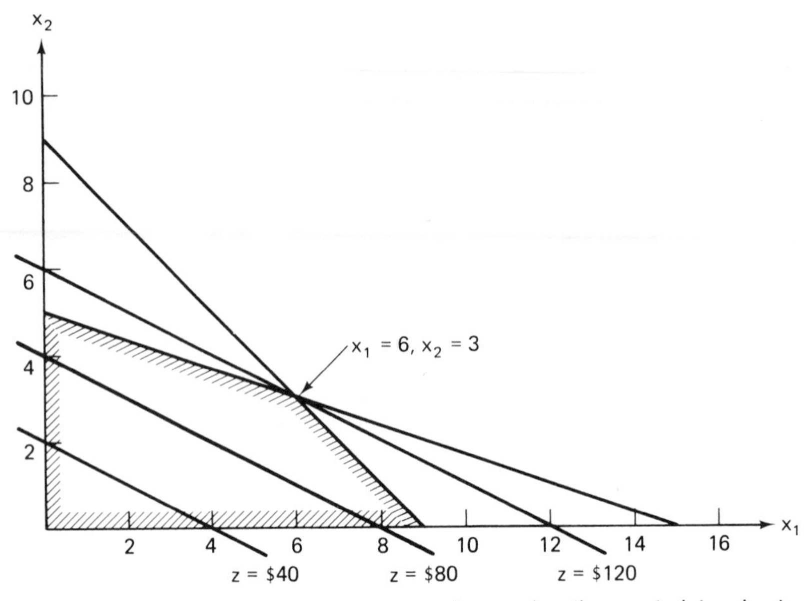

FIGURE 15.8 Constant-profit lines superimposed on the constraint region to give the solution at $x_1=6$, $x_2=3$, and $z=\$120$.

If $x_2=5$, then $x_1=0$, and the profit is

$$z=10(0)+20(5)=\$100$$

which is less than the previous value of $z=\$120$ found in the graphical method.

Although this example may be a somewhat trivial problem in linear programming, the L.P. approach has been applied to a significant number of real-life problems. For the interested reader, Whitehouse and Wechsler [17] provide a survey of linear programming application studies.

As indicated by the techniques and examples in this section, problems in steady-state optimal control do not use feedback to check on the value of the objective function. This is assumed to be unnecessary in our definition of steady-state optimal control. We next turn our attention to control systems whose objective is optimization but whose output must be monitored in order for the objective to be reached. This type of system is referred to as adaptive control.

## 15.4 ADAPTIVE CONTROL[2]

In Section 9.4, the subject of adaptive control machining was presented as a logical extension of numerical control. Applications of adaptive control are not limited to the machining process. Entire volumes have been written on

[2]This section is adapted largely from Groover [7].

the subject of adaptive control theory and methodology ([4], [6]). In this section, we survey some of the principles and techniques, supplementing the discussion with several examples.

Adaptive control has attributes of both feedback systems and optimal control systems. Like a feedback system, measurements are taken on certain process variables. Also like an optimal system, an overall measure of performance is used. In adaptive control, this measure is called the index of performance (IP). The feature that distinguishes adaptive control from the other two types is that an adaptive system is designed to operate in a time-varying environment. It is not unusual for a system to exist in an environment that changes over the course of time. If the internal parameters or mechanisms of the system are fixed, as is the case in a feedback control system, the system might operate quite differently in one type of environment than it would in another. For example, the controls of an airplane cause different effects at sea level in subsonic flight than during supersonic flight at an altitude of 60,000 ft. An adaptive control system is designed to compensate for the changing environment by monitoring its performance and altering, accordingly, some aspect of its control mechanism to achieve optimal or near-optimal performance. The term "environment" is used in a most general way and may refer to the normal operation of the process. For example, in a manufacturing process, the changing environment may simply mean the day-to-day variations that occur in tooling, raw materials, air temperature and humidity (if these have any influence on the process operation), and so on.

An adaptive system is different from a feedback system or an optimal system in that it is provided with the capability to cope with this time-varying environment. The feedback and optimal systems operate in a known or deterministic environment. If the environment changes significantly, these systems might not respond in the manner intended by the designer.

On the other hand, the adaptive system evaluates the environment. More accurately, it evaluates its performance within the environment and makes the necessary changes in its control characteristics to improve or, if possible, to optimize its performance. The manner of doing this involves three functions which characterize adaptive control and distinguish it from other modes of control.

## Three functions of adaptive control

To evaluate its performance and to respond accordingly, the adaptive controller is furnished with the capacity to perform the following three functions: identification, decision, and modification. It may be difficult, in any given adaptive control system, to separate out the components of the system that perform these three functions; nevertheless, all three must be present for adaptation to occur.

1. *Identification function.* This involves determining the current performance of the process or system. Normally, the performance quality of the system is defined by some relevant index of performance. The identification function is concerned with determining the current value of this performance measure by making use of the feedback data from the process. Since the environment will change over time, the performance of the system will also change. Accordingly, the identification function is one that must proceed over time more or less continuously. Identification of the system may involve a number of possible measurement activities. It may involve estimation of a suitable mathematical model of the process or computation of the performance index from measurements of process variables. It could include a comparison of the current performance quality with some desired optimal performance.

2. *Decision function.* Once the system performance is determined, the next function of adaptive control is to decide how the control mechanism should be adjusted to improve process performance. This decision procedure is carried out by means of a preprogrammed logic provided by the system designer. Depending on the logic, the decision may be to change one or more of the controllable inputs to the process; it may be to alter some of the internal parameters of the controller, or some other decision.

3. *Modification function.* The third adaptive function is to implement the decision. While the decision function is a logic function, modification is concerned with a physical or mechanical change in the system. It is a hardware function rather than a software function. The modification involves changing the system parameters or variables so as to drive the process toward a more optimal state.

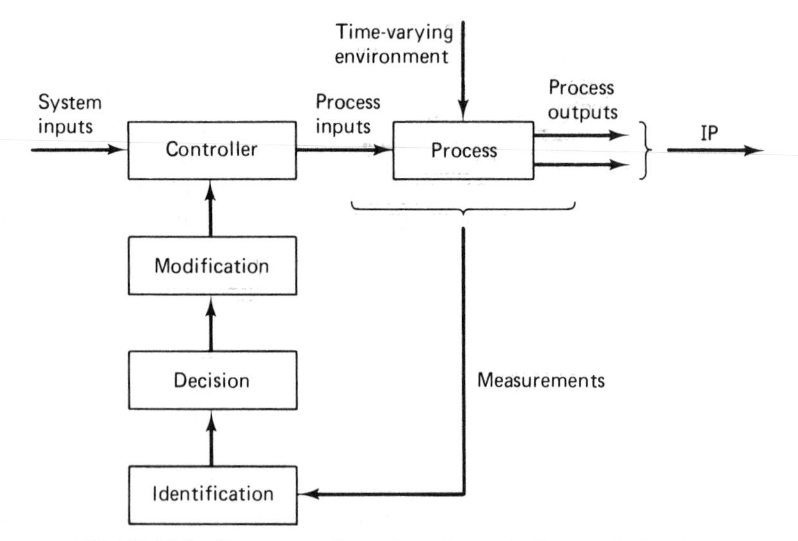


FIGURE 15.9 General configuration of an adaptive control system.

Figure 15.9 illustrates the sequence of the three functions in an adaptive controller applied to a hypothetical process. The process is assumed to be influenced by some time-varying environment. The adaptive system first identifies the current process performance by taking measurements of inputs and outputs. Depending on current performance, a decision procedure is carried out to determine what changes are needed to improve system performance. Actual changes to the system are made in the modification function.

## EXAMPLE 15.4

Before proceeding further, let us consider an example to illustrate several of the points presented up to now. First, we want to show the differences among the three types of control: feedback, optimal, and adaptive control. Second, we want to demonstrate the three functions of an adaptive control system: identification, decision, and modification.

This will be a simple numerical example to show how a hypothetical process might be controlled under a number of different circumstances. Consider the process shown in Figure 15.10, with transfer function = 5.0. That is, the output $y$ is related to the process input $x_p$ as

$$y = 5x_p$$

Let us suppose that the desired value of $y$ is 20. According to the foregoing relationship, it is obvious that the value of $x$ must be 4.0 in order for $y$ to equal 20. However, let us see how the problem might be perceived first as a feedback control problem, then as an optimal control problem, and finally as an adaptive control problem. We will ignore the presence of system dynamics. We could easily incorporate time into our model, but in doing so the mathematics would become more complex and we might lose track of our purpose, which is to show the differences between the three control strategies. Therefore, let us consider the problem without including time dynamics.

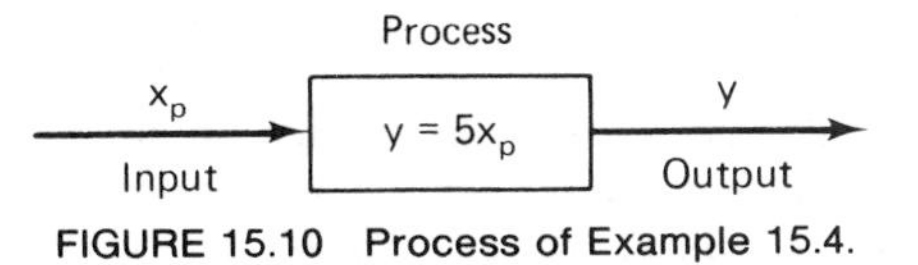


FIGURE 15.10 Process of Example 15.4.

1. *Feedback control.* Using feedback control to regulate the value of $y$, the block diagram of the system might appear as in Figure 15.11. The value of $y$ is fed back through some "measuring device" which produces a signal one-tenth the value of $y$. This signal is compared with the system input $x$, which has a value of 4.0. The difference between the two (the error, $e$) is fed into the proportional controller, which multiplies the signal with a "gain" of 2.0. The output of the controller is $x_p$, which drives the process.

By block diagram algebra, the following can be shown:

$$e = x - 0.1y = 4 - 0.1y$$

$$x_p = 2e$$

$$y = 5x_p$$

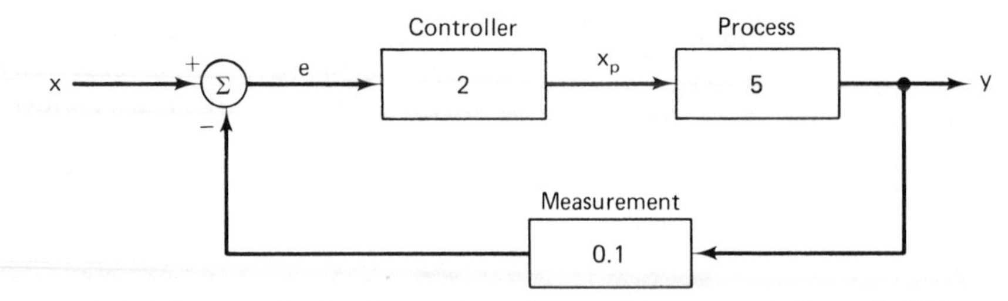


FIGURE 15.11 Feedback control used to regulate output $y$ in Example 15.4.

Therefore,

$$y=(5)(2)(4-0.1y)=40-y$$

This equation can be solved for $y$:

$$2y=40$$
$$y=20$$

The gain value for each block in the system is set to yield a steady-state value of $y$ at the desired level of 20.

However, we have so far ignored the possibility of a disturbance to the process. Let us assume that the process is suddenly disturbed, throwing the output $y$ to a new value of 25. Will the control system respond in such a way as to drive the system back to $y=20$?

Assuming that the system dynamics do not result in instability, we can show that the tendency will be for $y$ to resume its previous value. If $y=25$, then

$$e=4-0.1y=4-2.5=1.5$$

If $e=1.5$, then

$$y=(5)(2)e=15$$

It can be seen that a high vaue of $y$ will cause the error, $e$, to be low. When this error signal is fed through the controller and the process, the effect is to reduce the value of $y$. Likewise, a low value of $y$ will tend to be increased by the system. The system dynamics (time lags and second-order and higher-order effects) will determine the actual response of $y$ over time.

2. *Optimal control.* To solve the example as an optimal control problem, we will assume a least-squares-error objective function:

$$\text{minimize } z=(20-y)^2$$

The solution of the optimal control problem will consist of substituting the function $y-5x_p$ into the objective function, differentiating $z$ with respect to $x_p$, and setting the derivative to zero in order to find the value of $x_p$ that minimizes $z$:

$$z=(20-5x_p)^2$$
$$=400-200x_p+25x_p^2$$
$$\frac{dz}{dx}=0-200+2(25)x_p=0$$
$$x_p=4$$

The reader may recall that this is the same problem that was solved in Example 15.1.

3. *Adaptive control.* With this control strategy, the adaptive controller will try to monitor the performance and compensate for changes in environment.

Let us say in our example that the changes in the environment are manifested in an altered value of the process gain. Say that the gain has decreased from a value of 5 to a value of 4. We will now see how ineffective a feedback control system would be when used to try to maintain the value of $y$ at 20. As before,

$$e = x - 0.1y = 4 - 0.1y$$

$$x_p = 2e$$

$$y = 4x_p \quad \text{(the changed process)}$$

Therefore,

$$y = 8(4 - 0.1y) = 32 - 0.8y$$

$$1.8y = 32$$

$$y = 17.777$$

The steady-state value of $y$ is not maintained at the desired level of 20 by the feedback control strategy.

Under the optimal control policy, where the process was assumed to possess a gain = 5, the optimal value of $x_p$ was determined to be 4.0. Now that the process gain = 4, the value of $x_p = 4$ leads to the following:

$$y = 4x_p = 4(4) = 16$$

Steady-state optimal control, with no provision for monitoring the performance, misses the desired value even more than feedback control.

Let us proceed to see how adaptive control might be employed. The objective function will be the same as in the previous optimal control solution. However, it is termed the index of performance in adaptive control.

$$\text{minimize } z = (20 - y)^2$$

Measurements will be taken of both $x_p$ and $y$, the input and output of the process, and these signals will be fed back to the adaptive controller. Of course, it may be unnecessary to take actual measurements of the input, since this variable is determined directly by the controller unit.

The adaptive controller, as part of its logic function, will attempt to determine the mathematical function that exists between $x_p$ and $y$. This represents the identification function of adaptive control. Often, enough is known about the process that a good guess can be made as to the mathematical form of the function. In our case, we will assume that the suggested model form is

$$y = Kx_p$$

Now, using the measured values of $x_p$ and $y$, the value of $K$ can be calculated.

Consistent with our previous discussion, we will assume that the measured values of $x_p$ and $y$ are 4 and 16, respectively. Hence, the calculated value of $K$ would be

$$K = \frac{y}{x_p} = \frac{16}{4} = 4.0$$

Next, the adaptive controller will compute the value of $x_p$ to minimize the index of performance $z$. This would be the decision function. These computations would be based on the following preprogrammed steps, similar to the solution approach used in optimal control:

$$z = (20 - y)^2 = (20 - Kx)^2$$

$$= 400 - 40Kx + K^2x^2$$

$$\frac{dz}{dx} = 0 - 40K + 2K^2x = 0$$

$$2K^2x = 40K$$

$$Kx = 20$$

$$x = \frac{20}{K}$$

From the previous steps, $K$ had been computed to have a value of 4. Accordingly,

$$x_p = 5$$

Finally, as the last step in the adaptive control procedure, a modification would be made in the process input changing its value from 4 to the new value:

$$x_p = 5.0$$

As the process evolves over time and the value of the process gain continues to change unpredictably, the adaptive control system would make either continuous or periodic computations and adjustments in order to maintain the output $y$ at the value of 20. The sequence of activities is displayed in Figure 15.12.

We have considered a most elementary example. Yet it features the principal attribute for which adaptive control is ideally suited: a time-varying, unpredictable environment, manifested in changing values of process gain $K$. Neither feedback control nor optimal control are capable of providing a satisfactory response in this situation. The adaptive strategy possesses this capability to deal with an ever-changing environment.

If the process is not subjected to a time-varying and difficult-to-predict environment, a simpler form of strategy (e.g., feedback) would be more appropriate than adaptive control. It is therefore appropriate to examine whether a given process is a feasible candidate for adaptive control by considering whether the process is one that changes over time.

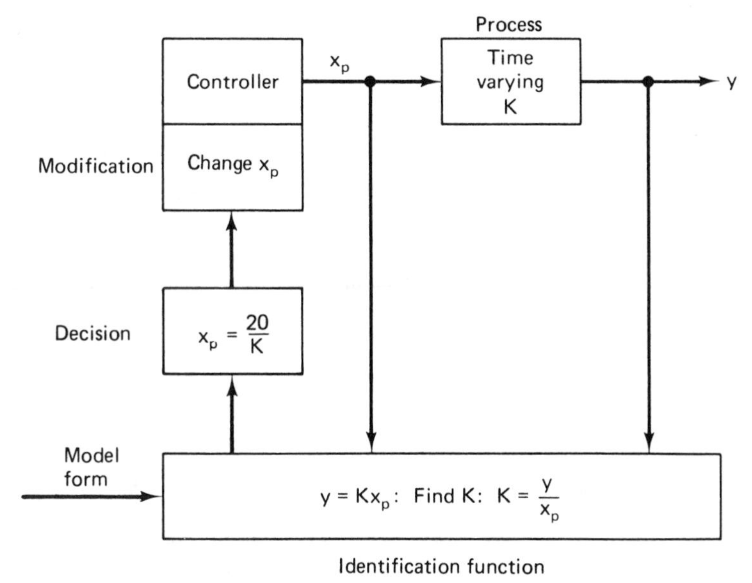


FIGURE 15.12 Adaptive control applied to Example 15.4.

## Why process changes occur

The following are several of the primary reasons why process changes might occur over time in an unpredictable manner:

1. *Environmental changes.* Some processes are affected by their environment. For example, some chemical processes are affected by the temperature and relative humidity in the air. Efforts are often made to maintain these properties at satisfactory values. In other cases, these factors cannot conveniently be controlled so other variables must be altered to compensate.

2. *Changes in raw materials.* It is not difficult to imagine the many possible variations that could occur in raw material inputs to a process. It is virtually impossible to control all the properties that define a material. The variations in properties might have an adverse effect on the operation of the process.

For example, considering a metal machining operation, the variations in raw materials include changes in hardness, strength, microstructure, chemistry, thermal properties, and so on. Each of these factors has an effect on machinability and each can be expected to vary to some extent from workpiece to workpiece or lot to lot. We have not even considered in the list above the possible variations in part size and geometry.

3. *Wear of components.* The moving parts of a piece of production equipment can be expected to wear over time. This will often cause misalignment of the equipment, for which compensating adjustments must be made.

In many manufacturing operations, some form of tooling is required which is used up or worn out during the process. Metal cutting is, of course, an example. An adaptive control system can sometimes be used to counteract the effects of wear.

4. *Failure of components.* Worse than the wear of some machine component is the actual failure of the component. This usually means that the entire system breaks down. For example, the final failure of the cutting tool in a machining operation means the process must be stopped and the tool changed. However, there are other systems or processes in which the failure of some minor component does not necessarily shut down the system. A production flow line with parallel stations might be an example. If one of the parallel stations breaks down, the others can maintain production. The throughput of the line may be reduced, but at least the line can be kept running.

In each of the preceding cases, the process has been changed in some manner. The various process changes that can occur over time are reflected as changes in the mathematical model of the process.

## Changes in process model

The identification function in adaptive control usually reduces to the problem of determining either the value of the index of performance or the mathematical model that describes the process. Even the index of performance, when it is the object of the identification procedure, is generally based on a mathematical model. For example, the index of performance for a machining process might be cost per cubic inch of metal, which would have to be formulated as a mathematical equation.

It is therefore of importance to consider how the mathematical model of the process can be affected by the changes occurring in the process over time. These effects can be cataloged into three basic types:

1. *Changes in the model coefficients.* In this case we assume that the structure or form of the mathematical model remains accurate, but the constants in the model become altered in value to reflect process changes.

As an illustration, the previous example used the model

$$y=5x_p$$

to express the relationship between output and input. The change that occurred in the hypothetical process of Example 15.4 was a shift in the value of the process gain from 5 to 4.

2. *Structural changes in model.* This is a more difficult case to deal with because the number of possible model forms is almost endless. Finding

the correct form, unless some limitations are placed on the possibilities, would be a very tedious task.

An example of a structural change in the model would be a change from the model

$$y = Bx \qquad \text{as above}$$

to

$$y = A + Bx$$

or

$$y = Bx^n$$

or some other form of equation.

3. *Changes in constraints.* In a practical problem, the variables usually have an allowable range of values. The upper and lower extremes of the range can be expressed as constraints within which the variables must operate. These constraint values represent part of the mathematical model describing the process. As changes occur to the process, these changes may be defined in terms of changes in the constraints.

For example, the amount of torque that a drill bit could sustain without breaking would depend on the size of the drill. The torque limit could be expressed as a constraint value. When the drill is changed, a new constraint value would become applicable.

## The identification problem

The previous two subsections have considered some of the possible reasons why a process may change over time, and how these changes might be simulated by changing the mathematical model of the process. If the changes over time are serious enough in terms of process performance, it is appropriate to consider adaptive control as a candidate for controlling the system.

In an adaptive control system the most difficult problem is usually the identification problem. Often, the solution of this problem is individualized. That is, the identification function is performed in a way that is unique for the given process. The types of identification problem can usually be placed in one of two categories.

1. *Where process identification is possible and feasible.* Enough information is known about the process that a mathematical model can be assumed with some reliability. Then, the identification function is concerned with using on-line measurements to determine the parameters of the model. In the more difficult case, the form of the mathematical model must be selected during identification. In any case, once the model and its parameters have been identified, the optimization procedure can be carried out.

2. *Where process identification is not possible or feasible.* In this situation, the process is poorly defined or a model of the process would be too complex. Hence, it is not feasible to make practical use of an assumed mathematical model. It is, however, possible to evaluate or estimate the performance quality of the process by means of measurements taken. Some sort of search procedure would be required to seek out the optimal point at which to operate the process.

We will consider methods of identification that fall into the first category in this section. The second category of identification problem, requiring a systematic search procedure, we will postpone until Section 15.5.

A variety of techniques have been proposed for solving the identification problem in adaptive control ([4], [6], [13]). Some of these techniques require a fairly high level of mathematical sophistication, and the interested reader is invited to seek out the references cited. Presented in the following subsections are two techniques for process identification:

1. Instantaneous approximation method.
2. Regression techniques.

These methods are relatively straightforward to use, yet effective in practice.

### Instantaneous approximation method

Many model identification techniques rely on observing the system over time in order to incorporate the dynamics of the system into the model. The instantaneous approximation method does not. However, the limitation of the method is that it can estimate only a single parameter in the model. Therefore, it is applicable only to those cases where only one of the process model parameters is expected to fluctuate over time. The other parameters in the model are assumed to remain at relatively constant values.

In this approximation technique, periodic measurements are taken of the input and output variables of the process. The measured values of these variables are inserted into the model of the process, along with the values of all parameters except the one to be estimated.

Then, the model is solved for the missing parameter value. For this method to yield satisfactory results, the model form must be known in advance and all parameter values must be fairly constant except the one to be estimated. Also, the system dynamics, if significant, must be included in the model.

**EXAMPLE 15.5**

To illustrate the instantaneous approximation method, assume a process whose mathematical model is given by

$$y = A + Bx$$

The process has been studied and it is known that the value of $A$ remains relatively constant at a value of 10, while the value of $B$ seems to fluctuate between 4.0 and 8.0. This variation takes place relatively slowly, but it makes it difficult to control the value of $y$. It is desired that the value of $B$ be estimated from measurements taken of $x$ and $y$.

If $x$ and $y$ are known from measurements made on the process and the parameter $A$ is assumed at a value of 10, then $B$ can be solved as follows:

$$B = \frac{y-A}{x} = \frac{y-10}{x}$$

For example, let us say that the values of $x$ and $y$ were measured as $x=3$ and $y=25$. Solving the foregoing equation for $B$ yields

$$B = \frac{25-10}{3} = \frac{15}{3} = 5.0$$

Having determined $B$, we have therefore identified the process model.

$$y = 10 + 5x$$

We can now make whatever adjustments are required in the input $x$ to produce the desired result in the output $y$.

## EXAMPLE 15.6

Suppose that the process model included a rate or derivative term, as follows:

$$\frac{dy}{dt} + y = A + Bx$$

where the rate of change of $y$ is expressed as the derivative $dy/dt$. We assume that the coefficient of this term is unity, but any other value could be assumed also. The instantaneous approximation method might be used to again find the current value of the varying parameter $B$, as long as all variables can be measured (including the derivative term), and as long as all other parameters in the model except $B$ remain at their assumed values. For example, suppose that the measured variables had values as follows:

$$y = 25$$

$$\frac{dy}{dt} = 6$$

$$x = 3$$

Solving for $B$, we obtain

$$B = \frac{dy/dt + y - A}{x}$$

$$= \frac{6+25-10}{3} = 7.0$$

The instantaneous approximation method produces a quick estimate of the unknown parameter value. It is often impractical because of the assumption that all other parameter values are constant and known. Most practical processes are more complex than that.

### Regression techniques

One of the several weaknesses of the instantaneous approximation method is that it presumes no error in the variable measurements. If error is present in the measurement of the process variables, the estimation of the unknown parameter value will contain error. Another limitation which was already mentioned is the fact that only one parameter can be approximated. Thus, the preceding method would be quite inadequate where more than one parameter was to be estimated, and where the variable measurements contained noise or error.

Regression techniques can be employed to overcome these problems. To use these methods, the process must be observed over a length of time, with measurements taken periodically of the process variables. A model form for the process must be assumed, but the parameters of the model will vary over time. We must assume that the variation in parameter values takes place slowly compared with our procedure of sampling the process and estimating the parameters. After a sequence of values of the process variables has been collected, a regression computation is made to determine the model parameter values at the current time. The whole procedure is repeated at regular intervals, depending on the rate at which the model parameters drift.

METHOD OF LEAST SQUARES. Regression analysis is typically performed by means of the least-squares technique. To illustrate the method of least squares in the simplest case, suppose that it was desired to determine the linear relationship between some dependent variable $y$ and the independent variable $x$. Table 15.1 shows seven pairs of $x$-$y$ values which will be used for demonstration purposes. These data may represent the measurements collected from some physical experiment, or the results of a calibration procedure on some sensing instrument, or the sampled data compiled on some manufacturing process. Our interest here is in the data rather than the source from whence they came.

The data are plotted in Figure 15.13. The linear relationship between $x$ and $y$ is determined by finding the straight line that best fits the set of data

TABLE 15.1 Hypothetical Data to Illustrate the Least-Squares Technique

| $i$ | 1 | 2 | 3 | 4 | 5 | 6 | 7 |
|---|---|---|---|---|---|---|---|
| $y_i$ | 0.5 | 1.2 | 1.3 | 1.1 | 1.8 | 2.2 | 2.7 |
| $x_i$ | 0.8 | 1.0 | 1.5 | 2.5 | 3.1 | 3.6 | 4.4 |

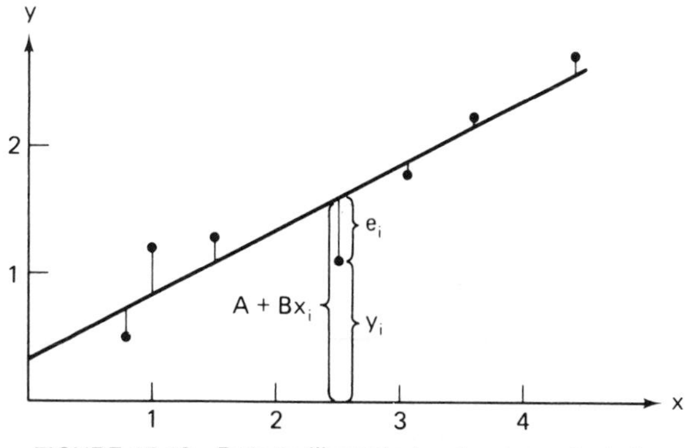


FIGURE 15.13 Data to illustrate least-squares technique.

presented in this figure. One way to do this would be to fit the line to the data "by eye." The trouble is, different analysts would probably come up with different lines. To overcome this variation in judgment, statisticians have developed the least-squares criterion. This criterion states that the "best fit" is the line that minimizes the sum of the squared errors between the data points and the line. To elucidate this statement, the equation for a straight line is

$$y = A + Bx \tag{15.7}$$

where $B$ = a constant representing the slope of the line
$A$ = a constant representing the intercept of the line with the $y$-axis (at $x = 0$)

One possible line that follows this general equation is shown in Figure 15.13. The vertical distance between the straight line and any given data point (defined by $x_i$, $y_i$ is termed the error $e_i$. That is,

$$e_i = y_i - (A + Bx_i) \tag{15.8}$$

Mathematically expressing the criterion of least squares, the objective is to find values of $A$ and $B$ that

$$\text{minimize} \sum_{i=1}^{n} e_i^2 = \sum_{i=1}^{n} [y_i - (A + Bx_i)]^2 \tag{15.9}$$

where $n$ is the number of points in the data set.

To find the values of $A$ and $B$ that satisfy this objective, the partial derivatives of Eq. (15.9) with respect to $A$ and $B$ can be found and set equal to zero.

$$2 \sum_{i=1}^{n} [y_i - (A + Bx_i)](-1) = 0$$

$$2 \sum_{i=1}^{n} [y_i - (A + Bx_i)](-x_i) = 0$$

These equations can be written in a more convenient form by separating the terms being summed as follows:

$$\sum_{i=1}^{n} y_i = An + B\sum_{i=1}^{n} x_i$$
$$\sum_{i=1}^{n} x_i y_i = A\sum_{i=1}^{n} x_i + B\sum_{i=1}^{n} x_i^2 \quad (15.10)$$

Equations (15.10) are called the *normal equations*. By solving these two equations simultaneously, the values of $A$ and $B$ can be computed which provide the "best fit" to the data by the least-squares criterion. The statistically minded reader may find objectionable the use of the symbols $A$ and $B$ in Eq. (15.10), since these are the same symbols used in Eq. (15.7). $A$ and $B$ in Eq. (15.7) represent the true values of the constants in the relationship between $x$ and $y$. The values calculated from Eq. (15.10) are estimates of these true values based on a sampling of data from the relationship.

## EXAMPLE 15.7

The method of least squares will be illustrated by means of the data presented in Table 15.1. To make use of Eqs. (15.10), it is convenient to construct a calculations table similar to the one shown in Table 15.2. In this table, the intermediate computations are performed to collect the terms used in the normal equations.
From Table 15.2,

$$\sum_{i=1}^{7} y_i = 10.8 \qquad \sum_{i=1}^{7} x_i = 16.9$$
$$\sum_{i=1}^{7} x_i y_i = 31.68 \qquad \sum_{i=1}^{7} x_i^2 = 52.07$$

Inserting these terms into Eqs. (15.10), we have

$$10.8 = 7A + 16.9B$$
$$31.68 = 16.9A + 52.07B$$

TABLE 15.2 calculations Table for Example 15.7

| $i$ | $y_i$ | $x_i$ | $x_i y_i$ | $x_i^2$ |
|---|---|---|---|---|
| 1 | 0.5 | 0.8 | 0.40 | 0.64 |
| 2 | 1.2 | 1.0 | 1.20 | 1.00 |
| 3 | 1.3 | 1.5 | 1.95 | 2.25 |
| 4 | 1.1 | 2.5 | 2.75 | 6.25 |
| 5 | 1.8 | 3.1 | 5.58 | 9.61 |
| 6 | 2.2 | 3.6 | 7.92 | 12.96 |
| 7 | 2.7 | 4.4 | 11.88 | 19.36 |
| Totals | 10.8 | 16.9 | 31.68 | 52.07 |

Solving these two equations for $A$ and $B$, we determine that $A=0.342$ and $B=0.497$. The resulting equation is

$$y=0.342+0.497x$$

This equation is shown in Figure 15.13 as the straight line through the data points.

It is usually desirable to examine some of the statistical characteristics of the regression equation which indicate how good the relationship is. Among these characteristics are the sample correlation coefficient and the standard error of estimate.

To calculate these two statistics, it is convenient to define several intermediate terms as follows:

$$S_{xx}=n\sum_{i=1}^{n}x_i^2-\left(\sum_{i=1}^{n}x_i\right)^2 \tag{15.11}$$

$$S_{yy}=n\sum_{i=1}^{n}y_i^2-\left(\sum_{i=1}^{n}y_i\right)^2 \tag{15.12}$$

$$S_{xy}=n\sum_{i=1}^{n}x_iy_i-\left(\sum_{i=1}^{n}x_i\right)\left(\sum_{i=1}^{n}y_i\right) \tag{15.13}$$

The *sample correlation coefficient* can now be defined as

$$r=\frac{S_{xy}}{\sqrt{S_{xx}S_{yy}}} \tag{15.14}$$

The sample correlation coefficient ranges in value between $-1.0$ and $+1.0$. It indicates the strength of the linear relationship between the sampled values of $x$ and $y$. If $r$ is close to $+1.0$, this indicates a strong positive correlation between $x$ and $y$. If $r$ is close to $-1.0$, it is an indication that a strong negative linear correlation exists between the two variables. A value of $r$ near zero indicates the absence of any strong correlation.

The *standard error of estimate* can be defined as

$$s_e=\sqrt{\frac{S_{xx}S_{yy}-(S_{xy})^2}{n(n-2)S_{xx}}} \tag{15.15}$$

The $s_e$ statistic for a regression equation gives the average value of the $e_i$ errors. It is not an arithmetic average, but rather a root-mean-square average corrected for the number of degrees of freedom. Although less convenient than Eq. (15.15), an alternative way of defining the standard error of estimate is given by

$$s_e = \sqrt{\frac{\sum_{i=1}^{n} e_i^2}{n-2}} = \sqrt{\frac{\sum_{i=1}^{n} [y_i - (A + Bx_i)]^2}{n-2}} \tag{15.16}$$

The squared term inside the radical sign shows the root-mean-square averaging process. The $n-2$ term is the number of degrees of freedom.

The sign of a "good fit" is when the value of $s_e$ is low relative to the values of $y$ in the data set. If $s_e = 0$, this indicates that the equation fits the data perfectly.

### EXAMPLE 15.8

We will demonstrate how these two statistics are calculated by using the data from Example 15.7. From Table 15.2, we have

$$\sum_{i=1}^{7} y_i = 10.8 \qquad \sum_{i=1}^{7} x_i = 16.9$$

$$\sum_{i=1}^{7} x_i y_i = 31.68 \qquad \sum_{i=1}^{7} x_i^2 = 52.07$$

The only additional term we need in order to make the calculations is $\sum_{i=1}^{7} y_i^2$:

$$\sum_{i=1}^{7} y_i^2 = (0.5)^2 + (1.2)^2 + \ldots + (2.7)^2$$

$$= 19.96$$

From Eqs. (15.11) through (15.13),

$$S_{xx} = 7(52.07) - (16.9)^2 = 78.88$$

$$S_{yy} = 7(19.96) - (10.8)^2 = 23.08$$

$$S_{xy} = 7(31.68) - (16.9)(10.8) = 39.24$$

From Eq. (15.14), the sample correlation coefficient is

$$r = \frac{39.24}{\sqrt{(78.88 \times 23.08)}} = 0.9197$$

This is close to $+1.0$ and indicates that the values of $x$ and $y$ are closely correlated. From Eq. (15.15), the standard error of estimate is

$$s_e = \sqrt{\frac{(78.88)(23.08) - (39.24)^2}{7(5)(78.88)}} = 0.3189$$

The average value of $y$ is $10.8/7 = 1.543$. The standard error of estimate is roughly 20% of this average $y$.

The reader can visually interpret these calculated statistics in Figure 15.13.

MULTIPLE REGRESSION. Most industrial processes are too complex to be modeled by the simple linear relation of Eq. (15.7). There are often two or more input variables which determine the process output variable of interest. Also, the dynamics of the process usually play an important role in its behavior. Accordingly, these additional features of the process must be taken into account in the identification function. If the model is linear (or if it is nonlinear but can be transformed into an equivalent linear form), the least-squares method can be extended to determine the parameters of the model. The regression of some dependent variable $y$ on more than a single independent variable is called *multiple regression*.

To illustrate how multiple regression might be used in process identification, suppose that the model for a certain industrial process is the following:

$$K\frac{dy}{dt} + y = A + Bx_1 + Cx_2 \tag{15.17}$$

where $dy/dt$ = rate of change of the output variable $y$
$x_1$ and $x_2$ = two process input variables

We can take measurements of all variables, including the derivative term, but there is random noise in each of the signals. We will assume that the noise is normally distributed about the mean value of the variable. In operating the process, we must make routine changes in the inputs $x_1$ and $x_2$ and we then observe the effect on the output $y$ and its rate of change $dy/dt$. One final assumption is this: that the rate at which the model parameters, $K$, $A$, $B$, and $C$, change is slow relative to our ability to perform the estimating procedure.

To carry out the model identification scheme, we first take measurements of the four variables $y$, $dy/dt$, $x_1$, and $x_2$ at discrete intervals, making minor changes in $x_1$ and $x_2$ to observe the effects on $y$ and $dy/dt$. This will provide a set of values over time in the format shown in Table 15.3. In effect, we are sampling the process at discrete points in time ($t$, $t+1$, $t+2$, etc.). With the data from Table 15.3, the least-squares computation can be performed to determine current estimates of the four parameters, $K$, $A$, $B$, and $C$.

To carry out the least-squares computations, Eq. (15.17) must be rearranged into the following form:

$$y = A + Bx_1 + Cx_2 - K\frac{dy}{dt}$$

Since the sign of the fourth term will take care of itself in the least-squares

TABLE 15.3 Data Set Format Used in Multiple Regression Computation

| *Sampling time* | $\frac{dy}{dt}$ | $y$ | $x_1$ | $x_2$ |
|---|---|---|---|---|
| $t$ | $\frac{dy}{dt}(t)$ | $y(t)$ | $x_1(t)$ | $x_2(t)$ |
| $t+1$ | $\frac{dy}{dt}(t+1)$ | $y(t+1)$ | $x_1(t+1)$ | $x_2(t+1)$ |
| $t+2$ | $\frac{dy}{dt}(t+2)$ | $y(t+2)$ | $x_1(t+2)$ | $x_2(t+2)$ |
| . | . | . | . | . |
| . | . | . | . | . |
| . | . | . | . | . |
| $t+n$ | $\frac{dy}{dt}(t+n)$ | $y(t+n)$ | $x_1(t+n)$ | $x_2(t+n)$ |

calculations, it is more convenient to write this equation as

$$y = A + Bx_1 + Cx_2 + K\frac{dy}{dt} \tag{15.18}$$

Now the dependent variable $y$ can be regressed on $x_1$, $x_2$, and $dy/dt$. Statisticians might object to this on the grounds that the variable $dy/dt$ is not independent of $y$. While this objection is valid from a statistical viewpoint, the least-squares method is nevertheless an appropriate and convenient calculation procedure for estimating the unknown parameters in the linear model.

The approach to this problem is similar to the procedure used for the case of $y = A + Bx$. The objective is to minimize the sum of the squared error terms:

$$\text{minimize} \sum_{i=1}^{n} e_i^2 = \sum_{i=1}^{n} \left[ y_i - (A + Bx_1 + Cx_2 + K\frac{dy}{dt} \right]^2$$

From this the "Normal Equations" can be derived by taking the partial derivatives with respect to $A$, $B$, $C$, and $K$, and setting each equal to zero. The resulting Normal Equations are:

$$\begin{aligned}
\Sigma y &= nA + B\Sigma x_1 + C\Sigma x_2 + K\Sigma \frac{dy}{dt} \\
\Sigma x_1 y &= A\Sigma x_1 + B\Sigma x_1^2 + C\Sigma x_1 x_2 + K\Sigma x_1 \frac{dy}{dt} \\
\Sigma x_2 y &= A\Sigma x_2 + B\Sigma x_1 x_2 + C\Sigma x_2^2 + K\Sigma x_2 \frac{dy}{dt} \\
\Sigma y\frac{dy}{dt} &= A\Sigma \frac{dy}{dt} + B\Sigma x_1 \frac{dy}{dt} + C\Sigma x_2 \frac{dy}{dt} + K\Sigma \left(\frac{dy}{dt}\right)^2
\end{aligned} \tag{15.19}$$

The normal equations for other model forms can be derived in the same manner. Problem 15.15 asks the reader to derive the normal equations for the model $y = A + Bx_1 + Cx_2$. Computer software packages are available for the least-squares computations in multiple regression. The reader may therefore ask why we need bother with these manual calculations. The answer is that it is often necessary to program this type of computation into the software used for specific process monitoring and control applications.

## EXAMPLE 15.9

Let us proceed through the calculations for the multiple regression case. Suppose that sampled data have been collected for a manufacturing process that can be modeled by an equation of the form of Eq. (15.17). The data are collected and summarized in Table 15.4. The sampling interval is 1 min.

TABLE 15.4 Hypothetical Data Set for Example 15.9

| *Sampling time (min)* | $y$ | $x_1$ | $x_2$ | $\frac{dy}{dt}$ |
|---|---|---|---|---|
| 1 | 22.3 | 1.1 | 3.0 | 1.0 |
| 2 | 23.5 | 1.2 | 3.2 | 1.3 |
| 3 | 24.8 | 1.2 | 3.4 | 1.1 |
| 4 | 25.9 | 1.3 | 3.6 | 1.4 |
| 5 | 26.2 | 1.3 | 3.5 | 0.4 |
| 6 | 26.5 | 1.4 | 3.5 | 0.2 |
| 7 | 26.3 | 1.4 | 3.4 | 0.2 |
| 8 | 25.8 | 1.3 | 3.3 | 0.7 |

The resulting equation for this sampling period (calculated using a computer statistical package) is

$$y = 3.172 + 2.389x_1 + 5.763x_2 - 0.771\frac{dy}{dt}$$

Rewriting this in the form of Eq. (15.17) yields

$$0.771\frac{dy}{dt} + y = 3.172 + 2.389x_1 + 5.763x_2$$

The sampling of the process would be repeated at suitable intervals to recalculate the parameter values. In this way, any drift in the parameters would be identified.

There are several aspects of this regression technique that ought to be mentioned. First, the number of data sets required depends on statistical and practical considerations. More data sets will usually give more accurate estimates of the parameter values. Second, a major disadvantage of the regression method discussed here is that the process must be perturbed in order to assess the effects of changes in the controllable input variables. If no changes occurred in $x_1$ and $x_2$, the parameter values could not be estimated. Third, it is important to acknowledge the effects that system dynamics play in the operation of the process. All the significant rate terms should be included in the process model.

### Practical problems with adaptive control

Although adaptive control is the appropriate way to solve the problem in which the system is confronted with unpredictable environmental changes, there are some difficulties associated with its use. The following discussion explores most of these difficulties, some of which are not exclusively associated with adaptive control but apply to other control systems as well.

1. *Complexity of the system.* Adding an adaptive control loop will increase the complexity of the analysis problem. The use of AC may turn a relatively simple linear system into a complex nonlinear system. Accordingly, this difficulty should be considered and adaptive control should be applied only where it represents the most feasible way of dealing with the problem.

2. *Difficulty with the identification function.* There are several problems associated with the system identification function in adaptive control.
   a. *Definition of index of performance.* The performance of the system can be no better than the measure used to evaluate performance. There are clearly difficulties involved when an attempt is made to characterize the overall performance of a process or system with a single index of performance. Despite these difficulties, the use of adaptive control requires that process performance be combined into one measure. It is therefore important that the index of performance selected be truly representative of overall performance.
   b. *Sensor problems.* One of the biggest problems in applying adaptive control to the machining process is the difficulty in obtaining accurate and reliable measurements of the process variables. Without such measurements, the index of performance cannot be accurately assessed and the process cannot be accurately identified.
   c. *Identification under normal operating conditions.* The identification function should be performeed under normal operating conditions. Tests required to identify the process model (transfer function of the process) should not disturb the routine operation of the process. This is often very difficult. Among the model identification methods explored in this section, the regression techniques require that the process be disturbed during the measurement procedure. Also, the search techniques discussed in Section 15.5 require perturbations of the process to carry out the optimum seeking strategy. These disturbances must be considered as costs associated with the identifica-

tion function. The benefits of adaptive control must more than compensate for these costs.

3. *System stability.* As a consequence of the preceding, the system may become unstable. This means that the system has a tendency to get out of control, possibly to cause damage to itself.

4. *System cost.* The attachment of adaptive control to the process will obviously have a cost. The improvement in process performance must be greater in value than the associated cost.

## 15.5 ON-LINE SEARCH STRATEGIES

When model identification is infeasible, it may still be possible to make a determination of the index of performance for the process. This can either be done through direct observation of the IP or by calculating the IP from measurements of the process variables that determine its value. Even though we can evaluate the index of performance, the problem still remains that the relationship between the IP and the process inputs is unknown. Therefore, we cannot directly determine the values of these inputs which will optimize the system performance. In this type of situation, we must resort to some form of search procedure.

The general strategy in any of the search techniques is to make adjustments to the controllable input variables and observe the effects on the output variables. Based on the effects, decisions are made to systematically change the inputs so as to improve process performance.

The common search techniques encompass a variety of approaches, ranging from pure trial and error to gradient strategies. The scope of this book restricts us from examining the entire range. We will focus on the gradient strategies. There is little evidence to suggest that these search techniques are widely applied in discrete-parts manufacturing. The applications are more prevalent in the continuous process industries.

### Desirable properties of a search strategy

A search strategy is a procedure of logical computations used to adjust the process inputs in order to try to improve the index of performance. By repeating the logical procedure, the search strategy tends to move toward the optimum value of the IP. As mentioned above, there are many different search strategies. One strategy may be appropriate for some search problems and inappropriate for others. Because the nature of the search problem is that we are dealing with an unknown process, it is often difficult to decide in advance which search strategy would work best. The success of the search sometimes reduces to a matter of luck on the part of the programmer selecting the strategy. The general criteria used to judge the effectiveness of a search strategy are the following:

1. *Speed of arriving at the optimum.* It is desirable for the strategy to arrive at the optimum IP value in the minimum number of steps. In process control, this is especially important when the industrial process is subjected to frequent shifts and a new optimum set of operating conditions must continually be sought.

2. *Simplicity of the strategy.* This is desirable from the viewpoint of the operating personnel who must supervise the process. If the strategy is complicated, it will be difficult to understand by those using it. This may result in the operator's overriding the strategy. Simplicity of the strategy is also an advantage in programming the strategy on the control computer.

3. *Capability to deal with difficult or unusual search problems.* Some search strategies work well on certain problems, while other strategies are more suited to other situations. A desirable search strategy is one that is versatile enough to cope with a variety of different search problems.

4. *Stopping criteria.* When the optimum has been reached, the search should be terminated. Because of the stochastic nature of many manufacturing processes, the optimum is often disguised by the presence of random noise. This makes it difficult to identify the optimum operating conditions. One criterion used to judge a search strategy is its ability to discern this stopping point.

## Some basic definitions

In order to explain the operation of any of the search techniques, it is necessary to establish certain basic definitions.

RESPONSE SURFACE. Perhaps the most fundamental concept required to understand how a search strategy works is the concept of the response surface. A response surface is a graphical representation of the index of performance (or other dependent variable) as a function of the input variables. For two inputs, $x_1$ and $x_2$, the response surface can be plotted very conveniently as shown in Figure 15.14. The plot reads something like a geological survey map. The contour lines are lines of constant IP value. In Figure 15.14, the value of each contour line is identified.

In concept a response surface can be defined mathematically as

$$z = f(x_1, x_2) \tag{15.20}$$

where $z$ = index of performance or dependent variable
$x_1, x_2$ = inputs or independent variables on which $z$ is functionally dependent

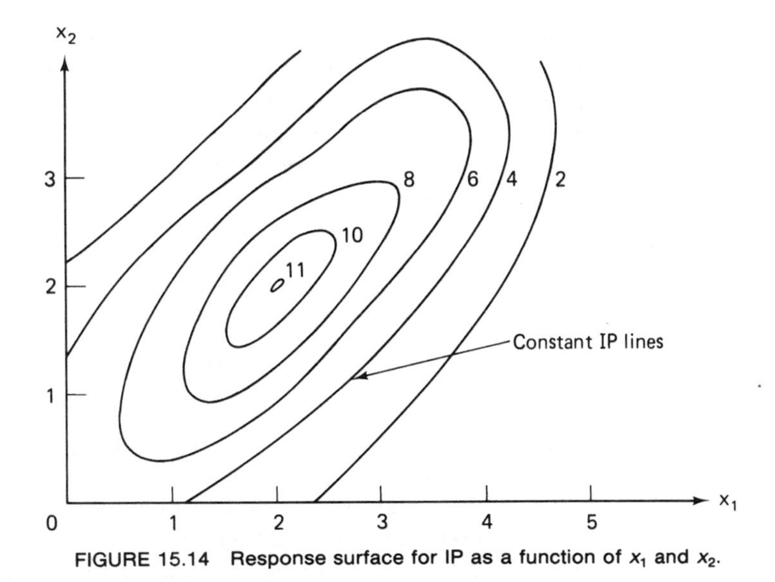


FIGURE 15.14 Response surface for IP as a function of $x_1$ and $x_2$.

Although a response surface can be mathematically defined in principle, it turns out that this mathematical definition is often very difficult to realize in practice for many industrial processes. In this section, we will use mathematical functions to define and illustrate certain terms relating to search strategies. However, the reader should recognize that real manufacturing processes cannot often be defined so precisely. If they could, there would be little need to apply search techniques for finding the optimum. We could use the methods of Sections 15.3 and 15.4 to determine optimum operating conditions.

OPTIMUM POINT. The optimum point on a response surface is the combination of $x_1$ and $x_2$ values at which the index of performance is optimized. In Figure 15.14, the optimum (maximum) IP value is slightly greater than 11.0 and its location is approximately $x_1=2.0$ and $x_2=2.0$. For a maximization problem, the optimum point is at the peak or summit of the response surface. For a minimization problem, the optimum point is located at the deepest point in the valley of the response surface.

UNIMODALITY. Search strategies rely on the assumption that the response surface is unimodal. What this means is that there is only one peak in the response surface. The objective of the search strategy is to find that single peak. If more than one peak exists, the strategy might seek out a peak that is not the highest one. Hence, the true optimum point would be neglected.

GRADIENT. Many search strategies are based on the use of gradients. The gradient is a vector quantity whose components are along the axes of the independent variables ($x_1$ and $x_2$). The magnitude of each component is equal to the partial derivative of the index of performance with respect to the corresponding independent variable. Our interest will be limited to two independent variables, although the concept of the gradient applies to $n$-dimensional response surfaces. For two inputs, $x_1$ and $x_2$, the components of the gradient are defined as

$$G_{1p} = \left.\frac{\partial z}{\partial x_1}\right|_p \qquad G_{2p} = \left.\frac{\partial z}{\partial x_2}\right|_p \tag{15.21}$$

where $G_{1p}, G_{2p}$ = components of the gradient in the $x_1$ and $x_2$ directions, respectively; these components must be evaluated at a particular location on the response surface and this location is identified as point $p$

$z$ = index of performance, a function of $x_1$ and $x_2$

The two components add together to form the gradient

$$\mathbf{G}_p = iG_{1p} + jG_{2p} \tag{15.22}$$

where $i$ and $j$ represent unit vectors parallel to the $x_1$ and $x_2$ axes. The gradient points in the direction of the steepest slope. Moving in the direction of steepest slope is a reasonable strategy to reach the top of the response surface. This is why many search strategies are based on the use of gradients.

The magnitude of the gradient is a scalar quantity given by

$$M_p = \left[\left(\left.\frac{\partial z}{\partial x_1}\right|_p\right)^2 + \left(\left.\frac{\partial z}{\partial x_2}\right|_p\right)^2\right]^{1/2} \tag{15.23}$$

Again, the magnitude of the gradient is defined at a particular point $p$ on the $x_1 - x_2$ surface.

The direction of the gradient is a unit vector defined as

$$\mathbf{D}_p = \frac{1}{M_p}\mathbf{G}_p \tag{15.24}$$

It is sometimes more convenient to work with the direction of the gradient rather than the gradient itself because the length of the direction vector does not vary. Its length is always 1 unit.

The definitions of gradient, magnitude, and direction given above can all be extended to response surfaces with more than two independent variables. The visualization is more difficult, but the concepts are identical in multidimensional space.

TRAJECTORY. Whether or not the search strategy makes use of gradients to find its way, the trajectory is the sequence of moves followed by the strategy to seek out the optimum. An efficient search will exhibit a fairly straight line trajectory from the starting point to the optimum point.

### EXAMPLE 15.10

We will illustrate several of the definitions presented in this subsection. Suppose that the response surface for some hypothetical process is given by

$$Z = 24x_1 + 22x_2 - x_1^2 - 0.5x_2^2$$

The variable $z$ represents perhaps the yield of the process as a function of two imputs, $x_1$ and $x_2$. The objective is to maximize the yield. The current operating conditions are at $x_1 = 5$ and $x_2 = 6$. Let us determine the values of the gradient itself, and the magnitude and direction of the gradient. From Eq. (15.21), the components of the gradient are

$$G_{1p} = \left.\frac{\partial z}{\partial x_1}\right|_p = 24 - 2x_1 = 24 - 2(5) = 14$$

$$G_{2p} = \left.\frac{\partial z}{\partial x_2}\right|_p = 22 - x_2 = 22 - 6 = 16$$

The gradient, according to Eq. (15.22), is

$$\mathbf{G}_p = 14i + 16j$$

The magnitude of the gradient is given by Eq. (15.23):

$$M_p = \sqrt{14^2 + 16^2} = 21.26$$

and the direction of the gradient can be determined from Eq. (15.24):

$$D_p = \frac{14i + 16j}{21.26} = 0.658i + 0.753j$$

All of these quantities are defined at the point $p(x_1 = 5, x_2 = 6)$. The terms would have different values at different $x_1$, $x_2$ locations.

To demonstrate the usefulness of the gradient in pointing the way toward the optimum, we can determine the actual location of the optimum by the same approach used in Example 15.2 of Section 15.3. Setting the partial derivatives equal to zero and solving for $x_1$ and $x_2$, we obtain

$$\frac{\partial z}{\partial x_1} = 24 - 2x_1 = 0 \qquad x_1 = 12$$

$$\frac{\partial z}{\partial x_2} = 22 - x_2 = 0 \qquad x_2 = 22$$

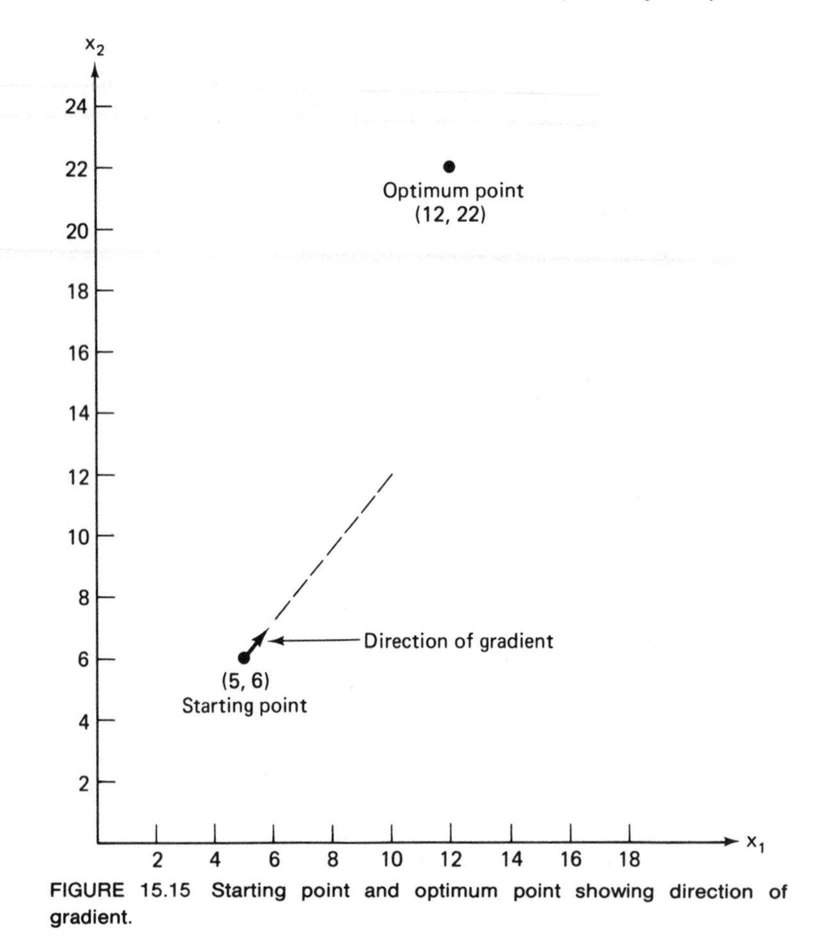


FIGURE 15.15 Starting point and optimum point showing direction of gradient.

In Figure 15.15, this optimum point is plotted along with the direction of the gradient at $x_1=5$ and $x_2=6$. It can be seen that the gradient direction is pointed roughly toward the optimum.

## Gradient search strategies

The most familiar gradient search strategy is called the *method of steepest ascent*. This method begins by estimating the gradient at the current operating point. It then moves the operating point to a new position in the direction of the gradient. The gradient is determined again at the new position in anticipation of the next move toward the optimum. The cycle of gradient determination and step move is repeated until the optimum point is achieved.

DETERMINING THE GRADIENT. In a practical problem, the mathematical equation for the response surface is not usually known. Accordingly, the gradient cannot simply be found by employing Eq. (15.21). Instead, the slope of the response surface is determined by making several exploratory moves centered around the current operating point. The exploratory moves are arranged in the form of a factorial experiment. That is, a square of experimental points is established around the current operating point, as illustrated in Figure 15.16. At each point the value of the index of performance is determined. Then the gradient components are estimated by means of the following equations:

$$G_{1p} = \frac{(z_2+z_3)-(z_1+z_4)}{2\Delta x_1}$$
$$G_{2p} = \frac{(z_2+z_4)-(z_1+z_3)}{2\Delta x_2} \tag{15.25}$$

where $z_1, z_2, z_3, z_4$ = values of the performance index at the four experimental points

$\Delta x_1$ = difference in the independent variable $x_1$ separating the experimental points

$\Delta x_2$ = difference in the independent variable $x_2$ separating the experimental points

The reason for sequencing the exploratory points 1, 2, 3, and 4 as shown in Figure 15.16 is to reduce the effect of any drift in the process. The values of $\Delta x_1$ and $\Delta x_2$ must be decided according to two opposing factors. First, Eqs. (15.25) approximate the true partial derivatives of Eqs. (15.21) more accurately as $\Delta x_1$ and $\Delta x_2$ become smaller. On the other hand, if experimental error is present in the measurement of $z$ (and it invariably is present in most process situations), the separation between experimental points must be large enough to overcome the effect of the error. Also, the exploratory moves may have to be repeated to average the errors. Judgment must be used by the analyst in order to decide on the values of $\Delta x_1$ and $\Delta x_2$ as well as the number

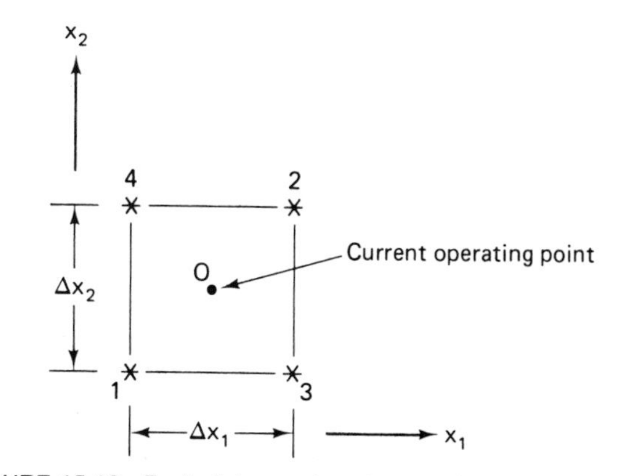


FIGURE 15.16 Factorial experiments to estimate the gradient.

of experimental replications. We shall leave the statistical analysis of the errors to other volumes, such as Box and Draper [2], if the reader wishes to pursue the subject.

### EXAMPLE 15.11

Let us compare the use of Eqs. (15.25) with Eqs. (15.21) in the previous Example 15.10. The equation for the response surface was

$$z = 24x_1 + 22x_2 - x_1^2 - 0.5x_2^2$$

The current operating point is $x_1 = 5$ and $x_2 = 6$, which yields $z = 209$. For convenience we will use $\Delta x_1 = 2$ and $\Delta x_2 = 2$, which means that the four test points and corresponding $z$ values are:

| *Test point* | $x_1$ | $x_2$ | $z$ |
|---|---|---|---|
| 1 | 4 | 5 | 177.5 |
| 2 | 6 | 7 | 237.5 |
| 3 | 6 | 5 | 205.5 |
| 4 | 4 | 7 | 209.5 |

The components of the gradient are calculated by Eqs. (15.25)

$$G_{1p} = \frac{(237.5 + 205.5) - (177.5 + 209.5)}{2(2)} = 14$$

$$G_{2p} = \frac{(237.5 + 209.5) - (177.5 + 205.5)}{2(2)} = 16$$

From Eq. (15.22), the gradient is

$$\mathrm{G}_p = 14i + 16j$$

For this response surface, the set of exploratory moves has provided the exact value of the gradient at the point $x_1 = 5$, $x_2 = 6$.

STEP MOVES. Exploratory moves are used for the purpose of determining the gradient. Once the gradient has been determined, a step move is made to the new operating point. The step move is taken in the direction of the gradient. The input variables $x_1$ and $x_2$ are incremented in proportion to the components of the direction vector.

$$\begin{aligned} \text{new } x_1 &= \text{old } x_1 + C\frac{G_{1p}}{M_p} \\ \text{new } x_2 &= \text{old } x_2 + C\frac{G_{2p}}{M_p} \end{aligned} \qquad (15.26)$$

where $C$ is a scalar quantity that determines the size of the step move. The use of the constant $C$ in Eqs. (15.26) means that the length of the step move is the same for every cycle. This occurs in spite of the fact that the gradient components will change in value both relatively and absolutely with every cycle.

STOPPING CRITERIA. The search continues until the optimum is reached. At the optimum value of the index of performance, the gradient has a value of zero. It would be sheer coincidence if a step move were to land exactly on the optimum point. A more likely occurrence is for the strategy to "overshoot" the optimum. When this happens, it can be identified by the fact that the next gradient changes direction very abruptly, perhaps heading in roughly the opposite direction from previous step moves.

When the vicinity of the optimum is found, it is usually beneficial to reduce the size of the step move. This is accomplished by reducing the value of the constant $C$ in Eqs. (15.26). In the beginning of the search, a large step size would be used to speed convergence to the optimum. The final resolution of the optimum must be achieved with smaller step moves. A reasonable criterion for stopping the search would be when repeated step moves produce no significant improvement in the index of performance.

## EXAMPLE 15.12

Let us continue Example 15.11 through the first two cycles of the method of steepest ascent. We will use a step move of length 5 units. That is, $C$ in Eqs. (15.26) equals 5.

*Cycle 1:* From the previous example, the gradient at the starting point has components

$$G_{1p} = 14 \quad \textit{and} \quad G_{2p} = 16$$

The magnitude of the gradient is 21.26. By application of Eqs. (15.26), we get the new operating point:

$$\text{new } x_1 = 5.0 + (5.0)(0.658) = 8.290$$

$$\text{new } x_2 = 6.0 + (5.0)(0.753) = 9.765$$

The index of performance can be calculated to be

$$z = 24(8.29) + 22(9.765) - (8.29)^2 - 0.5(9.765)^2 = 331.6$$

This value compares with the IP at the starting point ($x_1 = 5, x_2 = 6$):

$$z = 24(5) + 22(6) - (5)^2 - 0.5(6)^2 = 209$$

The search has led us to an improved IP value.

*Cycle 2:* Since the response surface is mathematically defined for this hypothetical example, we could use either Eq. (15.21) or (15.25). For ease of computation, we will use Eq. (15.21).

$$G_{1p} = 24 - 2x_1 = 24 - 2(8.290) = 7.420$$

$$G_{2p} = 22 - x_2 = 22 - 9.765 = 12.235$$

The magnitude of the gradient = 14.309.

$$\text{new } x_1 = 8.290 + (5.0)\frac{7.420}{14.309} = 10.883$$

$$\text{new } x_2 = 9.765 + (5.0)\frac{12.235}{14.309} = 14.040$$

The index of performance has a value at this point of $z = 353.1$.

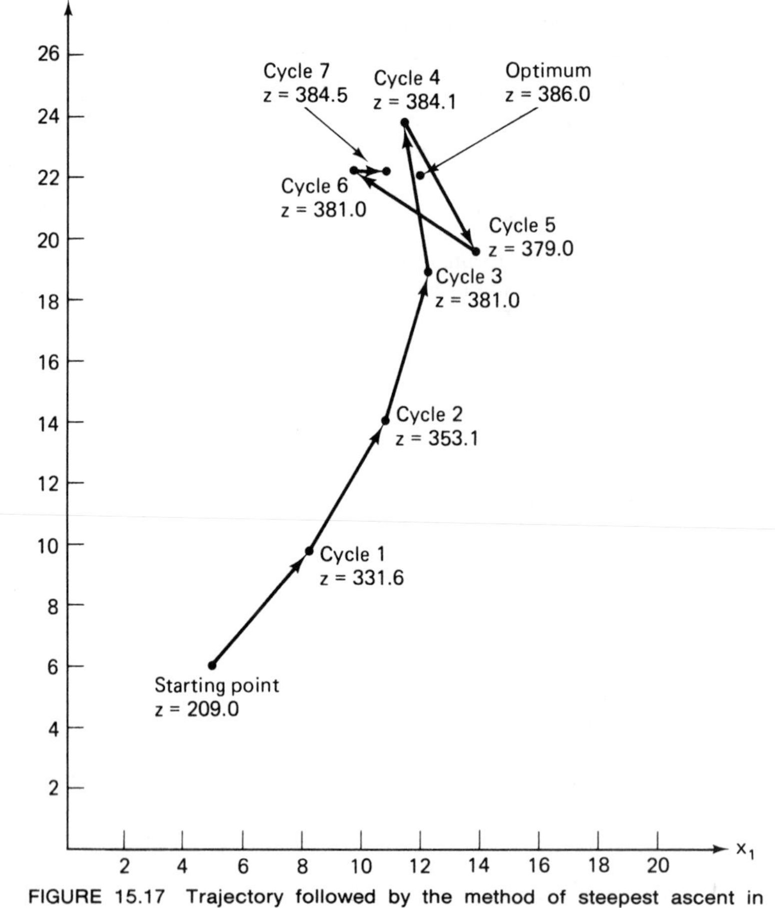


FIGURE 15.17 Trajectory followed by the method of steepest ascent in Example 15.12.

A continuation of the method of steepest ascent produces the trajectory shown in Figure 15.17. The $z$ values are indicated for each step move in the search. The reader will note that after cycle 4 ($x_1=11.498$, $x_2=23.799$), the search begins to oscillate. The gradient direction changes in cycles 5 and 6, and this is usually a tipoff that the vicinity of the optimum has been located. The reader will also note that the IP does not improve significantly. In fact, it tends to fluctuate in value. The search is stepping back and forth across the optimum. When this happens, the step size should be reduced. This is done in cycle 7 (the length of the step move is reduced from $C=5.0$ to $C=1.0$). The result is that the next move brings the search to within a very small error (0.39%) of the optimum IP value.

Other Gradient Strategies. There are other gradient search strategies in addition to the method of steepest ascent. A close cousin is the *optimum gradient method.* The procedure of the optimum gradient method is as follows:

1. The strategy begins with a determination of the index of performance and its gradient at the starting point.
2. A step move is made in the direction of the gradient.
3. The index of performance is determined at the new operating point. If the current index of performance is greater than the previous IP, take another step move in the direction of the previous gradient.
4. Repeat step 3 until no further improvement in IP results. When the current IP is less than the previous IP, determine the gradient at the previous point. Go to step 2 in the procedure.

The advantage of the optimum gradient method is that in most search problems it will not be necessary to make exploratory moves to find the gradient after every step move. Since there is a time and cost associated with each exploratory move, the optimum gradient method will be less time-consuming and less expensive to operate in most search problems.

## EXAMPLE 15.13

To illustrate how the optimum gradient method works, we will use the response surface from the previous examples in this section.

$$z=24x_1+22x_2-x_1^2-0.5x_2^2$$

Starting from the point $x_1=5$, $x_2=6$, the gradient was found in Example 15.10 to be

$$\mathbf{G}_p=14i+16j$$

To compare the optimum gradient method with the method of steepest ascent, the same step-move size of 5 units will be used from Example 15.12. The first step move in the optimum gradient method will be identical to the initial move in the method of steepest ascent (see Example 15.12).

*Cycle 1:*

$$\text{new } x_1 = 5.0 + 5(0.658) = 8.290$$
$$\text{new } x_2 = 6.0 + 5(0.753) = 9.765$$

At the starting point the index of performance had a value $z = 209$, and after the first move the new IP value is $z = 331.6$. Since the new IP value is greater than the previous one, we would take another step move in the same direction (step 3 of the procedure).

*Cycle 2:*

$$\text{new } x_1 = 8.290 + 5(0.658) = 11.580$$
$$\text{new } x_2 = 9.765 + 5(0.753) = 13.530$$

The new IP value is

$$z = 24(11.58) + 22(13.53) - (11.58)^2 - 0.5(13.53)^2 = 349.9$$

Another step move is taken in the original direction since the IP has been further increased.

*Cycle 3:*

$$\text{new } x_1 = 11.58 + 5(0.658) = 14.870$$
$$\text{new } x_2 = 13.53 + 5(0.753) = 17.295$$

The corresponding $z = 366.7$.

*Cycle 4:*

$$\text{new } x_1 = 14.87 + 5(0.658) = 18.16$$
$$\text{new } x_2 = 17.295 + 5(0.753) = 21.06$$

The new IP value is $z = 347.6$.

Since the latest index of performance value is less than the previous value, we would go back to the previous point and determine the gradient as provided in step 4 of the procedure. The previous operating point was $x_1 = 14.87$, $x_2 = 17.295$. The gradient components are

$$G_{1p} = 24 - 2(14.87) = -5.74$$
$$G_{2p} = 22 - (17.295) = 4.705$$

The magnitude of the gradient is

$$M_p = 7.422$$

Therefore, using this point ($x_1 = 14.87$, $x_2 = 17.295$) to proceed with our search in the revised direction, we get:

*Cycle 5:*

$$\text{new } x_1 = 14.87 + 5\left(\frac{-5.74}{7.422}\right) = 11.003$$

$$\text{new } x_2 = 17.295 + 5\left(\frac{4.705}{7.422}\right) = 20.465$$

The index of performance is $z=383.8$, which is greater than the previous point, where $z=366.7$.

*Cycle 6:*

$$\text{new } x_1 = 11.003 + 5(-0.773) = 7.136$$

$$\text{new } x_2 = 20.465 + 5(0.634) = 23.635$$

The corresponding $z=361.0$. The reduction in index of performance means that we would redetermine the gradient components at the previous point ($x_1=11.003$, $x_2=20.465$).

$$G_{1p} = 24 - 2(11.003) = 1.994$$

$$G_{2p} = 22 - 20.465 = 1.535$$

$$M_p = 2.516$$

The reader might observe that the direction of the gradient has virtually reversed itself from the previous value. This is the signal that the search has overstepped the optimum point. A reduction in the step size is therefore advisable. We shall (arbitrarily) reduce the length of the step move from 5.0 to 1.0 and proceed.

*Cycle 7:*

$$\text{new } x_1 = 11.003 + 1\left(\frac{1.994}{2.516}\right) = 11.796$$

$$\text{new } x_2 = 20.465 + 1\left(\frac{1.535}{2.516}\right) = 21.075$$

The corresponding IP value is $z=385.5$.

Additional cycles will cause the search to begin to oscillate back and forth across the summit of the response surface.

From Example 15.10, the reader will recall the actual location of the optimum to be at $x_1=12$ and $x_2=22$. The corresponding IP value is $z=386$. The optimum gradient method has led us to within 0.13% of the maximum IP on this response surface. Of course, in a real search problem in manufacturing, the true optimum index of performance would not be known.

The trajectory followed by this search procedure is illustrated in Figure 15.18.

For further discussion on search techniques, the reader is advised to seek out several of the references: [2], [9], [18], and [19].

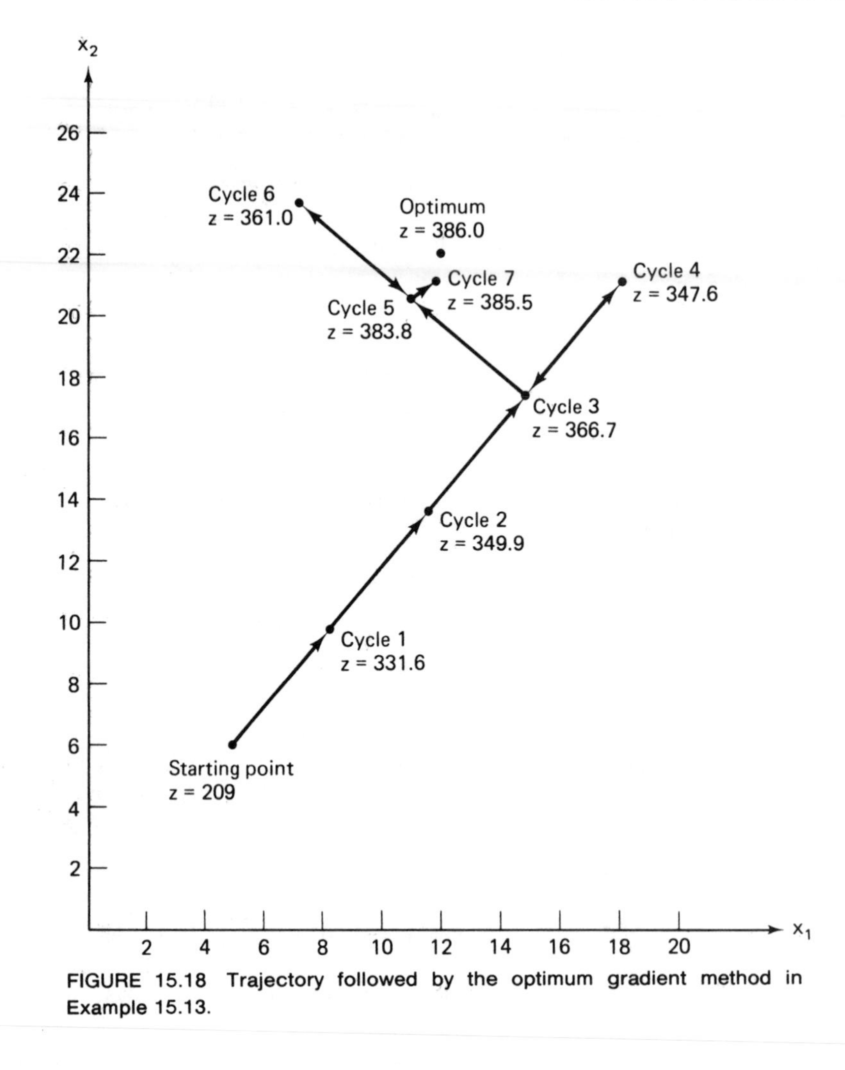


FIGURE 15.18 Trajectory followed by the optimum gradient method in Example 15.13.

## REFERENCES


[1] BELLMAN, R. E., and DREYFUS, S. E., *Applied Dynamic Programming*, Princeton University Press, Princeton, N.J., 1962.

[2] BOX, G. E. P., and DRAPER, N. R., *Evolutionary Operation*, John Wiley & Sons, Inc., New York, 1969.

[3] CASSELL, D. A., *Introduction to Computer-Aided Manufacturing in Electronics*, Wiley–Interscience, New York, 1972.

[4] DAVIES, W. D. T., *System Identification for Self-Adaptive Control*, Wiley–Interscience, London, 1970.

[5] DRAPER, N. R., and SMITH, H. *Applied Regression Analysis*, John Wiley & Sons, Inc., New York, 1966.

[6] EVELEIGH, V. W., *Adaptive Control and Optimization Techniques*, McGraw-Hill Book Company, New York, 1967.

[7] GROOVER, M. P., "Adaptive Control and Adaptive Control Machining," *Educational Module*, MAPEC, Copyright Purdue Research Foundation, 1977. All rights reserved.

[8] HARRISON, T. J (Editor), *Minicomputers in Industrial Control*, Instrument Society of America, Pittsburgh, Pa., 1978.

[9] IDLESOHN, J. M., "10 Ways to Find the Optimum," *Control Engineering*, June, 1964, pp. 97–100.

[10] LEE, T. H., ADAMS, G. E., and GAINES, W. M., *Computer Process Control: Modeling and Optimization*, John Wiley & Sons, Inc., New York, 1968.

[11] PRESSMAN, R. S., and WILLIAMS, J. E., *Numerical Control and Computer-Aided Manufacturing*, John Wiley & Sons, Inc., New York, 1977.

[12] SAVAS, E. S., *Computer Control of Industrial Processes*, McGraw-Hill Book Company, New York, 1965.

[13] SMITH, C. L. *Digital Computer Process Control*, Intext Educational Publishers, Scranton, Pa., 1972.

[14] TAHA, H. A., *Operations Research, An Introduction*, 2nd ed., Macmillan Publishing Co., Inc., New York, 1976.

[15] TIPNIS, V. A., "Development of Mathematical Models for Adaptive Control Systems," *Proceedings*, 13th Annual Meeting of the Numerical Control Society, March, 1976, Cincinnati, Ohio, pp. 149–156.

[16] WAGNER, H. M., *Principles of Operations Research*, Prentice-Hall, Inc., Engelwood Cliffs, N.J., 1970.

[17] WHITEHOUSE, G. E., and WECHSLER, B. L., *Applied Operations Research*, John Wiley & Sons, Inc., New York, 1976.

[18] WILDE, D. J., *Optimum Seeking Methods*, Prentice-Hall, Inc., Englewood Cliffs, N.J., 1964.

[19] WILDE, D. J., and BEIGHTLER, C. S., *Foundations of Optimization*, Prentice-Hall, Inc., Englewood Cliffs, N.J., 1967.

## PROBLEMS

**15.1.** Consider some manufacturing or industrial process with which you are familiar. For this process, classify the process variables into the following categories:

(a) Input variables

i. Controllable input variables

ii. Uncontrollable input variables

iii. Fixed variables

(b) Output variables
    i. Measurable output variables
    ii. Performance evaluation variable

Discuss the general effect of each of the input variables on the measurable output variables. (If there is a large number of input variables of the three types, limit your discussion to those variables which are the most important.) How is the performance evaluation variable determined from the measurable output variables?

**15.2.** For the process considered in Problem 15.1, what are the practical constraints on the values the variables can assume?

**15.3.** In Problem 15.1, can a mathematical model of the process be determined to express:

(a) The relationship between the controllable input variables and the measurable output variables?
(b) The relationship between the measurable output variables and the performance evaluation variable?

According to your answers to (a) and (b) and your knowledge of the process, what type of control strategy would be most appropriate for the process?

**15.4.** Consider a process similar to the one shown in Figure 15.4, except that the transfer function for the process is given by

$$\frac{y(s)}{x(s)} = \frac{37.2}{s+5.8}$$

The desired steady-state value of the output $y$ is 10.0. Set up the problem as a steady-state optimal control problem with objective function:

$$\text{minimize } z = (10-y)^2$$

Solve for the required step input $x$ that will achieve this objective in the steady state.

**15.5.** The performance of a certain manufacturing process is determined by two input variables, $x_1$ and $x_2$. The performance evaluation variable $z$ represents the process yield, which is to be maximized. The relationship between yield $z$ and the two inputs $x_1$ and $x_2$ has been determined as follows:

$$z = 0.523 + 0.020x_1 - 0.0012x_1^2 + 0.17x_2 - 0.025x_2^3$$

The constraints on the values of $x_1$ and $x_2$ are as follows: $2.2 < x_1 < 14.0$, $0.7 < x_2 < 2.0$. Within these constraints, what values of $x_1$ and $x_2$ will maximize the yield? What is the maximum yield that can be expected from the process?

**15.6.** The measurable output variables for a new experimental welding process are $y_1$ and $y_2$. These outputs are regulated by means of two inputs to the process, which can be manipulated as desired. The two inputs are $x_1$ and $x_2$. The relationships between inputs and outputs are as follows:

$$y_1 = 1.6 + 2.2x_1 + 3.7x_2$$

$$y_2 = 7.1x_2$$

The most appropriate performance evaluation variable, $z$, for the welding process is the linear speed of the welding head, since this is a measure of the production rate. The objective in the process is to maximize this performance variable. In the research on the process, it has been determined that $z$ is related to the two outputs $y_1$ and $y_2$ as follows:

$$z = 3.0y_1 - 0.40y_1^2 + 1.8y_2 - 1.208y_2^2$$

Determine the values of $x_1$ and $x_2$ that maximize the welding head speed $z$.

**15.7.** The performance $z$ of a process is related to the inputs $x_1$ and $x_2$ by the following equation:

$$z = 25x_1 - 2x_1^2 + 41x_2 - 5x_2^2 + 4x_1x_2$$

Find the values $x_1$ and $x_2$ that maximize $z$. What is the optimum $z$ value?

**15.8.** Solve the following linear programming problem graphically:

$$\text{maximize } z = 10x_1 + 3x_2$$

subject to

$$6x_1 + 4x_2 \leqslant 24$$
$$4x_1 + 8x_2 \leqslant 40$$
$$10x_1 + 2x_2 \leqslant 30$$
$$x_1 \leqslant 0, \quad x_2 \leqslant 0$$

**15.9.** One of the production shops at Special-T Company makes two products, each of which requires three manufacturing operations. The company can sell all it can make of these two products. In fact, it is considering a proposal to increase production capacity. For the time being, however, the constraints on the operations and other data are given in the following table:

| | *Production time required Per unit (h/unit)* | | | | *Selling* |
|---|---|---|---|---|---|
| | | | | *Cost* | *Price* |
| *Product* | *Oper 1* | *Oper 2* | *Oper 3* | *($)* | *($)* |
| 1 | 1.2 | 2.3 | 4.5 | 80 | 95 |
| 2 | 2.3 | 6.8 | 1.9 | 110 | 130 |
| Hours available | 24.0 | 32.0 | 24.0 | | |

(a) Determine the product mix that maximizes profit.

(b) If production capacity were to be increased, to which operations should the additional capacity be allocated? Is there surplus capacity on any operation?

**15.10.** Assume a process whose mathematical model is given by

$$y = A + Bx$$

The value of $A$ remains constant throughout process operation at $A = 50$.

However, the value of $B$ varies through a range from $B=10$ to 20. In order to optimize the process the value of $B$ must be measured. It has been decided that the instantaneous approximation method is to be used. If $y$ and $x$ can be measured with sufficient accuracy, what is the value of $B$ if $y=135$ and $x=5$.

**15.11.** Assume that a process can be modeled by an equation of the form

$$y = Ax + B/x$$

The value of $A$ remains constant throughout process operation at $A=32$. However, the value of $B$ fluctuates slowly because of gradual environmental changes.

(a) Use the instantaneous approximation method to determine the value of $B$ if $y$ and $x$ have been recently measured at values $y=95.0$ and $x=2.1$.

(b) Using the value of $B$ determined in part (a), determine the value of $x$ that will minimize $y$, which is a measure of product cost.

(c) What new value of $y$ will result if the change in $x$ from part (b) is implemented?

**15.12.** Consider the welding process from Problem 15.6. It has been found in applications that the process model is sensitive to variations in work material and atmospheric conditions. The input/output relationships can be modeled by equations of the form

$$y_1 = 1.6 + Ax_1 + 3.7x_2$$

$$y_2 = Bx_2$$

Fortunately, the four variables $y_1$, $y_2$, $x_1$, and $x_2$ can all be measured during the operation. During the welding of a special high-strength alloy in an enclosed vessel, the following values were measured:

$$y_1 = 4.05 \qquad x_1 = 0.91$$

$$y_2 = 0.87 \qquad x_2 = 0.13$$

As in Problem 15.6, the objective is to maximize the objective function

$$z = 3.0y_1 - 0.40y_1^2 + 1.8y_2 - 1.208y_2^2$$

(a) From the variable measurements, determine the unknown parameter values, $A$ and $B$.

(b) What changes should be made in the inputs $x_1$ and $x_2$?

**15.13.** The tool-chip thermocouple is a thermocouple whose two dissimilar metals are the cutting tool and the chip. The cutting temperature in a machining operation can be monitored by measuring the emf output at the tool-chip interface (this interface constitutes the junction of the two dissimilar metals of the thermocouple). It has been proposed to monitor cutting temperature as part of an adaptive control machining system. The data given below were taken during the calibration of the tool-chip thermocouple.

| Tool-chip thermocouple (mV) | 11.5 | 12.7 | 14.0 | 15.3 | 17.1 | 18.2 | 18.9 |
|---|---|---|---|---|---|---|---|
| Temperature (°C) | 552 | 601 | 670 | 742 | 839 | 878 | 930 |

(a) Use the least-squares method to determine the calibration equation

$$y = A + Bx$$

where $y$ = temperature, °C
$x$ = emf output of the tool-chip thermocouple, mV

(b) What are the correlation coefficient and the standard error of estimate for the equation?

**15.14.** The model for a certain component of a manufacturing process is given by

$$y = Ax^n$$

However, the values of $A$ and $n$ are observed to change slowly over time because of unpredictable environmental factors. Determining these two parameters is a necessary condition for control of the process. Both the independent variable $x$ and the dependent variable $y$ can be measured during the operation. Over a 15-min observation period, the following data have been monitored on $x$ and $y$:

| $x$ | 50.1 | 52.0 | 54.9 | 57.2 | 53.8 | 59.7 |
|---|---|---|---|---|---|---|
| $y$ | 606 | 621 | 642 | 659 | 635 | 675 |

Determine by the least-squares method the values of $A$ and $n$ in the model above. (*Hint:* By a logarithmic transformation, the equation $y = Ax^n$ can be converted into

$$\ln y = \ln A + n \ln x$$

which has the standard linear equation form for using the least squares.)

**15.15.** Derive the normal equations for the linear equation

$$y = A + Bx_1 + Cx_2$$

**15.16.** The process model for a certain component of an industrial operation is

$$K\frac{dy}{dt} + y = A + Bx$$

where the parameters $K$, $A$, and $B$ vary over time. During operation, the values of $dy/dt$, $y$, and $x$ can all be measured. During a particular 7-min interval the following values have been sampled:

| *Time* | *dy/dt* | *y* | *x* |
|---|---|---|---|
| 1 | 4.0 | 11.0 | 3.0 |
| 2 | 3.1 | 18.2 | 4.0 |
| 3 | 1.9 | 22.4 | 4.5 |
| 4 | 1.0 | 25.5 | 4.7 |
| 5 | −0.2 | 26.4 | 4.4 |
| 6 | −0.9 | 25.9 | 4.0 |
| 7 | −1.0 | 24.8 | 3.8 |

Use the normal equations derived in Problem 15.15 to determine the current values of the parameters $K$, $A$, and $B$ in the model above. *Notes:* (1) The same sort of rearrangement of the equation illustrated in Example 15.9 must be used in this problem; (2) the computations must be carried out with a high degree of precision.

**15.17.** The response surface for the index of performance as a function of two input variables is given by the following equation:

$$z = 25x_1 - 2x_1^2 + 41x_2 - 5x_2^2 + 4x_1x_2$$

(a) Determine the gradient at the point defined by $x_1 = 2$, $x_2 = 2$.
(b) Determine the magnitude of the gradient at the same point.
(c) Determine the direction of the gradient.

**15.18.** For the same response surface from Problem 15.17, solve for the gradient, the magnitude of the gradient, and the direction of the gradient at the point

$$x_1 = 15, \; x_2 = 10.$$

**15.19.** For the same response surface from Problem 15.17, solve for the gradient, the magnitude, and the direction at the point $x_1 = 17.25$, $x_2 = 11.0$.

**15.20.** The current operating point for a particular long-run machining job is: cutting speed = 200 ft/min (1.016 m/s) and feed rate = 0.010 inch/rev (0.254 mm/rev). At these conditions, the unit cost of the operation = \$2.00. A series of exploratory moves have been made to determine whether improvements in cutting conditions can be made. The results given in the accompanying table have been recorded (these data represent the averages of several replications).

| *Speed* | | *Feed* | | *Unit cost* |
|---|---|---|---|---|
| *ft/min* | *m/s* | *in./rev* | *mm/rev* | |
| 250 | 1.27 | 0.012 | 0.305 | \$2.22 |
| 250 | 1.27 | 0.008 | 0.203 | 1.96 |
| 150 | 0.762 | 0.012 | 0.305 | 2.01 |
| 150 | 0.762 | 0.008 | 0.203 | 1.78 |

(a) With data such as these, there is a large difference in the units scale for speed versus the units scale for feed rate. This results in problems during calculation and interpretation of the gradient. To overcome these problems, the data can be coded. The objective of the data coding is to approximately equalize the scale of the two variables, in this case speed and feed. For example, in the U.S. Customary units, the speed in ft/min could be divided by 20 and the feed in in./rev could be multiplied by 1000. The problem is not so severe in the SI units. Here the scale could be leveled by multiplying feed in mm/rev by 4.0. Rewrite the table by coding both the U.S.C. and SI variables as indicated.

(b) Determine the gradient components in either the (coded) U.S.C. and SI units system.

(c) If the step move to be taken in the method of steepest ascent is defined by $C=2.0$ in Eqs. (15.26), determine the new operating conditions.

**15.21.** Suppose that the response surface for a certain manufacturing process was defined by the equation

$$z = 17x_1 + 27x_2 - x_1^2 - 0.9x_2^2$$

Determine the approximate optimum operating point using the method of steepest ascent. The starting point of the search should be $x_1=2$, $x_2=3$, and the step size should be $C=4.0$.

**15.22.** Solve Problem 15.21 using the optimum gradient method.

part 

# Systems for Manufacturing Support

# Production Systems at the Operations Level

## 16.1 OVERVIEW

In Parts II, III, and IV of the book, we considered automated production systems. Our focus in Part II was on flow line systems for high-volume production. In Part III, numerical control systems used to automate the job shop and batch manufacturing were described. In Part IV, we discussed computer-aided manufacturing with emphasis on systems for process monitoring and control.

Now, in Part V, we will consider various systems for manufacturing support. These systems distinguish themselves from those previously considered by being more software-oriented rather than hardware-oriented. Transfer lines, NC machines, and process control systems are concerned with automation hardware—direct automation of the manufacturing operations. When we speak of systems for manufacturing support, we are referring to systems that interface with the manufacturing operations in an indirect way. Most of these systems rely on the use of the computer for practical implementation. Accordingly, they fall within a scope that we have defined in Chapter 10 as CAM for manufacturing support.

Part V is divided into two chapters. The current chapter deals with manufacturing support systems and concepts at the level of the production processes. At this level, we are concerned with making the individual operations as efficient as possible. Chapter 17 is concerned with systems that support manufacturing at the plant level. At this higher level, our interest is in managing the plant as a responsive, productive, and efficient enterprise.

Production systems for manufacturing support at the operations level include the following topics:

1. Computer-generated time standards.
2. Machinability data systems.
3. Cutting conditions optimization.

This list attempts to represent a variety of different techniques that have been developed to solve industrial problems at the operations level.

## 16.2 COMPUTER-GENERATED TIME STANDARDS

Work measurement can be defined as the development of a time standard to indicate the value of a work task. According to a recent survey conducted by *Industrial Engineering* magazine and Patton Consultants, Inc., 95% of the manufacturing firms responding to the survey use work measurement [15]. There are various purposes for which work measurement is used. Among these are

Wage incentives.
Estimating and job costing.
Production scheduling and capacity planning.
Measurement of worker performance.

The techniques used to establish time standards in manufacturing include the following:[1]

1. Direct time study.
2. The use of standard data.
3. Predetermined time standard systems.
4. Estimates based on previous experience.
5. Work sampling.

[1] The reader unfamiliar with these terms should consult the standard texts in work measurement, such as Mundel [10] or Niebel [11].

By far the most often used work measurement technique is direct time study. According to the survey, 46% of all standards are set by this method.

Direct time study involves the direct observation of the task, timing the elements of the work cycle with a stopwatch, rating the performance of the operation (also called "leveling"), applying the necessary allowances (for personal delay, machine cycle, etc.), and calculating the standard time for the job. Among the disadvantages of the direct time study method are:

The performance rating is often disputed by the worker.

The time standard cannot be set until after the job is in production. This often means wasted time in the shop until after the standard has been set.

There tends to be variability in the standards among time study analysts.

Much time is required by the time study analyst in taking the study and calculating the standard.

Even though many of the jobs done in a given plant are similar, the stopwatch time study procedure must nevertheless be applied to all new jobs. (This, of course, depends on the terms of the labor–management agreement.)

In recent years, a number of computer packages have been commercially introduced for setting time standards. The objective of these computer packages is to reduce the time required in the development of work standards, and to overcome some of the problems associated with direct time study. In addition, these systems provide an opportunity to generate a manufacturing data base for storing information not only on operation time standards, but also standard cost data, tooling information, job instructions, and so on.

The computerized systems are based on the use of standard data stored in computer files. The term *standard data*, when applied in the context of work measurement, refers to previously determined time values corresponding to particular work elements or groups of work elements. These elements of various types comprise all manual production elements in the factory. For example, in a turret lathe operation, the workpiece must be loaded onto the machine and unloaded. The time required to load (or unload) the piece depends on the weight and configuration of the part, how it is to be fixtured in the turret lathe, and so on. Standard data can be developed to indicate the time required for the loading element as well as other work elements in the operation cycle. The individual element times depend on the attributes of the job (workpiece weight, size, machine tool, etc.). The computer stores these elemental times either in a data file or in the form of a mathematical formula. To use the package, the time study analyst must analyze the job to be timed by dividing it into its elements and specifying the attributes of the job for each element. The computer then retrieves from the file or calculates the element times, sums the times, and applies the necessary allowances to determine the standard time for the total cycle.

Among the advantages of using a computerized system for generating time standards are the following:

1. Reduction in time required by the time study analyst to set the standard. A reduction of approximately 20% was reported by one consulting firm [7] in total manpower required for methods analysis and maintenance. In another application, a 50% reduction in clerical effort was reported compared to manual methods of applying standard data [16].
2. Greater accuracy and uniformity in the time standards.
3. Ease of maintaining the methods and standards file when engineering and methods changes occur.
4. Elimination of the controversial performance rating (leveling) step.
5. Time standards can often be set before the job gets into production.
6. Improved manufacturing data base for production planning, scheduling, forecasting manning requirements, tool control, and so on.

**EXAMPLE 16.1**[2]

There are several commercially available computerized standards packages. We will use the 4M DATA System as an example of these packages. 4M stands for "Micro-Matic Methods and Measurement."

The 4M DATA System is a computerized work measurement system available from the MTM Association as a method for applying MTM-1 predetermined time standards. The MTM-1 motions are elementary body motions that are used in all manual production operations. About 90% of the manual work performed in the typical industrial situation consists of the motions REACH, GRASP, MOVE, POSITION, and RELEASE. These motions can be combined into GET and PLACE motion groups. Special symbols for these two motion groups convey the necessary data to the 4M computer program for calculating the proper standard time values. In addition to the GET and PLACE motion groups, the computer program also recognizes all other MTM-1 motions with their corresponding variables. The methods/time study analyst can develop the time standard for a given manual operation cycle by calling the required elements in the 4M DATA package.

To illustrate the use of the system, let us examine the GET motion group. The format of the GET symbol is

GXX xx

[2] This example is based on reference [9].

G stands for GET. The next two digits, XX, are numeric and represent the category of grasp and degree of control over the object. The definitions of these digits are presented in Table 16.1. The final digits, xx, give the distance of the GET reach, expressed in inches. The symbols for other motion elements are based on a similar format and coding system. The reader familiar with MTM will recognize the general construction of the elements.

TABLE 16.1 GET Symbol in 4M DATA System

GXX xx = format of GET symbol
G = GET
XX = category of grasp and degree of control
- 01-No grasp time, A Reach*
- 02-No grasp time, B Reach*
- 03-No grasp time, E Reach*
- 11-Pickup grasp, A Reach*
- 12-Pickup grasp, B Reach*
- 13-Pickup grasp, D Reach*
- 20-Regrasp
- 30-Transfer grasp
- 41-Search and select, greater than 1.0 $in.^3$
- 42-Search and select, 0.01 to 1.0 $in.^3$
- 43-Search and select, less than 0.01 $in.^3$

xx = Distance of reach in inches.

* Definitions of A Reach, B Reach, etc.:

A Reach = Reach to object in fixed location or to object in other hand or on which other hand rests

B Reach = Reach to single object in location which may vary slightly from cycle to cycle

D Reach = Reach to a very small object or where accurate grasp is required

E Reach = Reach to indefinite location to get hand in position for body balance or next motion or out of the way

Figure 16.1 shows the general input format for a manual operation cycle. The two columns of symbols are for the left-and right-hand motions. The 4M program accepts these data and performs a series of calculations and data processing steps (apply allowances, uncoding the symbols for printing, computation of method improvement indices, etc.). It then prints out the analysis and results according to a number of different options. One possible format is shown in Figure 16.2 (this is for a different operation cycle than that shown in Figure 16.1). The standard time is given in hours and minutes per cycle. The MAI stands for Motion Assignment Index and is a measure of how completely the two hands are utilized during the cycle. An index of 50% would mean effective use of only one hand, while 100% would indicate complete utilization of both hands. The other indices (RMB, GRA, POS, and PROC) represent additional guides that might indicate possible methods improvements.

MMMM DATA

MICRO-MATIC METHODS & MEASUREMENT

Keypunch on ALL cards

study no. 105

| H 1 | oper. no. 01 | oper. descr. ASSEMBLE COMPENSATORS AND MAGNETS | mo 06 | day 02 | yr. 72 |
|---|---|---|---|---|---|
| H 2 | dept. no. | dwg. & item | part or apparatus METER FRAME | | |
| H 3 | mech. no. | tooling ASSY FIXTURE | analyst TRAINING FILM | Lng. MTM | prac. (Y) N · allow 08 · micro (Y) N · print A B C |

| ELEM NO. | LINE NO. | LEFT HAND MOTIONS LITERALS | ROTATION | DIST | WGT | LEAD NO | ROTATION | DIST | WGT | RIGHT HAND MOTIONS LITERALS | PROCESS (min x 10) | FREQUENCY | DIR. CODE |
|---|---|---|---|---|---|---|---|---|---|---|---|---|---|
| 6 7 | 8 9 | 10–31 | 32–35 | 36–37 | 38–39 | 40 | 41–44 | 45–46 | 47–48 | 49–70 | 71–74 | 75–78 | 79–80 |
| 0 | | PLACE COMPONENTS IN CAVITY | | | | | | | | | | | E |
| | 1 | ASSEMBLY TO TABLE | P03 | 20 | | | G12 | 22 | | FRAME FROM CONTAINER | | | B |
| | 2 | LARGE COMPENSATOR | G4R | 15 | | | P13R | 22 | | FRAME IN FIXTURE | | | B |
| | 3 | COMPENSATOR IN CAVITY | P131 | 07 | | | G4R | 04 | | SMALL COMPENSATOR | | | B |
| | 4 | LARGE MAGNET | G4R | 07 | | | P131 | 07 | | COMPENSATOR IN CAVITY | | | B |
| | 5 | MAGNET IN CAVITY | P131 | 08 | | | G4R | 08 | | SMALL MAGNET | | | B |
| | 6 | FRAME | G1T | 04 | | | P131 | 08 | | MAGNET IN CAVITY | | | B |
| | 7 | | | | | | | | | | | | |
| | 8 | | | | | | | | | | | | |
| | 9 | | | | | | | | | | | | |
| | 0 | | | | | | | | | | | | |
| | | | | | | | | | | | | | E |
| | 1 | | | | | | | | | | | | |
| | 2 | | | | | | | | | | | | |

Keypunch element headings

FIGURE 16.1 Input format for a namual operation using the 4M DATA system. (Reprinted by permission of MTM Association.)

SUMMARY OPERATOR INSTRUCTIONS DATE 06/02/72

STUDY NO. 112 OPERATION NO. 09 ASSEMBLE BRACKETS TO RING 02/10/77
DEPT. NO. 30 IDENTIFICATION 20W1301 WATT-HOUR METER
MACH. NO. TOOLING PRESS DIE ANALYST TRAINING FILM

| | | FREQ. | LH | RH | TOTAL |
|---|---|---|---|---|---|
| 01 | PLACE RING ON FIXTURE | | 669 | 849 | 905 |
| 02 | ASSEMBLE RIVETS, NUT, BRACKET | 2.0000 | 1904 | 2394 | 3348 |
| 03 | REMOVE RING. PLACE BRACKET, NUT IN FIXTURE. REPLACE RING | | 779 | 921 | 1062 |
| 04 | ASIDE COMPLETED ASSEMBLY. PLACE BRACKET, NUT IN FIXTURE | | 349 | 301 | 467 |

TOTAL 5782 MU RMB 44 PERCENT
STANDARD TIME .00624 HOURS/ .37440 MIN. MAI 70 PERCENT GRA 22 PERCENT PROC 0 PERCENT
CYCLES PER HOUR 160.3 POS 34 PERCENT

**FIGURE 16.2** One possible output format of the 4M DATA system. (Reprinted by permission of MTM Association.)

## 16.3 MACHINABILITY DATA SYSTEMS

Machinability can be defined as "the relative ease with which a metal can be machined by an appropriate cutting tool."[3] There are a number of criteria that are typically used to assess machinability in a given operation, including:

Tool life. A long tool life is desirable.

Surface finish. The workpiece finish must satisfy the design specification.

Dimensional stability. Workpart dimensional tolerances must be achieved.

Cutting forces, power consumption, and temperatures. Good machinability implies low forces, power, and temperatures in the operation.

Ease of chip disposal.

Economic criteria, such as minimum cost or maximum production rate.

To obtain these desirable performance criteria, the machining operation must be carried out using the appropriate machine, cutting tool, speed, feed, and depth of cut. The machine is specified on the route sheet for the part by the process planner. The selection of cutting tool must be based on such factors as type of operation (turning, milling, drilling, etc.), whether a roughing or finishing operation is used, work material type, and the experience of the individual who selects the tool. Specification of cutting tool includes definition of both tool geometry and tool material.

Finally, after the machine tool and cutting tool have been selected, success in the operation depends on the proper choice of cutting conditions—speed, feed, and depth of cut. Depth of cut is usually predetermined by the workpiece geometry and operation sequence. Therefore, the problem reduces to one of determining the proper speed and feed combination. Machinability data systems are basically intended to solve this problem.

### Definition of the problem

Stated precisely, the objective of a machinability data system is to select

1. Cutting speed, and
2. Feed rate

given that the following characteristics of the operation have been defined:

[3]M. P. Groover, "A Survey on the Machinability of Metals," *Tech. Paper MR*76-269, Society of Manufacturing Engineers, Dearborn, Mich., 1976, p. 1.

1. Type of machining operation.
2. Machine tool.
3. Cutting tool.
4. Workpart.
5. Operating parameters other than feed and speed.

The magnitude of the problem can best be appreciated by contemplating the multitude of different parameters that are included within these five operation characteristics. A partial list of the parameters is presented in Table 16.2.

The methods of solving the speed/feed selection problem are these:

1. Experience and judgment of process planner, foreman, or machine operator.
2. Handbook recommendations.
3. Computerized machinability data systems.

TABLE 16.2 Characteristics of a Machining Operation

1. *Type of Machining Operation*
   a. Process type—turning, facing, drilling, tapping, milling, boring, grinding, etc.
   b. Roughing operation vs. finishing operation.
2. *Machine Tool Parameters*
   a. Size and rigidity
   b. Horsepower
   c. Spindle speed and feed rate levels
   d. Conventional or NC
   e. Accuracy and precision capabilities
   f. Operating time data
3. *Cutting Tool Parameters*
   a. Tool material type (high-speed steel, cemented carbide, ceramic, etc.)
   b. Tool material chemistry or composition
   c. Physical and mechanical properties (hardness, wear resistance, etc.)
   d. Type of tool (single point, drill, milling cutter, etc.)
   e. Geometry (nose radius, rake angles, relief angles, number of teeth, etc.)
   f. Tool cost data
4. *Workpart Characteristics*
   a. Material—basic type and specific grade
   b. Hardness and strength of work material
   c. Geometric size and shape
   d. Tolerances
   e. Surface finish
   f. Initial surface condition of workpiece
5. *Operating Parameters Other Than Feed and Speed*
   a. Depth of cut
   b. Cutting fluid, if any
   c. Workpiece rigidity
   d. Fixtures and jigs used

Relying on the experience and judgment of any individual is the least systematic approach and carries the greatest risk. The risk lies in the potential loss of the individual who has acquired the needed experience and judgment over many years in the shop. Personal judgment is also undesirable because it usually has no scientific foundation. Cutting conditions derived from personal experience are not based on economic criteria.

Handbook recommendations are compiled from the experiences of more than one person. Handbooks of machinability data are generally developed from a systematic analysis of large quantities of machining data. The cutting recommendations are often based on designed laboratory experiments whose objective is to determine optimum speeds and feeds. The best known of these handbooks is the *Machining Data Handbook* [8].

Although the handbook approach represents a definite improvement over personal judgment, it often suffers from several drawbacks when applied to a particular company's machining environment. First of all, handbook recommendations often tend to be conservative, meaning that the suggested feeds and speeds are based on worst-case conditions. Second, handbooks are intended to be general guides and often do not coincide with the particular product line and machine tools of a given shop. Third, the use of handbooks is not compatible with the objective of many firms to automate the process planning function using a computerized manufacturing data base.

### Computerized machinability data systems

To overcome these difficulties, efforts have been directed to the development of computerized machinability data systems. These efforts date back to the early 1960s and are continuing today. Some of the systems have been developed by individual firms to meet their own specific requirements. The importance of these systems has grown with the increase in the use of NC machines and the economic need to operate these machines as efficiently as possible. The importance of computerized machinability data systems will continue to grow as CAM becomes more common, and with it the development of integrated manufacturing data bases.

Computerized machinability data systems have been classified into two general types by Pressman and Williams ([13],[14]):

1. Data base systems.
2. Mathematical model systems.

We shall discuss the two types below.

DATA BASE SYSTEMS. These systems require the collection and storage of large quantities of data from laboratory experiments and shop experience. The data base is maintained on a computerized storage file that can be accessed either by a remote terminal (CRT or teletype) or in a batch mode for a more permanent printout of cutting recommendations. An example of a typical printout is shown in Figure 16.3.

To collect the machinability data required for a data base system, cutting experiments are performed over a range of feasible conditions. These experiments are most commonly conducted in the laboratory. However, many data base machinability systems allow for shop data to be entered into the files also. For each set of conditions, computations are made to determine the cost of the operation. Not only is the total cost per piece calculated, but the cost components that make up the total are also calculated. Examples of these costs are illustrated in Figure 16.3.

The computations are based on the traditional concept in machining economics that the total cost per piece is composed of elements as given in the following equation:

$$C_{pc} = C_o T_m + C_o T_h + \frac{T_m}{T}(C_t + C_o T_{tc}) \tag{16.1}$$

where $C_{pc}$ = cost per workpiece, \$/piece
$C_o$ = cost to operate the machine tool (labor, machine, and applicable overhead), \$/min
$T_m$ = machining time, min
$T_h$ = workpiece handling time, min
$T$ = tool life, min
$C_t$ = cost of tooling, \$/cutting edge
$T_{tc}$ = tool change time, min

The equations used in many programs are more complicated than this,[4] but the basic machining cost concept is embodied in Eq. (16.1).

The organization of a typical data base system consists of files of data not unlike the arrangement of Table 16.2. To access the data on these files, the user would have to enter certain descriptive data that would identify the type of machining operation, work material, tooling, and so on. The printout would consist of a listing of the machining recommendations corresponding to the input data.

MATHEMATICAL MODEL SYSTEMS. These systems go one step further than the data base systems. Instead of simply retrieving cost information on operations that have already been performed, the mathematical model systems attempt to predict the optimum cutting conditions for an operation. The prediction is generally limited to optimum cutting speed, given a certain feed

[4] For example, the General Electric System and the Machinability Data Center System, both described in reference [12], use expanded versions of Eq. (16.1).

COST AND PRODUCTION RATE FOR TURNING
THROWAWAY CARBIDE TOOLS

| DATA SET NO | WORK MATERIAL | HARDNESS | TOOL MATL | CUT SPD F/M | FEED IN/REV | TOOL LIFE MIN | FEED COST $ | RAPD TRAV $ | LOAD UNLD $ | SET UP $ | INDX INST $ | HLDR DEPR $ | INSERT COST $ | TOTAL COST $/PC | PROD RATE PC/HR |
|---|---|---|---|---|---|---|---|---|---|---|---|---|---|---|---|
| 16 | AISI 4340 | 515 | C-8 | 300 | 0.0100 | 8 | 0.87 | 0.04 | 0.34 | 0.15 | 0.04 | 0.00 | 0.11 | 1.58 | 6.1 |
| 17 | AISI 4340 | 515 | C-8 | 200 | 0.0100 | 30 | 1.30 | 0.04 | 0.34 | 0.15 | 0.01 | 0.00 | 0.04 | 1.91 | 4.8 |
| 18 | AISI 4340 | 515 | C-8 | 150 | 0.0100 | 48 | 1.74 | 0.04 | 0.34 | 0.15 | 0.01 | 0.00 | 0.03 | 2.33 | 3.9 |
| 19 | VASJET1000 | 52RC | C-8 | 162 | 0.0100 | 15 | 1.61 | 0.04 | 0.34 | 0.15 | 0.04 | 0.00 | 0.11 | 2.32 | 4.0 |
| 20 | VASJET1000 | 52RC | C-8 | 133 | 0.0100 | 30 | 1.96 | 0.04 | 0.34 | 0.15 | 0.02 | 0.00 | 0.07 | 2.60 | 3.5 |
| 21 | VASJET1000 | 52RC | C-8 | 110 | 0.0100 | 45 | 2.37 | 0.04 | 0.34 | 0.15 | 0.02 | 0.00 | 0.05 | 2.99 | 3.0 |
| 22 | VASJET1000 | 52RC | C-8 | 95 | 0.0100 | 60 | 2.74 | 0.04 | 0.34 | 0.15 | 0.01 | 0.00 | 0.04 | 3.36 | 2.7 |
| 23 | 250 MARAGE | 53RC | C-3 | 345 | 0.0100 | 5 | 0.75 | 0.04 | 0.34 | 0.15 | 0.06 | 0.00 | 0.16 | 1.53 | 6.6 |
| 24 | 250 MARAGE | 53RC | C-3 | 315 | 0.0100 | 15 | 0.82 | 0.04 | 0.34 | 0.15 | 0.02 | 0.00 | 0.05 | 1.45 | 6.4 |
| 25 | 250 MARAGE | 53RC | C-3 | 275 | 0.0100 | 35 | 0.94 | 0.04 | 0.34 | 0.15 | 0.01 | 0.00 | 0.02 | 1.53 | 5.9 |

**FIGURE 16.3 Typical output format of computerized machinability data system. (Reprinted by permission from Field and Ackenhausen [2].)**

rate. The definition of optimal is based on either the objectives of minimizing cost or maximizing production rate.

A common mathematical model to predict optimum cutting speed relies on the familiar Taylor equation for tool life,

$$VT^n = C \tag{16.2}$$

where $V$ = surface speed, ft/min or m/s

$T$ = tool life, min

$C$ and $n$ = constants

By combining Eqs. (16.1) and (16.2), and accounting for the fact that machining time $T_m$ is inversely proportional to cutting speed, $V$, the equation for minimum cost cutting speed can be derived.

$$V_{\min} = \frac{C}{\left(\dfrac{1-n}{n}\,\dfrac{C_o T_{tc} + C_t}{C_o}\right)^n} \tag{16.3}$$

In a similar way, the cutting speed that yields maximum production rate can also be derived:

$$V_{\max} = \frac{C}{\left(\dfrac{1-n}{n} T_{tc}\right)^n} \tag{16.4}$$

Equations (16.3) and (16.4), or equations similar to these, are used in the predictive-type machinability data systems to determine recommendations that approximate optimal cutting conditions.

The potential weakness of the mathematical model systems lies in the validity of the Taylor tool life equation. Equation (16.2) is an empirical equation[5] derived from experimental data that contain random errors. These random variations tend to distort the accuracy of the minimum cost and maximum production equations. Also, there are dangers in extrapolating the Taylor equation beyond the range over which the experimental data were collected.

The two types of computerized machinability data systems (data base system and mathematical model system) can be combined into one system.

### EXAMPLE 16.2

A computerized machinability data base system was developed for the Abex Corporation by Zimmers ([12],[17],[18]). It is principally a data base system.

Five data files are used in the Abex System to store the required machinability data needed in the computations.

[5] For definition of the term "empirical equation," see Section 13.2.

MACHINABILITY ANALYSIS – COST SUMMARY AND CUTTING CONDITION DETAILS

PART NO. 03189-T OPER NO. 100 DATE 03/

MATERIAL DESCRIPTION – 4340B 0.86 HP/CU. IN.

OPERATION NAME – TURN

CUT DETAILS – DIAMETER – 3.5000 LENGTH – 19.0000 DEPTH – 0.100

FINISH – 125. RMS

CALCULATIONS USING SHOP DATA

| CUT SPEED RPM | CUT SPEED SFPM | FEED RATE IPR | FEED RATE IPM | TOOL LIFE MIN. | WORK DIAM IN. | TOOL DESCRIPTION OP****TH**RAK** **MATL**NR***LA | SIZE TOLER IN. | SURF FIN URMS | TOOL COST $/100 | TIP COST $/100 | MACH COST $/100 | TOTAL COST $/100 | PROD RATE HRS/100 | PARTS PER T.L. |
|---|---|---|---|---|---|---|---|---|---|---|---|---|---|---|
| 448. | 470. | 0.010 | 4.48 | 15. | 4.000 | 11 78 0103P0600 | 0.005 | 125. | 0.19 | 5.35 | 63.54 | 125.61 | 9.15 | 3.54 |
| 381. | 400. | 0.010 | 3.81 | 30. | 4.000 | 11 78 0103P0600 | 0.005 | 125. | 0.11 | 3.14 | 74.66 | 128.33 | 9.71 | 6.02 |
| 343. | 360. | 0.010 | 3.43 | 45. | 4.000 | 11 78 0103P0600 | 0.005 | 125. | 0.08 | 2.32 | 82.96 | 133.52 | 10.38 | 8.13 |
| 310. | 325. | 0.010 | 3.10 | 60. | 4.000 | 11 78 0103P0600 | 0.005 | 125. | 0.07 | 1.93 | 91.90 | 140.96 | 11.25 | 9.79 |
| MINIMUM COST CONDITIONS, SHOP DATA | | | | | | | | | | | | | | |
| 512. | 470. | 0.010 | 5.12 | 15. | 3.500 | 11 78 0103P0600 | 0.005 | 125. | 0.19 | 5.35 | 63.54 | 125.61 | 9.15 | 3.54 |

| MACH OVHD $/MIN | NO. CUTS | APPR TO WRK IN. | RPDTRV COST $/100 | LOAD TIME MIN. | LOAD COST $/100 | SETUP TIME MIN. | SETUP COST $/100 |
|---|---|---|---|---|---|---|---|
| 0.15 | 1. | 4.0 | 4.05 | 2.3 | 34.49 | 21.0 | 3.14 |

MACHINE GROUPS – GROUP NO. MACHINE DESCRIPTION MACH HP

126-0 WARNER AND SWASEY NO. 2A TURRET LATHE 15 10.500

| | | | | |
|---|---|---|---|---|
| 1 | RPM– 321. | FPR– 0.0110 | IPM– 3.53 | HRS/100– 9.49 |
| 2 | RPM– 439. | FPR– 0.0110 | IPM– 4.82 | HRS/100– 7.00 |
| 3 | RPM– 600. | FPR– 0.0110 | IPM– 6.59 | HRS/100– 5.18 |

**FIGURE 16.4** Output format of Abex computerized machinability data system. (Reprinted by permission from Zimmers [17].)

1. Machine tool data (DB100). Machine identification, horsepower, and available speeds and feeds.
2. Base cutting conditions (DB200). Base cutting conditions for certain types of materials for HSS and cemented carbide tools.
3. Experimental data (DB300). Results of laboratory tests to supplement file DB200. Data contained in the experimental data bank include operation type, tooling, work material, cutting conditions, observed surface finish, and tool-life values.
4. Shop-generated data (DB400). Results of successful experience in production. Data contained in DB400 are similar to those of DB300 except that they originated in the shop rather than the laboratory.
5. Plant cost data (DB500). Cost parameters needed to compute cost per piece and optimum conditions.

Input to the program consists of data on the workpart, depth of cut, surface finish requirements, and other data related to the operation. (Development of the process sequence must precede the use of the machinability data program.) During operation of the program, the computer performs a series of data file searches and calculations to determine the recommended cutting conditions. A typical printout is shown in Figure 16.4. If shop data (DB400) are available, the recommendations are based on this file. Otherwise, experimental data (DB300) are used, and if no applicable data are available in DB300, recommended speeds and feeds are based on the DB200 file.

## 16.4 CUTTING CONDITIONS OPTIMIZATION

Machinability data systems of the type discussed in Section 16.3 are useful for providing general recommendations on speeds and feeds for machining. As indicated, some systems attempt to obtain optimum cutting speed values for a given feed rate. However, with the many variables indicated in Table 16.2, no general machinability system can anticipate all the possible job variations that may exist in the shop. For a particular job, there may be a potential for significant improvements over the recommended conditions. This section outlines one approach that might be used to realize these potential improvements ([5], [6]).

The scheme is based on an off-line search procedure similar to the strategies discussed in Section 15.5. An off-line search procedure refers to the case where process variables are measured at the end of a cut or sequence of cuts. An on-line search is one in which process variables are measured during the operation.

The cutting conditions optimization procedure uses an index of performance (IP) to guide the search. Either cost or production rate can be used as the index of performance. Feed and speed are the independent variables of interest in the machining process. The index of performance can be viewed as a response surface with feed and speed as the coordinate axes. The procedure is roughly based on "evolutionary operations" [1], and can be outlined as follows:

1. A starting point is selected based on recommended cutting conditions for the tool, work, depth, and so on. A point is defined in terms of the speed and feed values. The starting point becomes the current operating point.

2. Place test points about the current operating point in a square pattern as illustrated in Figure 16.5. These points represent a set of exploratory moves to determine which direction will improve the index of performance.

3. Determine the value of the index of performance at each of the test points. This is accomplished by performing machining operations at each of the speed/feed combinations represented by the points. For each set of

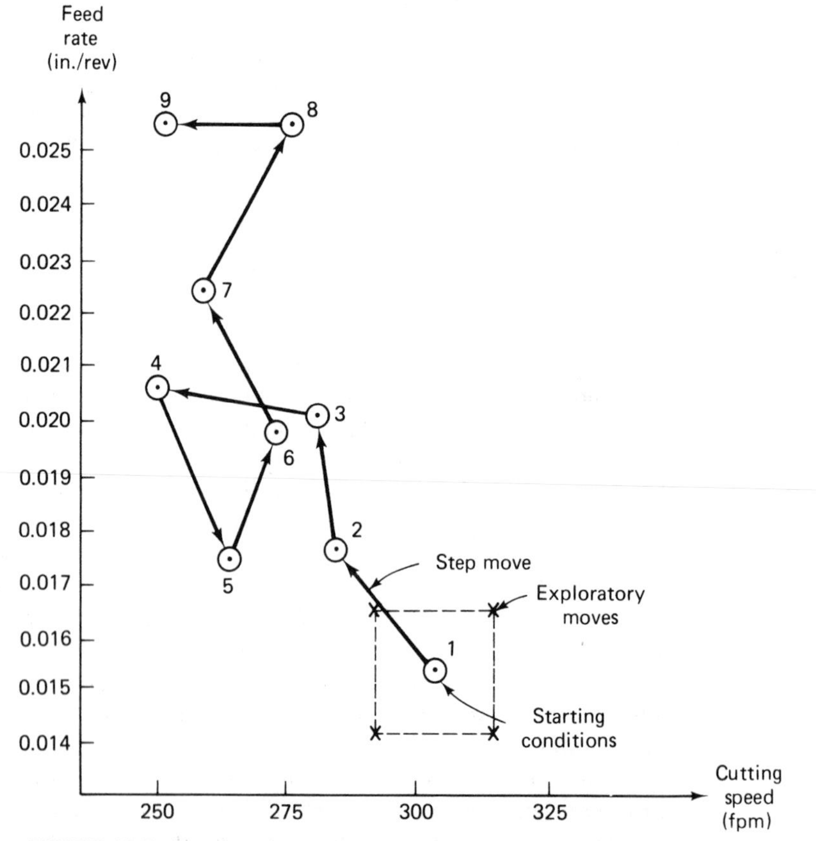


FIGURE 16.5 Results of machining conditions search trajectory in turning operation.

CUTTING CONDITIONS OPTIMIZATION

Calculation Procedure: Turning Cost per Piece

1. Known Information

   Price of Insert + Toolholder Depreciation ($) P = ________

   Total Cutting Edges on Tool Insert E = ________

   Machine Operator's Rate ($/min.) MR = ________

   Tool Changing Time (min./cutting edge) TCT = ________

   Work Changing Time (min./workpiece) WCT = ________

2. Machining Time per Piece = MT

   Speed V = ________ sfpm Feed, f = ________ ipr

   Work Length L = ________ in. Diameter D = ________ in.

   Machining Time $MT = \frac{3.142\ DL}{12\ fv}$ = ________ min.

3. Number of Workpieces per Cutting Edge = N

   It is usually the case that one cutting edge will last for more than a single workpiece. Therefore, N will have a value greater than 1. There are two alternative methods to determine N.

   a. Exact Method: Count the number of workpieces until tool change or tool failure. N = workpiece count.

   b. Approximate Method: N can be approximated by means of the fraction of the tool used up per workpiece.

      WF = Wear level at tool failure WF = ________

      WFC = Wear rate per workpiece WPC = ________

      $N = \frac{WF}{WPC}$ = workpieces/cutting edge (approximate)

4. Tool cost per cutting edge = TC

   TC = P/E = ________ $/cutting edge

5. Maching cost per workpiece, COST

   $COST = MR(MT + WCT) + \frac{1}{N}(MR \times TCT + TC)$ ________ $/piece

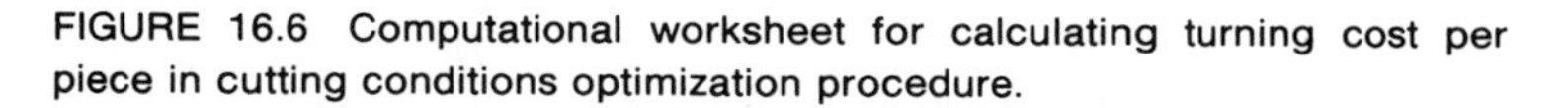

FIGURE 16.6 Computational worksheet for calculating turning cost per piece in cutting conditions optimization procedure.

conditions, the necessary data must be acquired to determine the IP value. A worksheet for organizing the data and making the IP calculation is illustrated in Figure 16.6.[6] Either tool wear must be measured or the number of workpieces until tool failure must be counted to obtain the tool cost allocation.

[6] The symbols used in Figure 16.6 are not consistent with the symbols used elsewhere in this book.

4. Determine the speed and feed change components. Figure 16.7 illustrates the procedure.

5. Take a move of prescribed length in the direction indicated by the gradient components. This will define a new operating point which should be superior to the preceding point.

6. Repeat steps 2 through 5 until: (a) an optimum point has been reached, or (b) a further improvement in the IP will result in unsatisfactory surface finish, chatter, excessive horsepower consumption, or some other undesirable effect.

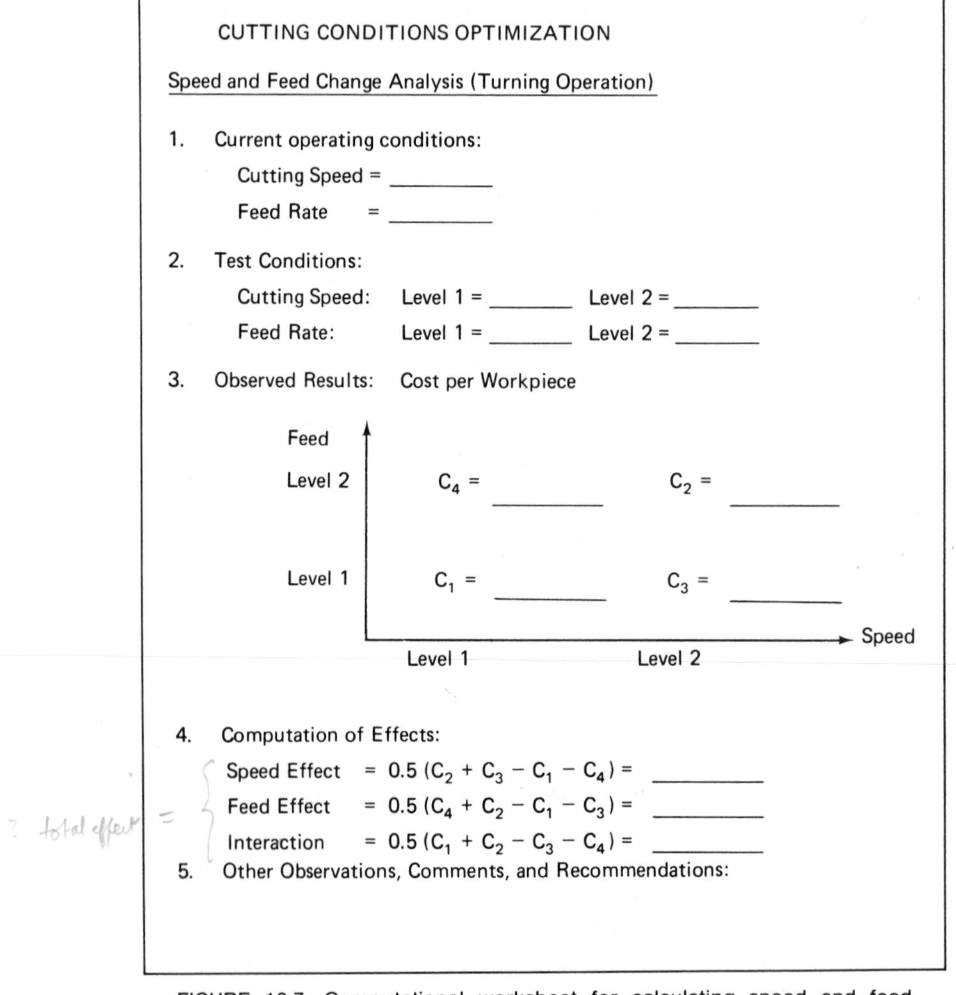

CUTTING CONDITIONS OPTIMIZATION

Speed and Feed Change Analysis (Turning Operation)

1. Current operating conditions:
   - Cutting Speed = ________
   - Feed Rate = ________
2. Test Conditions:
   - Cutting Speed: Level 1 = ________ Level 2 = ________
   - Feed Rate: Level 1 = ________ Level 2 = ________
3. Observed Results: Cost per Workpiece



4. Computation of Effects:
   - Speed Effect $= 0.5\,(C_2 + C_3 - C_1 - C_4) =$ ________
   - Feed Effect $= 0.5\,(C_4 + C_2 - C_1 - C_3) =$ ________
   - Interaction $= 0.5\,(C_1 + C_2 - C_3 - C_4) =$ ________
5. Other Observations, Comments, and Recommendations:

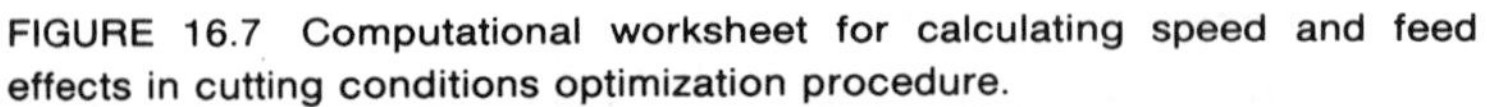

FIGURE 16.7 Computational worksheet for calculating speed and feed effects in cutting conditions optimization procedure.

Figure 16.5 shows the progress of one application of the cutting conditions optimization procedure in a machining job. It can be seen that cycle 4 resulted in a "false move." One of the difficulties in the application of the procedure arises from the variability inherent in the machining process. A number of replications must be taken at each test point to average out these random variations. If this painstaking procedure is not followed, the probability of a "false move" in the wrong direction is increased.

The cutting conditions optimization scheme is impractical for short-run jobs. Application of the procedure requires a considerable amount of machining. This limits its use to production jobs of fairly long duration.

## REFERENCES

[1] Box, G. E. P., and Draper, N. R., *Evolutionary Operations*, John Wiley & Sons, Inc., New York, 1969.

[2] Field, M., and Ackenhausen, A. R., *Determination and Analysis of Machining Costs and Production Rates Using Computer Techniques*, Report No. AFMDC 68-1, Machinability Data Center, Cincinnati, Ohio, 1968.

[3] French, J. H., "Incentive Rates by Computer," *Industrial Engineering*, September, 1974, pp. 16–20.

[4] Groover, M. P., "A Survey on the Machinability of Metals," *Tech. Paper MR76-269*, Society of Manufacturing Engineers, Dearborn, Mich., 1976.

[5] Groover, M. P., Johnson, R. J., and Gunda, A. M., "Determination of Machining Conditions by a Self-Adaptive Procedure," *Final Report*, National Science Foundation, Grant GK30418, 1975.

[6] Groover, M. P., Gunda, A. M., and Johnson, R. J., "Determination of Machining Conditions by a Self-Adaptive Procedure," *Proceedings*, Fourth North American Metalworking Research Conf., 1976, pp. 267–271.

[7] Lindgren, L. H., and Murphy, R. D., "Computerized Standard Data—A Powerful New Tool for Manufacturing Engineering," *CAD/CAM and the Computer Revolution*, Society of Manufacturing Engineers, Dearborn, Mich., 1974, pp. 240–257.

[8] *Machining Data Handbook*, 2nd ed., Machinability Data Center, Cincinnati, Ohio, 1972.

[9] MTM Association, "4M DATA: Computerized Work Measurement System," Publication 3233-77, Fairlawn, N.J.

[10] Mundel, M. E., *Motion and Time Study*, 5th ed., Prentice-Hall, Inc., Englewood Cliffs, N.J., 1978.

[11] Niebel, B. W., *Motion and Time Study*, 6th ed., Richard D. Irwin, Inc., Homewood, Ill., 1976.

[12] Parsons, N. R. (Editor), *N/C Machinability Data Systems*, Society of Manufacturing Engineers, Dearborn, Mich., 1971.

[13] PRESSMAN, R. S., and WILLIAMS, J., "A New Approach to the Prediction of Machinability Data for Numerical Control," *NC/CAM, The New Industrial Revolution, Proceedings*, 13th NC Society Conference, 1976, pp. 421–435.

[14] PRESSMAN, R. S., and WILLIAMS, J., *Numerical Control and Computer-Aided Manufacturing*, John Wiley & Sons, Inc., New York, 1977, Chapter 9.

[15] RICE, R. S., "Survey of Work Measurement and Wage Incentives," *Industrial Engineering*, July, 1977, pp. 18–31.

[16] WEAVER, R. F., KOLLMAR, J. J., and BOEPPLE, E. A., "Developing Standards by Computer," *Industrial Engineering*, January, 1978, pp. 26–31.

[17] ZIMMERS, E. W., "Practical Applications of Computer Augmented Systems for Determination of Metal Removal Parameters and Production Rate Standards in a Job Shop," *Tech. Paper MS71-136*, Society of Manufacturing Engineers, Dearborn, Mich., 1971.

[18] ZIMMERS, E. W., "A Computer Assisted System for the Generation of Machining Parameters in a Job Shop," Ph.D. dissertation, Lehigh University, 1973.

## PROBLEMS

**16.1.** For a certain tool/work material combination, the Taylor tool life relationship, Eq. (16.2), has parameters $C = 929$ and $n = 0.27$. These values apply to a turning operation carried out at a feed rate of 0.010 in./rev and a depth of cut of 0.070 in. In addition, the following data are given:

$$C_o = \$0.30/\text{min}$$
$$C_t = \$0.50/\text{cutting edge}$$
$$T_{tc} = 2.0 \text{ min}$$

Compute the cutting speed that will yield
(a) Minimum cost per piece.
(b) Maximum production rate.

**16.2.** In the operation of Problem 16.1, the machining time per workpiece can be determined from the following equation:

$$T_m = 2000/V \qquad (T_m \text{ in minutes})$$

where $V$ is the cutting speed in fpm. Compute the cost per piece for the results of Problem 16.1, part (a), given that the workpiece handling time is $T_h = 3.0$ min.

**16.3.** Using the data provided in Problems 16.1 and 16.2, write a computer program to perform the necessary calculations and print out results in the general format of Figure 16.3. The work material is 4140, with a hardness of Rockwell C25. The tool material is Grade 350. Use the following cutting speeds to construct the table: 350, 400, 450, 500, 550, 600, and 650 fpm. The individual cost components should be separated as given by the three terms of Eq. (16.1).

**16.4.** In an application of the cutting conditions optimization method, the current operating conditions are: cutting speed = 250 fpm and feed rate = 0.010 in./rev. Test conditions used were as follows:

| | |
|---|---|
| Cutting speed: | level 1 = 220 fpm |
| | level 2 = 280 fpm |
| Feed rate: | level 1 = 0.008 in./rev |
| | level 2 = 0.012 in./rev |

The observed results were as follows:

$C_1 = \$1.03/\text{workpiece}$

$C_2 = \$0.99/\text{workpiece}$

$C_3 = \$1.07/\text{workpiece}$

$C_4 = \$0.95/\text{workpiece}$

Compute the speed and feed effects as indicated by the format of Figure 16.7. What changes in cutting conditions would you recommend based on the results?

chapter 17

# Production Systems at the Plant Level

## 17.1 PRODUCTION PLANNING AND CONTROL

In Chapter 16, we described several systems for manufacturing support at the level of the individual production operations. In the present chapter, we turn our attention to systems for manufacturing support at the plant level. Supporting the aggregate activity at the plant level means production planning and control. The purpose of this chapter is to cover some of the important systems concepts in modern production planning and control.

The traditional functions of production planning and control in a manufacturing firm were discussed in Section 2.3. These traditional functions include ordering raw materials and purchased components, production scheduling, dispatching, and expediting. For plants with large volumes of parts and complex assembled products, the work involved in executing these functions can be formidable. Even with computerized systems, the amount of human effort required to plan and control factory operations can be significant, particularly in areas such as dispatching and expediting. The types of problems commonly encountered in the planning and management of plant operations are the following:

1. *Plant capacity problems.* Production falls behind schedule due to lack of manpower and equipment resources. This results in excessive overtime, delays in meeting delivery schedules, customer complaints, backordering, and so on.

2. *Suboptimal production scheduling.* The wrong jobs are scheduled because of a lack of clear order priorities, inefficient scheduling rules, and the ever-changing status of jobs in the shop. As a consequence, production runs are interrupted by jobs whose priorities have suddenly increased, machine setups are increased, and jobs that are on schedule fall behind.

3. *Long manufacturing lead times.* The manufacturing lead time is the time required to process a job through the shop. In an attempt to compensate for problems 1 and 2 production planners allow extra time to produce the order. The shop becomes overloaded, order priorities become confused, and the result is excessively long manufacturing lead times.

4. *Inefficient inventory control.* At the same time that total inventories are too high for raw materials, work-in-progress, and finished products, there are stockouts that occur on individual raw-material items needed for production. High total inventories mean high carrying costs, while raw material stockouts mean delays in meeting production schedules.

5. *Low work center utilization.* This problem results in part from poor scheduling (excessive product changeovers and job interruptions), and from other factors over which plant management has limited control (equipment breakdowns, strikes, reduced demand for products, etc.).

6. *Process planning not followed.* This is the situation in which the regular planned routing is superseded by an ad hoc process sequence. It occurs for instance because of bottlenecks at work centers in the planned sequence. The consequences are longer setups, improper tooling, and less efficient processes.

7. *Errors in engineering and manufacturing records.* Bills of materials are not current, route sheets are not up to date with respect to the latest engineering changes, inventory records are inaccurate, and production piece counts are incorrect.

8. *Quality problems.* Quality defects are encountered in manufactured components and assembled products, thus causing delays in the shipping schedule.

All of these problems give rise to the need for better systems to plan and control plant operations. There are differences among writers, practitioners, and companies in the terminology used to describe these systems. IBM uses the term "communications oriented production information and control system—COPICS" [8] to identify the group of system elements. George Plossl integrates the various system concepts under the name "manufacturing control" [11]. Computer-Aided Manufacturing-International, Inc., calls its development effort in this area the "factory management project" [12]. All these terms represent integrated information systems designed to reduce the problems described above of planning and managing plant operations. The basic ingredients and framework for a modern production planning and control system are pictured in the flow diagram of Figure 17.1. These ingredients include:

1. *Master production schedule.* The master schedule was defined in Section 2.3 as a listing of the products to be manufactured, when they are to be delivered, and in what quantities. It is developed from customer orders and forecasts of future demand.

2. *Engineering and manufacturing data base.* This data base comprises all the information needed to fabricate the components and assemble the products. It includes the bills of material (assembly lists), part design data (either as engineering drawings or some other suitable format), process route sheets, and so on. Ideally, these data should be contained in some master file to avoid duplication of records and to facilitate update of the files when engineering changes are made.

3. *Material requirements planning (MRP).* MRP involves determining when to order raw materials and components for assembled products. It can also be used to reschedule orders in response to changing production priorities and demand conditions. The term *priority planning* is now widely used in

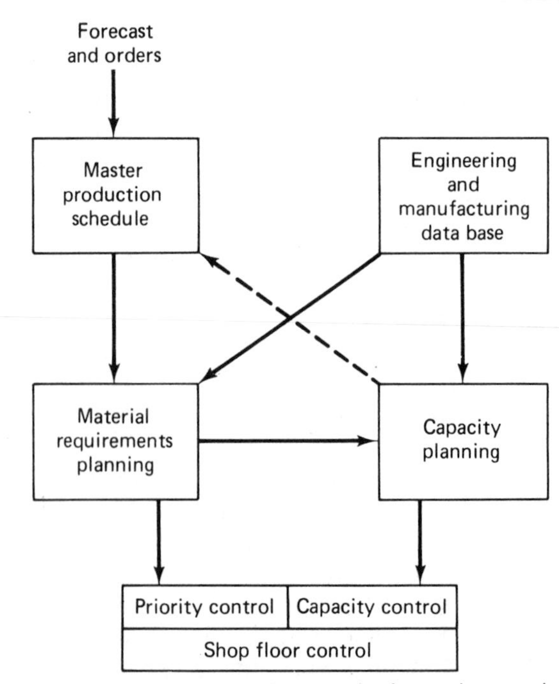


FIGURE 17.1 Basic elements and framework of a modern production planning and control system.

describing computer-based systems for time-phased planning of raw materials, work-in-progress, and finished goods.

4. *Capacity planning.* As we have previously defined it in Chapter 2, production is the transformation of raw materials into finished goods by the use of the firm's human and equipment resources. MRP is concerned with the planning of materials and components. Capacity planning, on the other hand, is concerned with determining the labor and equipment resources needed to meet the production schedule.

Capacity planning will often necessitate a revision in the master production schedule. It would be infeasible, and counterproductive in all likelihood, to develop a master schedule that exceeds plant capacity. Therefore, the master schedule is checked against available plant capacity to make sure that the schedule can be realized. If not, either the schedule or plant capacity must be adjusted to be brought into balance.

5. *Shop floor control.* The term "shop floor control" refers to a system for monitoring the status of production activity in the plant and reporting the status to management so that effective control can be exercised. Other terms are found in the literature to describe some of the activities included under shop floor control. *Priority control* is concerned with keeping priorities for work-in-progress at the appropriate levels to reflect changes in the status of final product priorities. *Capacity control* is concerned with adjustments that can be made in labor and equipment usage to maintain production rates consistent with the master schedule.

In the following sections, we will describe three of these production planning and control topics: material requirements planning, capacity planning, and shop floor control.

## 17.2 MATERIAL REQUIREMENTS PLANNING (MRP)[1]

### What is MRP?

Material requirements planning is a computational technique that converts the master schedule for end products into a detailed schedule for the raw materials and components used in the end products. The detailed schedule identifies the quantities of each raw material and component item. It also tells when each item must be ordered and delivered so as to meet the master schedule for the final products.

MRP is often thought of as a method of inventory control. While it is an effective tool for minimizing unnecessary inventory investment, MRP is also useful in production scheduling and purchasing of materials.

[1] Recently the term MRP is being used to reference Manufacturing Resource Planning [14]. It indicates an overall system for planning and controlling all of the resources of a manufacturing firm.

The concept of MRP is relatively straightforward. What complicates the application of the technique is the sheer magnitude of the data to be processed. The master schedule provides the overall production plan for final products in terms of month-by-month delivery requirements. Each of the products may contain hundreds of individual components. These components are produced out of raw materials, some of which are common among the components. For example, several parts may be produced out of the same sheet steel. The components are assembled into simple subassemblies. Then these subassemblies are put together into more complex assemblies—and so forth, until the final product is assembled together. Each production and assembly step takes time. All of these factors must be incorporated into the MRP computations. Although each separate computation is uncomplicated, the magnitude of all the data to be processed is so large that the application of MRP is virtually impossible unless carried out on a digital computer.

Material requirements planning is based on several concepts that are implicit in the description above but not explicitly defined. These concepts are discussed in the paragraphs below.

INDEPENDENT VERSUS DEPENDENT DEMAND. This distinction is fundamental to MRP. *Independent* demand means that demand for a product is unrelated to demand for other items. End products and spare parts are examples of items whose demand is independent. Independent demand patterns must usually be forecasted.

*Dependent* demand means that demand for the item is related directly to the demand for some other product. The dependency usually derives from the fact that the item is a component of the other product. Not only component parts, but raw materials and subassemblies are examples of items which are subject to dependent demand.

While demand for the firm's end products must often be forecasted, the raw materials and component parts should not be forecasted. Once the delivery schedule for the end products is established, the requirements for components and raw materials can be calculated directly. For example, even though the demand for automobiles in a given month can only be forecasted, once that quantity is established we know that five tires will be needed to deliver the car (don't forget the spare).

MRP is the appropriate technique for determining quantities of dependent demand items. These items constitute the inventory of manufacturing: raw materials, work-in-progress, component parts, and subassemblies. That is why MRP is such a powerful tool in the planning and control of manufacturing inventories.

LEAD TIMES. The lead time for a job is the time that must be allowed to complete the job from start to finish. In manufacturing there are two kinds of lead times: ordering lead times and manufacturing lead times. An *ordering* lead time for an item is the time required from initiation of the purchase requisition to receipt of the item from the vendor. If the item is a raw material that is stocked by the vendor, the ordering lead time should be relatively short, perhaps a few weeks. If the item must be fabricated by the vendor, the lead time may be substantial, perhaps several months.

*Manufacturing lead time* is a term we have previously defined and used. It is the time needed to process the part through the sequence of machines specified on the route sheet. It includes not only the operation times but also the nonproductive dead time that must be allowed.

In MRP, lead times are used to determine starting dates for assembling final products and subassemblies, for producing component parts, and for ordering raw materials.

COMMON USE ITEMS. In manufacturing, the basic raw materials are often used to produce more than one component type. Also, a given component may be used on more than one final product. For example, the same type of steel rod stock may be used to produce screws on an automatic screw machine. Each of the screw types may then be used on several different products. MRP collects these common use items from different products to effect economies in ordering the raw materials and manufacturing the components.

### Inputs to MRP

As indicated previously, MRP converts the master production schedule into the detailed schedule for raw materials and components. For the MRP program to perform this function, it must operate on the data contained in the master schedule. However, this is only one of three sources of input data on which MRP relies. The three inputs to MRP are:

1. The master production schedule and other order data.
2. The bill-of-material file, which defines the product structure.
3. The inventory record file.

Figure 17.2 presents a diagram showing the flow of data into the MRP processor and its conversion into useful output reports. The three inputs are described in the paragraphs below.

MASTER PRODUCTION SCHEDULE. The master schedule is a listing of what end products are to be produced, how many of each product are to be produced, and when they are to be ready for shipment. The general format of a master production schedule is illustrated in Figure 17.3. Manufacturing firms generally work toward monthly delivery schedules. However, in Figure 17.3, the master schedule uses weeks as the time periods (for purposes of an example which will be developed later).

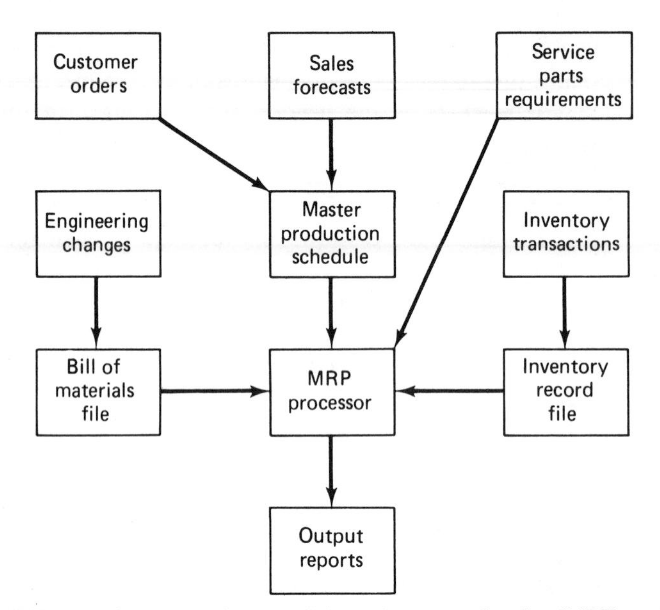


FIGURE 17.2 Structure of a material requirements planning (MRP) system.

| Week number | 6 | 7 | 8 | 9 | 10 |
|---|---|---|---|---|---|
| Product P1 | | | 50 | | 100 |
| Product P2 | | 70 | 80 | 25 | |
| etc. | | | | | |

FIGURE 17.3 Master production schedule for products P1 and P2 showing weekly delivery quantities.

The master schedule must be based on an accurate estimate of demand for the firm's products together with a realistic assessment of its production capacity. (Capacity planning is discussed in Section 17.3.)

Product demand that makes up the master schedule can be separated into three categories. The first consists of firm customer orders for specific products. These orders usually include a specific delivery date which has been promised to the customer by the sales department. The second category is forecasted demand. Based on statistical techniques applied to past demand, estimates provided by the sales staff, and other sources, the firm will generate a forecast of demand for its various product lines. This forecast may constitute the major portion of the master schedule. The third category is demand for individual component parts. These components will be used as repair parts and are stocked by the firm's service department. This third category is often excluded from the master schedule since it does not represent demand for end products. In Figure 17.2 we show it as feeding directly into the MRP processor.

BILL-OF-MATERIALS FILE. In order to compute the raw material and component requirements for end products listed in the master schedule, the product structure must be known. This is specified by the bill of materials, which is a listing of component parts and subassemblies that make up each product. Putting all these assembly lists together, we have the bill-of-materials file (BOM).

The structure of an assembled product can be pictured as shown in Figure 17.4. This is a relatively simple product in which a group of individual components make up two subassemblies, which in turn make up the product. The product structure is in the form of a pyramid, with lower levels feeding into the levels above. We can envision one level below that shown in Figure 17.4. This would consist of the raw materials used to make the individual components. The items at each successively higher level are called the parents of the items in the level directly below. For example, subassembly S1 is the parent of components C1, C2, and C3. Product P1 is the parent of subassemblies S1 and S2.

The product structure must also specify how many of each item are included in its parent. This is accomplished in Figure 17.4 by the number in parentheses to the right and below each block. For example, subassembly S1 contains four of component C2 and one each of components C1 and C3.

Figure 17.4 is not the way the bill of materials would be recorded in a company's files. Figure 17.5 illustrates a more likely format for the product structure which is illustrated graphically in Figure 17.4.

INVENTORY RECORD FILE. This is also called the *item master file* in a computerized inventory system. The types of data contained in the inventory record for a given item are illustrated in Figure 17.6. The file is divided into three segments:

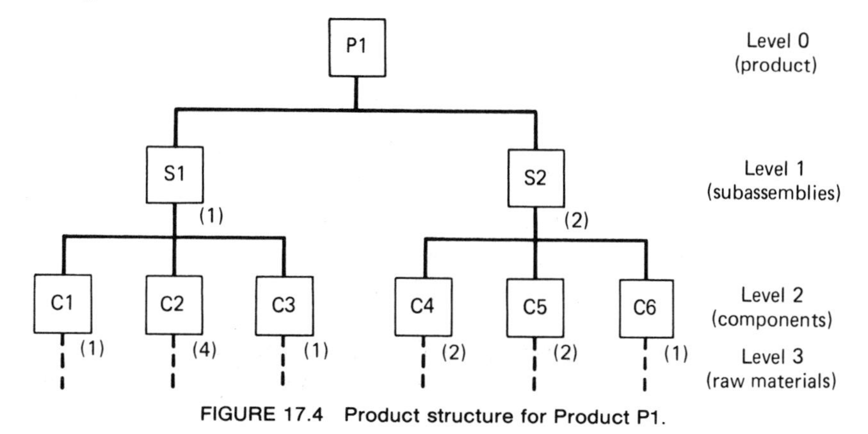


FIGURE 17.4 Product structure for Product P1.

```
P1
      S1  (1)
      S2  (2)
S1
      C1  (1)
      C2  (4)
      C3  (1)
S2
      C4  (2)
      C5  (2)
      C6  (1)
```

FIGURE 17.5 Parts list and product structure for product P1 (alternative format to Figure 17.4).

| | | | | | | | | | | | | |
|---|---|---|---|---|---|---|---|---|---|---|---|---|
| ITEM MASTER DATA SEGMENT | Part No. | Description | | Lead time | | Std. cost | | Safety stock | | | | |
| | Order quantity | Setup | | Cycle | | Last year's usage | | Class | | | | |
| | Scrap allowance | Cutting data | | Pointers | | Etc. | | | | | | |
| INVENTORY STATUS SEGMENT | Allocated | | Control balance | Period | | | | | | | | Totals |
| | | | | 1 | 2 | 3 | 4 | 5 | 6 | 7 | 8 | |
| | Gross requirements | | | | | | | | | | | |
| | Scheduled receipts | | | | | | | | | | | |
| | On hand | | | | | | | | | | | |
| | Planned-order releases | | | | | | | | | | | |
| SUBSIDIARY DATA SEGMENT | Order details | | | | | | | | | | | |
| | Pending action | | | | | | | | | | | |
| | Counters | | | | | | | | | | | |
| | Keeping track | | | | | | | | | | | |

FIGURE 17.6 Record of an inventory item. (Reprinted by permission from Orlicky [10].)

1. Item master data segment.
2. Inventory status segment.
3. Subsidiary data segment.

The first segment gives the item's identification (by part number) and other data, such as lead time, cost, and order quantity. The second segment (inventory status segment) provides a time-phased record of inventory status. In MRP it is important to know not only the current level of inventory, but also the future changes that will occur against the inventory status. Therefore, the inventory status segment lists the gross requirements for the item, scheduled receipts, on-hand status, and planned-order releases. The third file segment (subsidiary data segment) contains miscellaneous information pertaining to purchase orders, scrap or rejects, engineering change actions, and so on.

It is important that the inputs to the MRP processor be kept current. The bill-of-materials file must be maintained by feeding any engineering changes that affect the product structure into the BOM. Similarly, the inventory record file is maintained by inputing the inventory transactions to the file. The structure of these updating procedures is shown in Figure 17.2.

## How MRP works

The material requirements planning processor operates on the data contained in the master schedule, the bill-of-materials file, and the inventory record file. The master schedule specifies a period-by-period list of final products required. The BOM defines what materials and components are needed for each product. The inventory record file contains information on the current and future inventory status of each component. The MRP program computes how many of each component and raw material are needed by "exploding" the end product requirements into successively lower levels in the product structure. Referring to the master schedule in Figure 17.3, 50 units of product P1 are specified in the master schedule for week number 8. Now referring to the product structure in Figure 17.4, 50 units of P1 explodes into 50 units of subassembly S1 and 100 units of S2, and the following numbers of units for the components:

C1: 50 units.
C2: 200 units.
C3: 50 units.
C4: 200 units.
C5: 200 units.
C6: 100 units.

The quantities of raw materials for these components would be determined in a similar manner.

There are several factors that complicate the MRP parts and materials explosion. First, the component and subassembly quantities given above are gross requirements. Quantities of some of the components and subassemblies may already be in stock or on order. Hence, the quantities that are in inventory or scheduled for delivery in the near future must be subtracted from gross requirements to determine net requirements for meeting the master schedule.

A second complicating factor in the MRP computations is manifested in the form of lead times: ordering lead times and manufacturing lead times. The MRP processor must determine when to start assembling the subassemblies by offsetting the due dates for these items by their respective manufacturing lead times. Similarly, the component due dates must be offset by their manufacturing lead times. Finally, the raw materials for the components must be offset by their respective ordering lead times. The material requirements planning program performs this lead-time-offset calculation from data contained in the inventory record file (see Figure 17.6).

A third factor that complicates MRP is common use items. Some components and many raw materials are common to several products. The MRP processor must collect these common use items during the parts explosion. The total quantities for each common use item are then combined into a single net requirement for the item.

Finally, a feature of MRP that should be emphasized is that the master production schedule provides time-phased delivery requirements for the end products, and this time phasing must be carried through the calculations of the individual component and raw material requirements.

## EXAMPLE 17.1

To illustrate how MRP works, let us consider the requirements planning procedure for one of the components of product P1. The component we will consider is C4. This part happens to be used also on one other product: P2 (see the master schedule of Figure 17.3). However, only one of item C4 is used on each P2 produced. The product structure of P2 is given in Figure 17.7. Component C4 is made out of raw material M4. One unit of M4 is needed to produce 1 unit of C4. The ordering and manufacturing lead times needed to make the MRP computations are as follows:

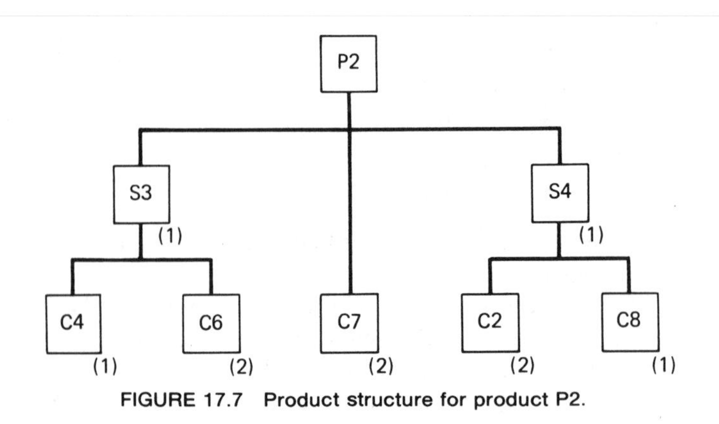

FIGURE 17.7 Product structure for product P2.

P1: assembly lead time = 1 week.

P2: assembly lead time = 1 week.

S2: assembly lead time = 1 week.

S3: assembly lead time = 1 week.

C4: manufacturing lead time = 2 weeks.

M4: ordering lead time = 3 weeks.

The current inventory and order status of item M4 is shown in Figure 17.8. There are no stocks or orders for any of the other items listed above.

The solution is presented in Figure 17.9. The delivery requirements for products P1 and P2 must be offset by the 1-week assembly lead time to obtain the planned order releases. Since the subassemblies to make the products must be ready, these order release quantities are "exploded" into requirements for subassemblies S2 (for P1) and S3 (for P2). These net requirements are then offset by the 1-week lead time and combined (in week 6) to obtain the gross requirements for part C4. Net requirements are equal to gross requirements for P1, P2, S2, S3, and C4 because of no on-hand inventory and no planned orders. We see the effect of current and planned stocks in the time-phased inventory picture for M4. The on-hand stock of 50 plus the scheduled receipts of 40 are used to meet the gross requirements of 70 units of M4 in week 3. Twenty units remain after meeting these requirements, which can be applied to the gross requirements of 280 units of M4 in week 4. Net requirements in week 4 are therefore 280 − 20 = 260 units. With an ordering lead time of 3 weeks, the order release for the 260 units must be planned for week 1.

| Period | | 1 | 2 | 3 | 4 | 5 | 6 | 7 |
|---|---|---|---|---|---|---|---|---|
| Item: RAW MATL. M4 | | | | | | | | |
| Gross Requirements | | | | | | | | |
| Scheduled Receipts | | | | 40 | | | | |
| On Hand | 50 | | | 90 | | | | |
| Net Requirements | | | | | | | | |
| Planned Order Releases | | | | | | | | |

FIGURE 17.8 Initial inventory status of material M4 in Example 17.1.

## Output reports

The material requirements planning program generates a variety of outputs that can be used in the planning and management of plant operations. These outputs include:

1. Order release notices, to place orders that have been planned by the MRP system.

| Period | | 1 | 2 | 3 | 4 | 5 | 6 | 7 | 8 | 9 | 10 |
|---|---|---|---|---|---|---|---|---|---|---|---|
| Item: PRODUCT P1 | | | | | | | | | | | |
| Gross Requirements | | | | | | | | | 50 | | 100 |
| Scheduled Receipts | | | | | | | | | | | |
| On Hand | 0 | | | | | | | | | | |
| Net Requirements | | | | | | | | | 50 | | 100 |
| Planned Order Releases | | | | | | | | 50 | | 100 | |
| Item: PRODUCT P2 | | | | | | | | | | | |
| Gross Requirements | | | | | | | | 70 | 80 | 25 | |
| Scheduled Receipts | | | | | | | | | | | |
| On Hand | 0 | | | | | | | | | | |
| Net Requirements | | | | | | | | 70 | 80 | 25 | |
| Planned Order Releases | | | | | | | 70 | 80 | 25 | | |
| Item: SUBASSBY S2 | | | | | | | | | | | |
| Gross Requirements | | | | | | | | 100 | | 200 | |
| Scheduled Receipts | | | | | | | | | | | |
| On Hand | | | | | | | | | | | |
| Net Requirements | | | | | | | | 100 | | 200 | |
| Planned Order Releases | | | | | | | 100 | | 200 | | |
| Item: SUBASSBY S3 | | | | | | | | | | | |
| Gross Requirements | | | | | | | 70 | 80 | 25 | | |
| Scheduled Receipts | | | | | | | | | | | |
| On Hand | | | | | | | | | | | |
| Net Requirements | | | | | | | 70 | 80 | 25 | | |
| Planned Order Releases | | | | | | 70 | 80 | 25 | | | |
| Item: COMPONENT C4 | | | | | | | | | | | |
| Gross Requirements | | | | | | 70 | 280 | 25 | 400 | | |
| Scheduled Receipts | | | | | | | | | | | |
| On Hand | | | | | | | | | | | |
| Net Requirements | | | | | | 70 | 280 | 25 | 400 | | |
| Planned Order Releases | | | | 70 | 280 | 25 | 400 | | | | |
| Item: RAW MATL. M4 | | | | | | | | | | | |
| Gross Requirements | | | | 70 | 280 | 25 | 400 | | | | |
| Scheduled Receipts | | | | 40 | | | | | | | |
| On Hand | 50 | | | 90 | 20 | | | | | | |
| Net Requirements | | | | -20 | 260 | 25 | 400 | | | | |
| Planned Order Release | | 260 | 25 | 400 | | | | | | | |

FIGURE 17.9 MRP solution to Example 17.1.

2. Reports showing planned orders to be released in future periods.
3. Rescheduling notices, indicating changes in due dates for open orders.
4. Cancellation notices, indicating cancellation of open orders because of changes in the master schedule.
5. Reports on inventory status.

The above outputs of the MRP system are called primary outputs by Orlicky [10]. In addition, secondary output reports can be generated by the MRP system at the user's option. These reports include:

1. Performance reports of various types, indicating costs, item usage, actual versus planned lead times, and so on.
2. Exception reports, showing deviations from schedule, orders that are overdue, scrap, and so on.
3. Inventory forecasts, indicating projected inventory levels (both aggregate inventory as well as item inventory) in future periods.

### Benefits of material requirements planning

Many advantages are claimed for a well-designed MRP system. Among the benefits reported by users are the following:

Reduction in inventory. Some users claim as much as a 40% reduction.
Improved customer service.
Quicker response to changes in demand.
Reduced setup and product changeover costs.
Better machine utilization.
Improved capability to respond to changes in the master schedule.
Aid in developing the master schedule.

As a result of the general recognition of these benefits, MRP applications have grown dramatically since the late 1960s. Over 1000 companies are currently using computerized MRP systems.

## 17.3 CAPACITY PLANNING

### Plant capacity and capacity planning

The master schedule defines the production objectives of the firm in terms of what products, how many, and when. A realistic master schedule must be compatible with the production capabilities and limitations of the plant which will produce the products. Accordingly, the firm must know its production capacity and must plan for changes in its capacity to meet the changing production requirements specified in the master schedule.

PLANT CAPACITY. The term "plant capacity" is used to define the maximum rate of output that the plant can produce under a given set of assumed operating conditions. The assumed operating conditions refer to the number of shifts (one, two, or three shifts per day), number of days of plant operation per week, employment levels, whether or not overtime is included in the definition of plant capacity, and so on. For a continuous chemical plant, the equipment may have to be operated 24 hours per day, 7 days per week. For an automobile assembly factory, capacity is typically defined in terms of one shift per day but with overtime allowances included.

Capacity for a production plant is traditionally measured in terms of output units of the plant. Examples would be tons of steel for a steel mill, number of automobiles for a car assembly plant, and barrels of oil for a refinery. When the output units of a plant are nonhomogeneous, input units may be more appropriate for measuring plant capacity. A job shop, for instance, may use labor hours or available machine hours to measure capacity.

CAPACITY PLANNING. Capacity planning is concerned with determining what labor and equipment capacity is required to meet the current master production schedule as well as the long-term future production needs of the firm. Capacity planning is typically performed in terms of labor and/or machine hours available.

The function of capacity planning in the overall production planning and control system was shown in Figure 17.1. The master schedule is transformed into material and component requirements using MRP. Then these requirements are compared with available plant capacity over the planning horizon. If the schedule is incompatible with capacity, adjustments must be made either in the master schedule or in plant capacity. The possibility of adjustments in the master schedule is indicated by the dashed line in Figure 17.1.

Capacity adjustments can be accomplished either in the short term or the long term. Capacity planning for short-term adjustments would include decisions on such factors as the following:

1. Employment levels. Employment in the plant can be increased or decreased in response to changes in capacity requirements.
2. Number of work shifts. Increasing or decreasing the number of shifts per week.
3. Labor overtime hours or reduced workweek.
4. Inventory stockpiling. This would be used to maintain steady employment during temporary slack periods.

5. Order backlogs. Deliveries of product to customers would be delayed during busy periods.

6. Subcontracting. Letting of jobs to other shops during busy periods, or taking in extra work during slack periods.

Capacity planning to meet long-term capacity requirements would include the following types of decisions:

1. Investing in new equipment. Investing in more productive machines to meet increased production requirements, or investing in new types of machines to match future new product designs.
2. New plant construction.
3. Purchase of existing plants from other companies.
4. Closing down or selling off existing facilities which will not be needed in the future.

## Capacity planning model

Estimates of plant capacity and capacity requirements can be developed from the production models derived in Section 2.6. For job shop and batch production, capacity calculations must be performed on each machine or group of similar machines in a section of the shop. For flow line production, capacity calculations are carried out treating the line as a section or work center. Two types of capacity problem will be considered in the paragraphs below:

1. Assessing current production capacity.
2. Determining future capacity requirements.

Examples will be used to illustrate the two problem areas.

ASSESSING CURRENT CAPACITY. The following symbols will be used: Let PC be the production capacity of a given work center or group of work centers in a section of the factory. Production capacity will be measured in terms of the number of good pieces produced per week. Let $W$ be the total number of work centers in the section of the factory being considered. A work center is a worker–machine production system capable of producing at a rate $R_p$ pieces per hour. The term $R_p$ was defined by Eqs. (2.10) and (2.11) for a job shop or batch production. These terms also apply to flow line operation, except that the setup time becomes negligible in Eq. (2.10) for mass-produced items. Each work center operates for $H$ hours per shift. $H$ is an average that excludes time for machine breakdowns and repairs, maintenance, machine interference, operator delays, and so on. Provision for setup time on the machine is allowed in $R_p$. Let $S_w$ be the number of shifts per week for the work center(s). The fraction scrap loss is given by $q$.

These parameters can be utilized to calculate the production capacity for the shop section as follows:

$$PC = WS_wHR_p(1-q) \tag{17.1}$$

Use of $R_p$ and $q$ assumes that the items processed through the section are fairly homogeneous.

### EXAMPLE 17.2

The turret lathe section has six machines, all devoted to the production of similar parts. The section operates 10 shifts per week. Nominal hours per shift are eight, but machine utilization is about 80% for each shift, owing to machine breakdowns and operator delays. Average production rate is 17 pieces per hour, and the scrap loss rate averages 3%. Determine the production capacity of the turret lathe section.

From the data provided:

$W =$ 6 machines  $R_p =$ 17 pieces/h
$S_w =$ 10 shifts  $q =$ 0.03
$H =$ 8(80%) = 6.4 h

Hence, Eq. (17.1) gives production capacity as

$$PC = 6(10)(6.4)(17)(1-0.03) = 6332 \text{ pieces/week}$$

This type of computation can be made for each section or work center in the shop to ensure a balanced flow of work in the shop. Bottleneck work centers can be identified. Future workload requirements can be compared to current capacity.

Determining Capacity Requirements. The previous equation can be rearranged to determine the number of work centers needed to meet a certain weekly production rate requirement. Let $D_w$ be the weekly demand rate for good pieces. Replacing PC by $D_w$ in Eq. (17.1) and rearranging for the required number of work centers, we obtain

$$W = \frac{D_w}{S_wHR_p(1-q)} \tag{17.2}$$

Since it is probably difficult to adjust the number of work centers $W$ on a short-term basis, we can solve for the number of work center hours per week as an alternative to Eq. (17.2).

$$S_wH = \frac{D_w}{WR_p(1-q)} \tag{17.3}$$

Equations (17.2) and (17.3) suggest three ways of adjusting capacity up or down to meet changing production requirements.

1. Change the number of work centers in the section of the shop ($W$).
2. Change the number of shifts per week ($S_w$).
3. Change the number of hours per shift ($H$): overtime, for example.

## EXAMPLE 17.3

A weekly demand rate of 7000 pieces per week is forecasted for the turret lathe section of Example 17.2. Recommend two possible alternatives for meeting this capacity requirement.

According to Eq. (17.2), the number of work centers (turret lathes) required to meet the new production rate is

$$W = \frac{7000}{(10)(6.4)(17)(0.97)} = 6.63 \text{ work centers}$$

Since turret lathes cannot be divided into fractional units, seven lathes would be required. Considering the lead time to acquire a machine tool, this alternative is likely to be infeasible in the short term.

According to Eq. (17.3), the number of hours per work center per week required to meet the capacity requirement using six machines would be

$$S_w H = \frac{7000}{6(17)(0.97)} = 70.75 \text{ h/week}$$

This is slightly greater than the 64 hours per week available. According to the data given, only 80% of the shift hours are productive. To achieve 70.75h//week of production time, each work center would have to be operated 70.75/0.80=88.4 h/week. This increase over the current 80 h/week could be achieved either by operating additional shifts, or by authorizing overtime for each of the current 10 shifts.

In cases where production rates differ significantly for different workparts, it is necessary to alter Eqs. (17.2) and (17.3). Let the subscript $i$ be used to identify products. $D_{wi}$ is the weekly demand for product $i$. Similarly, $R_{pi}$ and $q_i$ are the hourly production rate and scrap rate, respectively. Then, Eq. (17.2) becomes

$$W = \frac{1}{S_w H} \sum \frac{D_{wi}}{R_{pi}(1-q_i)} \tag{17.4}$$

where the summation is carried out over the various workparts processed through the work centers in the shop section for which $W$ is to be determined.

By the same reasoning, Eq. (17.3) becomes

$$S_w H = \frac{1}{W} \sum \frac{D_{wi}}{R_{pi}(1-q_i)} \tag{17.5}$$

## EXAMPLE 17.4

Three products are to be processed through a certain type of work center, and it is desired to determine how many of the work centers will be required to meet demand. Pertinent data are given in the following table:

| *Product* | *Weekly demand* | *Hourly production rate* | *Scrap rate(%)* |
|---|---|---|---|
| 1 | 400 | 10 | 5 |
| 2 | 1000 | 30 | 2 |
| 3 | 2200 | 70 | 10 |

A total of 65 h of production will be available per week for each work center.

Equation (17.4) can be used to figure the number of work centers required. For product 1,

$$\frac{D_w}{R_p(1-q)} = \frac{400}{10(1-0.05)} = 42.11 \text{ h}$$

For product 2,

$$\frac{D_w}{R_p(1-q)} = \frac{1000}{30(1-0.02)} = 34.01 \text{ h}$$

For product 3,

$$\frac{D_w}{R_p(1-q)} = \frac{2200}{70(1-0.1)} = 34.92 \text{ h}$$

Summing for the three products,

$$\sum \frac{D_{wi}}{R_{pi}(1-q_i)} = 111.04 \text{ h}$$

This represents the total number of production hours required per week to meet demand. Since each work center can provide 65 hours of production time per week, the number of work centers is

$$W = \frac{111.04}{65} = 1.71 \text{ work centers}$$

This would be rounded up to 2 work centers.

Computer programs can be written to perform capacity calculations similar to those represented by the example problems given above. IBM's Communications-Oriented Production Information and Control System [8] contains a module that performs capacity requirements planning computations.

## 17.4 SHOP FLOOR CONTROL

In this section we consider two related problems that perplex production managers everywhere: operation scheduling and acquiring up-to-date information on the progress of orders in the factory. The second of these problems has been called "shop floor control" [7].

### Operation scheduling

The master production schedule gives the timetable for end product deliveries. This is translated into material and component requirements by using MRP, and checked against plant production capacity by means of capacity planning. The next link in this planning chain is operation scheduling.

*Operation scheduling* is concerned with the problem of assigning specific jobs to specific work centers on a weekly, daily, or hourly basis. The end products specified in the master schedule consist of components, each of which is manufactured by a sequence of processing operations. Operation scheduling involves the assignment of start dates and completion dates to the batches of individual components and the designation of work centers on which the work is to be performed. The scheduling problem is complicated by the fact that there may be hundreds or thousands of individual jobs competing for time on the limited number of work centers. These complications are compounded by unforeseen interruptions and delays such as machine breakdowns, changes in job priority, worker absenteeism, and strikes.

The objectives of an operation scheduling system are to assign jobs to work centers so as to:

1. Meet the required delivery dates for completion of all work on the jobs.
2. Minimize in-process inventory. This is accomplished by minimizing the aggregate manufacturing lead time.
3. Maximize utilization of machines and manpower resources.

There are a variety of scheduling methods used in production. Different methods are appropriate, depending on whether the factory is engaged in job shop operations, batch production, or mass production. To describe the operation scheduling problem, let us consider the typical job shop or batch production situation. Jobs are to be processed through the appropriate sequence of work centers. A *job* is defined as a single part or a batch of parts. A job could also consist of groups of components to be assembled. Operation scheduling can be described as consisting of the following two steps:

1. Machine loading.
2. Job sequencing.

To process the jobs through the factory, the jobs must be assigned to work centers. Since the total number of jobs exceeds the number of work centers, each work center will have a queue of jobs waiting to be processed. Allocating the jobs to the work centers is referred to as *machine loading* or *shop loading*. Ten jobs may be the loading for a particular work center during the next week. The unanswered question is: In what sequence will the 10 jobs be processed? Answering this question is the problem in job sequencing.

*Job sequencing* involves determining the order in which to process the jobs through a given work center. To accomplish this, priorities are established among the jobs in the queue. Then the jobs are processed in the order of their relative priorities. Among the priority rules that have been used are the following:

Highest priority is given to the job with the earliest due date.

Highest priority goes to the job with the lowest time remaining/work remaining ratio (the so-called "critical ratio").

Highest priority goes to the job with the "shortest processing time."

Jobs are processed on a first come first served basis.

When a job is completed at one work center, it enters the queue at the next work center in its process routing. That is, it becomes part of the machine loading for the next work center, and the priority rule determines its sequence of processing among those jobs.

It is beyond the scope of this section to provide a comprehensive survey of operation scheduling techniques. Our purpose has been to explain to the unfamiliar reader what the problem is and how it relates to the overall production planning and control system. References cited at the end of the chapter give a more detailed coverage of the subject ([1], [2], [5]).

### Shop floor control

Many of the production problems discussed at the beginning of this chapter occur because management loses control over the jobs in the factory. Even with the best material requirements planning systems, capacity planning, and production scheduling methods, the plant will not operate as an efficient system unless control is maintained over the activities in the shop. The basis of control is accurate, timely information. Shop floor control is concerned with the generation of such information based on data collected from factory operations.

The shop floor control subsystem described in PICS [7][2] will be used for illustration purposes. This version of shop floor control consists of two functional modules:

[2]The shop floor control functions in PICS [7] are replaced in COPICS [8] by manufacturing activity planning, order release, and plant monitoring and control activities.

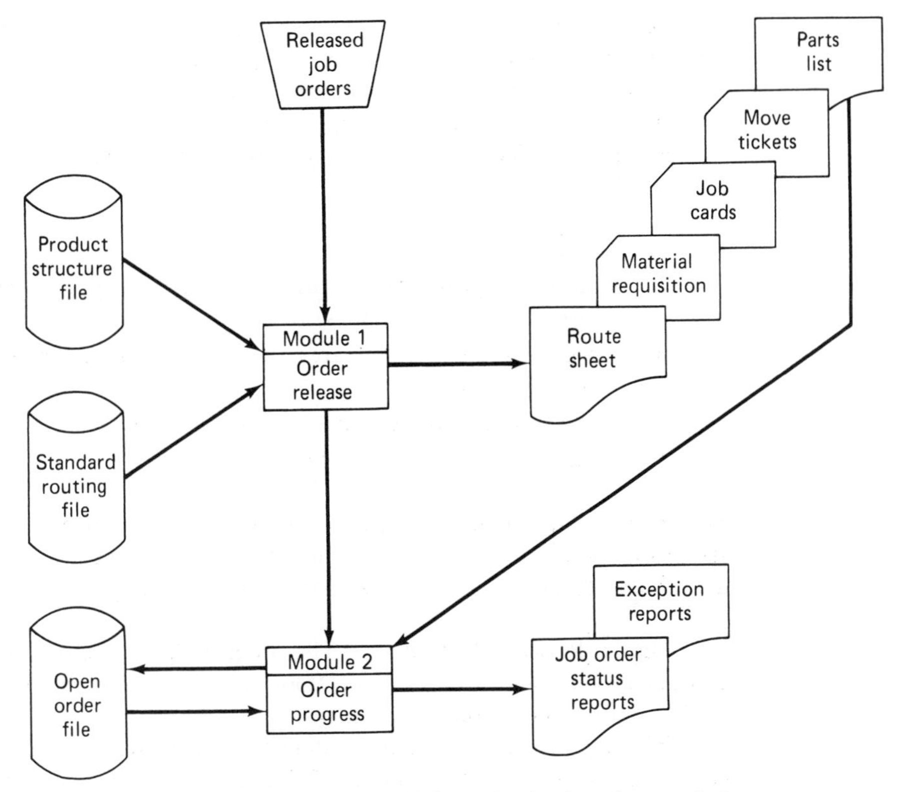


FIGURE 17.10 Flow of information in shop floor control.

1. Release of new orders.
2. Order progress reporting.

The purpose of the order release module is to provide the necessary documents for identification of jobs in the factory. The function of the second module is to report on the status of these jobs as they progress through the factory. The flow of information in shop floor control is depicted in Figure 17.10. A highly computerized information system is assumed by this flow diagram.

ORDER RELEASE MODULE. Information on released job orders is made available from the MRP system and capacity planning. This information includes the item number, the order quantity, and dates for start and finish of the order. These data constitute the input to the order release module whose function is to prepare the job documents. To accomplish this function, the standard routing file and the product structure file must be called. The standard routing file contains the sequence of operations (and work centers) used to process the job through the shop. The product structure file is used for assembly operations to identify the components needed in fabricating the various subassemblies as well as the final product.

From the various inputs, the order release module prepares a series of documents that will accompany the job through the shop. These documents are referred to collectively as the *shop packet*. The shop packet for each job consists of:

1. Route sheet.
2. Material requisitions—to draw the necessary raw materials (or components for assemblies) from stock.
3. Job cards—enough job cards to report the labor for each operation on the route sheet.
4. Move tickets—to move the parts between work centers.
5. Parts list—for assembly jobs.

The shop packet moves with the job through the sequence of processing or assembly operations. It comprises the documentation necessary to complete the job and to monitor the status of the job in the shop.

ORDER PROGRESS MODULE. The order progress module is designed to accept data collected from the shop floor and to generate reports that can be used to assist production management.

The mechanism by which the data are collected from the shop may be manual or automated. The subject of data collection systems was previously covered in Chapter 12.

The various documents contained in the shop packet provide the means to identify the job. When a machine operator completes a particular process specified in the route sheet, relevant data to indicate the completion are entered into the order progress module. The data would include such items as piece count, scrappage (if any), operator identification number, operation number, work centers, and time of completion. The order progress module maintains a file of the transactions reported on each of the uncompleted jobs. This file is called the *open order file*, and it contains the latest status of each job order in the factory. The types of reports that can be generated from the open order file include the following:

1. Job order status reports. Detailed reports that show the progress of each job through the shop.

2. Exception reports. These are designed to pinpoint deviations from the production schedule, overdue jobs, inconsistent piece counts (e.g., an increase in the number of pieces in the order), and other exception information.

These reports can be generated daily to achieve better control over jobs in the plant. Depending on the design of the shop floor control system, the information in the open order file might also be obtained on an inquiry basis. The various types of data that can be collected from factory operations are described in Chapter 12.

Substantial improvements in plant level operations can be achieved by successful implementation of shop floor control, MRP, and other production planning and control systems. The interested reader may wish to consult references [8], [10], and [13] for more detailed discussions of these topics than we have provided in this chapter.

## REFERENCES

[1] ADAM, E. E., and EBERT, R. J., *Production and Operations Management*, Prentice-Hall, Inc., Englewood Cliffs, N.J., 1978.

[2] BAKER, K. R., *Introduction to Sequencing and Scheduling*, John Wiley & Sons, Inc., New York, 1974.

[3] BONSACK, R. A., "Computer-Based Manufacturing Planning and Control Systems," *Production and Inventory Management*, Second Quarter, 1976, pp. 94–117.

[4] "Capacity Planning and Control Study Guide," *Production and Inventory Management*, First Quarter, 1975, pp. 1–16.

[5] CHASE, R. B., and AQUILANO, N. J., *Production and Operations Management*, rev. ed., Richard D. Irwin, Inc., Homewood, Ill., 1977.

[6] DEVARAJAN, D. A., "Care and Feeding of MRP," *Industrial Engineering*, August, 1978, pp. 22–27.

[7] International Business Machines, *Production Information and Control System*, Publication GE20-0280-2.

[8] International Business Machines, *Communications Oriented Production Information and Control System*, Volume I, Publication G320-1974.

[9] OLSEN, E. D., "Shop Control System Far Cry from Job Jar Approach," *Industrial Engineering*, December, 1978, pp. 16–20.

[10] ORLICKY, J., *Material Requirements Planning*, McGraw-Hill Book Company, New York, 1975.

[11] PLOSSL, G. W., *Manufacturing Control*, Reston Publishing Company, Inc., Reston, Va., 1973.

[12] STAUFFER, R. N., "Getting a Handle on Manufacturing Management," *Manufacturing Engineering*, February, 1979, p. 75.

[13] WIGHT, O. W., *Production and Inventory Management in the Computer Age*, CBI Publishing Company, Inc., 1974.

[14] WIGHT, O. W., Correspondence dated May 22, 1979.

## PROBLEMS

**17.1.** Using the master schedule of Figure 17.3 and the product structures in Figures 17.4 and 17.7, determine the time-phased requirements for component C6. The raw material used in component C6 is M6. Two units of C6 are obtained from every unit of M6. Lead times are as follows:

P1: assembly lead time = 1 week.

P2: assembly lead time = 1 week.

S2: assembly lead time = 1 week.

S3: assembly lead time = 1 week.

C6: manufacturing lead time = 2 weeks.

M6: ordering lead time = 2 weeks.

Assume that the current inventory status for all of the items above is: units on hand = 0, units on order = 0. The format of the solution should be similar to that presented in Example 17.1.

**17.2.** Solve Problem 17.1 if the current inventory and order status for S3, C6, and M6 is:

S3: inventory on hand = 2, on order = 0.

C6: inventory on hand = 5, on order = 10 due for delivery in week 2.

M6: inventory on hand = 10, on order = 50 due for delivery in week 2.

**17.3.** The following data apply to a cluster of NC turret lathes in one section of the shop:

Number of machines = 4.

Number of shifts per day = 2.

Number of days per week = 5.

Nominal hours per day = 8.

Machine utilization during one shift averages 70%.

From these data compute the turret lathe capacity in terms of production hours per week.

**17.4.** For the turret lathe section of Problem 17.3, the average production rate is 15 parts per hour for the current product mix and average batch size. This rate includes an allowance for setup time. It is anticipated that the production requirements for next year will average 3600 parts/week. The scrap rate has historically averaged about 5%. The future product mix and batch size are expected to remain unchanged.

(a) Compute the average number of production hours needed per week from the section to meet this requirement.

(b) Propose at least two alternative methods for meeting this new production requirement.

**17.5.** Five workpart types are to be processed through the NC lathe section, which contains four NC machines. Workpart data are given in the accompanying table:

| *Product* | *Weekly demand* | *Hourly production rate* | *Scrap rate (%)* |
|---|---|---|---|
| 1 | 1200 | 10 | 0 |
| 2 | 400 | 20 | 0 |
| 3 | 2000 | 15 | 4 |
| 4 | 2500 | 30 | 2 |
| 5 | 1700 | 17 | 2 |

If the work centers operate 6.85 hours per shift, how many shifts must the plant operate per week to achieve the required capacity?

**17.6.** There are nine machines in the automatic lathe section of the NC machine shop. The setup time on an automatic lathe averages 6 h. Average lot size for parts processed through the section is 90. Production cycle time averages 8 min. Three operators run the machines. Under the shop work rules, each operator is permitted to be assigned up to three machines. However, an average of 15% of the machine time is lost as a result of machine breakdowns. In addition to the machine operators, there are two setup workers who perform setups exclusively. These people are kept busy the full shift. The section runs one 8-h shift per day, 6 days/week. Scrap losses are negligible.

The production control manager claims that the production capacity of this section should be 1836 pieces/week. However, the actual output of the section averages only 1440 pieces/week.

(a) What is the problem in this section?

(b) Recommend a solution.

# Group Technology and Integrated Manufacturing Systems

chapter 18

# Group Technology

## 18.1 INTRODUCTION

The modern industrial environment poses a myriad of problems and challenges in discrete parts manufacturing. Among these problems are the following:

The increasing demand toward customized products with special options and features to meet particular needs of the customer.

The growing trend toward production in small lot sizes. Some experts estimate that in coming years 75% of manufactured parts will be in small lot sizes (50 or fewer parts). This compares with 25% to 35% now.

The demand for higher reliability of products and closer tolerances for the components that go into these products.

The need to process a wider variety of work materials including metals with high strength/weight ratios, plastics, ceramics, and composite materials.

The conventional job shop organization (process type layout as discussed in Chapter 2) is turning out to be inefficient and outdated.

The growing need to integrate the activities of design and manufacturing.

One of the modern concepts in manufacturing that holds the promise of meeting these challenges is group technology. In this chapter we discuss three related topics:

1. Group technology (GT).
2. Parts classification and coding.
3. Automated process planning.

All these techniques are gaining acceptance in U.S. industry today. The applications are growing, and there is a widespread belief among manufacturing people that the use of group technology will increase significantly in future years.

*Group technology* is a manufacturing philosophy in which similar parts are identified and grouped together to take advantage of their similarities in manufacturing and design. Similar parts are arranged into part families. For example, a plant producing 10,000 different part numbers may be able to group the vast majority of these parts into 50 or 60 distinct families. Each family would possess similar design and manufacturing characteristics. Hence, the processing of each member of a given family would be similar, and this results in manufacturing efficiencies. These efficiencies are achieved by arranging the production equipment into machine groups, or cells, to facilitate work flow. In product design, there are also advantages obtained by grouping parts into families. These advantages lie in the classification and coding of parts.

Parts classification and coding is concerned with identifying the similarities among parts and relating these similarities to a coding system. Part similarities are of two types: *design attributes* (such as geometric shape and size), and *manufacturing attributes* (the sequence of processing steps required to make the part). While the processing steps required to manufacture a part are usually closely correlated with the part's design attributes, this is not always the case. Accordingly, classification and coding systems are often devised to allow for differences between a part's design and its manufacture. The reason for using a coding scheme is to facilitate retrieval for design and manufacturing purposes. In design, for example, a designer faced with the task of developing a new part can use the design-retrieval system to determine if a similar part is already in existence. A simple change in an existing part would be much less time-consuming than designing from scratch. In manufacturing, the coding scheme can be used in an automated process planning system.

Automated process planning is often called *computer-aided process planning* (CAPP) because the computer must be utilized. CAPP involves the automatic generation of a process plan (or route sheet) to manufacture the part. The process routing is developed by recognizing the specific attributes of the part in question and relating these attributes to the corresponding manufacturing operations. The use of an automated process planning system must be preceded by an appropriate parts classification and coding system.

Group technology, parts classification and coding, and automated process planning are all interrelated. Group technology is the underlying manufacturing concept, but some form of parts classification and coding is almost a necessity in order to implement GT and automated process planning.

## 18.2 PART FAMILIES

A *part family* is a collection of parts which are similar either because of geometric shape and size or because similar processing steps are required in their manufacture. The parts within a family are different, but their similarities are close enough to merit their identification as members of the part family. Figures 18.1 and 18.2 show two part families. The two parts shown in Figure 18.1 are similar from a design viewpoint but quite different in terms of manufacturing. The 13 parts shown in Figure 18.2 might constitute a parts family in manufacturing, but their geometry characteristics do not permit them to be grouped as a design parts family.

One of the big manufacturing advantages of grouping workparts into families can be explained with reference to Figures 18.3 and 18.4. Figure 18.3 shows a process-type layout for batch production in a machine shop. The various machine tools are arranged by function. There is a lathe section, milling machine section, drill press section, and so on. During the machining of a given part, the workpiece must be moved between sections, with perhaps the same section being visited several times. This results in a significant amount of material handling, a large in-process inventory, usually more setups than necessary, long manufacturing lead times, and high cost. Figure

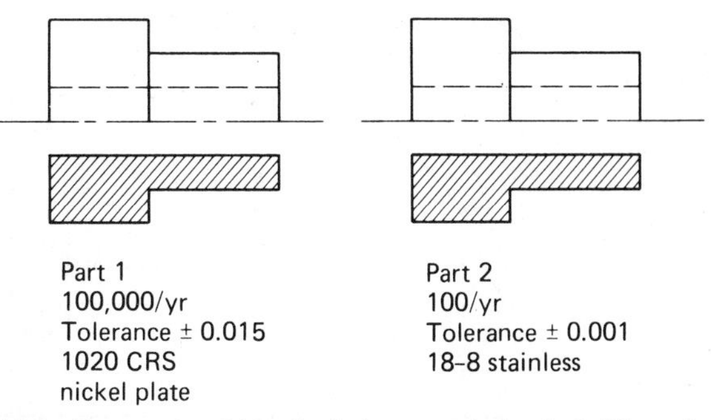


FIGURE 18.1 Two parts of identical shape and size but different manufacturing requirements.

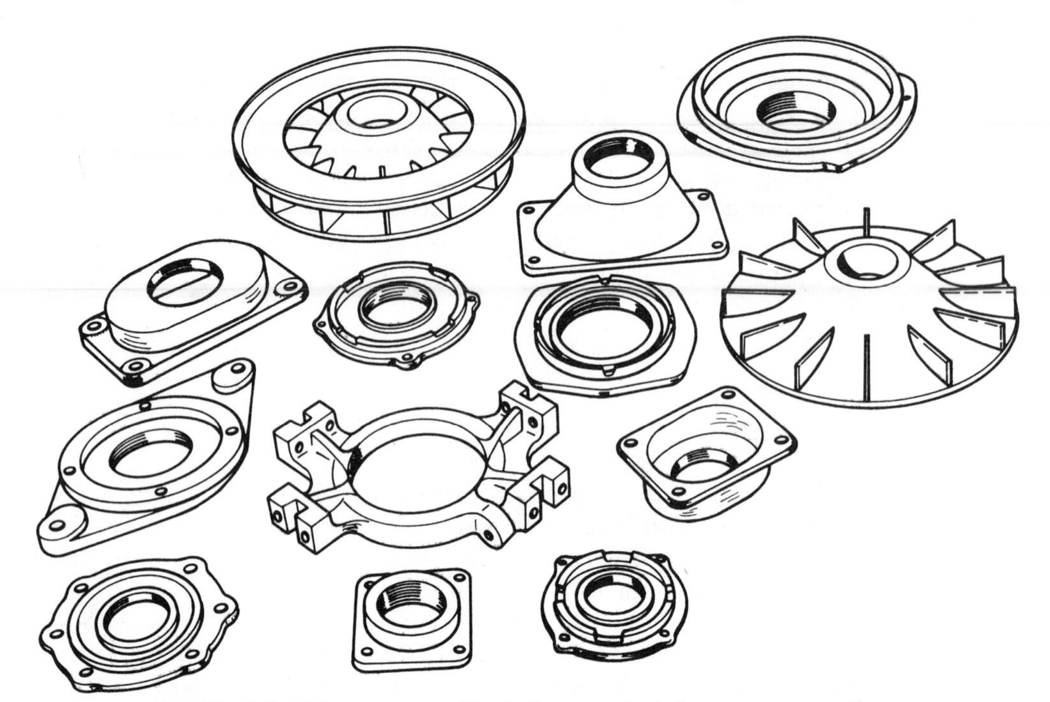

FIGURE 18.2 Thirteen parts with similar manufacturing process requirements but different design attributes.

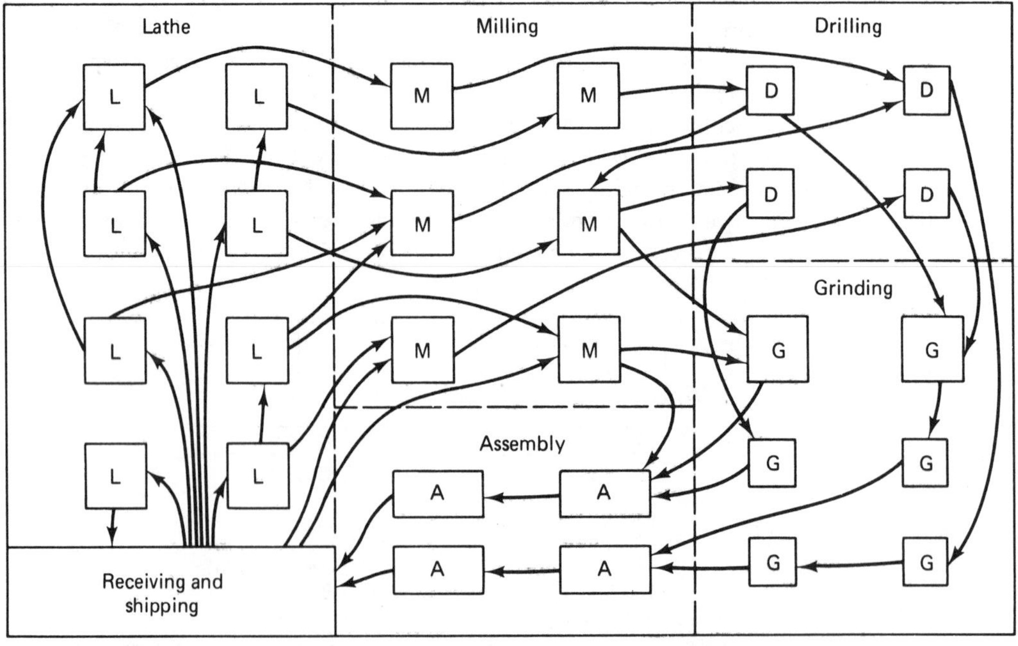


FIGURE 18.3 Process-type layout.

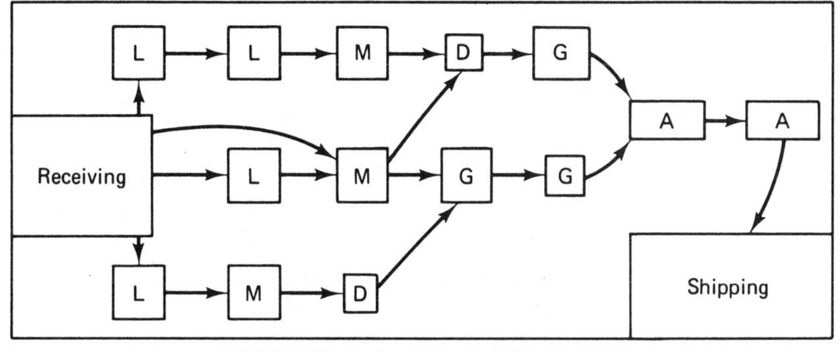


FIGURE 18.4 Group technology layout.

18.4 shows a production shop of equivalent capacity, but with the machines arranged into cells. Each cell is organized to specialize in the manufacture of a particular part family. Advantages are gained in the form of reduced workpiece handling, lower setup times, less in-process inventory, and shorter lead times. Some of the manufacturing cells can be designed to form production flow lines, with conveyors used to transport workparts between machines in the cell.

The biggest single obstacle in changing over to group technology from a traditional production shop is the problem of grouping parts into families. There are three general methods for solving this problem. All three methods are time-consuming and involve the analysis of much data by properly trained personnel. The three methods are:

1. Visual inspection.
2. Classification and coding by examination of design and production data.
3. Production flow analysis (PFA).

The *visual inspection* method is the least sophisticated and least expensive method. It involves the classification of parts into families by looking at either the physical parts or photographs and arranging them into similar groupings. Although this method is generally considered to be the least accurate of the three, one of the first major success stories of GT in the United States made the changeover using the visual method. This was the Langston Division of Molins Machine Company, located in Cherry Hill, N.J. [5].

The second method involves classifying the parts into families by examining the individual design and/or manufacturing attributes of each part. The classification results in a code number that uniquely identifies the part's attributes. This classification and coding may be carried out on the entire list of active parts of the firm, or some sort of sampling procedure may be used to establish the part families. For example, parts produced in the shop during a certain given time period could be examined to identify part family categories. The trouble with any sampling procedure is the risk that the sample may be unrepresentative of the entire population. The method of parts classification and coding seems to be the most commonly used method today. A number of classification and coding systems are described in the literature ([3], [11]), and there are several commercially available packages being sold to industrial concerns. We will discuss parts classification and coding more thoroughly in Section 18.3.

The third method, *production flow analysis*, makes use of the information contained on route sheets rather than part drawings. Workparts with identical or similar routings are classified into part families. Production flow analysis will be discussed in more detail in Section 18.4.

## 18.3 PARTS CLASSIFICATION AND CODING

As mentioned previously, the three methods of identifying part families all require a significant investment in time and manpower. The most time-consuming and complicated of the three methods is parts classification and coding. Many systems have been developed throughout the world, but none of them has been universally adopted. One of the reasons for this is that a classification and coding system should be custom-engineered for a given company or industry. One system may be best for one company while a different system is more suited to another company.

The major benefits of a well-designed classification and coding system for group technology have been summarized as follows by Ham [4]:

1. It facilitates the formation of part families and machine cells.
2. It permits quick retrieval of designs, drawings, and process plans.
3. It reduces design duplication.
4. It provides reliable workpiece statistics.
5. It facilitates accurate estimation of machine tool requirements and logical machine loadings.
6. It permits rationalization of tooling setups, reduces setup time, and reduces production throughput time.
7. It allows rationalization and improvement in tool design.
8. It aids production planning and scheduling procedures.
9. It improves cost estimation and facilitates cost accounting procedures.
10. It provides for better machine tool utilization and better use of tools, fixtures, and manpower.
11. It facilitates NC part programming.

### Types of classification and coding systems

Although it would seem from the foregoing list that nearly all departments in the firm can benefit from a good parts classification and coding system, the two main functional areas that use the system are design and manufacturing. Accordingly, parts classification systems fall into one of three categories:

1. Systems based on part design attributes.
2. Systems based on part manufacturing attributes.
3. Systems based on both design and manufacturing attributes.

The types of design and manufacturing workpart attributes which are typically included in classification schemes are listed in Table 18.1. It is clear that there is a certain amount of overlap between the design and manufacturing attributes of a part.

TABLE 18.1 Design and Manufacturing Part Attributes Typically Included in a Group Technology Classification System

| | |
|---|---|
| *Part design attributes* | |
| Basic external shape | Major dimensions |
| Basic internal shape | Minor dimensions |
| Length/diameter ratio | Tolerances |
| Material type | Surface finish |
| Part function | |
| *Part manufacturing attributes* | |
| Major process | Operation sequence |
| Minor operations | Production time |
| Major dimension | Batch size |
| Length/diameter ratio | Annual production |
| Surface finish | Fixtures needed |
| Machine tool | Cutting tools |

The parts coding scheme consists of a sequence of numerical digits devised to identify the part's design and manufacturing attributes. Coding schemes for parts classification can be of two basic structures:

1. *Hierarchical structure.* In this code structure, the interpretation of each succeeding symbol depends on the value of the preceding symbols.

2. *Chain-type structure.* In this type of code, the interpretation of each symbol in the sequence is fixed. It does not depend on the value of the preceding symbol.

For example, consider a two-digit code, such as 15 or 25. Suppose that the first digit stands for the general part shape. The symbol 1 means round workpart and 2 means flat rectangular geometry. In a hierarchical code structure, the interpretation of the second digit would depend on the value of the first digit. If preceded by 1, the 5 might indicate some length/diameter ratio, and if preceded by 2, the 5 might be interpreted to specify some overall length. In the chain-type code structure, the symbol 5 would be interpreted the same way regardless of the value of the first digit. For example, it might indicate overall part length, whether the part is rotational or flat rectangular. The advantage of the hierarchical code structure is that more information can be contained in the code. Some parts classification and coding systems use a combination of the hierarchical and chain-type structures.

The number of digits required can range from 6 to 30. Coding schemes that include only design characteristics require fewer digits, maybe 12 or fewer. Most modern classification and coding systems incorporate both design and manufacturing data into the code. To adequately accomplish this, code numbers with 20 to 30 digits may be needed.

Two classification systems will be discussed in the following subsections: the Opitz System and the MICLASS System. The Opitz Classification System was one of the earlier systems and is relatively simple to understand. The MICLASS System is relatively sophisticated but provides more features and options (such as a computer-automated process planning package).

### The Opitz classification system

This parts classification and coding system was developed by H. Opitz of the University of Aachen in West Germany. It represents one of the pioneering efforts in the group technology area and is probably the best known of the classification and coding schemes.

The Opitz Coding System uses the following digit sequence:

12345 6789 ABCD

The basic code consists of nine digits, which can be extended by adding four more digits. The first nine digits are intended to convey both design and manufacturing data. The general interpretation of the nine digits is indicated in Figure 18.5. The first five digits, 12345, are called the "form code" and describe the primary design attributes of the part. The next four digits, 6789, constitute the "supplementary code." It indicates some of the attributes that would be of use to manufacturing (dimensions, work material, starting raw workpiece shape and accuracy). The extra four digits, ABCD, are referred to as the "secondary code" and are intended to identify the production operation type and sequence. The secondary code can be designed by the firm to serve its own particular needs.

The complete coding system is too complex to provide a comprehensive description here. Opitz wrote an entire book on his system [11]. However, to obtain a general idea of how the Opitz System works, let us examine the first five digits of the code, the form code. The first digit identifies whether the part is a rotational or a nonrotational part. It also describes the general shape

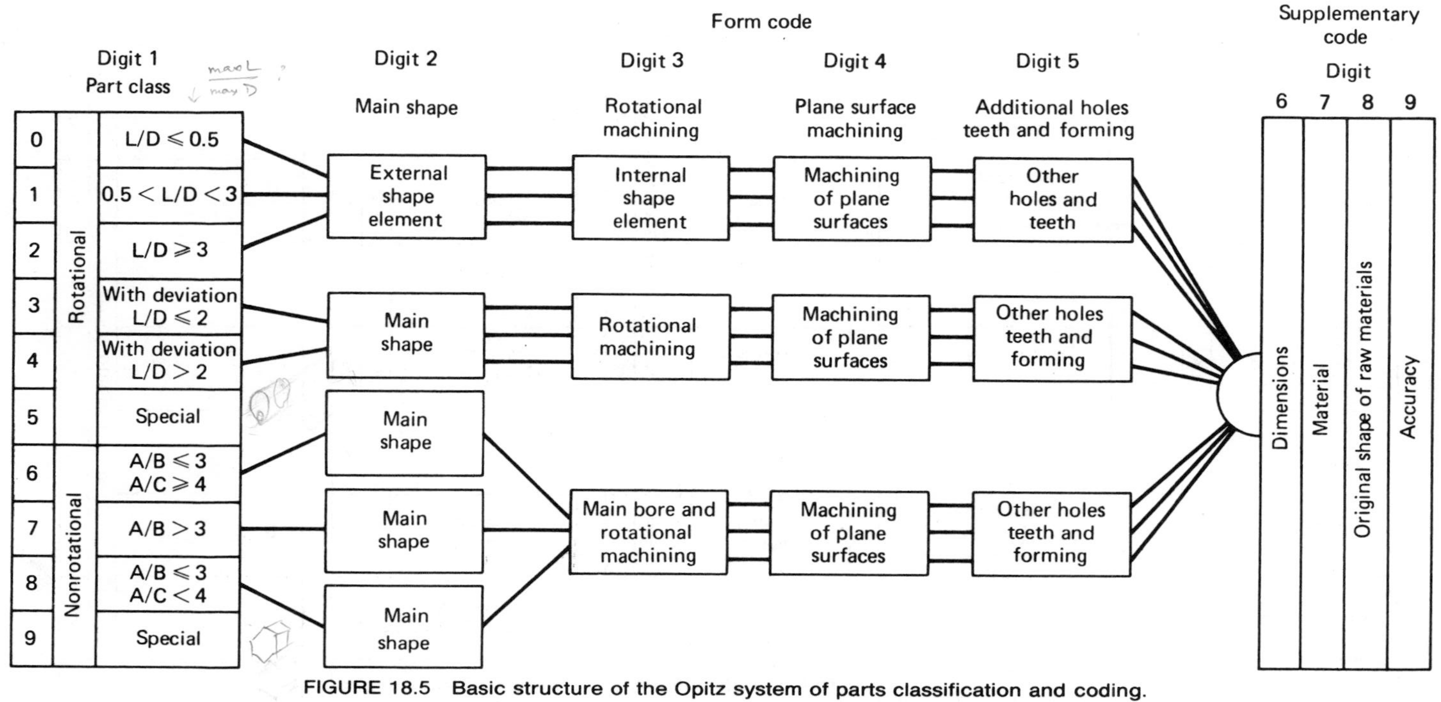


FIGURE 18.5 Basic structure of the Opitz system of parts classification and coding.

| | Digit 1 | | Digit 2 | | Digit 3 | | Digit 4 | | Digit 5 |
|---|---|---|---|---|---|---|---|---|---|
| | | Part class | | External shape, external shape elements | | Internal shape, internal shape elements | Plane surface machining | | Auxiliary holes and gear teeth |
| 0 | Rotational parts | L/D ⩽ 0·5 | | Smooth, no shape elements | | No hole, no breakthrough | No surface machining | No gear teeth | No auxiliary hole |
| 1 | Rotational parts | 0·5 < L/D < 3 | Stepped to one end | No shape elements | Smooth or stepped to one end | No shape elements | Surface plane and/or curved in one direction, external | No gear teeth | Axial, not on pitch circle diameter |
| 2 | Rotational parts | L/D ⩾ 3 | Stepped to one end or smooth | Thread | Smooth or stepped to one end | Thread | External plane surface related by graduation around a circle | No gear teeth | Axial on pitch circle diameter |
| 3 | Rotational parts | | Stepped to one end or smooth | Functional groove | Smooth or stepped to one end | Functional groove | External groove and/or slot | No gear teeth | Radial, not on pitch circle diameter |
| 4 | Rotational parts | | Stepped to both ends | No shape elements | Stepped to both ends | No shape elements | External spline (polygon) | No gear teeth | Axial and/or radial and/or other direction |
| 5 | Rotational parts | | Stepped to both ends | Thread | Stepped to both ends | Thread | External plane surface and/or slot, external spline | No gear teeth | Axial and/or radial on PCD and/or other directions |
| 6 | Nonrotational parts | | Stepped to both ends | Functional groove | Stepped to both ends | Functional groove | Internal plane surface and/or slot | With gear teeth | Spur gear teeth |
| 7 | Nonrotational parts | | | Functional cone | | Functional cone | Internal spline (polygon) | With gear teeth | Bevel gear teeth |
| 8 | Nonrotational parts | | | Operating thread | | Operating thread | Internal and external polygon, groove and/or slot | With gear teeth | Other gear teeth |
| 9 | Nonrotational parts | | | All others | | All others | All others | | All others |

FIGURE 18.6 Form code (digits 1 through 5) for rotational parts in the Opitz System. Part classes 0, 1, and 2.

and proportions of the part. We will limit our survey to rotational parts possessing no unusual features, those with code values 0, 1, or 2. See Figure 18.5 for definitions. For this general class of workparts, the coding of the first five digits is given in Figure 18.6. An example will demonstrate the coding of a given part.

### EXAMPLE 18.1

Given the part design of Figure 18.7, define the "form code" (the first five digits) using the Opitz System.

The overall length/diameter ratio, $L/D = 1.5$, so the first digit code = 1. The part is stepped on both ends with screw thread on one end, so the second digit code would be 5. The third digit code is 1 because of the through-hole). The fourth and fifth digits are both 0, since no surface machining is required and there are no auxiliary holes or gear teeth on the part. The complete form code in the Opitz System is 15100. To add the supplementary code, we would have to properly code the sixth through ninth digits with data on dimensions, material, starting workpiece shape, and accuracy.

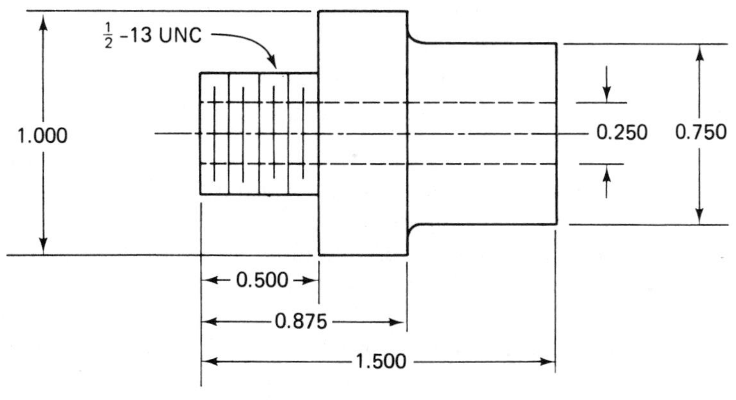


FIGURE 18.7 Workpart of Example 18.1.

## The MICLASS system

MICLASS stands for Metal Institute Classification System and was developed by TNO, the Netherlands Organization for Applied Scientific Research. It was started in Europe about 5 years before being introduced in the United States around 1974. The MICLASS system was developed to help automate and standardize a number of different design, manufacturing, and management functions, including:

Standardization of engineering drawings.

Retrieval of drawings according to classification number.

Standardization of process routing.

Automated process planning.

Selection of parts for processing on particular groups of machine tools.

Machine tool investment analysis.

The MICLASS classification number can range from 12 to 30 digits. The first 12 digits are a universal code that can be applied to any part. Up to 18 additional digits can be used to code data that are specific to the particular company or industry. For example, lot size, piece time, cost data, and operation sequence might be included in the 18 supplementary digits.

The workpart attributes coded in the first 12 digits of the MICLASS number are as follows:

| | |
|---|---|
| 1st digit | Main shape |
| 2nd and 3rd digits | Shape elements |
| 4th digit | Position of shape elements |
| 5th and 6th digits | Main dimensions |
| 7th digit | Dimension ratio |
| 8th digit | Auxiliary dimension |
| 9th and 10th digits | Tolerance codes |
| 11th and 12th digits | Material codes |

One of the unique features of the MICLASS system is that parts can be coded using a computer interactively. To classify a given part design, the user responds to a series of questions asked by the computer. The number of questions depends on the complexity of the part. For a simple part, as few as seven questions are needed to classify the part. For an average part, the number of questions ranges between 10 and 20. On the basis of the responses to its questions, the computer assigns a code number to the part. Because the system developer, TNO, is an international organization, the program was written to converse in any of four languages: English, French, German, or Dutch. Also, it can operate in either inches or metric, or both. An example will illustrate the use of the MICLASS computer system for parts classification.

### EXAMPLE 18.2

Figure 18.8 shows a rotational part to be classified and coded using MICLASS. Dimensions are given in inches. Figure 18.9 shows a copy of the user interrogation by the computer. Most of the questions require "Yes" or "No" answers determined from a user analysis of the part drawing. When the interrogation is completed, the computer prints the proper code number. For this part, the universal code (the first 12 digits) is 1271 3231 3144.

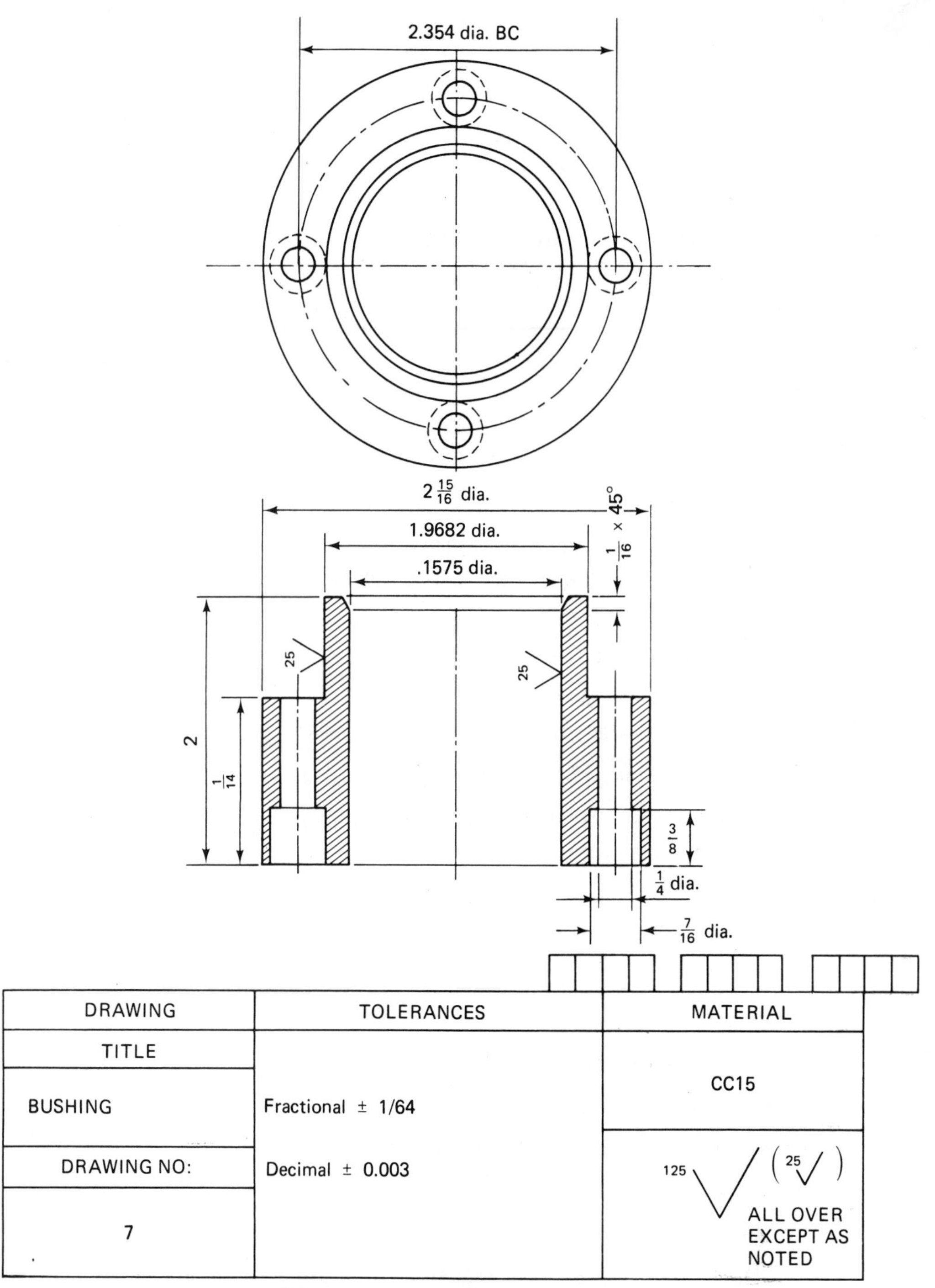


FIGURE 18.8 Workpart of Example 18.2.

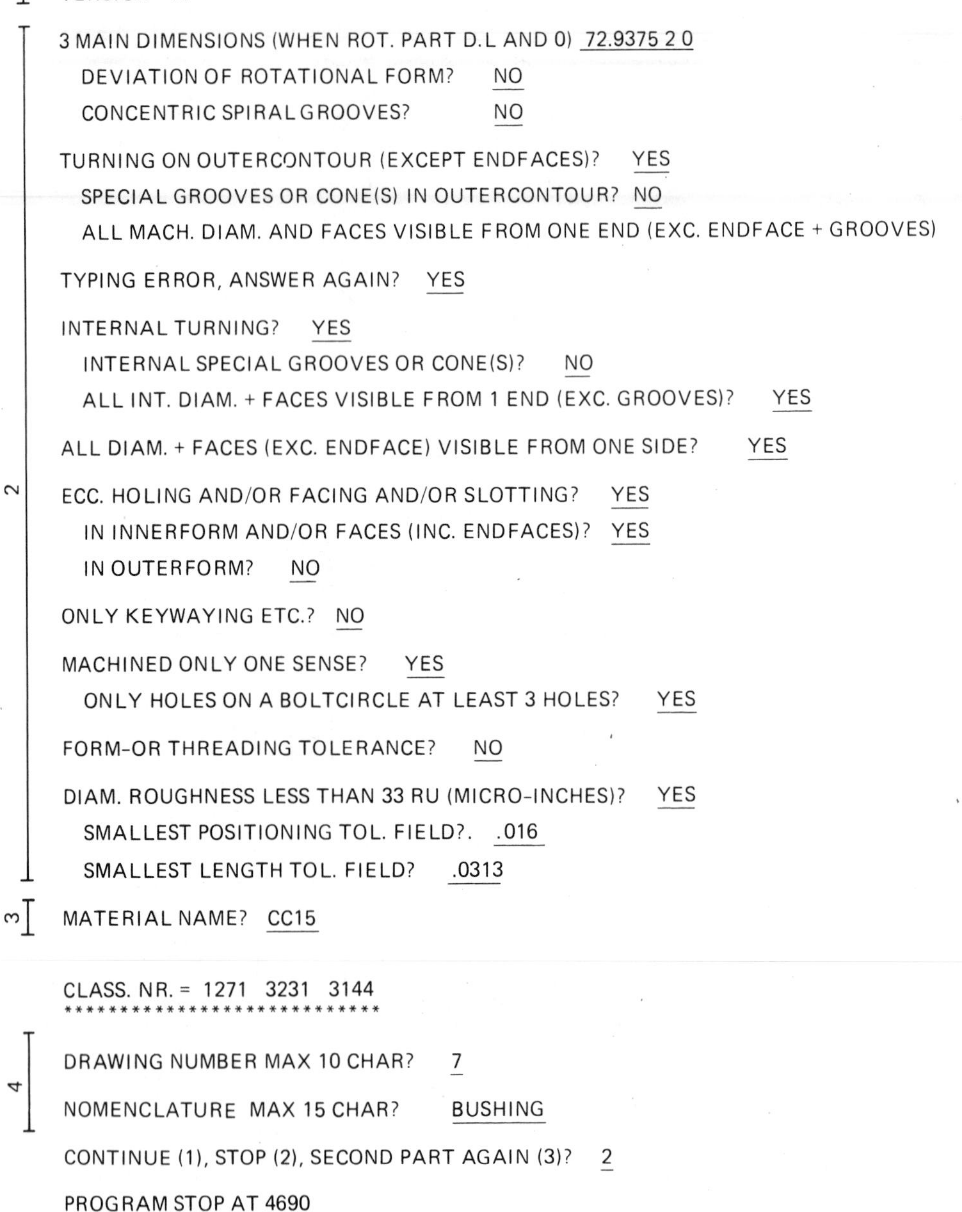


**FIGURE 18.9** Computerized MICLASS System determination of code for workpart of Figure 18.8.

## 18.4 PRODUCTION FLOW ANALYSIS

Production flow analysis (PFA) is a method for identifying part families and associated groupings of machine tools. It does not use a classification and coding system and it does not use part drawings to identify families. Instead, PFA is used to analyze the operation sequence and machine routing for the parts produced in the given shop. It groups parts with identical or similar routings together. These groups can then be used to form logical machine cells in a group technology layout. Since PFA uses manufacturing data rather than design data to identify part families, it can overcome two possible anomalies. First, parts whose basic geometries are quite different may nevertheless require similar or identical process routings. Second, parts whose geometries are similar may nevertheless require process routings that are quite different. However, the disadvantage of using production flow analysis is that it provides no mechanism for rationalizing the manufacturing routings. It takes the route sheets the way they are, with no consideration being given to whether the routings are optimal or consistent or even logical.

### PFA procedure

The procedure in production flow analysis can be organized into the following steps:

1. *Data collection.* The first step in the PFA procedure is to decide on the scope of the study and to collect the necessary data. The scope defines the population of parts to be analyzed. Should all the parts produced in the shop be included in the study, or should a representative sample be selected for analysis? Once the population is defined, the minimum data needed in the analysis are the part number and machine routing (operation sequence) for every part. These data can be obtained from the route sheets. Additional data, such as lot size, time standards, and annual production rate, might be useful for designing machine cells of the desired productive capacity.

2. *Sorting of process routings.* The second step is to arrange the parts into groups according to the similarity of their process routings. For a large number of parts in the study, the only practical way to accomplish this step is to code the data collected in step 1 onto computer cards. One possible format (highly simplified) for these cards is illustrated in Figure 18.10. The format allows space for the part number and a sequence of code numbers that identify particular machines in the routing. Table 18.2 presents a list of possible code numbers that might be used (again, highly simplified).

A sorting procedure would be used on the cards to arrange them into "packs." A pack is a group of parts with identical process routings. Some

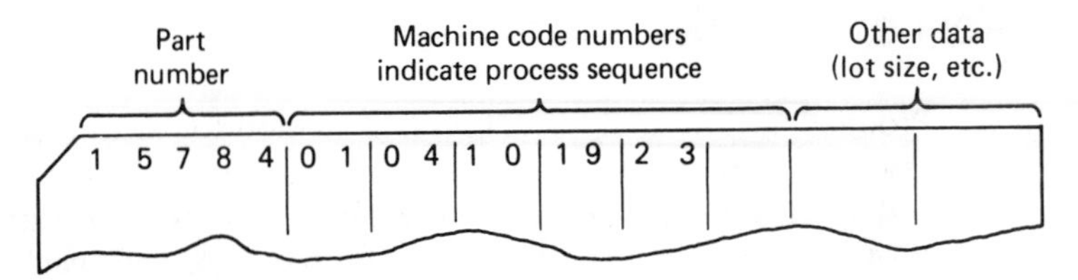


FIGURE 18.10 Card format for organizing process routing data in production flow analysis.

TABLE 18.2 Possible Code Numbers to Indicate Processes and Machines (Highly Simplified)

| *Process* | *Code* | *Process* | *Code* |
|---|---|---|---|
| Cutoff | 01 | Shaper | 13 |
| Lathe | 02 | Planer | 14 |
| Turret lathe | 03 | Broach | 15 |
| Chucker | 04 | Deburr | 16 |
| Drill manual | 05 | Polish | 17 |
| NC drill | 06 | Buff | 18 |
| Mill | 07 | Clean | 19 |
| Bore | 08 | Paint | 20 |
| Grind—surface | 09 | Plate | 21 |
| Grind—exterior cylinder | 10 | Assemble | 22 |
| Grind—interior cylinder | 11 | Inspect | 23 |
| Grind—centerless | 12 | Package | 24 |

packs will contain only one part number. Each pack is given a pack identification number or letter.

3. *PFA chart*. The processes used for each pack are next displayed graphically on a PFA chart. A simplified version of the PFA chart is shown in Figure 18.11. It is merely a plot of the process code numbers for all the packs which have been determined.

4. *Analysis*. This is the most subjective and most difficult step in production flow analysis, yet it is the crucial step in the procedure. From the pattern of data exhibited in the PFA chart, similar groups must be identified. This can be done by rearranging the data on the original PFA chart into a new pattern which brings together packs with similar routings. One possible rearrangement is shown in Figure 18.12. The different groupings are indicated within blocks. The machines identified together within the blocks of Figure 18.12 would be synthesized into logical machine cells.

Invariably, there will be packs (process routings) that do not fit into similar groupings. These parts can be analyzed to determine if a revised process sequence can be developed which fits into one of the groups. If not, these parts must continue to be manufactured through a conventional process-type plant layout.

Pack indentification codes

| Machine code numbers | A | B | C | D | E | F | G | H | I | J | K | L | M | N | O | P | Q | R | S | T |
|---|---|---|---|---|---|---|---|---|---|---|---|---|---|---|---|---|---|---|---|---|
| 01 | X | | X | X | X | X | | X | | | | X | X | X | X | | | X | X | |
| 02 | | | X | | | X | | | X | X | | X | | | | | | | | |
| 03 | X | | | X | | | | X | | | | | | | | | | X | X | X |
| 04 | | | | | X | | | | | | | | X | | | | | | | |
| 05 | | X | | | | | | | X | | X | | | | | X | | | X | |
| 06 | | | | | | | | | | X | | | | | | | X | | | |
| 07 | | X | | | | | X | | X | X | X | | | | | X | X | | | |
| 08 | | | | | | | X | | | X | | | | | | | | | | |
| 09 | | | | | | | X | | | X | | | | | | | | | | |
| 10 | X | | | | X | X | | | X | | | X | X | | | | | X | | |
| 11 | X | | X | | X | | | | | | | X | | | | | | | | |
| 12 | | | | X | | | | X | | | | | | X | X | | | | | X |
| 13 | | X | | | | | X | | | | | | | | | | X | | | |
| 14 | | | | | | | | | | | | | | | | | | | | |
| 15 | | X | | | | | | | | | | | | | | X | | | | |
| 16 | X | X | X | | X | X | | X | | | | X | X | X | | | X | X | | X |
| 17 | | | | | | | | X | | X | | | | X | | | | | | X |
| 18 | | | | X | | | | | | X | | | | | | | | | | |
| 19 | | | X | | | | | | X | | | X | X | | X | X | X | | | |
| 20 | | | | | | | | | X | | | | | | X | | X | | | |
| 21 | | | | | | | | | | | | X | X | | | X | | | | |
| 22 | X | | | X | | | | | X | | | | | | X | | | | | |
| 23 | X | | X | X | | X | | X | X | X | | | | X | X | | | | | X |
| 24 | X | | | X | | | | | X | | | | | | X | | | | | |

FIGURE 18.11 PFA chart (highly simplified).

## Comments on PFA

The weakness of production flow analysis is that the data used in the analysis are derived from production route sheets. The process sequences from these route sheets have been prepared by different process planners, and these differences are reflected in the route sheets. The routings may contain processing steps that are nonoptimal, illogical, and unnecessary. Consequently, the final machine groupings that result from the analysis may be suboptimal. Notwithstanding this weakness, PFA has the virtue of requiring less time to perform than a complete parts classification and coding procedure. It therefore provides a technique that is attractive to many firms for making the changeover to a group technology machine layout.

Pack identification code

| Machine code numbers | F | C | L | R | A | M | E | K | Q | G | P | B | J | S | H | N | T | D | O | I |
|---|---|---|---|---|---|---|---|---|---|---|---|---|---|---|---|---|---|---|---|---|
| 01 | X | X | X | X | X | X | X | | | | | | | X | X | X | | X | X | |
| 02 | X | X | X | | | | | | | | | | X | | | | | | | X |
| 03 | | | | X | X | | | | | | | | | X | X | | X | X | | |
| 04 | | | | | | X | X | | | | | | | | | | | | | |
| 05 | | | | | | | | X | | | X | X | | X | | | | | | X |
| 06 | | | | | | | | | X | | | | X | | | | | | | |
| 07 | | | | | | | | X | X | X | X | X | X | | | | | | | X |
| 08 | | | | | | | | | | X | | | X | | | | | | | |
| 09 | | | | | | | | | | X | | | X | | | | | | | |
| 10 | X | | X | X | X | X | X | | | | | | | | | | | | | X |
| 11 | | X | X | | X | | X | | | | | | | | | | | | | |
| 12 | | | | | | | | | | | | | | | X | X | X | X | X | |
| 13 | | | | | | | | | X | X | | X | | | | | | | | |
| 14 | | | | | | | | | | | | | | | | | | | | |
| 15 | | | | | | | | | | | X | X | | | | | | | | |
| 16 | X | X | X | X | X | X | X | | X | | | X | | | X | X | X | | | |
| 17 | | | | | | | | | | | | | X | | X | X | X | | | |
| 18 | | | | | | | | | | | | | X | | | | | X | | |
| 19 | | X | X | | | X | | | X | | X | | | | | | | | X | X |
| 20 | | | | | | | | | X | | | | | | | | | | X | X |
| 21 | | | X | | | X | | | | | X | | | | | | | | | |
| 22 | | | | | X | | | | | | | | | | | | | X | X | X |
| 23 | X | X | | | X | | | | | | | | X | | X | X | X | X | X | X |
| 24 | | | | | X | | | | | | | | | | | | | X | X | X |

FIGURE 18.12 Rearranged PFA chart, indicating possible machine groups.

## 18.5 MACHINE CELL DESIGN

Whether part families and machine groups have been determined by parts classification and coding or by production flow analysis, the problem of designing the machine cells must be solved. The current section will consider some of the aspects of this important problem in group technology.

### The composite part concept

Part families are defined by the fact that their members have similar design and manufacturing attributes. The composite part concept takes this part family definition to its logical conclusion. It conceives of a hypothetical part

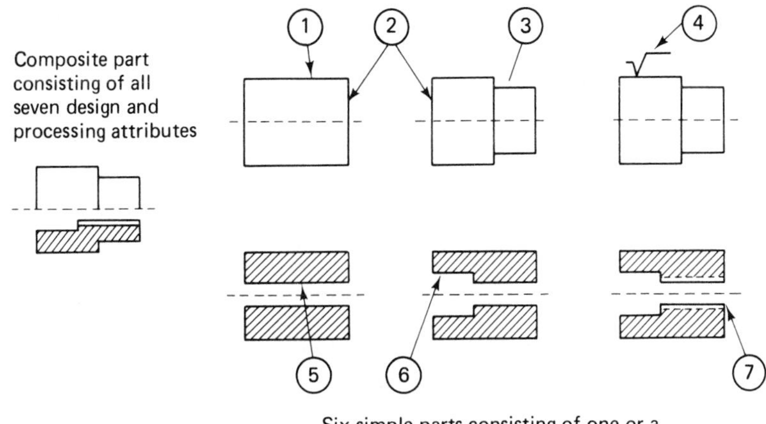


FIGURE 18.13 Composite part concept.

that represents all of the design and corresponding manufacturing attributes possessed by the various individuals in the family. Such a hypothetical part is illustrated in Figure 18.13. To produce one of the members of the part family, operations are added and deleted corresponding to the attributes of the particular part design. For example, the composite part in Figure 18.13 is a rotational part made up of seven separate design and manufacturing features. These features are listed in Table 18.3.

A machine cell would be designed to provide all seven machining capabilities. The machines, fixtures, and tools would be set up for efficient flow of workparts through the cell. A part with all seven attributes, such as the composite part of Figure 18.13, would go through all seven processing steps. For part designs without all seven features, unneeded operations would simply be canceled.

In practice, the number of design and manufacturing attributes would be greater than seven, and allowances would have to be made for variations

TABLE 18.3 Design and Manufacturing Attributes of the Composite Part in Figure 18.13

| *Number* | *Design and manufacturing attribute* |
|---|---|
| 1 | Turning operation for external cylindrical shape |
| 2 | Facing operation for ends |
| 3 | Turning operation to produce step |
| 4 | External cylindrical grinding to achieve specified surface finish |
| 5 | Drilling operation to create through hole |
| 6 | Counterbore |
| 7 | Tapping operation to produce internal threads |

in overall size and shape of parts in the part family. Nevertheless, the composite part concept is useful for visualizing the machine cell design problem.

### Types of cell designs

The organization of machines into cells can follow one of three general patterns:

1. Single machine cell.
2. Group machine layout.
3. Flow line cell design.

The single machine approach can be used for workparts whose attributes allow them to be made on basically one type of process, such as turning or milling. For example, the composite part of Figure 18.13 could be produced on a conventional turret lathe with the exception of the cylindrical grinding operation (step 4).

The group machine layout is a cell design in which several machines are used together, with no provision for conveyorized parts movement between the machines. The cell contains the machines needed to produce a certain family of parts, and the machines are organized with the proper fixtures, tools, and operators to efficiently produce the parts family.

The flow line cell design is a group of machines connected by a conveyor system. Although this design approaches the efficiency of the automated flow lines discussed in Part II of this book, the limitation of the flow line layout is that all the parts in the family must be processed through the machines in the same sequence. Certain of the processing steps can be omitted, but the flow of work through the system must be in one direction.

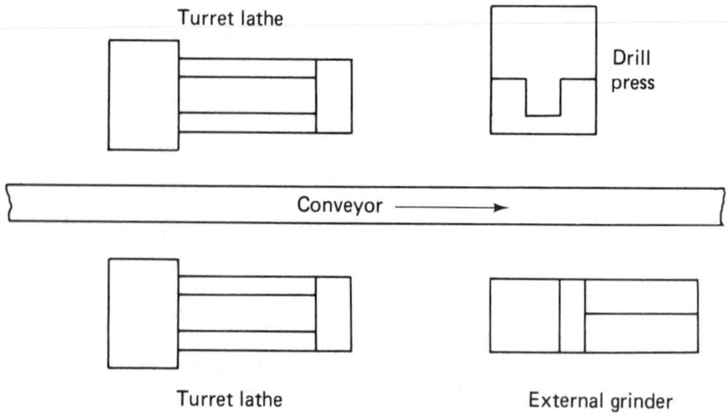


FIGURE 18.14 Flow line cell design.

Reversal of work flow is accommodated in the more flexible group machine layout, but not conveniently in the flow line configuration.

One possible flow line cell design for producing the parts family of Figure 18.13 (the composite part) is illustrated in Figure 18.14.

## 18.6 BENEFITS OF GROUP TECHNOLOGY

The problems that have prevented the widespread application of group technology in the United States include the following:

The problem of identifying part families among the many components produced by a plant.

The expense of parts classification and coding.

Rearranging the machines in the plant into the appropriate machine cells.

The general resistance that is commonly encountered when changeover to a new system is contemplated.

When these problems are solved and group technology is applied, benefits are typically realized in the following areas:

Design.

Tooling and setups.

Materials handling.

Production and inventory control.

Process planning.

Employee satisfaction.

### Product design benefits

In the area of product design, the principal benefit derives from the use of a parts classification and coding system. When a new part design is required, the engineer or draftsman can devote a few minutes to figure the code of the required part. Then the existing part designs that match the code can be retrieved to see if one of them will serve the function desired. The few minutes spent searching the design file with the aid of the coding system may save several hours of the designer's time. If the exact part design cannot be found, perhaps a small alteration of the existing design will satisfy the function.

Another advantage of GT is that it promotes design standardization. Design features such as inside corner radii, chamfers, and tolerances are more likely to be standardized with GT.

### Tooling and setups

Group technology also tends to promote standardization of several areas of manufacturing. Two of these areas are tooling and setups.

In tooling, an effort is made to design group jigs and fixtures that will accommodate every member of a parts family. Workholding devices are designed to use special adapters which convert the general fixture into one that can accept each part family member.

The machine tools in a GT cell do not require drastic changeovers in setup because of the similarity in the workparts processed on them. Hence, setup time is saved, and it becomes more feasible to try to process parts in an order so as to achieve a bare minimum of setup changeovers.

### Materials handling

Another advantage in manufacturing is a reduction in the workpart move and waiting time. The group technology machine layouts lend themselves to efficient flow of materials through the shop. The contrast is sharpest when the flow line cell design is compared to the conventional process-type layout (Figures 18.3 and 18.4).

### Production and inventory control

Several benefits accrue to a company's production and inventory control function as a consequence of group technology.

Production scheduling is simplified with group technology. In effect, grouping of machines into cells reduces the number of production centers that must be scheduled. Grouping of parts into families reduces the complexity and size of the parts scheduling problem. And for those workparts that cannot be processed through any of the machine cells, more attention can be devoted to the control of these parts.

Because of reduced setups and more efficient materials handling within machine cells, manufacturing lead times and work-in-process are reduced. Estimates are that throughput time may be reduced by as much as 60% and in-process inventory by 50% [1].

### Process planning

Proper parts classification and coding can lead to an automated process planning system. This topic will be discussed in Section 18.7. Even without an automated process planning system, reductions in the time and cost of process planning can still be accomplished. This is done through standardization. New part designs are identified by their code as belonging to a certain parts family, for which the general process routing is already known.

### Employee satisfaction

The machine cell often allows parts to be processed from raw material to finished state by a small group of workers. The workers are able to visualize their contributions to the firm more clearly. This tends to cultivate an improved worker attitude and a higher level of job satisfaction.

Another employee-related benefit of GT is that more attention tends to be given to product quality. Workpart quality is more easily traced to a particular machine cell in group technology. Consequently, workers are more responsible for the quality of work they accomplish. Traceability of part defects is sometimes very difficult in a conventional process-type layout, and quality control suffers as a result.

## 18.7 AUTOMATED PROCESS PLANNING

Automated process planning involves the use of a computer to generate the operation sequence (route sheet) based on input data relative to the workpart. Because the computer is used, the procedure is also called computer-aided process planning, or CAPP. The CAPP systems which have thus far been developed rely on two prerequisites:

1. Some form of parts classification and coding system.
2. Standard process plans for the part families produced by the plant.

### Standardization of manufacturing process plans

It is traditionally the task of the manufacturing engineers in an organization to determine the process plans for new part designs that are produced in the plant. The process plan is documented on a route sheet that lists the sequence of work centers and operations required in the part's manufacture. It is the manufacturing engineer's responsibility to develop an optimal routing for each new part. However, for a number of reasons, "stray" routings are not uncommon in most plants. Different process planners have different opinions about what makes the best routing. New machine tools in the factory render old routings less than optimal. Machine breakdowns force the planner to use temporary alternate routings, and these become the documented routings even after the machine is repaired. For these and other reasons, a significant proportion of the total number of process plans may be nonoptimal.

In one case cited in reference [16], a total of 42 different routings had developed for various sizes of a relatively simple part called an "expander sleeve." There were a total of 64 different sizes and styles, each with its own part number. The 42 routings included 20 different machine tools in the shop. The reason for this absence of process standardization was that many different individuals had worked on the parts: 8 or 9 manufacturing engineers, 2 planners, and 25 NC part programmers. Upon analysis, it was determined that only two different routings through four machines were needed to process the 64 part numbers.

A standard machine routing does not constitute a standard process plan. A variety of (similar) parts can be processed through the same sequence of machines, but the operations required at each machine may be quite distinct from one part to the next. The complete process plan must document the operations to be performed as well as the sequence of machines through which the part must be routed.

To develop standardized process plans, individual part families have to be identified, where part families are distinguished by the sequence of operations required. Several part families may have identical machine routings, but a unique production part family has the same (or nearly the same) list of operations. To define the individual part families, some form of parts classification and coding is required. The coding of the part is then used to identify the standard routing and operation sequence.

### How automated process planning works

A number of CAPP systems have been developed. These include MIAPP, which is one of the MICLASS modules [6], the CAPP systems by Computer-Aided Manufacturing-International [9], and some systems by individual companies [10]. One of the problems that complicates the development of a universal automated process planning system is that different companies use various levels of detail in specifying their process plans. Some companies use only a very basic routing; others list out the operations in great detail.

Rather than explaining how one particular system works, the general operation of an idealized automated process planning system will be described.

The flow of information in computer-aided process planning would proceed as illustrated in Figure 18.15. The user might initiate the procedure by entering the part code number at a computer terminal. The CAPP program then searches the part family matrix file to determine if a match exists. If so, both a standard machine routing and a standard operation sequence are available in the computer storage files. The process plan formatter prepares the desired document in the proper form. If an exact match cannot be found, the computer may be programmed to search the machine routing file and the operation sequence file for a similar code number. It would note where the deviations in the process sequence occur so that the user could edit the process plan to be compatible with the new part design. The machine routing file is distinguished from the operation sequence file in Figure 18.15 to emphasize that the machine routing may apply to a range of distinct part families. Editing of the operation sequence is more likely to be required. Once the operation sequence for a new part code

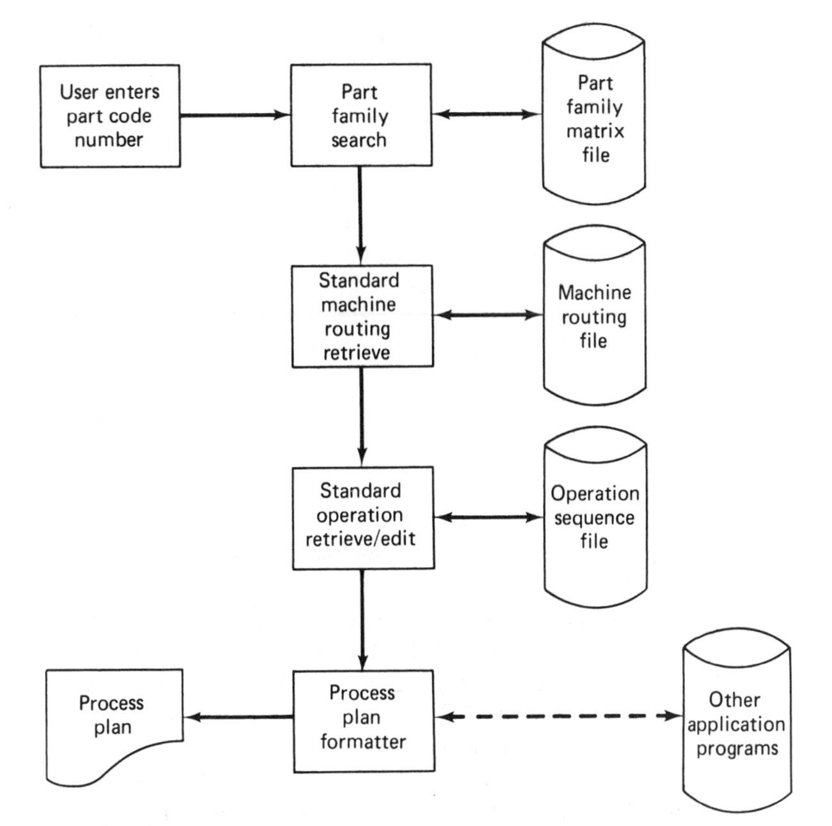


FIGURE 18.15 Information flow in computer-aided process planning.

number has been entered, it becomes the standard process for future parts of the same classification.

The process plan formatter may make use of other application programs. These could include programs to compute standard costs, setup times, and work standards (see Section 16.2). Standard cost programs might be used by the product design department to compare costs of alternative part designs. Presumably, the lowest-cost part among equivalent functional designs would be selected to minimize total product costs.

## Benefits of automated process planning

Among the benefits derived from computer-automated process planning are the following:

1. *Process rationalization and standardization.* The discipline of establishing optimal standardized process plans is actually a precondition for automated process planning. Standardized process plans should result in lower manufacturing costs and higher product quality.

2. *Increased productivity of process planners.* One system was reported to increase productivity of process planners by 600% [10].

3. *Reduced turnaround time.* The process planner working with the computer system can prepare the route sheet for use in less time compared to manual preparation.

4. *Improved legibility.* The computer-prepared document is neater and easier to read than are manually written route sheets.

5. *Incorporation of other application programs.* The CAPP program can be designed to operate in conjunction with cost estimating programs, work standards packages, and other systems to automate many of the time-consuming manufacturing support functions.

## REFERENCES

[1] ABOU-ZEID, M. R., "Group Technology," *Industrial Engineering*, May, 1975, pp. 32–39.

[2] DEVRIES, M. F., HARVEY, S. M., and TIPNIS, V. A., *Group Technology*, Publication No. MDC 76-601, Machinability Data Center, Cincinnati, Ohio, 1976.

[3] GALLAGHER, C. C., and KNIGHT, W. A., *Group Technology*, Butterworth & Co. Ltd., London, 1973.

[4] HAM, I., "Introduction to Group Technology," *Technical Report MMR76-03*, Society of Manufacturing Engineers, Dearborn, Mich., 1976.

[5] HOLTZ, R. D., "GT and CAPP Cut Work-in-Process Time 80%," *Assembly Engineering*, Part 1: June, 1978, pp. 24–27; Part 2: July, 1978, pp. 16–19.

[6] HOUTZEEL, A., "The Many Faces of Group Technology," *American Machinist*, January, 1979, pp. 115–120.

[7] HOUTZEEL, A., and SCHILPEROORT, B. A., "A Chain Structured Part Classification System (MICLASS) and Group Technology," *NC/CAM—The New Industrial Revolution*, Proceedings, 13th Annual Meeting and Technical Conference of the NC Society, 1976, pp. 383–400.

[8] HOUTZEEL, A., *Classification and Coding*, TNO, Organization for Industrial Research, Inc., Waltham, Mass.

[9] LINK, C. H., "Computer Aided Process Planning (CAPP)," *Tech. Paper MS77-314*, Society of Manufacturing Engineers, Dearborn, Mich., 1977.

[10] MCNEELY, R. A., and MALSTROM, E. M., "Computer Generates Process Routings," *Industrial Engineering*, July, 1977, pp. 32–35.

[11] OPITZ, H., *A Classification System to Describe Workpieces*, Pergamon Press, Oxford, England.

[12] OPITZ, H., and WIENDAHL, H. P., "Group Technology and Manufacturing Systems for Medium Quantity Production," *International Journal of Production Research*, Vol. 9, No. 1, 1971, pp. 181–203.

[13] PHILLIPS, R. H., and ELGOMAYEL, J., "Group Technology Applied to Product

Design," *Educational Module*, Manufacturing Productivity Educational Committee (MAPEC), Purdue University, 1977.

[14] SHARMA, S. C., "A Critical Study of the Classification and Coding Systems in Manufacturing Companies," *M.S. Thesis*, Lehigh University, 1978.

[15] TNO, *An Introduction to MICLASS*, Organization for Industrial Research, Inc., Waltham, Mass.

[16] TNO, *The 1978 MICLASS Users' Meeting*, Organization for Industrial Research, Inc., Waltham, Mass., 1978.

## PROBLEMS

**18.1.** Develop the form code (first five digits) in the Opitz System for the part illustrated in Figure P18.1.

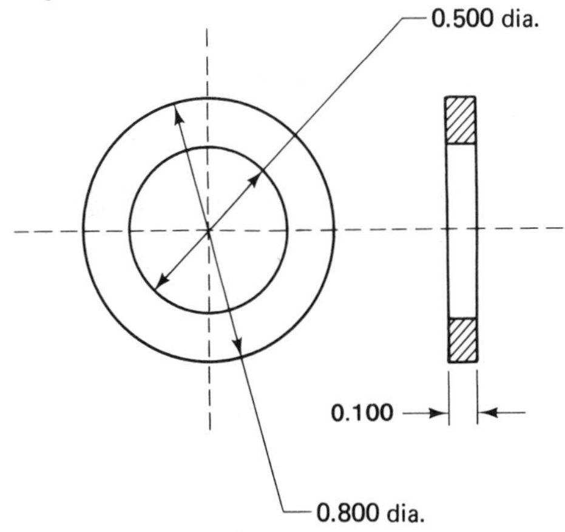


FIGURE P18.1

**18.2.** Develop the form code (first five digits) in the Opitz System for the part illustrated in Figure P18.2.

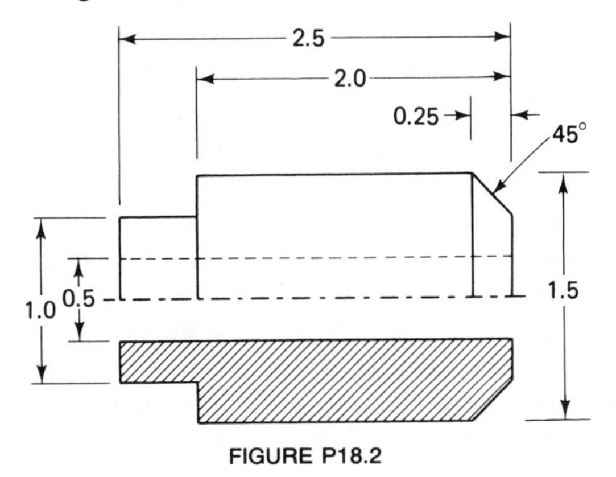


FIGURE P18.2

# Flexible Manufacturing Systems

## 19.1 INTRODUCTION

Flexible manufacturing systems incorporate many individual automation concepts and technologies into a single production system. These include:

- Automatic materials handling between machines.
- Numerical control machine tools and CNC.
- Computer control over the materials handling system and machine tools (DNC).
- Group technology principles.

We have discussed each of these concepts and technologies as separate topics throughout this book. Their integration into a single automated system is an appropriate subject on which to conclude the text.

A flexible manufacturing system (FMS) consists of a group of processing stations (usually NC machines) connected together by an automated workpart handling system. It operates as an integrated system under computer control. The FMS is capable of processing a variety of different part types simultaneously under NC program control at the various workstations.

Human labor is used to carry out the following functions to support the operation of the FMS:

Load raw workparts onto the system.

Unload finished parts from the system.

Change tools and tool setting.

Equipment maintenance and repair.

In addition, the operators may be required to input data, change part programs, and perform other tasks related to the computer control system.

The workparts are loaded and unloaded at a central location in the FMS. Pallets are used to transfer workparts between machines. Once a part is loaded onto the handling system it is automatically routed to the particular workstations required in its processing. For each different workpart type, the routing may be different, and the operations and tooling required at each workstation will also differ. The coordination and control of the parts handling and processing activities is accomplished under command of the computer. One or more computers can be used to control a single FMS. The computer functions can be categorized as follows:

CNC—computer control at the individual machines.

DNC—direct numerical control of all the machine tools in the FMS. Both CNC and DNC functions can be incorporated into a single FMS.

Computer control of the materials handling system.

Monitoring—collection of production-related data such as piece counts, tool changes, and machine utilization.

Supervisory control—functions related to production control, traffic control, tool control, and so on.

The supervisory computer can be programmed to determine alternative routings for parts when one of the machines in the FMS breaks down.

## 19.2 PROBLEMS SOLVED BY AN FMS

Flexible manufacturing systems are designed to fill the gap between high-production transfer lines and low- production NC machines. The relative position of the FMS concept is illustrated in Figure 19.1. Transfer lines are very efficient when producing parts in large volumes at high output rates. The limitation on this mode of production is that the parts must be identical. The highly mechanized lines are inflexible and cannot tolerate variations in part design. A changeover in part design requires the line to be shut down and

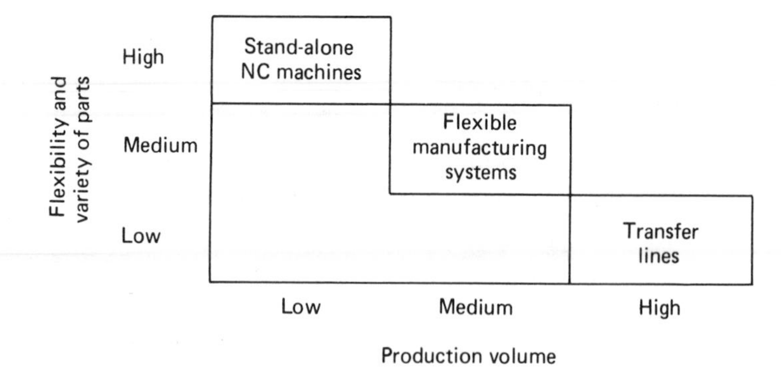


FIGURE 19.1 Application characteristics of the flexible manufacturing system concept.

retooled. If the design changes are extensive, the line may be rendered obsolete. On the other hand, stand-alone NC machines are ideally suited for variations in workpart configuration. Numerically controlled machine tools are appropriate for job shop and small batch manufacturing because they can be conveniently reprogrammed to deal with product changeovers and part design changes. In terms of manufacturing efficiency and productivity, a gap exists between the high-production-rate transfer machines and the highly flexible NC machines. This gap includes parts produced in midrange volumes (200 to 20,000 units/year). The parts are of fairly complex geometry, and the production equipment should be flexible enough to handle a limited variety of part designs. Transfer lines are not suited to this application because they are inflexible. NC machines are not suited to this application because their production rates are too slow. The solution to this mid-volume production problem is the flexible manufacturing system.

The following is a listing of the application problems the FMS concept is intended to solve:

1. *Production of families of workparts.* A flexible manufacturing system is designed to handle a variety of workpart designs. The versatility of the FMS is not as great as for a stand-alone NC machine. It applies the group technology concept for the manufacture of several different part families on the same series of machines.

2. *Random launching of workparts onto the system.* Random launching means that any workpiece among the part families handled by the FMS can be introduced to the system without downtime for setup. When the part is launched onto the line, it is identified to the computer control system, which then routes it to the proper machines in the system. The only limitation is that the workstations must be equipped in advance with the tooling required to process the part. Also, the FMS is designed to process various part configurations simultaneously by using its workstations concurrently. If

only parts of the same type are loaded onto the system, certain workstations (the ones required for the particular part) will tend to be fully utilized, while other workstations will be underutilized. Accordingly, it is desirable to maintain a mixed product launching schedule on the FMS. Within this requirement, various combinations of production rates and volumes can be processed through the system to meet specified product demands.

3. *Reduced manufacturing lead time.* Most workparts require processing in batches through several different work centers. There is setup time and waiting time at each of the work centers (see Section 2.6 on job shop production modeling). With flexible manufacturing systems, the nonoperation time is drastically reduced between successive workstations on the line. Also, setup time is minimized in the FMS operation. The setup for a traditional production machine consists of two main elements: tooling setup and workpart setup. Tooling setup means collecting the required tools from the tool crib and setting them in the machine. Workpart setup involves adjusting the workholding fixture, getting the raw materials ready, and so on. The tooling for a FMS is preset off-line. The tool setup for a particular workstation consists of loading the preset tools required for the job into the tool drum at that station. Each tool drum may be capable of holding up to 60 or more cutting tools. The workpart setup is performed external to the FMS. Since pallets are used to transport parts from station to station, the setup consists of adapting the pallet to the particular part for holding it and properly registering it at each workstation. This setup is accomplished in the load/unload area before the part and pallet are launched onto the system. Fixtures are designed that can adapt to various part configurations within a part family. Several different adaptible fixtures may be required to handle the various part families. The parts are clamped in place onto the fixture, and the fixture is attached to the pallet. The FMS has a large number of these pallets (the pallets may outnumber the workstations by 5 to 1), so some pallets can be off the system being loaded or unloaded, while other pallets are being used on the system. These features of the flexible manufacturing system allow the processing lead time to be significantly reduced.

4. *Reduced in-process inventory.* The float of parts on the FMS is limited. In fact, too many parts loaded on the system tends to cause congestion. Interference between the pallets tends to hinder fast, efficient handling of parts through the system.

5. *Increased machine utilization.* Most NC machines may operate at about 50% utilization or less. Because of minimum setup times, efficient workpart handling, simultaneous workpart processing, and other features, the utilization of a flexible manufacturing system may run as high as 80%. Downtime on a FMS is typically due to the following reasons:

Scheduled maintenance.

Scheduled tool changeovers.

Tooling problems (failures and adjustments).

Electrical failures.

Mechanical problems (e.g., oil leaks).

These are basically the same types of problems that plague the operation of a stand-alone NC machine.

6. *Reduced direct and indirect labor.* In the typical operation of many NC machines, one machine operator is used per machine. In the operation of an FMS, the entire manning may consist of three or four direct labor personnel for 6 to 10 workstations (one workstation is equivalent to one NC machine). Hence, the ratio of direct labor to machines is reduced. Indirect labor is reduced compared to job shop operation through automated materials handling rather than manual parts handling between stations.

7. *Better management control.* Since throughput time on the FMS is substantially reduced, parts do not have the opportunity to "get lost" in the shop. This results in better information and control of parts moving through the plant.

## 19.3 SYSTEM COMPONENTS

The equipment in a flexible manufacturing system consists of three main categories:

1. Machine tools.
2. Material handling system.
3. Computer control system.

These system components will be examined in this section.

### Machine tools

The machine tools used in a FMS depend on the processing requirements to be accomplished by the system. These processing needs have tended to divide flexible manufacturing systems into two distinct types [2]:

1. Dedicated FMS.
2. Random FMS.

The two different types result in different machine tool requirements.

The dedicated system is designed to meet known specific machining applications. For example, the Army contract to Chrysler Corporation for the XM-1 tank results in well-defined manufacturing requirements over a known

FIGURE 19.2 Duplex multiple spindle head indexer used as module on FMS (Courtesy Kearney & Trecker Corp.)

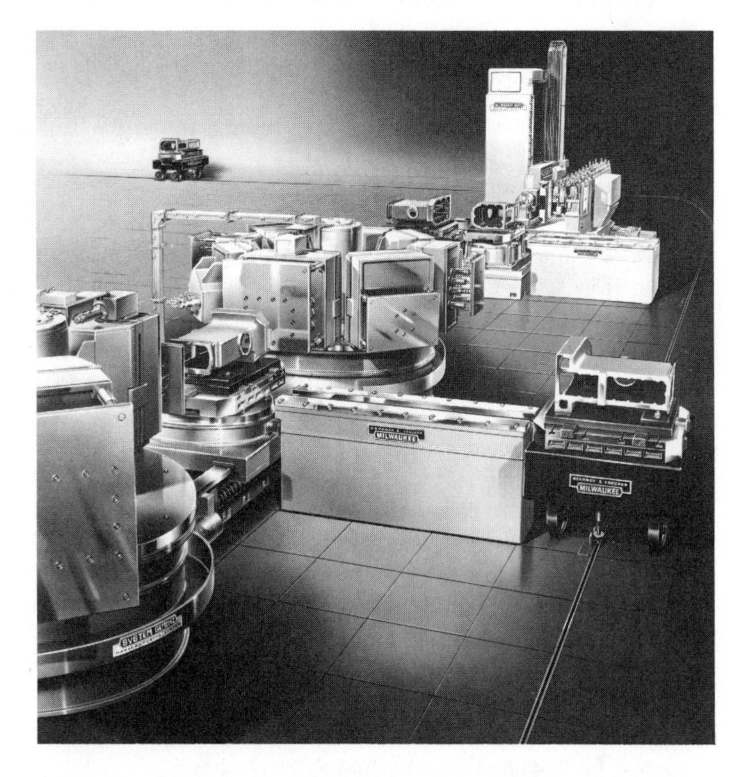


FIGURE 19.3 Machine tool modules and in-floor towline cart transfer system combined into FMS concept. (Courtesy Kearney & Trecker Corp.)

time horizon. These requirements translate into predetermined machining applications. In this type of case, the FMS can be designed as a dedicated system to meet a limited variety of processing needs. Special machine tools are used on the dedicated FMS. Specialization is also built into the tool-head changers. The heads are designed for the particular tools needed for the job.

The random FMS is designed to handle a greater variety of parts in random sequence. Standard NC machines are used to handle the product mix, which is not defined completely at the time the FMS is installed. New part designs may have to be handled by the system in future years. Consequently, the machine tools must possess a significant degree of flexibility. Four- and five-axis machining centers would be a typical choice. Such machines are designed to perform a wide variety of machining operations on a variety of workpart geometries.

In practice, a given flexible manufacturing system often tends to be a hybrid of the two types, incorporating both special machines and standard NC machines. Examples of modules used on flexible manufacturing systems are illustrated in Figures 19.2 and 19.3.

### Material handling system

The workpart handling system must satisfy the following design requirements:

1. *Random, independent movement of palletized workparts between workstations in the FMS.* The term "random" means that parts must be able to flow from any one station to any other station. Palletized means that the parts are mounted in pallet fixtures. The pallets must be able to move independently of each other to minimize interference and maximize workstation utilization. There are limitations to the degree of independence that can be achieved on a given material handling system.
2. *Temporary storage or banking of workparts.* The number of parts in the system will exceed the number of workstations. In this way, each machine in the system will have a queue of parts waiting to be processed. This helps to maximize machine utilization.
3. *Convenient access for loading and unloading workparts.* This includes both manual loading and unloading at the load/unload station, as well as automatic loading and unloading at the individual processing stations. Since a given FMS may have machines on both sides of the handling system, such a system must be capable of loading and unloading from either side.
4. *Compatible with computer control.* The computer control system will be discussed in the next subsection.
5. *Provision for future expansion.* The material handling system should be designed to be expandable on a modular basis.
6. *Adherence to all applicable industrial codes.* (Safety, noise, etc.).

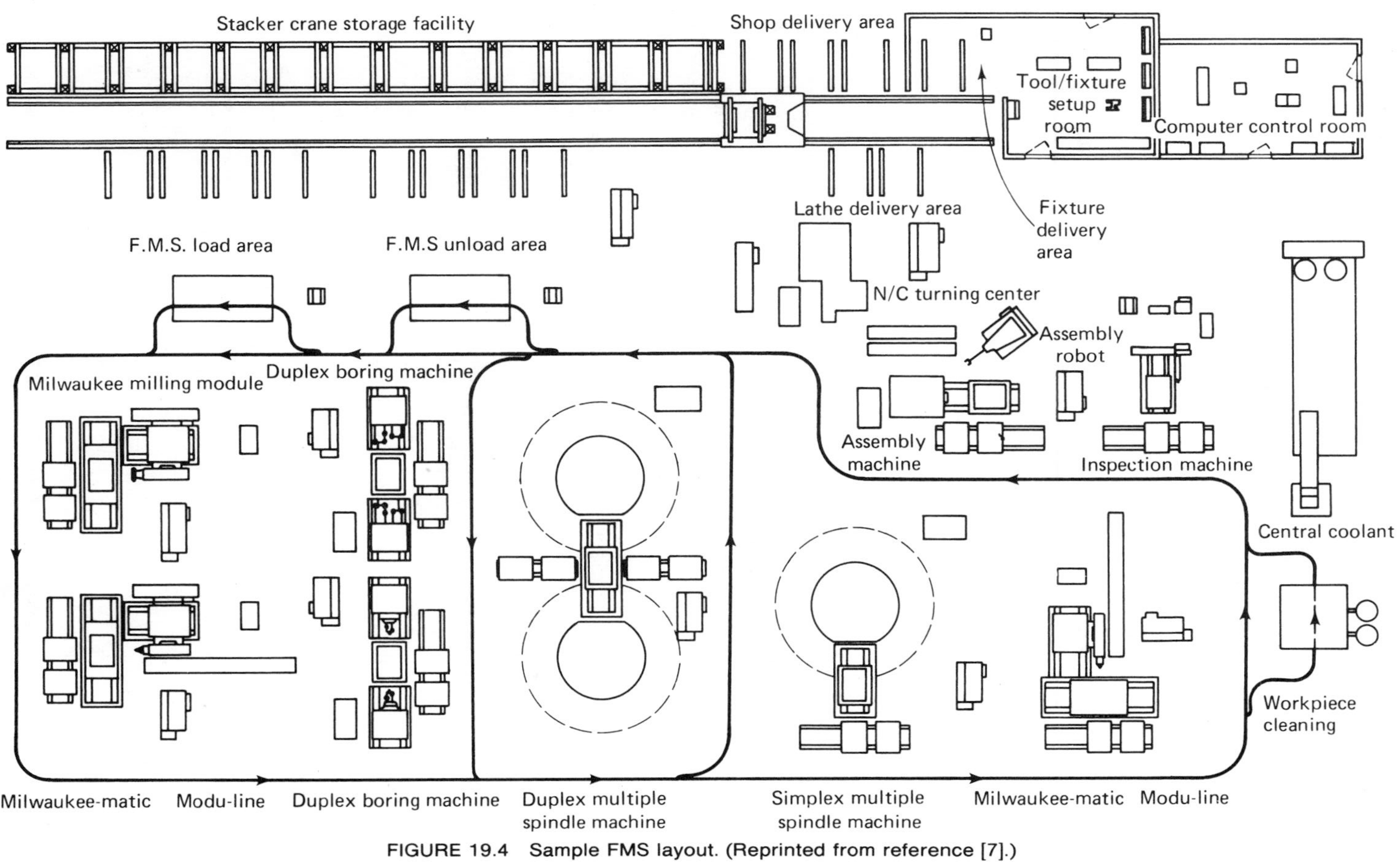


FIGURE 19.4 Sample FMS layout. (Reprinted from reference [7].)

7. *Access to machine tools.* The workpiece handling system should provide unobstructed floor-level access to the individual workstations on the line.

8. *Operation in shop environment.* The system must operate reliably in the presence of metal chips, cutting fluids, oil, and dirt which exist in the plant.

There are various ways to satisfy these design requirements. Workpart handling systems adaptable to the FMS concept include power roller conveyors, power and free overhead conveyors, shuttle conveyors, and floor "towline" systems. Industrial robots can also be used for parts transfer in integrated production systems [8]. The towline work transport system is illustrated in Figure 19.3. Floor plan diagrams, showing two alternative relationships of machine tools to handling system, are presented in Figures 19.4 and 19.5.

The workpiece handling system serves two functions. The first is the movement of parts between work centers. The second function is to interface with the individual work centers. This involves locating the workparts in the proper orientation for processing at the stations. It may also include banking of workparts at the workstations. Two separate workhandling systems may be required to best satisfy both functions. This is illustrated in the Kearney & Trecker FMS concept shown in Figure 19.3. The primary workhandling system is an under-the-floor towline system which pulls a series of carts between the different workstations. The carts have four wheels which roll on the floor surface. Slots in the floor define the permissible pathways for the carts. These paths may branch and merge in a manner similar to that depicted in Figure 19.4. The layout of the pathways depends on the design of the FMS. Guide pins at the front and back of the carts engage the slot in the floor to follow the correct path. The front guide pin engages a moving chain in the slot, which propels the cart.

The secondary workhandling system is the shuttle system at each machine. The shuttles are designed to transfer palletized parts to and from

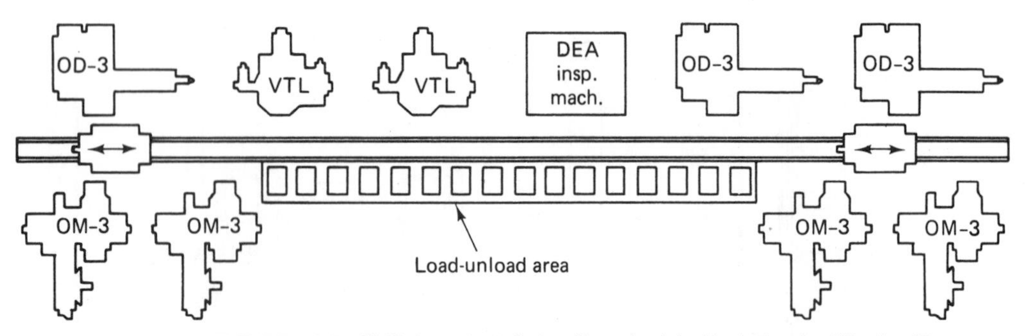


FIGURE 19.5 FMS layout at Caterpillar plant in East Peoria, Illinois. (Reprinted from reference [3].)

the towline carts, and to move them onto and off the machine tool table for accurate registration with the cutting tools. If the processing cycle at the workstation is relatively short, the shuttle system may also be required to maintain an inventory of parts ready for machining. Both the towline cart and shuttle system are shown in Figure 19.3.

### Computer control system

The operation and control of a complex flexible manufacturing system requires the capabilities of a digital computer. In this subsection, we discuss the various functions which the computer performs, the data files needed to carry out these functions, and the various types of reports which the FMS computer can be programmed to prepare.

COMPUTER FUNCTIONS. The functions performed by the FMS computer control system can be grouped into the following eight categories:

1. *NC part program storage.* The programs for the parts that are machined at the various workstations on the line must be stored.
2. *Distribution of the part programs to the individual machine tools.* This must be accomplished in the correct (postprocessor) format for that machine.
3. *Production control.* This function includes decisions on part mix and rate of input of the various parts onto the system. These decisions are based on data entered into the computer, such as desired production rate per day for the various parts, numbers of raw workparts available, and number of applicable pallets.[1] The computer performs its production control function by routing an applicable pallet to the load/unload area and providing instructions to the operator to load the desired raw part. A *data entry unit* (DEU) is located in the load/unload area for communication between the operators and the computer.
4. *Traffic control.* The term *traffic control* refers to the regulation of the primary workpiece transport system which moves parts between workstations. This control is effected by dividing the transport system into zones. For example, the Kearney & Trecker towline system installed at the Allis-Chalmers Tractor Division in West Allis, Wisconsin, has 120 zones. A zone is a section of the primary transport system (towline chain, conveyor, etc.) which is individually controlled by the computer. By allowing only one cart or pallet to be in a zone, the movement of each individual workpart is controlled. The traffic controller operates the switches at branches and merging points, stops workparts at machine tool loading points, and moves parts to operator load/unload stations.

[1]The term *applicable pallet* refers to a pallet that is fixtured to accept a workpart of the desired type.

5. *Shuttle control.* This is concerned with the regulation of the secondary part handling systems at each machine tool. Each shuttle system must be coordinated with the primary handling system, and it must also be synchronized with the operations of the machine tool it serves.

6. *Work handling system monitoring.* The computer must monitor the status of each cart and/or pallet in the primary and secondary handling systems as well as the status of each of the various workpart types in the system.

7. *Tool control.* Monitoring and control of cutting tool status is an important feature of a FMS computer system. There are two aspects to tool control: Accounting for the location of each tool in the FMS and tool-life monitoring.

The first aspect of tool control involves keeping track of the tools at each station on the line. If one or more tools required in the processing of a particular workpart are not present at the workstation specified in the part's routing, the computer control system will not deliver the part to that station. Instead, it will determine an alternate machine to which the part can be routed, or it will temporarily "float" the part in the handling system. In the second case, the operator is notified via the data entry unit what tools are required in which workstation. The operator then manually loads the tools and notifies the computer accordingly. Any type of tool transaction (e.g., removal, replacement, addition) must be entered into the computer to maintain effective tool control.

The second aspect of tool control is tool-life monitoring. A tool life is specified to the computer for each cutting tool in the FMS. Then, a file is kept on the machining time usage of each tool. When the cumulative machining time reaches the life for a given tool, the operator is notified that a replacement is in order.

8. *System performance monitoring and reporting.* The FMS computer can be programmed to generate various reports desired by management on system performance. The types of reports will be discussed later in this subsection.

These computer functions can be accomplished by any of several different computer configurations. One computer can be used for all components of the FMS, or several different computers can be used. Up to three levels are practical in a given manufacturing system. CNC would be used for control of each individual machine tool. DNC would be appropriate for distribution of part programs from a central control room to the machines. And a third control level would concern itself with production control, the operation of the work handling system, tool control, and generation of management reports.

Data Files. To exercise control over the FMS, the computer relies on data contained in files. The principal data files required for a flexible manufacturing system are of the following six types:

1. *Part program file.* The part program for each workpart processed on the system is maintained in this file. For any given workpart, a separate program is required for each station that performs operations on the part.

2. *Routing file.* This file contains the list of workstations through which each workpart must be processed. It also contains alternate routings for the parts. If a machine in the primary routing is down for repairs or there is a large backlog of work waiting for the machine, the computer will select an alternate routing for the part to follow.

3. *Part production file.* A file of production parameters is maintained for each workpart. It contains data relative to production rates for the various machines in the routing, allowances for in-process inventory, inspections required, and so on. These data are used for production control purposes.

4. *Pallet reference file.* A given pallet is fixtured only for certain parts. The pallet reference file is used to maintain a record of the parts that each pallet can accept. Each pallet in the FMS is uniquely identified and referenced in this file.

5. *Station tool file.* A file is kept for each workstation, identifying the codes of the cutting tools stored at that station. This file is used for tool control purposes.

6. *Tool-life file.* This data file keeps the tool-life value for each cutting tool in the system. The cumulative machining time of each tool is compared with its life value so that a replacement can be made before complete failure occurs.

System Reports. Data collected during monitoring of the FMS can be summarized for preparation of performance reports. These reports are tailored to the particular needs and desires of management. The following categories are typical:

1. *Utilization reports.* These are reports that summarize the utilization of individual workstations as well as overall average utilization for the FMS.

2. *Production reports.* Management is interested in the daily and weekly quantities of parts produced from the FMS. This information is provided in the form of production reports which list the required schedule together with actual production completions. One possible format for the production report is illustrated in Figure 19.6.

3. *Status reports.* Line supervision can call for a report on the current status of the system at any time. A status report can be considered an instantaneous "snapshot" of the present condition of the FMS. Of interest to supervision would be status data on workparts, machine utilization, pallets, and other system operating parameters.

PART SUMMARY
SHIFT 1 PAGE 01 02/13 17:35:44
02/13 09:28 TO 02/13 17:35 8.1 HRS.

| | P/S | SCHED. | COMPL. | DIFF. | PCT. |
|---|---|---|---|---|---|
| 268923 | 1 | 7 | 6 | −1 | −14.3 |
| 268923 | 2 | 7 | 9 | 2 | 28.6 |
| 268315 | 1 | 7 | 8 | 1 | 14.3 |
| 268315 | 2 | 7 | 8 | 1 | 14.3 |
| 268315 | 3 | 7 | 7 | 0 | 0.0 |
| 267171 | 1 | 7 | 5 | −2 | −28.6 |
| 267171 | 2 | 7 | 9 | 2 | 28.6 |

FIGURE 19.6 Production shift report, indicating FMS performance. (Reprinted from reference [7].)

4. *Tool reports.* These reports relate to various aspects of tool control. Reported data might include a listing of missing tools at each workstation. Also, a tool-life status report can be prepared at the start of each shift, similar to the one illustrated in Figure 19.7. This listing shows that several of the tools have been used well beyond their anticipated lives and are in need of replacement.

TOOL LIFE STATUS FOR STATION 3 PAGE 01 07/16 11:40'22

| TOOL NUMBER | EST. LIFE | ACT. USE | PCT. OF EST. |
|---|---|---|---|
| 00003 | 500.0 | 600.6 | 120.1 |
| 00165 | 50.0 | 3.0 | 6.0 |
| 00166 | 200.0 | 3.0 | 1.5 |
| 00173 | 91.5 | 22.0 | 24.0 |
| 10011 | 800.0 | 0.0 | 0.0 |
| 10014 | 135.0 | 3.9 | 2.9 |
| 10017 | 225.0 | 236.0 | 104.9 |
| 10020 | 300.0 | 59.0 | 19.7 |
| 10021 | 115.0 | 59.0 | 51.3 |
| 10023 | 210.0 | 0.0 | 0.0 |
| 10025 | 142.0 | 118.0 | 83.1 |
| 10027 | 100.0 | 0.0 | 0.0 |
| 10032 | 300.0 | 0.0 | 0.0 |
| 10034 | 999.0 | 0.0 | 0.0 |
| 10035 | 320.0 | 112.0 | 35.0 |
| 10036 | 380.0 | 1188.7 | 312.8 |
| 10037 | 35.0 | 85.0 | 242.9 |
| 10056 | 600.0 | 6.0 | 1.0 |
| 10057 | 400.0 | 118.0 | 29.5 |
| 10061 | 400.0 | 472.0 | 118.0 |
| 10062 | 350.0 | 472.0 | 134.9 |
| 10064 | 65.0 | 91.0 | 140.0 |

FIGURE 19.7 Tool life status report for one machine tool in an FMS. (Reprinted from reference [7].)

## 19.4 APPLICATION EXAMPLES

The features of some existing installations will be described to illustrate typical FMS applications.

### EXAMPLE 19.1

One of the first FMS installations in the United States was at the Roanoke, Virginia, plant of the Tool and Hoist Division of Ingersoll-Rand Company. The system was installed by Sundstrand in 1970. It consists of two five-axis OM-3 Omnimil Machining Centers, two four-axis Machining Centers, and two four-axis OD-3 Drilling Machines. The machines are equipped with 60-tool drums and automatic tool changers and pallet changers. A powered roller conveyor system is used for the primary and secondary workpart handling systems. Three operators plus foreman run the line three shifts. Up to 140 different part numbers can be run on the system. Part-size capability ranges up to a 3-ft cube. Production quantities for the various part numbers processed on the line range from 12 per year up to 20,000 per year. Cast-iron and aluminum castings are machined on the line, including hoist cases, motor cases, and so on. Control for the FMS is provided by an IBM 360/30 computer. A view of the system is presented in Figure 19.8.

FIGURE 19.8 Overview of the Ingersoll-Rand FMS in Roanoke, Virginia. (Courtesy White-Sundstrand Machine Tool, Inc.)

### EXAMPLE 19.2

Figure 19.5 shows the machine and conveyor layout for the flexible manufacturing system installed at the East Peoria (Illinois) plant of the Caterpillar Tractor Company. Installation of this system was completed in 1973. The line includes four Sundstrand five-axis OM-3's, three Sundstrand four-axis OD-3 Drilling Machines, two Giddings

FIGURE 19.9 Sundstrand Omniline concept at Caterpillar plant in East Peoria, Illinois. (Courtesy White-Sundstrand Machine Tool, Inc.)

and Lewis Vertical Turret Lathes, and one DEA coordinate-axis inspection machine. Parts movement is provided by a shuttle car system, as shown in Figure 19.5. A total of 21 conventional machine tools would have been required to achieve production equivalent of the FMS. Labor requirements are five operators, plus foreman on three shifts.

Workparts are cast-iron crankcase housings and covers ranging up to 900 pounds in weight. Six part numbers are machined and assembly work is also performed on the line. Production rate is 550 assemblies per month. Figure 19.9 illustrates the Sundstrand Omniline concept at Caterpillar.

## EXAMPLE 19.3

One of the most recent FMS installations is at the Avco-Lycoming plant in Williamsport, Pennsylvania. Figure 19.10 shows the machine and transfer system layout. The system was designed and built by Kearney & Trecker. Figure 19.3 illustrates the general type of in-floor towline cart transfer system and machine tools used in the FMS at Avco-Lycoming. The system is used to machine aluminum crankcase halves for aircraft engines. Machine tools are:

(2) K&T simplex multispindle head indexers.
(1) K&T duplex multispindle head indexer (see Figure 19.2).
(9) K&T Model 4848 Moduline Machining Centers.

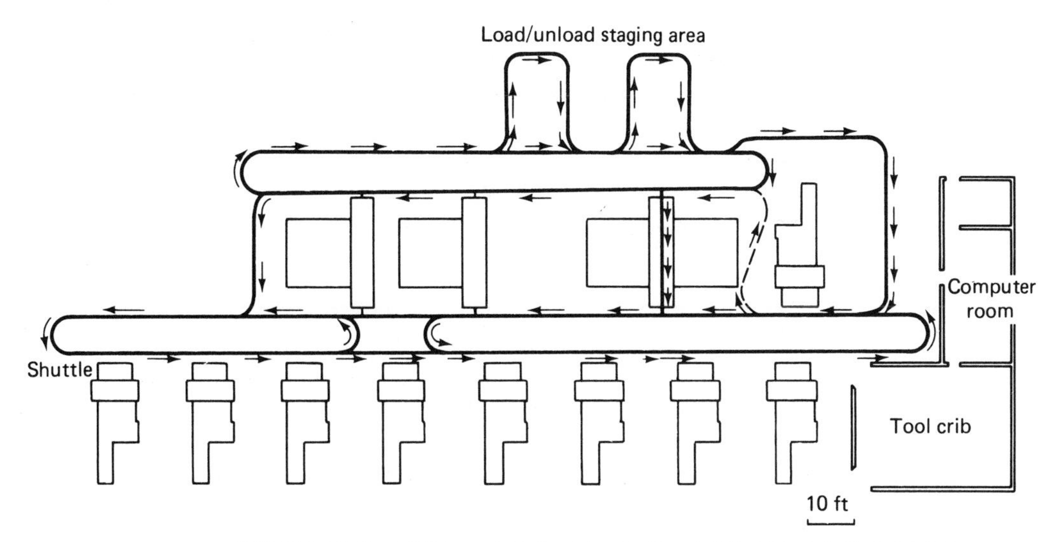


FIGURE 19.10 FMS layout at Avco-Lycoming in Williamsport, Pennsylvania. (Reprinted from reference [10].)

Two Model 70 Interdata minicomputers are used for control of the system. A total of 28 carts with pallets are driven at a maximum speed of 50 feet per minute to transport workparts between machines.

## 19.5 THE AUTOMATED FACTORY

The concept of the flexible manufacturing system can be expanded to plantwide operations. It is not difficult to visualize the combination of the many production systems concepts discussed in this book into an integrated computer automated factory. The term *unmanned factory* is sometimes used to describe such a plant [12]. However, this is a misnomer, at least in the foreseeable future, because certain functions in factory operations will still require human effort to perform. The present concept of the automated factory is similar to the highly integrated and computer-controlled plant found in the process industries of today (chemicals, petroleum refineries, basic metals, etc.). In such plants, a relatively small crew of perhaps five to ten persons manages the entire production operations. These individuals perform such functions as plant supervision, maintenance and repair, and so on. But they do not participate directly in the processing of the product. In the manufacturing industries, complex problems are encountered in achieving this same level of plant automation because of the difficulties in processing, assembling, and handling a diverse mix of discrete products. The future automated factory assumes that these technical problems can be solved, largely through the data handling, decision-making, and control capabilities of the computer. Hence, we are able to forecast the existence of a plant in which automatic machine tools will be interconnected by automatic work handling systems, delivering parts to and from automated warehouses. These activities will be directed under computer control by a crew of perhaps a dozen human operators and supervisors. If present trends continue, this computer-integrated factory will be realized by approximately the year 2000 [9].

## Features of the automated factory

An attempt will be made to describe some of the important features possessed by the modern automated factory of the year 2000. The principal functions that must be accomplished in a discrete-product manufacturing plant are: processing, assembly, inspection, parts transport, and storage. In addition, an information system is required to support these functions. Automation technology has been implemented in each of these individual functions in the factories of today. In the factory of tomorrow, these automated functions will be integrated into one consolidated system under computer control. Raw materials entering the factory will be automatically transferred to a temporary storage location, where they will await their turn for processing. The identification and location of each item in storage will be retained in computer files. According to a closely coordinated production schedule, the materials will be moved from storage to the production shop/warehouse interface by an automated stock picker system. From this interface, the plant work handling system will take over and transport the workparts through the proper sequence of machines for processing. Robots will be used to aid in this material handling function. The best routing for the workparts will be decided by the computer. As in the operation of a flexible manufacturing system, parts will be rerouted if a machine in the primary routing is down for repairs. When processing has been completed on individual parts, they will be routed to automatic assembly stations. By the year 2000, many workpart design practices will have been standardized to facilitate automatic assembly. After assembly, the finished products will be moved to an automated warehouse for temporary storage until shipment to customers.

The manufacturing plant of the foreseeable future will still require human workers. However, the structure of manual factory work will have changed dramatically by the year 2000. Tasks such as parts handling and counting, inspection work, and machine tending will either be eliminated or substantially reduced. Parts handling and counting will be performed automatically. Inspection work will be performed largely by industrial robots equipped with visual and tactile sensing capabilities. Checking of parts will probably be done on a 100% basis rather than by sampling. When workpart deviations are discovered, feedback data will be used to make adjustments in the process to correct the defects. The reliability of production machines will have been improved to reduce the need for machine tending. Computer monitoring systems will be used to ensure the satisfactory operation of the equipment.

If all these functions will have been automated, what useful work will human beings perform? The general level of work will be upgraded from manual and tedious tasks to functions requiring more skill and intellect. These functions will include management and supervision, planning, programming and operating the computer, engineering analysis and design, and maintenance and repair. The content of these jobs will have changed by the time of the computer automated factory of the future. New computer languages and techniques will be used to facilitate computer programming and NC part programming. Computer diagnosis of machine breakdowns will help the maintenance crew to pinpoint the cause of failure. Preventive maintenance schedules will be prepared by the computer to guide the maintenance crew. Manual labor may still be required in the shipping and receiving department, the interface between the automatic factory and the outside world.

## Social and economic forces

The trend toward the computer-automated factory seems unavoidable in modern industrialized societies such as the United States, Europe, and Japan. There are several social and economic factors tending to promote the development of such a factory. These include:

1. The need for greater productivity.
2. The desire to increase machine utilization.
3. The high cost of in-process inventory.
4. The need to respond quickly to customer demand—this means shorter manufacturing lead times.
5. The high cost of raw materials and energy—this means that both of these resources must be used efficiently.
6. The trend of the labor force away from production of goods and into the service sector. This means fewer workers in manufacturing, and higher wage rates for those who remain.
7. The demand by workers for more meaningful work.
8. Federal regulations regarding worker safety, noise, and so on.

This list includes many of the same factors presented in Chapter 1. Hence, we have come full circle. The same factors that were the driving force behind the development of transfer lines, NC, and other automated systems of today are also the impetus for further advances in production automation, culminating in the future realization of the computer-automated factory. On this note, we conclude the book.

## REFERENCES


[1] COOK, N. H., "Computer-Managed Parts Manufacture," *Scientific American*, February, 1975, pp. 22–29.

[2] CURTIN, F. T., "Manufacturing Systems Moving Stage Center in Dramatic Search for Better Productivity," *N/C Commline*, January/February, 1979, pp. 34–35.

[3] "DNC for Flexibility," *Production*, September, 1974, pp. 70–75.

[4] "DNC Lines Link Cutting to New Future," *Iron Age*, October 26, 1972.

[5] HUTCHINSON, G. K., "Production Capacity: CAM vs. Transfer Line," *Industrial Engineering*, September, 1976, pp. 30–35.

[6] HUTCHINSON, G. K., and WYNNE, B. E., "A Flexible Manufacturing System," *Industrial Engineering*, December, 1973, pp. 10–17.

[7] Kearney & Trecker Corp., *Understanding Manufacturing Systems* (A Series of Technical Papers), Vol. I, Milwaukee, Wis.

[8] MATTOX, J. E., "Industrial Robots and Integrated Manufacturing Systems," *N/C Commline*, January/February, 1979, pp. 41–45.

[9] MERCHANT, M. E., "The Inexorable Push for Automated Production," *Production Engineering*, January, 1977, pp. 44–49.

[10] "Our 'FMS' Will Do the Work of 67 Conventional Machine Tools," *Production*, April, 1978, pp. 66–69.

[11] Sundstrand Corp., *Omniline Systems* (marketing brochure), Belvidere, Ill.

[12] YOSHIKAWA, H., "Unmanned Machine Shop Project in Japan," *Advances in Computer-Aided Manufacture*, North-Holland Publishing Co., Amsterdam, 1977, pp. 3–22.

# Answers to Selected Problems

## CHAPTER 2

**2.1.** 141 hrs
**2.2.** 0.122 pc/hr, 1.034 pc/hr, 4.054 pc/hr, 5.724 pc/hr
**2.3.** 33.5 hrs
**2.4.** 20 min, 3.0 pc/hr
**2.5.** (a) 333.33 hrs, (b) 188.3 hrs, (c) 5.296 pc/hr

## CHAPTER 3

**3.1.** \$5674
**3.2.** \$949
**3.3.** \$1144
**3.4.** \$133,720
**3.5.** \$39,891
**3.6.** $UAC_m = \$32,441$  $UAC_a = \$27,933$  Select automatic
**3.7.** $UAC_m = \$33,578$  $UAC_a = \$35,194$  Select manual
**3.8.** \$1.203/pc

**3.9.** \$45.57/hr
**3.10.** \$21.59/hr
**3.11.** (a) −\$3321, (b) −\$221, (c) \$2879, (d) \$5979
**3.12.** 80,000 pc/yr:
(a) Conventional: \$50021; Automatic: \$50212
(b) Conventional: 4000 hrs; Automatic: 1333.33 hrs
(c) Conventional: \$5979; Automatic: \$5789
90,000 pc/yr:
(a) Conventional: \$53921; Automatic: \$50537
(b) Conventional: 4500 hrs; Automatic: 1500 hrs
(c) Conventional: \$9079; Automatic: \$12464
**3.13.** 71,328 pc/yr
**3.14.** (a) 20554 pc/yr, (b) 2055.4 hrs, (c) \$5335/yr
**3.15.** (a) 30831 pc/yr, (b) 3083.1 hrs, (c) −\$5598/yr
**3.16.** 72,354 pc/yr

## CHAPTER 4

**4.1.** 135° drive, 225° dwell
**4.2.** (a) 40 rpm, (b) 0.5 s, (c) 2400 pc/hr
**4.3.** (a) 22 s: 163.6 cycles/hr, 163.6 pc/hr
21 s: 171.4 cycles/hr, 165.8 pc/hr
20 s: 180 cycles/hr, 161.6 pc/hr
19 s: 189.5 cycles/hr, 149.5 pc/hr
18 s: 200 cycles/hr, 93.6 pc/hr
(b) 18.55 s: 194 pc/hr

## CHAPTER 5

**5.1.** (a) 0.513 min, (b) 117 pc/hr, (c) 0.649, (d) 0.351
**5.2.** $T_p = .533$ min, $R_p = 112.6$ pc/hr, $E = 0.625$, $D = 0.375$
**5.3.** \$.637/pc
**5.4.** \$.684/pc
**5.5.** 2400 pc/week
**5.6.** 42.4 hrs
**5.7.** (a) \$4821/week, (b) \$.482/unit, (c) Repair pays
**5.8.** (a) $F = .120$, (b) 90.9 pc/hr, (c) $E = 0.455$
**5.9.** (a) $F = .1136$, (b) 83.0 pc/hr, (c) $E = .468$
**5.10.** $D = 0.456$, $E = 0.544$, $R_p = 195.4$ pc/hr
**5.11.** (a) 1558 pcs, (b) 824.4 hrs, (c) 0.313, (d) 22.4 pc/hr
**5.12.** No
**5.13.** (a) 3058 pcs, (b) 0.363, (c) 76.4 pc/hr, (d) 0.0085
**5.14.** (a) 171.4 pc/hr, (b) 0.571, (c) 0.9415, (d) 0.0585, (e) 0.0857, (f) 0.3429

**5.15.** Current line: $C_{pc}=\$1.019/pc$
Proposed line: $C_{pc}=\$.788/pc$
Recommendation: Automate station number 5
**5.16.** Current $E_0=.61$, Max $E_\infty=.7576$
**5.17.** $E=.7145$, $R_p=57.16$ pc/hr
**5.18.** $E=.6967$, $R_p=55.73$ pc/hr
**5.19.** $E=.6996$, $R_p=55.97$ pc/hr
**5.20.** $E=.6854$, $R_p=54.83$ pc/hr

## CHAPTER 6

**6.1.** (a) 1.33 units/min, (b) 1.25 min, (c) .75 min/pc
**6.2.** 5.0 min: Not possible since $T_s=.833$ min/station
4.0 min: Possible since $T_s=.667$ min/station
3.0 min: Very likely since $T_s=.50$ min/station
**6.3.** (a) .582 min, (b) .5 m/min, (c) 1.72 pc/min, (d) .291 m/pc
**6.4.** (a) 11 stations, (b) $d=0.55\%$, (c) 1.403 pc/min
**6.5.** (a) 35 s, (b) $d=40\%$, (c) 1.285 pc/min, (d) 45%
**6.6.** (b) 5 stations, (c) $d=14\%$
**6.7.** (a) 5 stations, (b) $d=14\%$
**6.11.** (b) $d=20\%$
(c) Individual workers: UAC=$80,000/yr
Assembly line: UAC=$58,042/yr
Recommendation: Assembly line

## CHAPTER 8

**8.1.** 6111 rpm, 18.33 in./min
**8.2.** 2037 rpm, 24.44 in./min
**8.3.** 191 rpm, 6.11 in./min

## CHAPTER 9

**9.2.** NC: 400 s, Adaptive control: 250 s
**9.3.** (a) 26.67 min, (b) 8.33 min, (c) 68.8%, (d) 17.33 min
**9.5.** (a) $200/job, (b) 17.4%, (c) yes
**9.6.** Manual: UAC=$68,000/yr
Robots: UAC=$41,742/yr
Recommendation: Use the robots

## CHAPTER 11

**11.1.** 4096 levels, 0.000244, 0.01465 V, 0.00732 V

**11.2.** (a) 4 bits, (b) 6 bits, (c) 10 bits

**11.3.** 39.84 V

**11.4.** (a) $V(t)=36.25$, (b) $V(t)=36.25-2.5t$

**11.5.** $V(t)=36.25-3.125t-.625t^2$
First order hold: $V=33.75$ at 4th instant
Second order hold: $V=32.5$ at 4th instant

**11.6.** (a) 6°, (b) 300 pulses, (c) 75 pulses/s

**11.7.** (a) 200, (b) 111.11 pulses/s

## CHAPTER 13

**13.1.** $A=5.0$, $B=2.0$

**13.2.** $y=.5[1-\exp(-.4t)]$, $\tau=2.5$

**13.3.** (b) Probably second order—underdamped

**13.4.** $A\dfrac{d^2y}{dt_1^2}+B\dfrac{dy}{dt}+Cy=20$

**13.5.** $\dfrac{1}{5s+2}$

**13.6.** $\dfrac{1}{As^2+Bs+C}$

**13.7.** (a) $-2.68\pm j2.52$ (oscillatory), (b) 2.52 rad/s

**13.8.** (a) $\dfrac{.4}{s}$, (b) $\dfrac{1}{3s+7}$, (c) $\dfrac{1.2}{s^2+3s+2.5}$, (d) $\dfrac{s+3}{s^2+2s+8}$

**13.11.** (a) $\dfrac{C(A+B)}{1+CD(A+B)}$, (b) $\dfrac{AB/(1+A)}{1+CAB/(1+A)}$, (c) $\dfrac{B+A}{1+BC}$

**13.12.** (a) $\dfrac{5K}{s^2+7s+K}$, (b) $\dfrac{2s^2+s(17K+14)+24}{s^3+7.2s^2+(13.4+1.7K)s+2.4}$

**13.13.** $\dfrac{3}{s^2+9s+14.9}$

**13.14.** $\dfrac{s+2}{2s^3+20s^2+65s+66}$

**13.15.** (b) $\dfrac{15K(s+8)}{s(s+8)(s+6)+4.5K}$, (c) $\dfrac{15s(s+8)}{s(s+8)(s+6)+4.5K}$

**13.18.** (a) $y(s)=\dfrac{-1}{s+5}+\dfrac{1}{s+2}$, (b) $y(s)=\dfrac{.333}{s}-\dfrac{.833}{s+3}+\dfrac{.5}{s+1}$,
(c) $y(s)=\dfrac{5}{s+8}$

**13.19.** (a) $y(s)=\dfrac{-1}{(s+2)^2}+\dfrac{1}{s+2}$, (b) $y(s)=\dfrac{4}{7s^2}-\dfrac{4}{49s}+\dfrac{4}{49(s+7)}$

**13.23.** $y(t)=1.333+1.667\exp(-3t)$

**13.24.** $y(t)=1.28-t\exp(-2.5t)-1.28\exp(-2.5t)$

**13.25.** (a) 1.0, (b) 13/18, (c) 1/7

**13.26.** $K=6.25$

**13.29.** (a) Stable for all $K$
(b) Unstable for $K>160$
(c) Marginal stability at $K=0$
(d) Unstable for $K>26.25$

**13.30.** (a) $\dfrac{3K}{s(s+3)(s+7)}$, (b) $\dfrac{5K}{s(s+3)(s+7)+3K}$

**13.31.** (a) $\dfrac{4K}{(s^2+6s+10)(s+10)}$, (b) $\dfrac{20K(s+10)}{(s^2+6s+10)(s+10)+4K}$

**13.32.** Largest $K$ for stability is $K=70$
Largest $K$ for no complex roots is $K=4.199$

**13.33.** $K=3.285$

**13.34.** Stable at $K=4.0$

**13.35.** (a) No increase in $K$, (b) Two ways: (1) Increase input $x$ to $x=8.152$, or (2) Add gain block before summing point in $K=1.63$ Figure 12.28 with gain

**13.37.** (a) $K_2=3.333$, (b) $K_1=2.1$, (c) $y(t)=5-11.67\exp(-4t)+6.67\exp(-7t)$

## CHAPTER 14

**14.1.** DDC: $y(t)=6.594$ at $t=.2$ s
Analog: $y(t)=6.014$ at $t=.2$ s

**14.2.** DDC: $y(t)=6.269$

**14.3.** DDC: $y(t)=.363$ at $t=.1$ s
Analog: $y(t)=1.87$ at $t=.1$ s

**14.4.** 110 loops on DDC, 90 loops on analog

## CHAPTER 15

**15.4.** $x=1.559$

**15.5.** $x_1=8.333$, $x_2=1.506$, $z=.254$

**15.6.** $x_1=.80072$, $x_2=.10498$, $z=6.2955$

**15.7.** $x_1=17.25$, $x_2=11.0$, $z=441.125$

**15.8.** $x_1=2.571$, $x_2=2.143$, $z=\$32.14$

**15.9.** (a) $x_1=3.885$, $x_2=3.387$, $z=\$126.02$,
(b) Surplus capacity on operation 1

**15.10.** $B=17.0$

**15.11.** (a) $B=58.38$, (b) $x=1.3506$, (c) $y=86.44$

**15.12.** (a) $A=2.164$, $B=6.692$, (b) $x_1=.8032$, $x_2=.1113$

**15.13.** (a) $y=-42.21+51.137x$, (b) $r=.99886$, $s_e=7.51$

**15.17.** (a) $i25+j29$, (b) 38.29, (c) $i.653+j.757$

**15.18.** $i5+j1$, 5.099, $i.9806+j.196$

## CHAPTER 16

**16.1.** (a) 500 ft/min, (b) 589 ft/min
**16.2.** $2.54
**16.4.** Speed effect = $.04, decrease speed
Feed effect = − $.08, increase feed

## CHAPTER 17

**17.3.** 224 hrs
**17.4.** 252.63 hrs
**17.5.** 17.006 shifts

## CHAPTER 18

**18.1.** 00100
**18.2.** 11100

# Index

## A

B

D

## E



## F

T



U



V



W



Z